SPACE INFORMATION SYSTEMS
IN THE
SPACE STATION ERA

Proceedings of the AIAA/NASA International Symposium
on Space Information Systems
in the Space Station Era
Washington, DC and Greenbelt, MD, June 22–23, 1987

Organized and Operated by the
American Institute of Aeronautics and Astronautics (AIAA)

Sponsored by
National Aeronautics and Space Administration (NASA)

Edited by
Mireille Gerard and Pamela W. Edwards
AIAA

February 1988
American Institute of Aeronautics and Astronautics
370 L'Enfant Promenade, SW, Washington, DC 20024
(202) 646–7440

American Institute of Aeronautics and Astronautics
Washington, DC

Library of Congress Cataloging in Publication Data

AIAA/NASA International Symposium on Space Information Systems
in the Space Station Era (1987: Washington, DC and Greenbelt, MD)
Space information systems in the space station era.

   1. Space stations – Congresses. 2. Data transmission systems – Congresses. 3. Space
sciences – Information services – Congresses. I. Gerard, Mireille. II. Edwards, Pamela W. III.
American Institute of Aeronautics and Astronautics. IV. United States. National Aeronautics
and Space Administration. V. Title.
TL797.A35 1987   629.47'43   87-37417
ISBN 0-930403-36-3

Steering Committee

Co-Chairmen

ROBERT O. ALLER
Associate Administrator for Space
  Operations
NASA Headquarters

S. RICHARD COSTA
Director, Communications and Data
  Systems Division, Office of
  Space Operations
NASA Headquarters

DAVID R. McELROY
Group Leader
Lincoln Laboratory
Massachusetts Institute of
  Technology

MIREILLE GERARD
Administrator, Corporate and
  International Programs
American Institute of Aeronautics
  and Astronautics

DANA L. HALL
Acting Chief, Information Systems
  Management Division
NASA Headquarters

PETER KURZHALS
Director, Customer Integration
McDonnell Douglas Astronautics
  Company

JOHN J. QUANN
Deputy Director
NASA Goddard Space Flight Center

JOHN SAKSS
Chief, International Programs
  Support
NASA Headquarters

A. VILLASENOR
Program Manager
NASA Headquarters

Program Committee

Chairman
JOHN Y. SOS*
Chief, Systems Management Office
NASA Goddard Space Flight Center

WILLIAM C. COLLICOTT
Deputy Director, Satellite Data and
  Distribution
National Oceanic and Atmospheric
  Administration

EDWARD B. CONNELL*
Chief, Information Processing
  Division
NASA Goddard Space Flight Center

JAMES M. CONOVER*
Staff Member, TMIS Panel
McDonnell Douglas Astronautics
  Company

JAMES P. BIGHAM, JR.
SSIS Project Integration Manager,
  Space Station Project Office
NASA Johnson Space Center

PAUL EBERT*
Director
Communications Satellite
  Corporation

CLAUDE HONVAULT
Head, Systems and Project
  Support Department
European Space Agency

CHU ISHIDA
Systems Manager
National Space Development
  Agency of Japan

ALLAN JAWORSKI*
Manager, Advanced Technology
  Engineering
Ford Aerospace & Communications
  Corporation

GEORGES JEAMBRUN
Deputy Director, Operational
  Systems Division
Centre National d'Etudes Spatiales,
  France

KAZUO MATSUMOTO
Director, Tracking and Data
  Acquisition Department
National Space Development Agency
  of Japan

PETER PIOTROWSKI
Head, Data System Department
DFVLR, Federal Republic of
  Germany

EDWARD TAGLIAFERRI*
Director, Special Programs
Unisys

MICHAEL T. WARD*
Senior Systems Engineer
Ford Aerospace & Communications
  Corporation

***Organizers of sessions**

Administrative Committee

Chairman
PAMELA W. EDWARDS
Manager, Management Seminars
American Institute of Aeronautics
  and Astronautics

LYNN KRALOVICH
Symposium Coordinator
American Institute of Aeronautics
  and Astronautics

JANET K. WOLFE
Chief, Public Affairs
NASA Goddard Space Flight Center

HOWARD K. OTTENSTEIN
Chief, Public Services
NASA Goddard Space Flight Center

Table of Contents

Management 1: Standards and Protocols

Co-Chairmen: Dale L. Fahnestcok, NASA Goddard Space Flight Center and
        Hubertus Wanke, DFVLR

Systems 2: Integrated System Architectures

Co-Chairmen: Richard G. Schell, Computer Technology Associates and
        Bernard Curbelie, Centre National d'Etudes Spatiales

Technology 2: Software and Automation

Co-Chairmen: Howard L. Yudkin/Software Productivity Consortium and Rudolf J. Lauber/
University of Stuttgart, West Germany

Management 2: Plans and Approaches

Co-Chairmen: Robert E. Smylie/RCA American Communications, Inc. and
Claude Honvault/European Space Agency

Space Information Systems Workshop

Chairman: Michael J. Wiskerchen, Stanford University

Recording Secretary: Allen Jaworksi, Ford Aerospace / Communications Corporation

Chairman: C.E. (Tom) Velez, Computer Technology Associates, Inc

Recording Secretary: Thomas E. Ryan, NASA Goddard Space Flight Center

Preface

In the 1990s we will see a new era of space exploration and exploitation, exemplified by the international effort to launch a permanently manned Space Station. The Space Station will consist of a base constructed as a joint effort of the United States, Europe, and Japan; several related space platforms; and many ground elements distributed across the Earth. Other space activities during the 1990s include the Columbus program, JEM, the Space Telescope, and the Earth Observing System. Information systems will play a key role by tying the subsystems of these elements together and making them function productively. To perform this function, future space information systems must effectively support requirements for spacecraft control, manned operations, robotics, payload/sensor control, and data distribution.

We face many challenges in the development of these information systems. We must provide control systems that automate routine chores, increase the efficiency of both flight and ground operations, and improve the responsiveness of payload control functions. Data distribution systems must be built which press the edge of existing technology in high-bandwidth communications, processing, and data storage. Many of these systems must be engineered in the hostile environment of space, which poses major demands for information system component reliability, radiation-hardness, mechanical stability, tolerance to communications interruption, and weight minimization. We also face the challenge of coordinating these systems so that operational overhead is minimized and utilization of scarce resources is maximized. Moreover, our information systems must have the potential for expansion as we gain confidence in our space capabilities and see the full rewards of space exploration.

NASA and its partners, the European and Japanese Space agencies, have undertaken several major programs in preparation for the development of these systems. The 1987 AIAA/NASA International Symposium on Space Information Systems in the Space Station Era brought together many representatives of the government, industry, and academic community both from the United States and from abroad to share information and plan for these future efforts. Attendees included potential space information systems users, managers, systems architects, and implementers. These Proceedings contain papers presented at the Symposium and summaries of panel discussions in workshops held on the last day. It provides a panoramic view of the many topics presented at the Symposium, including system concepts, implementation approaches, operational scenarios, user interfaces, software environments, networking, data handling, and mission control systems. The workshop notes include specific recommendations for future space agency actions in the areas of customer support, technology growth, and achieving interoperability.

We would like to thank the many contributors to the Symposium, including the AIAA conference organization staff, authors, workshop chairpersons, recording secretaries, session chairpersons, panelists, and the AIAA Technical Committees supporting the Symposium. We would particularly like to thank the international members of the Program Committee whose efforts made this Symposium a truly international event, and the staff of Goddard Space Flight Center who devoted a significant amount to time and effort to provide many demonstrations of technologies under development.

KEYNOTE ADDRESS: Off and Running with the Space Station and a New Era in Information Systems

Andrew J. Stofan
Associate Administrator for Space Station
NASA Headquarters, June 22, 1987

Information systems are critical to the success of the Space Station and will push the state of art in technology. Hats off to James Beggs and President Reagan for the Space Station decision.

We have traveled a long preparatory course and, within the necessary budgetary constraints, attained a sound technical configuration for the Space Station. We have come up with a sound program. We also now have a largely new management structure in place; this flowing from the Phillips recommendation to NASA Administrator James Fletcher. Early on, we developed user involvement in the program — something we intend to keep "up front." And we have made significant progress on the international scene. I feel confident that by winter we will have memorandums of understanding signed with our three international partners, Canada, the European Space Agency (ESA), and Japan.

The process with Congress has gone well. In briefings and in hearings, we have worked very hard this past year with many individual Congressmen. The FY88 bill of $767 million has gone through the House and Senate Authorization Committees. Appropriations took a little money out of the program; but overall we have, I believe, strong support in Congress — a real vote of confidence.   The aerospace industry has supported the Space Station most effectively. All have been happy to see release of the RFPs. It took a long time to get them through the political process — the White House, the Office of Management and Budget (OMB), the National Security Council and the Office of Science and Technology Policy, plus all the staff offices in the White House. Just about everybody in Washington loves the Station, and they want to help it go ahead — and help me to define it.

The contract for technical and management information systems has been awarded. We expect to issue an RFP on "software support shortly." The RFP on the Flight Telerobotics Servicer is on the street. By mid-summer we expect to issue the program-support contract, and we are in the process of putting out an RFP for a what we call the crew emergency rescue vehicle. On the Phase C/D proposals, we expect to award contracts by the end of November. So, we have finally broken loose the program.

The Space Station is designed for evolution. We keep emphasizing this and are putting money up front in many areas — paying a premium in the short term — so that we can permit mods to the Station, allow it to grow for the long haul — to be in orbit 20 years and more, yet keep up to date technologically.

One challenge we must master is what I call "turnaround art." What we take up on each Shuttle has to be unloaded from its cargo bay and assembled by the astronauts within seven days. And then the next load to go up creates a virtually new spacecraft. That repeats again and again — something we have never done before. It will be a real technological challenge.

Another challenge will be software for the Station's information systems. We are pushing this state of the art. We are applying expert systems (AI) as much as the technology allows. We will be doing things like telescience, in which the scientist can operate his instruments directly from a laboratory on the ground. In this software control game, we have to succeed in making Station operations economical over a long period of time. We must bring in the latest technology not only to operate systems, but also to make them friendly to the user.

The information system has two major components. One we call the technical and management information system (TMIS): computers and software that allow us to communicate among all of the NASA centers, all of the contractors, and, we hope, all of our international associates. TMIS must permit the flow of data among all the elements. The second part is the Space Station Information System (SSIS) — the entire complex of integrated ground and airborne equipment that exercises software on the Space Station. These two elements go together to form the software system. Using this information system strategy, the people on board the Station should be able to pay attention to the operation of the Station when the Station tells them something has to be done. We also want the entire system transparent to the user, in the ground lab, pushing buttons.

The management system calls for high-volume, random-access, mass storage and efficient procedures — fault-tolerant and highly reliable. It must be a safe system, with software checks so that people can access it and get out without endangering anything that's going on in any other modules or any place in the Station. We will have challenges in controlling the software in such a diverse and widespread technical operation. But I think the capabilities in the technology will make that job do-able.

To sum up, the program is moving forward. We have a sound configuration; a strong contractor base, effective information concepts and contracts, and a sound Space Station program. We have the momentum now. We are letting the contracts, getting the people on board, and, by the end of November, will sign those final four contracts. We will finally be off and running in the great Space Station era and an equally great, and greatly needed, new information-systems era.

Space Information Systems Overview
Co-Chairmen: Robert O. Aller and Giovanni Mica

NASA SPACE INFORMATION SYSTEMS OVERVIEW

Dr. Dana L. Hall*
Space Station Program Office
NASA Headquarters
Washington, D.C.

Abstract

A major objective of NASA space missions is the gathering of information that when analyzed, compared, and interpreted furthers man's knowledge of his planet and surrounding universe. A space information system is the combination of data gathering, data processing, and data transport capabilities that interact to provide the underlying services that enable that advancement in understanding. Past space projects have been characterized by rather disjoint data systems that often didn't satisfy user requirements. NASA has learned from those experiences, however, and now is conceptualizing a new generation of sophisticated, integrated space information systems suitable to the wide range of near future space endeavors. This paper examines the characteristics of recent data systems and, based upon that characterization, outlines the scope and attributes of future systems. A description is offered of the information system for the Space Station Program as one real example of such advanced capabilities.

Introduction

The 1990's will see NASA beginning a new era of space application and exploitation. A major key to that advance is the provision of sophisticated, intelligent information systems that are able to perform mundane chores in a relatively autonomous manner while also offering flexible, adaptable higher level services amenable to a broad scope of applications. Such higher services might support activities as diverse as assembly of complex structures in space, dynamic experiment observation (with onsite, real time results processing), and interaction of multiple human participants with robots or other devices when those participants are distributed both in space and across many locations on the ground. Future space missions will often fly experiments and commercial applications generating hundreds of millions of bits of data each second. Some of that data will not be time critical and therefore could be stored for later (perhaps many weeks later) transfer to the customer. But frequently, particularly in situations of feedback and control or establishment of a scientific process, it will be critical to the success of the experiment to have near real time information systems services.

The information systems and supporting data technologies NASA and its industrial partners are now exploring are focused to accomplish challenges such as the above. These developments are based in large part on the lessons gleaned from past projects and are enabled by the startling advances being made in numeric computing, symbolic processing, mass data storage, and software coupled, of course, with the sharp decrease in required power, weight, and volume of the associated hardware.

This paper briefly reviews the experiences gained from previous space projects and applies those lessons to postulate the nature of future space information systems. The information system now being defined for the international Space Station Program is described as an example of one such near term system. Outlined as well are some of the technical and management challenges associated with developing and operating such all encompassing, integrated information capabilities.

Characteristics of Past Missions

The missions of the 1960's were characterized by sensors which provided a passive scan of all events in the field of view, spacecraft systems which could only be turned on and off, a single principal investigator who collected and reduced the data, and little or no need for correlating the data with that of other observations. The data systems built for such missions were simple and unique. Data rates were on the order of a few thousand bits per second and there was little in the way of real time interaction between the investigator and the instrument in space.

The supporting data systems evolved during the 1970's into a more complex arrangement of capabilities in response to the increased diversity of the space program. Since NASA had little experience to draw from and the supporting technology base was also new and largely untried, these data systems proved to have many limitations and considerable inconsistencies. The state of technology and the premium placed on size, weight, power, and reliability during the early years of space exploration dictated that the onboard instrumentation, including the data hand-ling and telemetry equipment, be as simple as possible. The ground system was assigned the role to make up for that onboard simplicity by providing the system intelligence needed for all of the complicated functions and compensating for any deficiencies.

Space data systems of the late 1970's can be characterized by the following assessment taken from the NASA End-to-End Data System (NEEDS) research program: [1]

- Ad hoc data system architectures developed mission by mission with unique software often force-fitted to meet previously selected or available, usually outdated hardware;

- Very limited system growth and adaptability;

- Software written in low level or machine languages, poorly documented (if at all), and supported by inadequate tools and insufficiently trained personnel;

- Poor customer service as evidenced by lengthy data delivery delays, difficult to understand data annotations and unique product formats, and uncoordinated, difficult to locate and access data bases.

* Member, AIAA

Unfortunately, the flight rate during the 1980's has been severely reduced compared to the heady trends that had been forecasted during the previous decade. Those missions that have flown, such as the latter Landsats, Spacelab, astronomy telescopes, and of course, the Space Shuttle have begun to explore the benefits (and problems) associated with complex, high rate data streams supporting multiple users with diverse objectives. As the instruments have become more sophisticated, the realization has grown that previously acceptable data handling methods simply are not suitable to the new job at hand. Techniques such as time multiplexing of unlabeled data from multiple sources, separate computation and later merging of ancillary data, and instrument specific data formats worked well for relatively simple missions but do not have the inherent evolvability to satisfy more complex missions. Processes that are faster, more flexible, cheaper, and less error prone are needed. Missions projected to be flown in the 1990s will involve a significant increase in the number of sensors and much higher volumes and rates of data. These characteristics will exceed the capabilities of the present space data systems, even if only a reasonable

fraction of the forecasted missions occur. Therefore, NASA must conceive, define, and develop space information systems which can satisfy the significantly different requirements expected for the future. Figure 1 is a simple schematic illustrating the major functions that together comprise an end-to-end integrated information system.

New technologies such as "smart" sensors also are prompting reexamination of the procedures for acquisition, conditioning, annotating, transmitting, and distributing sensor data to the user. The dynamic operation of artificially intelligent devices cause variation in bandwidth required and other resources that are difficult to anticipate in advance and require adaptable data handling and processing systems, presentation of information on-line, and feedback that may depend upon information just learned from a preceding event.

The evolution of the requirements for space information systems is perhaps best illustrated by examining experience gained from recent Spacelab flights.

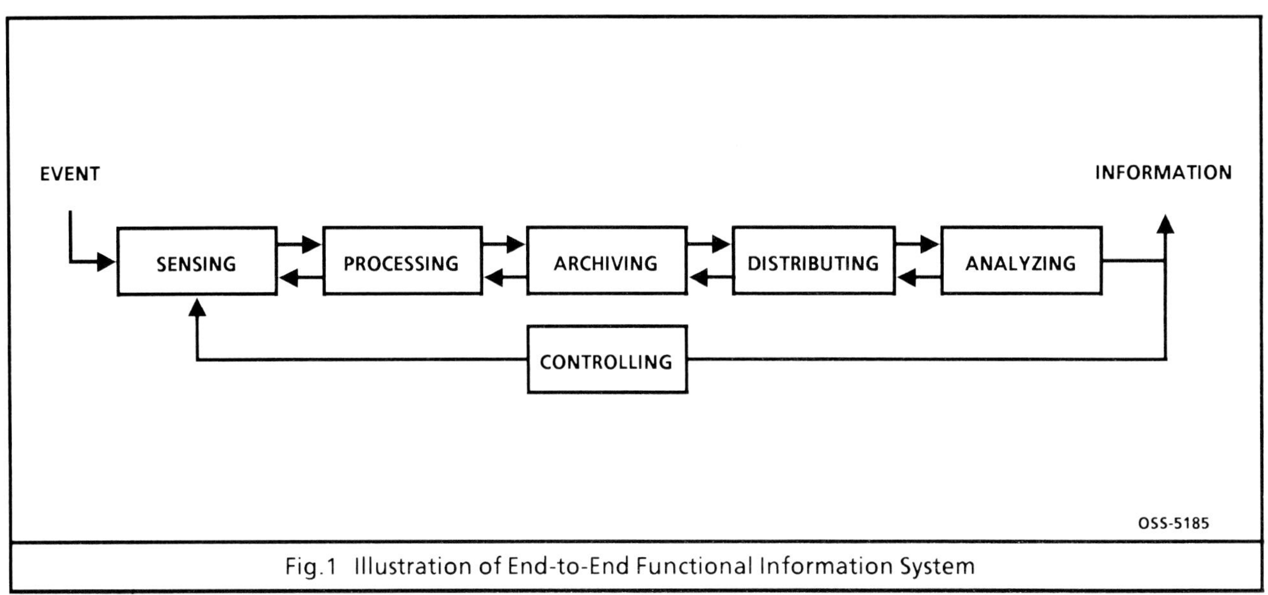

Fig.1 Illustration of End-to-End Functional Information System

Spacelab as a Model of the Future

The Spacelab missions provided hands-on experience with a mix of science experiments and users approximating what we will see in the 1990's. Valuable insight of the performance of the Spacelab data infrastructure was provided by a self-assessment team called the Spacelab End-to-End Data System (SEEDS) Working Group. That group suggested that future space information systems of the 1990s should reflect lessons and experiences such as the following:[2]

- The Spacelab data system requires users to learn and interact with several different data system interfaces as their instruments evolve through development, various levels of integration and test, actual flight use, and subsequent post-flight analysis. No standard user interface exists nor do standard methods for accessing support systems such as for ancillary data, crew availability, and mission planning and scheduling. The SEEDS Working Group recommended that NASA adopt a

uniform system control and operations language and user interface for payload test, operations, and data acquisition.

- Users are required to move along with their ground support equipment from center to center as the mission progresses; e.g., from instrument testing at the manufacturer facility or laboratory to integration and test at the Kennedy Space Center to flight operations at the Johnson Space Center (JSC). No capability exists in the realtime system at JSC for getting all needed data in electrical form for off-line analysis or to permit users to utilize home institution computing resources.

- Coordination has proven difficult between individuals performing different functions during a Spacelab mission. This in turn greatly reduces overall system effectiveness and thus ends up wasting precious mission time. Even the physical proximity of many of these teams at JSC has not alleviated the coordination difficulties.

- The SEEDS Working Group recommended that near-realtime, electronic access to ancillary data be provided to permit more timely processing of the scientific data into higher level products. At the time the SEEDS evaluation was conducted, there was delay of several months before ancillary data arrived to the investigator. Both the realtime and post-mission data analysis of many experiments were significantly and negatively affected by these conditions. In addition, the lack of realtime ancillary data in electronic format can lead to ineffective and cumbersome on-orbit operations because the investigator simply does not have enough information to make correct choices.

- The working group recommended that Spacelab instruments, flight subsystems, and onboard computers evolve to a packetized data format system. Insight gained from the many studies on spacecraft telemetry packetization has indicated that there are tremendous advantages to the users from implementing a packetized data system both onboard and on the ground. Sufficient experience has been gained to realize that such a data system is ideally suited   for complex, multi-user data type missions because data from various instruments can be separately packaged and then mixed, each packet is self-identifying, and the time-ordering of downlinked data (which is a problem because of the limited coverage of the TDRSS system) can be simplified by including a counter and identification code in the packet header.

- User command operations must be made easier and faster. The Spacelab mission forced users without access to current timeline planning data to go through many time consuming steps in order to specify, request, gain approval for, and transfer a simple command to their instruments. This was true even though most instrument commands were completely non-interactive with the rest of Spacelab and thus could cause no conflict with other instrument or subsystem needs.

The Space Station Information System

The lessons and experience of missions such as Spacelab are being assimilated into the Space Station Information System (SSIS) for the Space Station Program. The resulting system will be critical to the successful "useability" of the Program flight elements and the experiments and manufacturing processes those elements support. To be specific, the SSIS is defined as the integrated set of space and ground data and information networks which provide required data and information services to the flight crew, ground operations personnel, and the customer community. It includes as its elements not only flight element systems such as the onboard data management and communications and tracking systems, but also existing and planned institutional systems such as the NASA Communications System (NASCOM), the Tracking and Data Relay Satellite System (TDRSS), and the data and communications networks of the international partners. The SSIS is conceived to support the full range of users in all operations of their subsystems or experiments that involve data handling, processing, and/or storage regardless of where physically each user is located.[3]

Figure 2 illustrates the idea of networking the user community together and through such a hierarchy of local and wide area networks, providing access not only to each other and to their respective subsystems/ payloads but also to short term and long term storage facilities, data processing facilities, control centers, planning and other coordination activities, and all other space systems tasks that in some way require the exchange of information.

Figure 3 is a conceptual picture that indicates the all encompassing nature of these future space information systems. Note that the users are tied to a host of ground-based capabilities as well as to various kinds of payloads on potentially many different flight elements.

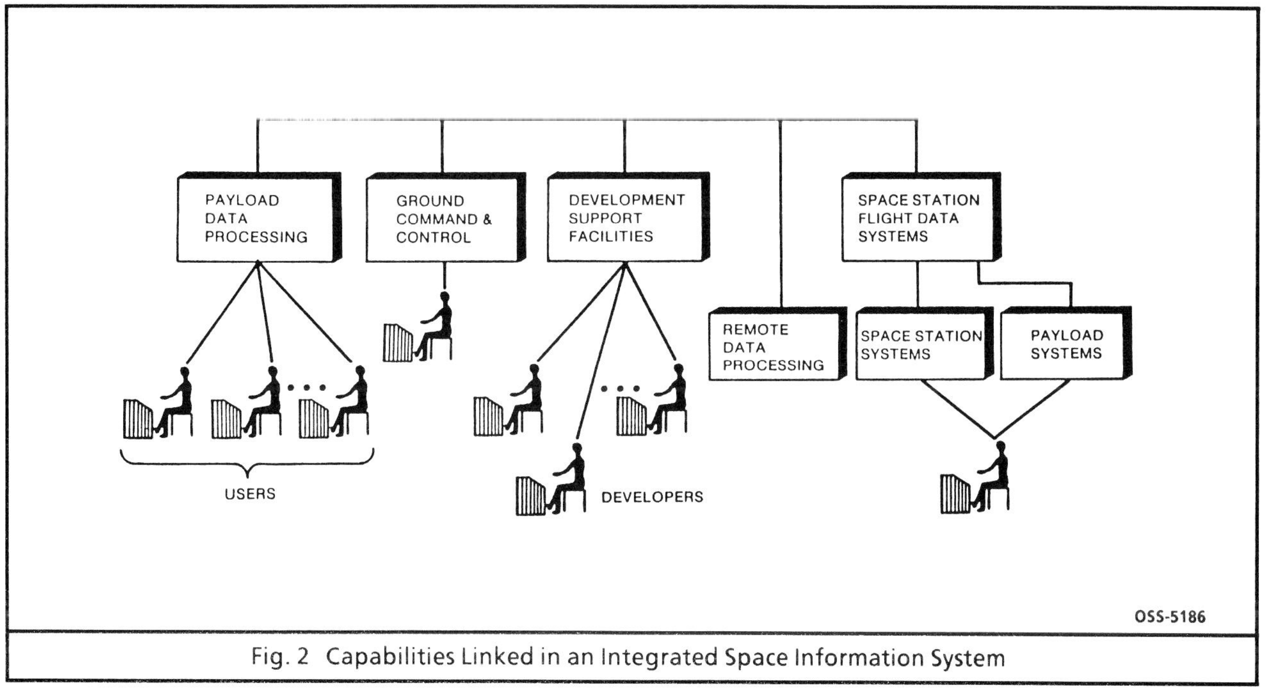

Fig. 2 Capabilities Linked in an Integrated Space Information System

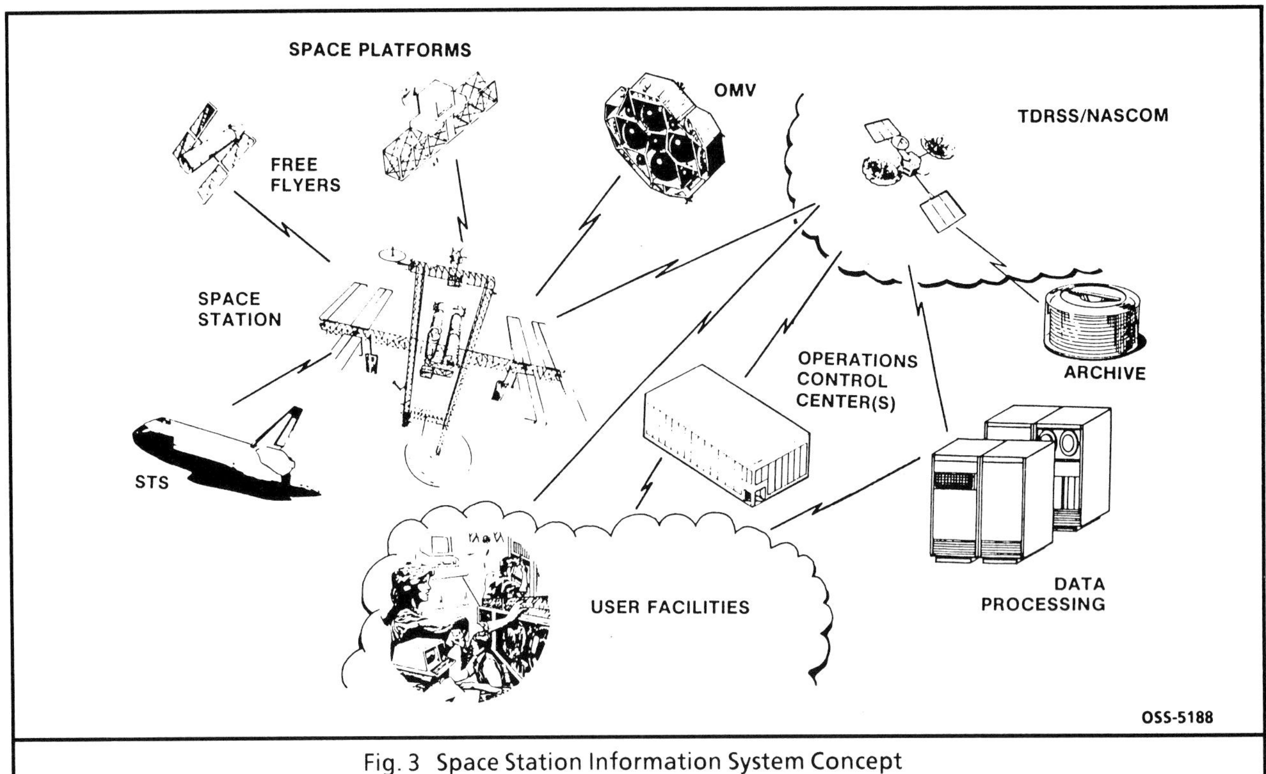

Fig. 3 Space Station Information System Concept

The end point of such an information system can be wherever the user is located. Sometimes that user will be a subsystem monitor located in a control center (whether human or an automated expert system). Other times the user might be a graduate student at a university conducting advanced research with a space located experiment. Alternatively, the user could be a crewperson aiding the conduct of a manufacturing process while in consultation with the ground home factory. The space information system end points will vary with time depending upon the desires of the users and the activities then underway.

Future space information systems like the SSIS will be characterized by autonomy in data structure, rapid data transport and reduction, universal data system standards and protocols, and high levels of fault tolerance and reliability.

The idea of data autonomy is that each data record of an event that can .be physically identified and separated in a complete and self-contained unit independent of other data units. Each such unit may include not only the data from the observation itself but also any ancillary parameters needed to transform the observation data into meaningful information. Such ancillary data typically might include the time the observation was made, the orbital location and attitude of the sensor at the data taking time, and other descriptors about the sensor or platform. Alternatively, the observation data packet stream might be interspersed with packets dedicated to ancillary data.

The concept of rapid transport and data reduction is based not only on good service to the user but also on cost. In general, the longer data and information remain undelivered in the system the more it costs the

system provider (e.g., NASA). Procedures which apply to reducing this time and obtaining information from the data include assessing which data has no useful information content or cannot be used from that which is useful as well as streamlining the delivery process via labeling and other rapid sorting aids.

The third major characteristic, that of standards and protocols, is perhaps most important and one of the more difficult problems for the information system designer. Most systems today are comprised of interconnecting computers and processing devices of different manufacturers. It is difficult to configure each element of the overall system so that data can be readily and reliably transferred and resources shared. This is due principally to the primitive nature of the systems interconnection standards that to date have emphasized only the physical level. To solve this problem, complete protocol standards that are agreed to by the hardware vendors and the users are needed. This includes not only the electrical link, but logical bit stream control, network control, traffic control end-to-end across several networks, dialog "sessions" control between two programs, "presentation" standards such as virtual terminal and virtual file storage, and "application-process" standards such as data base management, node management, remote job entry, and interactive user support. Included as well at the applications level are standard user interfaces such as screens and diagnostics. It should also be mentioned that standard interfaces (both in hardware and in software) permit more rapid sensor and subsystem development and integration. The SSIS definition team is investing much time in detailed assessment of the data standards concepts and products offered by external entities such as the International Standards Organization[4] and the Consultative Committee for Space Data Systems.[5]

Finally, the remaining example characteristic is system reliability and fault tolerance. Advanced information systems must demonstrate high responsiveness and reliability in correspondence to the increasing reliance on the systems at higher volumes and rates. Certain portions of the system such as those supporting flight safety must have very near 100 percent availability. It is possible to achieve this by designing the appropriate elements of the system so that if they fail, they fail soft and have modular redundancy; thus falling back perhaps to cold or hot backup elements. The question to be addressed, of course, is how much redundancy where in anticipation of what failures.

The SSIS now being defined for the Space Station Program tries to encompass these and other characteristics. Some of the major objectives of the SSIS are summarized below. These objectives are directly rooted in the lessons learned and experience gained from previous projects (like Spacelab) and research programs (such as the NEEDS) as mentioned earlier in this paper.

- Integrate the design, verification, and operation of flight element and institutional data/information system resources;

- Provide a "user friendly" environment (through techniques like a standard man-machine interface and minimal command checking constraints);

- Offer the user a private virtual circuit service to his destination whether that be in space or on the ground;

- Support a range of payload handling services in support of real time, quick look, archiving, data reduction, search and retrieval, and other functions;

- Provide core (ancillary) data where and when needed; and

- Utilize widely accepted commercial and international protocols and standards.

Much work has been invested by NASA and its supporting contractors in the definition and specification of the SSIS.

Some of the resulting proposals and ideas are discussed in other papers to be presented at this Symposium. [6,7] That work has resulted in a rather detailed understanding of the many challenges associated with such an advanced capability. On the technical list of challenges are items like the capture, distribution, and cataloging of the high rate (up to 300 Mbps) downlink flowing through the TDRSS. Methods being explored to help simplify that job include packetizing the data and assigning different kinds of data (perhaps differentiated by destination or type characteristics) into virtual channels. A virtual channel is nothing more than a common identification field in the packet header which can be used as a multiplexing and demultiplexing vehicle. In other words, the physical channel appears to be split into smaller logical pipelines.

Another challenge lies in the onboard buffering and storage of data usually in response to loss of TDRSS access (caused perhaps by zone of exclusion or, in the case of multiple flight elements served by one TDRS satellite, sharing of link support). Preliminary analyses indicate the onboard storage problem to be a particularly challenging one, especially if high rate users require no loss of data. Certainly one partial step to easing this challenge will be to do more data processing onboard. In the early stages, that processing will likely be limited to identifying and eliminating redundant or useless data such cloud obscured Earth images. We are closely monitoring the fast paced development of new storage technologies in industry; particularly optical media. It is reasonable to expect that the early years of the Space Station flight experience will rely on proven magnetic tape recorder technology and then upgrade to more sophisticated, higher capacity optical tape or disk devices as soon as possible in the latter 1990s.

Handling the users' command requests also is proving to be an interesting problem. As stated previously, it is a basic Space Station Program objective to provide a friendly, minimal interference environment to the user community. The problem with that objective when put in practice is protecting safety of the flight system and of other users from errant (unintentional or otherwise) commands. The most promising solution to this problem depends upon having sufficient margin in the onboard resources (such as power, data management, and thermal control) that each user can request and be given a time-varying envelope of resources for his application. This approach requires considerable advanced planning and coordination and onboard intelligence to detect and decide the corrective action when a resource boundary is being exceeded.

A fourth but certainly not least technical challenge is the sizing, selection, and flight qualification of a suitable family of processors and associated instruction set and support software. Although technology transparency is a Program goal, it will certainly be expensive and difficult to upgrade the flight processors whether standalone or embedded. The Program is investigating many candidate processors ranging from improved versions of the general purpose processors flying on the space shuttle to micro-VAX class machines to more exotic symbolic and numeric computers. The evolution and effective use of ground testbeds will play an important role in the establishment of the selection criteria and ultimate selection.

The SSIS in its entirety will transcend international and NASA-internal organizational boundaries. It requires commitment and involvement of resources beyond those controlled by the Space Station Program. Figure 4 shows the many organizations within and external to NASA that must cooperate in order for the concept of the SSIS to become a real capability. Although there are technical challenges in handling multi-megabit streams of data, processing user requests for service in real time etc., the real challenge in the SSIS in many opinions lies in its implementation management complexity. Detailed agreements are now being formulated between the respective organizations to specify development, funding, operating, and other responsibilities. The actual detailed definition and subsequent development will be aided by a hierarchy of working groups and configuration control boards.

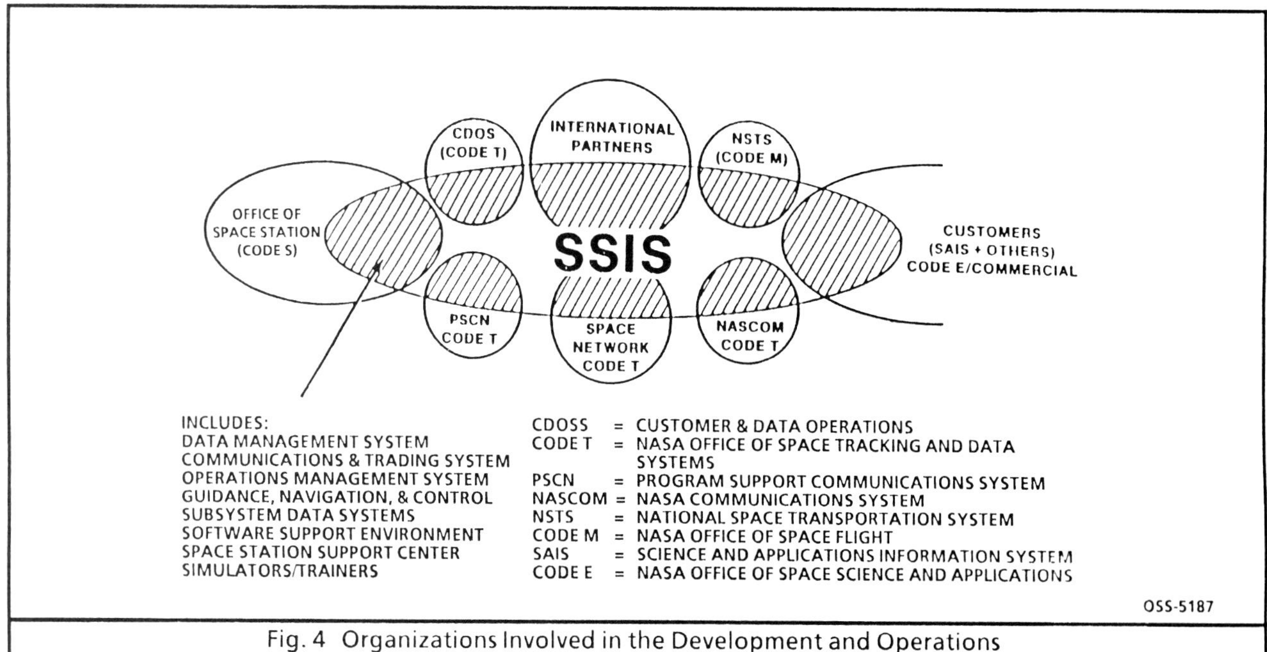

Fig. 4 Organizations Involved in the Development and Operations
of the Space Station Information System

Conclusion

Future space information systems will serve the user in the accomplishment of a wide range of activities. If designed and implemented properly, such capabilities will be as easy to use and flexible as current office automation systems. The only differences the user will see should be those resulting from the sheer physical separation and from the sharing of possibly scarce resources like crewperson time and space/ground link bandwidth. The success of the next generation of information systems depends upon learning from past space project experience and capitalizing on that knowledge and new technology. It is now entirely within our grasp to provide the user a very powerful, efficient, and effective space information system infrastructure. The system now being designed for the Space Station Program is one such example and should serve in turn as a springboard for subsequent information system advancements.

References

1. Obermann,R.M. and Austin, J.L., "Data Management Technology for Near-Term NEEDS", The MITRE Corporation, MTR-80W347, December 1980.

2. SPACELAB End-to-End Data System (SEEDS) Working Group Report, Office of Space Science and Applications, NASA Headquarters, October 24, 1984.

3. Space Station Information System Definition Document, Space Station Program Office, Washington, D.C., July 1986.

4. Information Processing Systems - Open Systems Interconnection - Basic Reference Model, Ref. No. ISO 7498-1984(E), International Organization for Standardization.

5. Consultative Committee for Space Data Systems, Space Station: Application of CCSDS Recommendations for Space Data Standards to the Space Station Information System (SSIS) Architecture, Green Book, October 1985, Office of Space Operations, NASA Headquarters.

6. Muratore,J., Bigham, J., Whitelaw,V., and Marker, W., "Space Station Information System Integrated Communications Concept", AIAA/NASA International Symposium on Space Information Systems in the Space Station Era, Washington D.C., June 1987.

7. Whitelaw, V., Marker, W., Muratore, J., and Bigham, J., "Interoperability in the Space Station Information System", AIAA/NASA International Symposium on Space Information Systems in the Space Station Era, Washington D.C., June 1987.

ESA'S FUTURE INTEGRATED SPACE DATA SYSTEM

Claude Honvault
European Space Operations Centre
Darmstadt, West Germany

Introduction

The European Space Agency is in the process of defining the components of its long term plan to be presented for approval to the Council of the Agency meeting at Ministerial level. This proposal will set up the basis for the work of the coming 13 years with the following objectives:

- participation in cooperative programmes as partners in the Space Station programme;
- eventual European independence of in-orbit infrastructure.

These two objectives have a strong impact on the definition of the programme, in particular on the facilities necessary to manage and utilise the large amount of data and information to be acquired, processed, transferred and archived from 1995 onwards.

This paper presents the status of the proposed solutions to such a problem, worked out at the level of the Executive, without prejudging the Agency's final decisions which will be taken at a later stage.

Overall Integrated Approach

The achievement of the objectives mentioned here above requires:

- the establishment of standards for electronic data exchange and processing;
- the implementation of an integrated system design approach for the ground facilities.

Standards for electronic data exchange and processing

Different levels of standardisation are required in line with the objectives of cooperation and independence:
- international standards;
- internal ESA standards.

International Standards

Since 1984, the Consultative Committee for Space Data Systems, with the assistance of numerous national and international space agencies, has been set up to define commonly agreed standards for telemetry, telecommand, data units, time datation, and localisation. The overall organisation of this committee is as shown in figure 1.

The CCSDS panels develop recommendations on technical issues of relevance to data exchange. This is done at several hierarchical levels, starting with concept papers and, after various

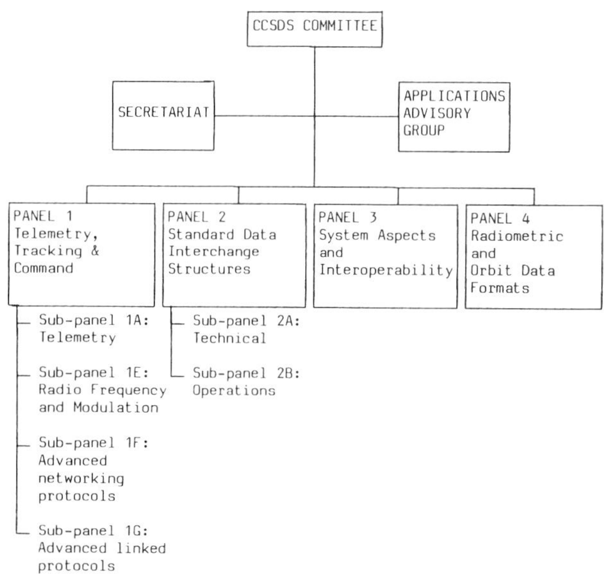

Figure 1

review stages, reaching a higher level of consensus until the final stage is reached: the issue of recommendations endorsed by the participating Agencies at the level of the committee, so-called "Blue Books" (previous issues are White or Red).

The rôle of the Applications Advisory Group is to convince as many programmes as possible to use the approved recommendations but it is up to each Space Agency to take the final decision of implementation. This implementation is a sine qua non step towards a possible transfer of data,

- from one element in-orbit to any station on the ground, or to any data relay satellite system, whoever the owner,

- from one data bank to any other without costly conversion processes.

Some of the results of the CCSDS work will be presented during this AIAA/NASA symposium.

Within the European Space Agency, coordination and supervision of activities in the area of combination of telecommunications and electronic data processing and applications to generation, transmission, acquisition, processing and storage of data (so-called "telematics") is considered necessary to keep pace with the latest state-of-the-art developments and trends in telematics, to acquire and operate cost-effective, coordinated telematic facilities that meet the Agency's current and foreseeable requirements. This coordination applies to telecommunications links and networks, electronic data exchange, end-to-end information systems, data acquisition technology and standards, computer-based automation processes, such as spacecraft check-out and electronic data processing.

For this purpose, specialised committees have been set up:

- Board for Software Standardisation and Control (BSSC);
- Standards for Telemetry and Telecommand Approval Board (STAB);
- Computer Standardisation Board (CSB);
- Committee for Operations and EGSE Standards (COES),

which are responsible for the coordination and control functions in their own fields and are grouped under the authority of the Telematics Supervisory Board (TSB), whose rôle is to:

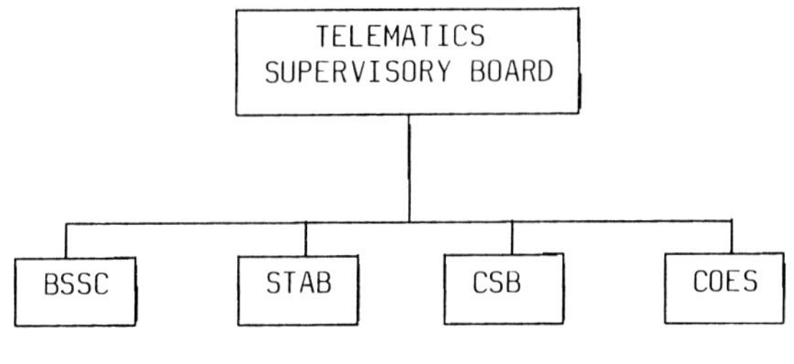

Figure 2

Internal ESA Coordination of
Data Management Activities

- survey the development of technology and international standards in the field of electronic information processing storage, and data exchange, assess the applicability to the Agency, and define the policy of the Agency in this field;

- review the existing telecommunications infrastructure of the Agency (both for voice and data communications) and formulate a policy for rationalisation, consistent with the technology and PTT Administrations' policy trends;

- establish a policy for the content of, and access to, data banks in the scientific, technical and administrative fields;

- coordinate the activities with the various internal Standardisation Boards and Committees in order to ensure coherence between the technical standards which will be applied on the ground and in space for the support of the

end-to-end data systems of future missions;

- establish telematic policies and, after approval by ESA's Director General (DG), will supervise their implementation;

- advise the Director General on the consequences and appropriateness (i.e. cost, schedule, risk, etc) of the implementation, in a given project or programme, of a new standard agreed by one of the specialised committees;

- review the performance of the specialist committees, audit their decisions and comment on their annual reports prior to their transmission to the Director General;

- review the specialised committees' terms of reference and, as appropriate, recommend changes.

This Board, chaired by the Director of Operations, is composed of representatives from ESA Programme and Support Directorates. In particular, it is in charge of approving for the Agency those standards generated by the different panels of the CCSDS before finalisation as Blue Books.

The coordination between CCSDS and TSB is ensured by the presence of identical experts sitting in both groups, panels and sub-committees.

<u>Integrated System Concept for the operations ground facilities of ESA's future programmes</u>

The above standards being considered as approved, it is necessary to build up a system concept which permits:

- the implementation of these standards;

- the combined operations between more than one element in orbit.

Indeed, the Long Term Plan of ESA includes the development of the Columbus programme elements:

- the Attached Pressurized Module (APM), permanently attached to the International Space Station;

- the Man-Tended Free Flyer (MTFF) space laboratory to be serviced by the Hermes spaceplane at the Space Station;

- the Polar Platform (PPF) to be serviced by the Shuttle;

- the Hermes spaceplane;

- the Coorbiting Platform, called Eureca-B, to be deployed and recovered by the Shuttle for refurbishment on-ground;

- the Data Relay Satellite System (DRSS).

None of these elements of the future European In-Orbit Infrastructure (IOI) can work independently, at least the DRSS is a permanent tool for all of them except the APM. For the APM it is planned that the transfer of data from space to ground take place via the U.S. TDRSS.

The PPF, the prime objective of which is to embark earth observation instruments, is complementary to the U.S. PPF. There is a strong interest on both sides of the Atlantic to receive, in quasi real-time, data from these two platforms. In addition, for the case of emergencies, there is also interest in using all communications resources available in space to ground systems. Since mid-1986, NASA and ESA have been studying the interoperability of U.S. and European relay satellite systems and Japan has decided to participate in the next meeting, which will take place in the first half of 1987. This coordination group is called SNIP (Space Network Interoperability Panel).

In the course of the preparation of the ESA Long Term Plan and in the context of the discussions concerning the coherence of the elements of the In-Orbit Infrastructure, the necessity was recognised for a more in-depth study into the configuration of the ground segment required to operate the space-based elements which are to be deployed. These ground facilities will represent considerable financial investment and their configuration will influence the level of such investments, as well as the yearly operating costs, for space facilities which are planned for a lifetime of up to 30 years. Since the scope of the ground segment for the IOI by far exceeds the capabilities of the present installations, there is the opportunity to develop a configuration which entirely meets the requirements and objectives identified, and which is not limited to existing concepts and infrastructure.

For this purpose, since the end of 1986 the European Space Agency's task has been to:

- analyse individual programmes and combined mission operations requirements;

- derive a functional task breakdown;

- identify the constraints (data processing power, data transmission capability, i.e. frequencies to be used, human resources, geographic sites, industrial capabilities);

- examine possible technical solutions and perform the necessary trade-off;

- determine interfaces between facilities;

- identify facilities which may be decentralised and those which must be co-located;

- identify the possibilities of national funding from certain Member States, in addition to the funding provided to the Agency's programmes;

- discuss and negotiate the integration of facilities developed with such national funding into the overall ground segment concept;

- for each individual programme, take into consideration political constraints and wishes from the participating states;

- establish criteria for the selection of the possible configurations;

and finally, to propose scenarii for the global configuration of facilities.

Unfortunately, there is no ideal single solution which will satisfy all requirements, and the final solution, whatever it might be, will be the result of compromise, depending upon the weighting factor of each criterion.

The following section describes the methodology applied for the definition of the possible scenarii.

<u>STUDY METHODOLOGY</u>

For a first analysis, the elements are arranged into a matrix with:

- mission or space element as rows;
- functions as columns.

Each matrix element then either represents a functional element of the ground system and its individual functions will be defined, or it is empty because the function is not required for the mission or element under consideration.

Each function is described in detail. For example: Mission Preparation includes all activities required to set up a functional operations system, from the early project inception until the start of the missions:

a) <u>Ground system preparation:</u>
   - operations requirements definition;
   - ground system design;
   - facilities procurement and integration;
   - staff recruitment;
   - preparation of operations procedures;
   - compilation of operations data bases;
   - training of operations staff;
   - scheduling of external support;
   - ground system configuration

b) <u>Ground system validation</u>
   - requirements validation;
   - design verification (against requirements);
   - facilities testing (against design);
   - end-to-end system validation;
   - back-up configuration validation.

c) <u>Mission specific functions</u>
   for Hermes (B_1):
   - landing sites configuration;
   - landing sites validation;
   - spaceplane configuration;
   - spaceplane validation;
   for manned missions (B_1, C_1, D_1):
   - crew training.

In order to establish a required or optimal grouping of different functional elements into facilities, where they will be co-located, the "connectivity" between pairs of functional elements is established based on two primary criteria:

- maximisation of safety (for man in space);
- minimisation of investment and operations cost, as far as compatible with the safety criterion.

A systematic evaluation of connectivities has been performed in the following areas:

  i) common safety requirements;
 ii) exchange of data/information;
iii) commonality of functions.

(i) and (iii) have an impact on the safety criterion. (ii) and (iii) have an impact on the cost criterion. This means that communications and data systems have a double impact on the connectivity analysis. The result of the connectivity study in the area of exchange of data/information is given in Figure 3 below.

When synthesizing:

- the results of the study;
- the financial considerations, in particular the possibility,
  -- to use and/or adapt facilities which will exist in certain countries;
  -- to use facilities which could be developed by certain Member States outside the ESA fundings;
- the political wishes of the major contributors to the large programmes,

the Agency proposes to develop a ground system based on the following concept:

Geographically decentralised element control centres, which will perform the following functions:

- configuration control of the element prior to the launch;
- generation of operations plans for the element control;
- generation of requirements and plans for the ground resources;
- preparation of ground dedicated facilities;
- mission direction during autonomous phases;
- flight control of the element;
- flight dynamics of the element;
- user interfaces;
- after mission: reconfiguration management, data processing of system parameters, performance evaluation, archiving of system data, generation of flight report, dedicated ground segment refurbishment;
- interfacing with the Combined Mission Control Centre described hereunder.

The necessary conditions which must be met to ensure operational safety for performing decentralised element control are:

- A centralised design authority exercising overall control of system integrity and verification throughout the development programme and the operational phase;

- Design and procurement must be oriented to meet the overall operational, safety and reliability requirements;

- Overall design must be completed and validated prior to the start of industrial procurement;

- In addition to the system assurance and verification function during operations, central control must be exercised over:

FUNCTION / MISSION	MISSION PREPARATION	MISSION PLANNING	DATA ACQUISITION DISTRIBUTION	SYSTEM CONTROL	PAYLOAD CONTROL	PAYLOAD EXPLOITATION	POST MISSION FUNCTION	QUALITY AND SAFETY ASSURANCE
COMBINED	A_1	A_2		A_4	-	-	A_7	A_8
HERMES	B_1	B_2	B_3	B_4	-	-	B_7	B_8
APM	C_1	C_2	C_3	C_4	C_5	C_6	C_7	C_8
MTFF	D_1	D_2	D_3	D_4	D_5	D_6	D_7	D_8
EURECA-B	E_1	E_2	E_3	E_4	E_5	E_6	E_7	E_8
PPF	F_1	F_2	F_3	F_4	F_5	F_6	F_7	F_8
DRS	G_1	G_2	G_3	G_4	G_5	-	G_7	G_8

<u>Figure 3</u>

-- the management, operation and allocation of the communication resources provided by the DRS system and the ground network;

-- combined operations utilising specific centralised facilities supplemented by the resources of the decentralised elements.

The Combined Mission Control Centre will:

- receive and evaluate processed operational data from:
  -- the Attached Pressurised Module Control Centre;
  -- the Man-Tended Free Flyer Control Centre;
  -- the Polar Platform Control Centre;
  -- the Hermes Spaceplane Control Centre.

- monitor the status of the terrestrial and space-based communications system, allocate these resources and control their use;

- receive and process medical and activity information concerning the mission specialists on board either Hermes or the Man-Tended Free Flyer;

- have direct and rapid access to extra-vehicular activity and tele-manipulator simulators;

- monitor and control proximity, docking and separation operation;

- monitor and control extra-vehicular activities;

- have direct and rapid access to rendezvous and docking simulators;

- compute attitude and position data for the Hermes and Man-Tended Free Flyer composite;

- organise the preparation of missions and ensur the operational qualification of the different teams with the equipment involved, i.e. by arranging the necessary simulations and rehearsals.

The implementation along these lines of the overall system can be depicted by the following diagram:

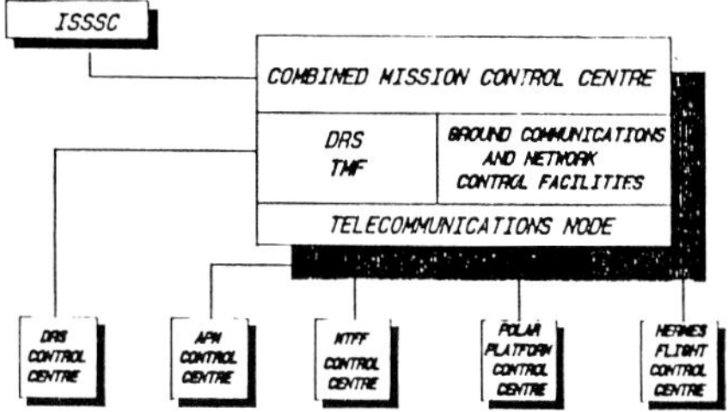

- *Concept requirements and overall design definition phase*
- *Procurement phase including the detailed design and implementation of each facility will be performed under the responsibility of an ESA Central Design Authority (located in ESOC)*
- *During the operations phase, ESOC will perform the overall system assurance and verification of all operations of ESA IOI elements.*

DRS: Data Relay Satellite
TMF: (DRS) Traffic Management Facility
ISSSC: International Space Station Support Centre
MTFF: Man-Tended Free Flyer
APM: Attached Pressurised Module

<u>Figure 4</u>

This very simplified diagram covers, in fact, a very complex Space Data System, as indicated in the figure 5 (next page). It is to be noted that the DRS system concept is also based on a decentralised system, whereby the downlink channels can be directly accessed by user terminals in the broadcast area and the uplinkings authorised by a unique Traffic Management Centre (see figure 6 following).

Figure 7 gives an overall view of the IOI Ground Segment Concept including the user centres and the engineering support necessary to the conduct of the operations in particular in case of difficulties on board the elements.

Figure 8 gives an overview of the communications requirements of the Hermes spaceplane, as they are identified today.

<u>CONCLUSION</u>

The previous chapters attempt to describe the approach taken by the European Space Agency to meet the challenge of the complex Space Data Systems of the next decade. The concept is still subject to discussions within the Executive of the Agency and with the Member States. The objective is to finalise it before the end of 1988 with detailed definition of each facility being completed by the end of 1989. This gives sufficient time to refine the system concept and to procure the equipment before a rather long validation period.

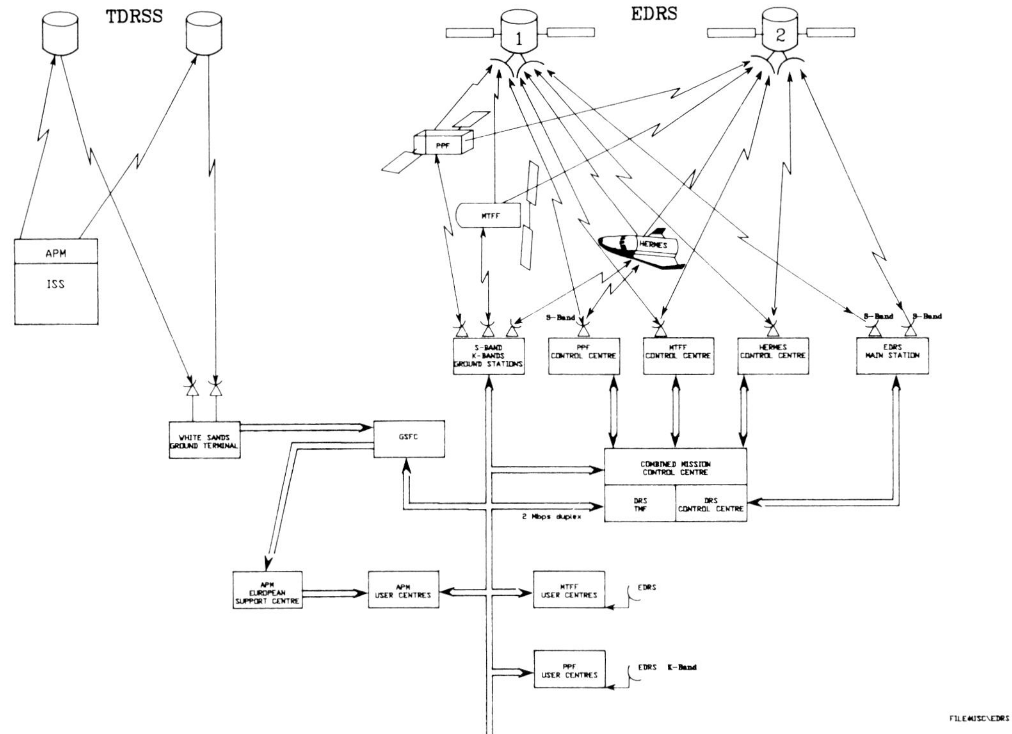

Figure 5

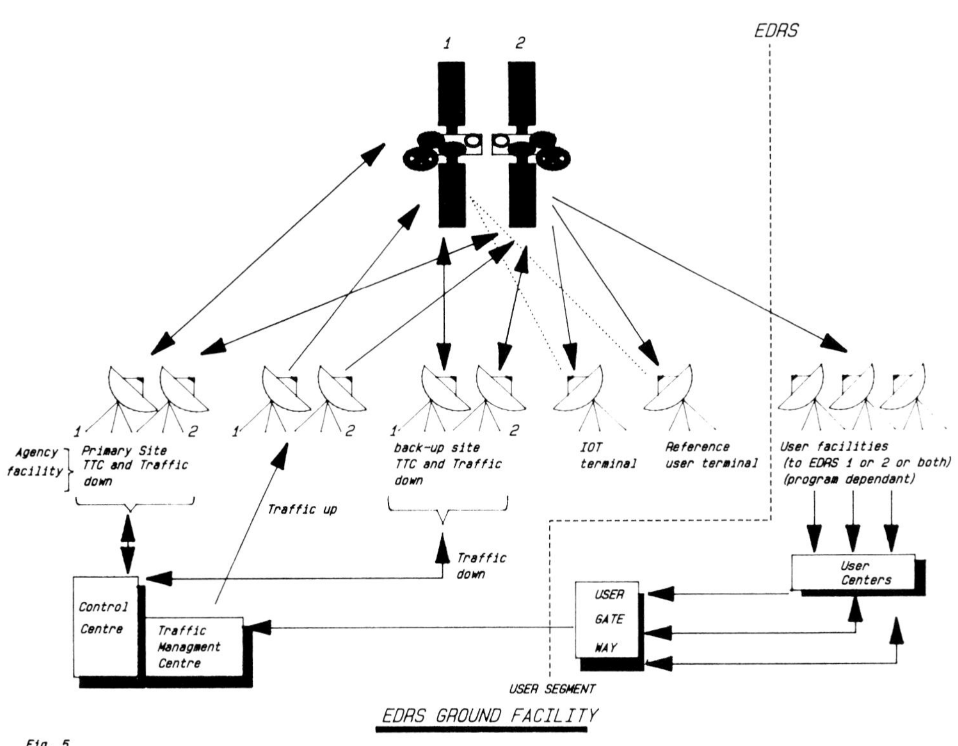

Figure 6

REF : DCIOIGOF
USA
CENTRALISED FACILITIES
ESA
DECENTRALISED FACILITIES
TDRSS NASCOM
ISS SUPPORT CENTRE
MISSION COORDINATION & PLANNING
HQ
CENTRAL MISSION CONTROL CENTRE
QUALITY ASSURANCE
ASTRONAUT'S OFFICE
INTEGRATED LOGISTICS SUPPORT
DRS TRAFFIC MANAGEMENT
CO-ORBITING PLATFORM CONTROL CENTRE
ESTEC
ESA
ESTEC
COMMUNICATIONS NODE
DRS CONTROL FACILITIES
PPF FLIGHT CONTROL CENTRE
APM FLIGHT CONTROL CENTRE
MTFF FLIGHT CONTROL CENTRE
HERMES FLIGHT CONTROL CENTRE
ASTRONAUT TRAINING FACILITIES
1
2
3
ARIANE V AND HERMES LAUNCH FACILITIES
LANDING SITES
KOUROU
PPF ENG. SUPPORT
APM ENG. SUPPORT
MTFF ENG. SUPPORT
HERMES ENG. SUPPORT
APM PAYLOAD INTEG. CENTRE
MTFF PAYLOAD INTEG. CENTRE
RESUPPLY
TRANSPORT
MAINT.
USER SUPPORT CENTRES
USER SUPPORT CENTRES
USER SUPPORT CENTRES
INSTRUMENT DEVELOP.
INSTRUMENT DEVELOP.
DETAILED CONFIGURATION OF THE IOI GROUND OPERATIONS FACILITIES

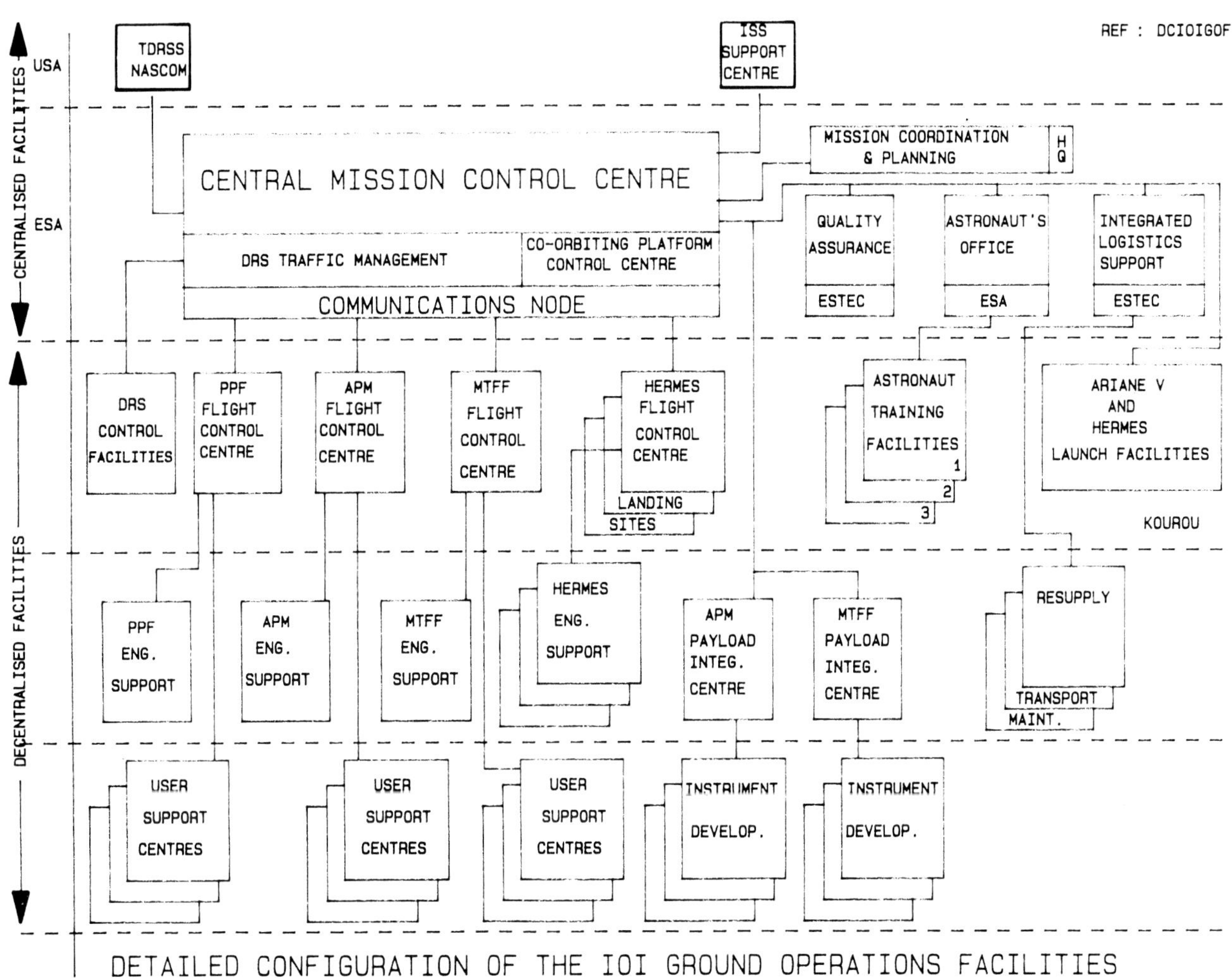

Figure 7

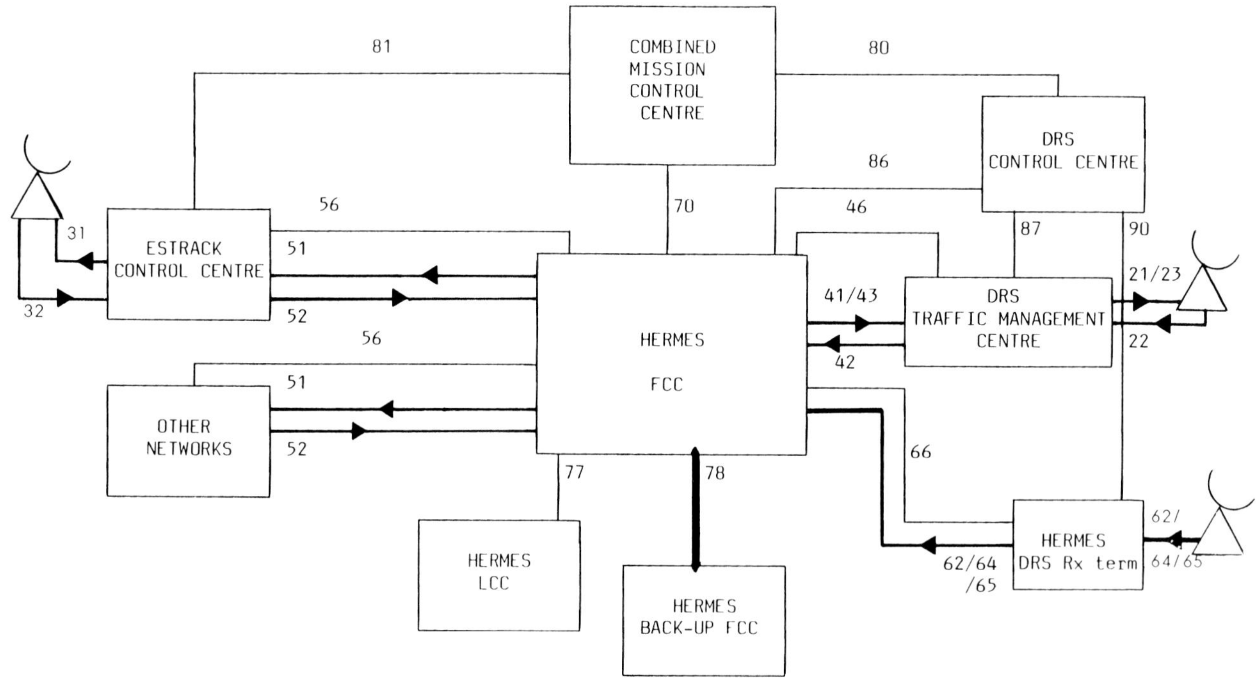

Figure 8

<u>Table 8.1</u> – Hermes Ground to Ground Communications description: Space Data Links

Code	Name		from	to	Mb/s peak rate	exp ber	main item	notes
21	Low Rate	FW	DRS/TMC	DRS/FEED	0.01	5	commands	pos. grouped with 46
41		FW	Hermes/FCC	DRS/TMC	0.01	5	commands	pos. grouped with 46
23	High Rate	FW	DRS/TMC	DRS/feed	0.30	5	commands	pos. grouped with 46
43		FW	Hermes/FCC	DRS/TMC	0.30	5	commands	pos. grouped with 46
22	Low Rate	RT	DRS/Feed	DRS/TMC	0.50		telemetry	
42		RT	DRS/TMC	Hermes/FCC	0.05	5	telemetry	
62		RT	H-DRS/Rx	Hermes/FCC	0.05	5	telemetry	redundant with 22+42
64	High Rate	RT	H-DRS/Rx	Hermes/FCC	10.00	5	telemetry	
65		RT	H-DRS/Rx	Hermes/FCC	34.00	4	television	? may be analog
31	Ground-Spaceplane		Network CC	Station	0.30	5	commands	
51			Hermes/FCC	Network CC	0.30	5	commands	
32			Station	Estrack CC	10.00	5	telemetry	
52	Spaceplane-Ground		Network CC	Hermes/FCC	10.00	5	telemetry	

<u>Table 8.2</u> – Hermes Ground to Ground communications description: Bi-directional links

Code	Name	from	to	
46	Traf. DRS	Hermes/FCC	DRS/TMC	CMCC operations
87	DRS-ops1	DRS/OCC	DRS/TMC	DRS operations
90	DRS-ops2	DRS/OCC	H-DRS/Rx	DRS operations
66	Herm-ops	Hermes/FCC	H-DRS/Rx	Hermes operations
56	Estrack ops	Hermes/FCC	Network CC	Hermes operations
70	CMCC ops	Hermes/FCC	CMCC	Combined Mission Operations
86	Hermes-DRS	Hermes/FCC	Hermes/LCC	Combined Mission Operations
77	Launch	Hermes/FCC	Hermes/LCC	Hermes operations
78	Back-up	Hermes/FCC	Hermes/FCC2	Hermes operations
80	DRS management	CMCC	DRS/CC	Combined Mission Operations
81	Network management	CMCC	Network CC	Combined Mission Operations

JAPANESE SPACE INFORMATION SYSTEM OVERVIEW

K. Matsumoto
Director of Tracking and Data Acquisition Department
National Space Development Agency of Japan

I. ABSTRACT

Japan's space development activities are implemented primarily by the National Space Development Agency (NASDA) and the Institute of Space and Astronautical Science (ISAS), in cooperation with other related organizations, and in accordance with the Space Development Program established by the Space Activities Commission, an advisory committee to the Prime Minister.

This paper briefly describes NASDA's present space flight control system and next generation system for H-II rocket and space station era with emphasis on system concept and implementation approaches.

Existing space flight control system consists mainly of S-band earth stations, tracking and control center and communications links. The system has successfully operated for about 20 NASDA satellites. The system also provides orbit determination support for ISAS satellites.

The preliminary conceptual study on the next generation system which is called space operations and data system (SODS) was conducted in house. The system will have sufficient capability to support operations of satellites launched by H-II rockets, Japanese Experimental Module of space station, and future space vehicles which will form space infrastructure. The SODS will not only satisfy Japanese future space operations needs but also contribute to the world as a part of world-wide space operations network realizing maximum data exchange and interoperability capabilities among space agencies.

II. INTRODUCTION

Present System

Present NASDA space flight control system consists of earth station network and satellite control system. The system has capability of VHF telemetry and telecommand, VHF/UHF one-way doppler tracking, and S-band telemetry, telecommand, ranging and one-way doppler tracking. Since 1981 satellite operation has been mostly executed by 2.1/2.2 GHz S-band system. NASDA S-band system covers full space operation S-band frequency range reflecting the conclusion of WARC '79. Thanks to support of other agency's networks such as NASA, ESA and CNES, the system has been able to extend its orbit visibility. On the other hand, the system could supplement the lack of orbit coverage of CNES network in the far east area receiving the telemetry data from SPOT-1. Satellite control is normally conducted in TACC (Tracking and Control Center) but TACSes (Tracking and Control Stations) also have satellite monitor and control functions to back up the system. Station operation is semi-automated but a small number of staff is assigned to do maintenance and keep reliability and

flexibility of operations. Dedicated mission operation stations such as earth observation data acquisition station are implemented separate from but in cooperation with the system.

Satellite control system in TACC consist of real-time satellite monitor and control system, communication interface equipment with agencies abroad, satellite data analysis and accumulation system, and flight dynamics system. The latter two systems are operated on the duplex main frame computers. On the other hand, the former two systems are operated on the same mini-computers. There are three kinds of work areas; Main Control Room (MCR), Mission Control Area (MCA) and Routine Phase Control Rooms (RPCRs). MCR is a focal place where directors can monitor the whole activities and make important decisions required to execute Launch and Early Orbit Phase (LEOP) operations. MCA is a focal place of satellite control and accommodates mini-computers, consoles and peripherals. One RPCR is assigned to each satellite and accommodates the same mini-computer system.

SODS Concept

Requirements for space flight control system evolves and swells year by year. The space operations system in Space Station era will not be realized by mere expansion of present system but require new system concept. The conceptual study on the space operations and data system (SODS) which can provide sufficient Japan's space operations capability in H-II and Space Station era, has revealed fundamental functions that is required to satisfy requirements for the SODS, namely
- Operations control function which can manage and control oprations of various space vehicles on orbit from the ground
- Communications network function which can connect various space vehicles on orbit with the operations control function on the ground
- Supporting information supply function which can supply useful information for effective SODS operations and space data utilization through communications network.

The three basic elements which realize above functions are as follows.
- Global communications network system which covers various altitude orbits world-wide
- Operations control system which provides space vehicles with sophisticated control or support effectively
- Engineering support and information system which executes planning, coordination and evaluation of the SODS activity. This also accummulates acquired data and effectively supplies necessary information to the payload experimenters and other data users.

The whole entity that consists of the combination of these indispensable operational elements is called the SODS. The basic structure of each basic element is described hereafter.

<u>Communications Network System</u>. This system consists of the following.
- Earth stations which connects space vehicle on-orbit with the ground network
- Data Relay and Tracking Satellite System (DRTSS) which provide continuous and high data rate communication links efficiently as a substitute for many earth stations
- Communications network which provides communications among earth stations, DRTSS, Operations control systems, Engineering support and information system, and other relevant facilities and organizations out of the SODS
- Network operations center which supervises and controls earth stations, DRTSS and communications network

<u>Operations Control System</u>. This System is defined as a collection of operational control centers for each space vehicle.

The system is classified into:
1) Type I system to control conventional satellites
2) Type II system to control manned system
3) Type III system to control over all on-orbit operations of space infrastructure such as randezvous-docking and re-entry.

<u>Engineering Support and Information System</u>.

This system consists of standardized computer file system and provides data service through communications network. The computer system accommodates data bases for SODS operation and user supports.

Typical data bases are:
1) Operations support data base
2) Space utilization data base

Fig. 1 shows SODS systematic structure.
Fig. 2 shows SODS elements and configuration with relevant outside entities.

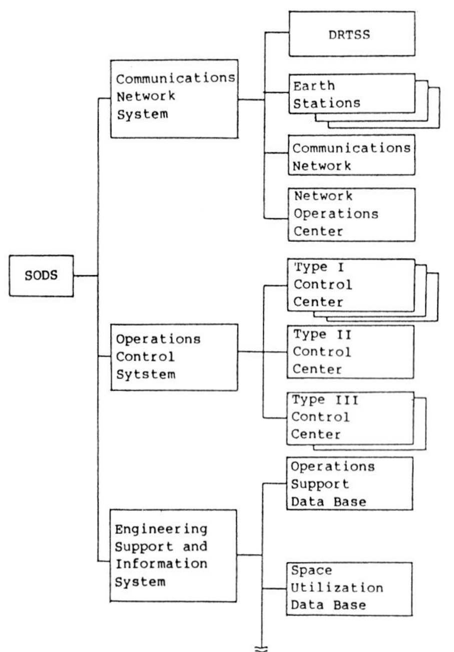

Fig. 1 Systematic structure of the SODS in early 2000s

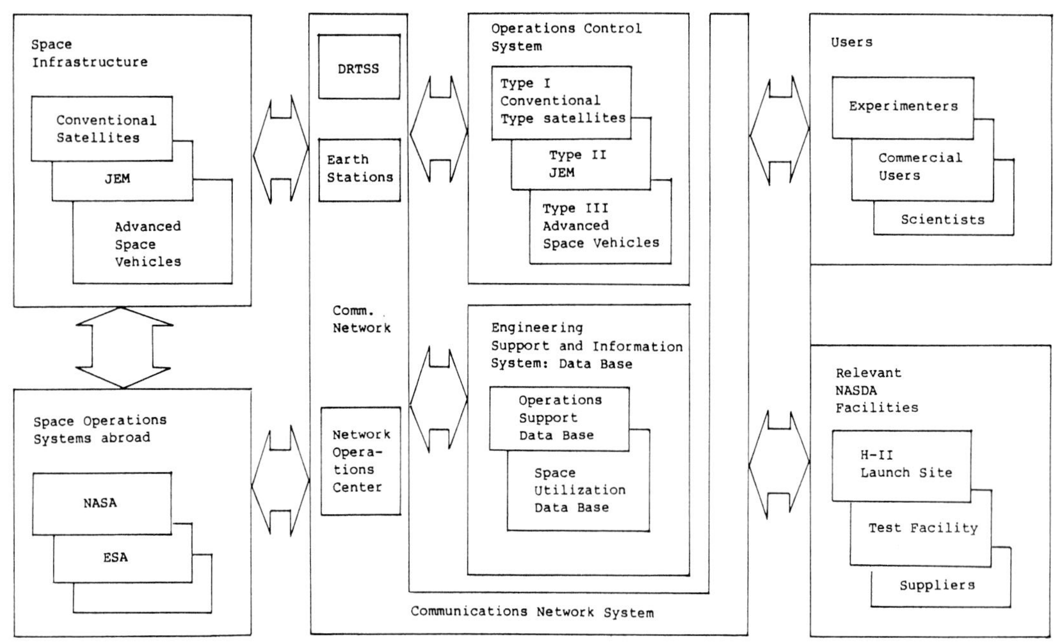

Fig. 2 Strawman structure of the SODS in early 2000s

III BACKGROUND

Current NASDA space flight control system has been evolved to meet each satellite requirements. This system has fully performed the required roles for NASDA space missions and it will also be capable of meeting requirement for coming several years.

However, in H-II and Space Station era, substantially different requirements are expected.

JEM Operation Requirement

JEM, which is fixed to NASA Space Station Base, provides facility for space experiments such as material processing. It is an international manned mission and needs functions to realize logistics activities such as transportation of crews and materials for experiment, and output substance for retrieval.

For this reason, continuous communication link for high data rate is required (initial stage: 20 -30 Mbps, mid term: around 100 Mbps). Therefore, communication network through NASA TDRS will be necessary.

For the next stage, as Space Station flourishes, development of NASDA own data relay satellite system is preferable to meet the increasing communication requirements. As the volume of data is enormous and complex operation is expected, sophisticated operation control and information sytems are needed.

Space Infrastructure Operation Requirement

In full scale space station era, Co-orbiting platform, polar orbiting platform, orbital servicing vehicle, space plane etc. emerge as low altitude orbit infrastructure. For geostationary orbit, orbital transfer vehicle and various geostationary platforms will be put into space. These vehicles are active and move around in space, so that monitoring and control from ground control center is required. Therefore, the facility to keep data communications and tracking capability via data relay satellite and earth stations without interruption is required. Particularly, control over randezvous-docking, remote manipulation, re-entry requires SODS addition of new capability - global coverage by DRTSS -.

H-II Rocket Launch Operation

H-II Rocket will be in operation from early 1990s. This rocket is capable of putting almost two tons of satellite up in to geostationary orbit. For the launch of satellites on commercial bases, it is necessary to minimize the launching cost by carrying multiple spacecrafts in a launch. In addition to launch operation, NASDA has to meet needs of LEOP operations of these spacecrafts.

Earth Observation

Earth observation mission requires high speed data reception, processing and distribution of large amount of data, so that it is considerable to be one of the major elements which may push data communications and data base capabilities up to the limit.

Summary of Requirements to the SODS

Above requirements are summerized as below:

1) SODS will facilitate operations of conventional type satellites, manned JEM and advanced space vehicles at the same time.
2) SODS will facilitate continuous and large capacity communications between those vehicles and operations control centers on the ground.
3) SODS will have capability to offer data service such as experiment data and (earth) observaiton data.
4) SODS will provide users and operators with necessary information for space vehicle operations and mission operations.
5) SODS will have capability of performing mutual cooperation with other agencies' space operations systems, especially with NASA and ESA.
6) SODS will promote standardization and realize total cost reduction along with improvement of efficiency of development.
7) SODS will be implemented adequately in step by step manner in order not to hamper daily operations being performed under existing system.

IV. CURRENT SPACE FLIGHT CONTROL SYSTEM

Current space flight control system in NASDA consists of network and mission control systems. Network function consists of earth station (TACS) network and network configuration control funciton in Tracking and Control Center (TACC) in Tsukuba Space Center. Mission Control System is also located in TACC.

Earth Station Network System

NASDA has three domestic ground stations (Katsuura, Okinawa and Masuda). During LEOP operation period NASDA has asked NASA/JPL DSN, CNES and/or ESA to support NASDA network in order to cover lack of visibility. The interconneciton with NASA and CNES is done by minicomputer gateway system in TACC which performs format and protocol conversion etc.

NASDA's earth stations are connected with TACC by dedicated communication lines which exchange telemetry, telecommand, facsimile data and verbal communication. Ranging data and station antenna prediction data are exchanged via NTT's packet switching system.

In addition to the standard telemetry receiving, telecommand transmitting and ranging functions, TACS has spacecraft monitoring and control function as a back up to the mission control function in TACC.

Each spacecraft monitoring and control function is performed by one pair of dedicated minicomputers located at TACS and TACC connected by one dedicated communication line. These systems also handle communication processing.

TACS has VHF, UHF, S-band transmission and receiving capabilities. The characteristics of antenna system are listed in Table 1, and Table 2 summarizes frequency range.

TACS facilities operation is monitored and controlled by station computer which enables semi-automated operation, and to reduce operation manpower.

Table 1. Characteristics of Antenna System of NASDA

Station	Frequency Band		Function	Antenna Type	Remarks
Masuda	VHF	148 MHz UP	CMD	Crossed Yagi	
	VHF/UHF	136 MHz DN 400 MHz DN	TLM/DOPP	18m Parabolic	
	S-band	2.1 GHz UP 1.7 GHz DN	TLM/CMD/RNG	10m Cassegrain	
		2.0 GHz UP 1.6 GHz DN	TLM/CMD		GMS mission
	USB(F)-1	2.0 GHz UP 2.2 GHz DN	TLM/CMD/RNG	18m Cassegrain	
	USB(F)-2	2.0 GHz UP 2.2 GHz DN	TLM/CMD/RNG	13m Cassegrain	
Katsuura	VHF	148 MHz UP	CMD	Crossed Yagi	
	VHF and USB(F)-1	136 MHz DN	TLM/DOPP	18m Parabolic Gregorian	
		2.0 GHz Up 2.2 GHz DN	TLM/CMD/RNG		
	VHF and USB	136 MHz DN	TLM/DOPP	18m Parabolic Gregorian	
		2.1 GHz UP 2.2 GHz DN	TLM/CMD/RNG		
	S-band	2.1 GHz UP 1.7 GHz DN	TLM/CMD/RNG	10m Cassegrain	
		2.0 GHz UP 1.6 GHz DN	TLM/CMD		GMS mission
	USB(F)-2	2.0 GHz UP 2.2 GHz DN	TLM/CMD/RNG	13m Cassegrain	
Okinawa	VHF	148 MHz UP	CMD	Crossed Yagi	
	VHF and USB(F)-1	136 MHz DN	TLM	18m Parabolic Gregorian	
		2.0 GHz UP 2.2 GHz DN	TLM/CMD/RNG		
	VHF and USB(F)-2	136 MHz DN	TLM	30m Parabolic	
		2.0 GHz UP 2.2 GHz DN	TLM/CMD/RNG	18m Gregorian	

GMS: Geostationary Meteorological Satellite

Table 2 Definition of frequency band and corresponding range

	VHF	UHF	S1	S2	USB	USB(F)
UP Link	148.0 ~ 149.9 MHz	---	2116.6 MHz	2034.2 MHz	2110.8 MHz	2025 ~ 2120 MHz
Down Link	136.0 ~ 137.9 MHz	400.0 ~ 401.9 MHz	1705 MHz	1694.0 MHz	2286.5 MHz	2200 ~ 2300 MHz

Control Center

Control Center in Tsukuba has responsibility of prime control of spacecraft operation and has three major functions: Network operation, spacecraft monitoring and control, and flight dynamics processing system commonly served for all projects.

Network operation performs operation resources scheduling and network configurations control and maintenance.

Spacecraft monitoring and control system has front end real time processing capabilities and back end support capabilities. Front end processor monitors spacecraft telemetry and transmits necessary commands to the spacecraft. Back end processor archives and retrieves data, and serves many types of trend analysis and display.

This system is a multi-satellite purpose one, however front end uses one computer and related facilities for each spacecraft. Computers and facilities of front end system are common and therefore they are interchangeable and can be used as backup of the other. Inter agency gateway system also utilizes this common system.

Flight dynamics support system also exists and processed in the back end system.

The Problem Area in the Current System

Current NASDA system has been implemented and modified for many years since 1974, therefore it is too old fashioned and is not adequate to the future system in space station era. From this standpoint the current system has the following problem area.
1) Co-existence of digital communications function in the spacecraft monitoring and control system
2) Old facilities require replacement
3) Small data service capability to the users
4) Lack of capability to handle large amount of high rate data
5) Excessive spacecraft monitoring & control capability in TACS results in increase of development and operations costs.
6) Limited coverage of domestic stations

V. SODS MAJOR ELEMENTS

In order to respond to the future requirements in H-II/Space Station era, the new Space Operation and Data System concept was studied. More detailed explanations of the major elements which appeared in system concept will be given in this section.

Communications Network System

DRTSS. DRTSS replaces many earth stations which support earth orbiting spacecrafts, and relays continuous high rate data between user spacecraft and ground feeder link terminal which is assumed to be located at Tsukuba Space Center. Two DRTSs will be located tentatively at 90°E and 170°W on geostationally orbit. This system will have a 90% orbit coverage for the spacecraft with 400km altitude.

DRTS has two communication capabilities with user spacecrafts, S-band low data rate link (max. several Mbps) and K-band high data rate link (max. several hundreds Mbps). The user spacecrafts need to have high gain and highly accurate pointing antenna for K-band communications, on the other hand need not for S-band communications.

The antenna pointing of DRTS shall be controlled at network operations center as is done for earth stations.

It should be also noted that DRTSS will play a key role for the success of missions of many user spacecrafts, the maximum compatibility with NASA and ESA system is highly desirable.

Earth Stations. The earth stations provide very reliable links between the spacecraft and the ground, and are mainly used in LEOP operations of the spacecraft. The link between spacecraft and a station is managed by the network operations center, and its facilities are planned to be remotelly controlled from the Network operations center. In order to extend the insufficient visibility of current ground station system during the LEOP operations, both the support from foreign agencies networks and the use of DRTSS must be taken into consideration.

Communications Network. The communications Network interconnects the DRTSS, the earth stations, operations control systems, data bases and data users, and plays following roles in SODS.
- to transfer space data between operations control centers and space vehicles
- to transfer data among data bases, operations control center and users
- to transfer data required to control and to monitor network itself
- to transfer management information within NASDA

The network of terrestrial and satellite communication circuits are leased from both the domestic and international carriers. The system is managed and operated comprehensively by the network operations center. The system is going to be implemented gradually according to the increase of needs. The system should have capability to commonly handle different types of data, this will enable efficient usage of communication circuit resources.

Network Operations Center. Network operations center manages and operates the earth stations, the DRTSS and the communications network. It also coordinates and conducts cross support with foreign space agencies like NASA and ESA.

Operations Control system

Type I Control Center. Type I control center manages and operates the conventional satellite missions from LEOP through routine phase. The mission operation planning, coordination with satellite users and evaluation of satellite performance are also conducted in this center. This type of centers will have capability to support simultaneous launch of multiple spacecrafts. In order to save operation cost, the facilities in the center must be much more standardized and adaptable to different missions than current system. The basic configuraiton and functions of the system are considered to be quite similar to the current system except operation performance and effectiveness.

Type II Control Center. The Type II control center manages and operates the early phase Japanese Experiment Module of the space station. In the scope of the SODS, two of this type centers, namely Tsukuba flight control center (TFCC) in Tsukuba and NASDA Johnson Branch (NJB), are planned to be settled. The NJB and the TFCC will be connected through the communications network system. The major role of the TFCC is to manage the experiments performed in the JEM, and to monitor and control the JEM in order to keep it in good conditions. Some portion of the JEM control function may be conducted at NJB, if very close coordination between NJB and the Space Station Support Center (SSSC) is required. The transmission of the several kinds of data such as telecommands, mission telemetry and house keeping data are exchanged through the TDRSS and the communications networks of the SODS. As a one of major relevant facilities of the TFCC, the JEM simulator is planned to be installed and connected to the TFCC in order to generate procedure for the space experiments and to investigate anomalies encountered in the operation. The data obtained in the JEM is processed and accumulated on the data base. The basic relations between the Type-II control center and the relevant facilities are shown in Fig. 3.

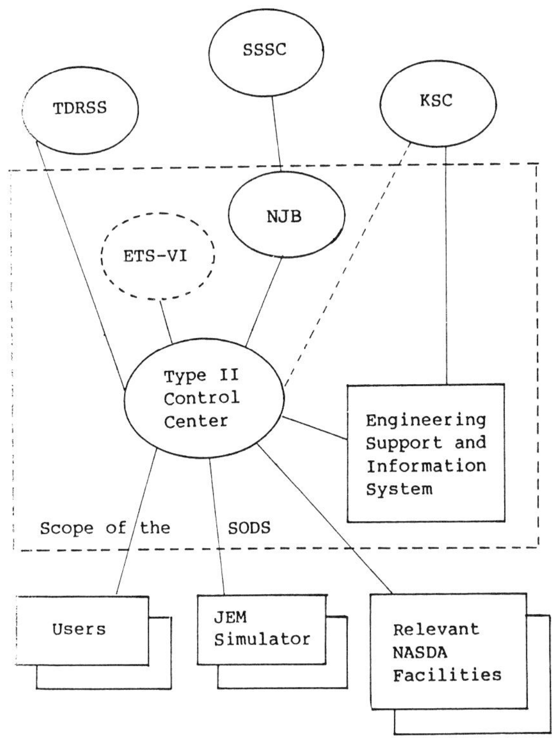

Fig. 3 Type II Control Center

Type III Control Center. The type III control center manages and controls the various types of space platforms simultaneously as an integrated flight mission, and enables sophisticated operations like randezvous-docking, various on-orbit operation, launch and retrieve operations where multiple space vehicles are involved in. For experiments performed in payloads, operations control function similar to the JEM control center will be required. In launch and re-entry phases, this center has interface with range control center or landing control center, and transfers operations control and necessary information to and from them.

When this center performs the above mentioned operations with JEM, the Type II control center and the Type III control center need to be connected or integrated. The basic relations between the Type III control center and relevant facilities are shown in Fig. 4.

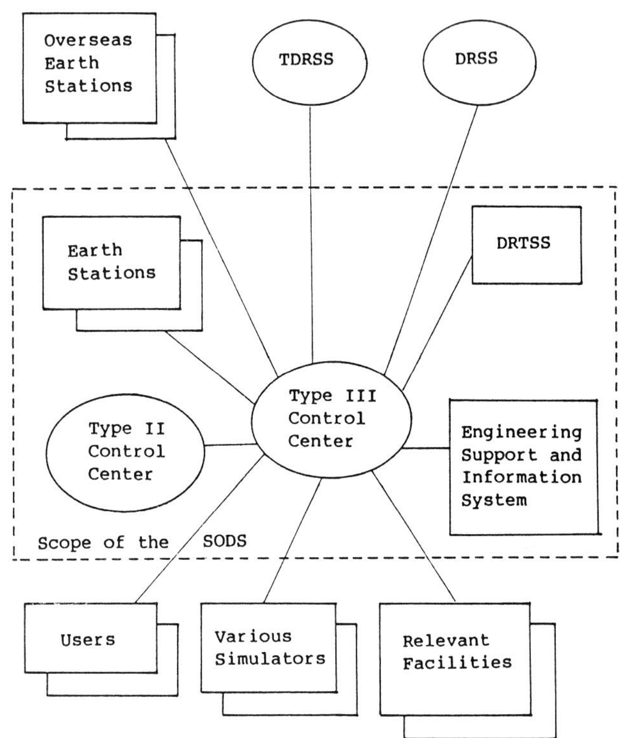

Fig. 4 Type III Control Center

Engineering Support and Information System. This system provides necessary information service for people involved in space operations to improve efficiency of operations and to promote space data utilization.
1) Operations support data base
  This data base consists of programs and integrated data files required for SODS operations planning, scheduling, evaluation of operation results and quick analysis in case of operations anomaly.
2) Space utilization data base
  This data base stores and provides observation data and space experiment data for vairous users, and supports data utilization such as sorting, retrieval, and engineering processing. This data base consists of the programs and integrated data files that are

required for planning, coordination and evaluation in support of space experiments performed on JEM.

Geographically distributed users can access to this data base through communications network system.

VI. IMPLEMENTATION APPROACHES

In SODS implementation, requirements from each project should be standardized, so that it makes minimize total cost in designing through operation phase. In order to meet above demand, it is insufficient only to pursuit efficiency of the SODS alone, but is very important to pursue the system which can realize maximum reliability and availability with minimum cost through maximizing interoperability among space agencies' systems. International activities such as Consultative Committee for Space Data Systems (CCSDS) and Space Network Interoperability Panel (SNIP) which coordinates interoperability of DRS systems are now seeking establishment of international standardization, interoperability and data exchange among various space data systems. In designing the SODS, NASDA will respect the results of those activities and other internationally recognized standards.

The SODS is tentatively planned to be implemented in accordance with the plan described below keeping steps with other programs.

Draft Implementation Plan

For SODS implementation, three main drivers are considered in developing the plan to meet projects need. These drives are:
o ETS-VI launch (Start of H-II era)
o JEM operations
o Operation of space infrastructure

Followings are principal activities to be performed in each phases;

1) Phase until ETS-VI Launch
 - To separate existing communications network from operations control systems and implement Type I control center.
 - To implement basic portion of the communications network which is compatible with world space agencies' communications networks.
 - To implement basic portion of the network operations center.
 - To implement further operation automated earth stations.
2) Phase until JEM Operational
 - To connect the communications network with NASA TDRSS and Johnson Space center.
 - To implement Type II control center
 - To develop communications experiment system between ETS-VI and JEM to conduct Preliminary experiment for DRTSS development.
 - To develop data bases for space utilization and operations support.
3) Phase until Operation of Space Infrastructure
 - To implement DRTSS
 - To implement Type III control centers for advanced space vehicles
 - To connect the communications network with ESA DRSS.

VII. CONCLUSION

This paper briefly presented Japanese next generation system which will fulfill Japanese space operation requirements in early 2000. NASDA will pursue to establish an integrated system that can keep appropriate autonomy and efficiency of Japanese space operation capability in cooperation with other agencies' space data systems. To realize this, NASDA recognizes further international cooperation and coordination in planning through operation phases are desirable.

Table 3. Development Schedule of the SODS

Fiscal Year	1986~7	1988	1989	1990	1991	1992	1993	1994~5
Common System				▲ BS-3a ERS-1	▲ BS-3b ▲ TFI △	△ ETS-VI	TBD	△ △ △ △ Space Station DRTS
1. Study	System Study							
2. Software		Design		Development and Integration		Operation		
3. Hardware			Design	Development and Adjustment		Operation		
Individual Ground System								
1. ETS-VI		Design		Develop.	Integration Operation			
2. Space Station			Design			Development	Integ.	Operation
3. DRTSS				Design		Development		Integration

Systems 1: Customer Needs
Co-Chairmen: Paul H. Smith and Hans E. Hoffman

EUROPEAN CUSTOMER NEEDS: A VIEW OF DIVERSE INDUSTRIAL
CONCERNS, INDUSTRIAL ISSUES, **INTOSPACE's** PLANS AND PROGRESS

Hans E.W. Hoffmann
Managing Director INTOSPACE GmbH.,
Hanover, W. Germany

HANS HOFFMANN: I represent one of these various users who were the subject of the discussion in the later part of this morning's panel, who are always talked about but obviously never appear alive. We talk about their needs and their requirements but we fail to ask them - themselves - what they really want from the aerospace people and I, therefore, in my contribution here this afternoon want to make a few remarks which come from the user world and which may serve as a background for the discussions going on here these three days and which will go on for a long time, I think, while we are designing the space station and are going to build it.

THE USER INDUSTRY IS THE NON-AEROSPACE INDUSTRY.

HOW ARE THEY INFORMED ABOUT THE OPPORTUNITIES OF SPACE UTILIZATION?

GENERALLY BY THE MEDIA.

IN EUROPE THERE EXISTS NO SYSTEM FOR AN IN-DEPTH INFORMATION BY THOSE WHO HAVE THE KNOW-HOW FOR THOSE WHO NEED TO KNOW.

THIS IS THE BACKGROUND OF THE INTOSPACE FOUNDATION.

The user industry, the real user industry is a non-aerospace industry. How are these users informed about opportunities of the space utilization? Up to now, generally by the media. In Europe, at least, there exists no system for an in-depth information by those who have until now the know-how and for those who need to know. That means those who have to know how are the aerospace industry, the governmental agencies and the universities. So the background of the foundation of INTOSPACE was to create a promotion organization to help the user industry to know more about space, the possibilities of space and to pull them on board our boat.

Speaking as a German but also generally as a European, the real user is the one who some day will invest his own money in space. He will make money with working in space and he will

pay taxes for the work and for the turnover he is doing by using technologies in space. This is for us the ultimate goal and one day the objective of a commercial utilization. That means we should try right from the beginning to get those private companies on board who some day may become the commercial users of the space station.

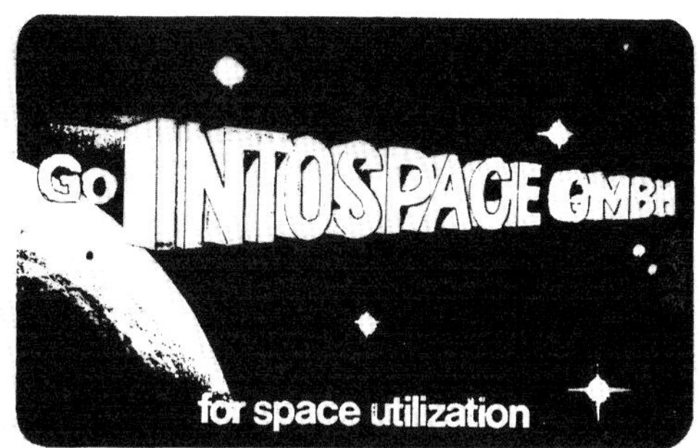

PURPOSE OF THE COMPANY

AN INTERNATIONAL COMPANY TO PROMOTE, INITIATE
AND SUPPORT MICROGRAVITY SPACE ACTIVITIES SUCH AS
RESEARCH, DEVELOPMENT AND PRODUCTION IN SPACE AS WELL AS TO
PROVIDE ASSISTANCE AND CONSULTANCY FOR SUCH
ACTIVITIES

The purpose of the INTOSPACE company on an international basis it to promote, initiate and support microgravity space activities such as research, development and production. Of course, now at this time it is only research, later on development and hopefully one day production in space.

PURPOSE OF INTOSPACE

o REDUCING CURRENT OBSTACLES TO SPACE UTILIZATION

o PROVIDING INFORMATION ABOUT SPACE OPPORTUNITIES FOR POTENTIAL USERS

o COORDINATING USER REQUIREMENTS WITH SPACE UTILIZATION

o IMPLEMENTING THE POLITICAL OBJECTIVES TO STIMULATE INDUSTRIAL APPLICATION OF ADVANCED RESEARCH

INTOSPACE supports industrial experiments
under microgravity and in vacuum, taking
advantage of existing carrier systems and
experimental equipment. We offer to the
industrial users preferred access to flight
opportunities and we try to have the best
relations to the responsible government agencies. INTOSPACE will help to reduce the current
obstacles to the commercial utilization and to
provide information about space opportunities
for potential users, to coordinate the user
requirements with the space utilization and,
last but not least, to implement the political
will of our government as I mentioned before.
Our target is to have one day a commercial
utilization of these very expensive facilities
which we are building and designing now.

We have ninety-four shareholders out of nine
European countries. We are a limited company,
according to German law, with a central office
in Hanover/Germany. At present we employ ten
people who are - except for myself who comes
from the space world - scientists and technical
people coming from those branches which are the
potential user branches.

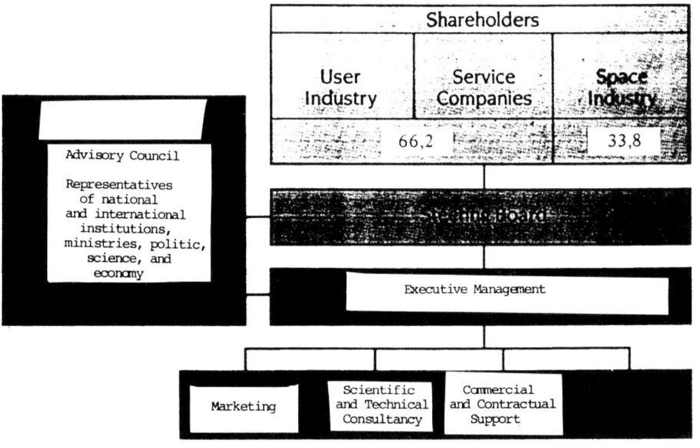

If you look at our shareholders, we have two
thirds non-aerospace and only one third aerospace companies. The original objective was to
get the user industry into this venture in a
majority position, that means in a position
carrying responsibility and taking initiatives
for the future.

If we talk to the user industry - what I am
doing very often now - the general impression
of the non-aerospace industry about space is
that space is a very risky business. Space is
very expensive. Space flight preparation takes
a very long time. Space is a government business.
The users are scientists. If you want to acquire
the technology which is gained by space flights,
you really do not need space. This shows that
there still is a big problem to be solved to
close the gap between the potential user industry
and our aerospace world.

RESULTS OF
A HEARING ON SPACE COMMERCIALIZATION
BEFORE THE U.S. HOUSE OF REPRESENTATIVES IN JUNE 1985

DIFFICULTIES :

(1) LIMITED AVAILABILITY OF CAPITAL	(2) INSITUTIONAL AND POLITICAL UNCERTAINTY
INVESTMENT IN SPACE IS UNPOPULAR	LONG-TERM GOVERMENT ASSURANCES
– LOSSES IN ILL-ADVISED INVESTMENT PROPOSALS – RELATIVELY LONG PAYBACK PERIODS	– CREATION OF THE INFRASTRUCTURE – CONTINUOUS SUPPORT FOR RESEARCH – TECHNOLOGY TRANSFER TO THE PRIVATE ENTERPRISE SECTOR (E.G. TECHNOLOGY CENTERS)
(3) HIGH COSTS OF TRANSPORTATION	(4) HIGH DEGREE OF RISK
– MORE FLIGHT OPPORTUNITIES – REGULAR LAUNCH SCHEDULE – PUBLIC PRICING POLICY	– DIFFICULTIES OF EXPERIMENT TESTING PRIOR TO AN EXPENSIVE SPACE FLIGHT – LONG PREPARATION TIME

I think we all agree when looking at commercial
utilization difficulties listed here under four
headlines:

We have the problem of limited availability of
capital. The private finance world is not yet
ready to invest money into space. There have been,
of course, ill-advised investment losses and
then there are the very unusual long payback
periods which are not effective for our present
banking systems.

We have institutional and political uncertain-
ties, this with respect to our political demo-
cratic system. The long-term investment of
industry cannot take the risk that after a four
years legislature period and following election,
there may be a new party coming into government
and introducing a new space policy which is
diverting from the policy of the last four years.
Industry needs continuity and stability. We have
to create an investment climate assuring this
stability.

We also need a system to transfer to the private
sector the know-how which has been accumulated
in the government agencies and at the universities having dealt already with the field of
space technology as well as with the field of
microgravity.

We have, of course - and I think we all should
be aware of this - the very high costs of transportation which at the moment represent a tremendous hurdle for the users. I need not speak
about the flight opportunities. We need flight
opportunities, very regular ones, we need flight
opportunities being frequent and reliable.

We need flight schedules which are a little bit
reliable like those of the European train service so that you know when you go to the station you can catch the next train and leave
according to the schedule.
And we need a reliable pricing policy being
valid over years, so when you start a long-term
investment and you finally get to fly, you know
what the price will be.

There is, of course, a risk. We all know this,
it is a special feature of space flight. And
there is still a very long preparation time
which is completely unusual for industry.
Chemical companies nowadays develop their products in three years and have, at the latest,
after five years the return of the investment
occurring. If we look at the present shuttle
manifest, our next Spacelab will fly in 1991,
that is four years from now. This means for the
experimenter to get the next chance for an
experiment within the period of four or five
years only. This is the time in which the normal
return of investment is already occurring for a
product which is started now in the industry,
the so-called user industry.

THE DIFFERENT STAGES LEADING TO COMMERCIAL UTILIZATION

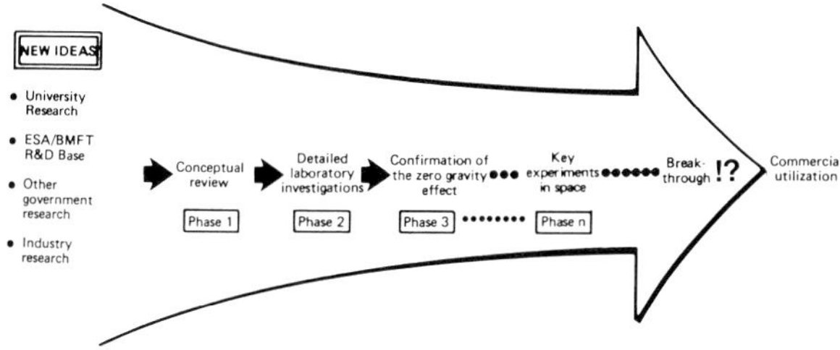

To counteract and counterargue the points of the
longtime development for space experiments, we
try to teach the potential user industry that it
takes a long time between the first idea and the
final commercial breakthrough. It needs a lot of
iterated loops, ground-tests first and then
space experiments to confirm the microgravity
effect. We can only hope that we may find soon
some potential users who will accept this new
mode of operation and learn to adopt in their
own internal strategy and planning that there
will be new product development lines taking
much more time than the usual products on the
ground.

OVERVIEW OF FLIGHT OPPORTUNITIES

Space System	TEXUS*	MAUS*	SPACELAB*	Reusable PLATFORMS SPAS*/EURECA°	Space Station COLUMBUS+
Duration of microgravity	6 minutes	up to 5 days	up to 10 days	up to 6 months	permanent
Type of mission	ballistic flight	orbital flight in the space shuttle		free-flying in low earth orbit	500 km orbit
Commercial Aspects	o reasonably priced o short access time o minimal technical and administrative complexity o suitable for small-scale commercial operation	o reasonably priced o short access time o independant of orbiter resources o suitable for small scale commercial operation	o wide range of experiments o more resources and services available than for TEXUS and MAUS o direct operation by payload specialist o suitable for medium-sized commercial operation	o very good micro-gravity level o advantageous price/payload ratio o suitable for large-scale commercial operation	o permanent operations o practically unlimited resources o manned and unmanned systems o very large commercial potential

* system available
° in development
+ in design

TEXUS = Abbreviation of the German project name, meaning technology experiments in zero gravity
MAUS = Abbreviation of the German project name, meaning self-contained materials science experiment in zero gravity

In principle, there are enough flight opportunities created in form of a variety of launch
vehicles. In Europe, a system has been established ranging from the very simple fall-tower,
sounding rockets, falling capsule systems to
the large number of shuttle-related flight
opportunities, get-away specials, the Spacelab,
SPAS and EUREKA and then, of course, the Space
Station. In this system we can offer the user
to begin very simple, very cheap, very fast,
make small experiments and work himself into the
more complicated, more expensive systems and
ending up in the space station. However, today
the availability is the problem as well as the
frequency of the flights. This has to be improved to become attractive for industrial experiments. Looking at the user world, we owe them
an answer in this field. We cannot offer them
space experiments and then when they ask for
the next flight opportunity, we have no answer.

FLIGHT OPPORTUNITIES

FALL TOWER	SPAS
PARABOLA FLIGHTS	EURECA
MIKROBA II	SPACEHAB
TEXUS	INDUSTRIAL SPACE FACILITY
MARIANNE	COLUMBUS
GAS/MAUS	HERMES
SHUTTLE (AFD)	LONG MARCH
SPACELAB	MIR AND OTHERS

There are, however, more flight opportunities.
We have very concrete offers from the People's
Republic of China for flights with their laun-
chers as well as from the Soviet Union. This
adds to the list of flight opportunities and
widens the variety and possibilities. Whether
they are applicable for commercial use, will
have to be seen. While waiting for the improvement in the launcher field, we continue our
preparation work. I think that the time is not

wasted because the information gap is so big
that we need this time in order to bring potential users up to the level so that they can
become real users and we can recommend them to
invest their own money.

MAIN DIRECTION OF ACTIVITIES

- INTENSIFY WITH PRESENT EXPERIMENTORS PRESENT FIELDS OF UTILIZA-
  TION WITH POSSIBILITIES TO IDENTIFY NEW FIELDS

- FIND NEW EXPERIMENTORS FOR PRESENT FIELDS OF APPLICATION

We try to intensify the work of our present
experimenters and to extend their activities
to new fields. Together with them, we try to
find new experiments for those fields which we
have already exploited.

OUR CUSTOMERS ARE MAINLY IN EUROPE. THEY

- PRODUCE WITH HIGH DEGREE OF ADDED VALUE

- ARE VERY HIGHLY RESEARCH AND DEVELOPMENT ORIENTED

- DEAL WITH MATERIALS NOT PRODUCEABLE UP TO NOW

- ARE INNOVATIVE

- ARE UNDER PRESSURE TO INNOVATE

For the time being, our customers are mainly
in Europe. They produce a high degree of added
value. They are highly research- and development
-oriented. They deal with products which are
not produceable up to now. They are under
pressure to innovate. The very important
industrial structure problems we have in Europe
bring certain customers to us when you look at
the ailing shipping yards and coal mining
industry in Europe and the very politically
debated field of nuclear energy. Space is one
of the promising future technology fields which
seems to be at least accessible without political problems and which also seems to have a real
future outlook.

IF WE WANT INDUSTRIAL MICROGRAVITY USERS:

- WE MUST GUARANTEE CONFIDENTIALITY LIKE IT EXISTS FOR INDUSTRIAL TERRESTRIAL
  RESEARCH AND DEVELOPMENT WORK

- WE MUST FURNISH ALL AVAILABLE INFORMATION FROM GOVERNMENTAL ORGANIZATIONS AND
  SCIENTIFIC SOURCES

- SCIENTIFIC MICROGRAVITATION UTILIZATION MUST NOT BECOME A SELF PURPOSE BUT IT
  MUST HELP INDUSTRY TO ENTER INTO THIS FIELD - ALSO IN ORDER TO JUSTIFY THE VERY
  LARGE PUBLIC INVESTMENTS

- PUBLIC FUNDING MUST ALSO BE MADE AVAILABLE TO INDUSTRY AT LEAST FOR THE STARTING
  PHASE

The first criterion of a user is the question of
confidentiality. If somebody invests his own
money, he wants to have the results for himself
- mainly - in a competitive environment.

The second criterion is the request for information what has been done before and what seems
to be possible. They want to have contact with
recognized scientists in their respective field
of activity. There is a great interest to have
a real scientific background of the microgravity
work which is intended to be done and in very
most cases the industrial companies do not have
enough scientific capacity to answer these
questions very clearly by themselves such that
investments can be made and recommended.

There is also a very great interest for contacts
with complementary industrial partners irre-
spective of their nationality. The industry is
very clearly profit-oriented and the passport
of the partner - be it an industrial or a scientific partner - plays no role whatsoever.
And then there is, of course, the question of
the finance which in this initial phase has to
be solved to reduce the burden of the very high
transport costs. There may be soon users ready
to finance their own experiment at a hundred
percent. But to pay the very expensive ticket
is a matter where the governments still must
come in. Here in this country, the joint endea-
vour agreement is a solution and there are similar solutions in practice by the European governments. We all recognize that in this early phase
it has to be done this way.

THE FIRST INDUSTRIAL USER CRITERION IS : CONFIDENTIALITY

THE SECOND INDUSTRIAL USER CRITERION IS: DETAILED INFORMATION OF WHAT HAS' BEEN DONE
                                          BEFORE AND WHAT SEEMS TO BE POSSIBLE

THE THIRD INDUSTRIAL USER CRITERION IS : CONTACT WITH REKNOWN SCIENTISTS IN THEIR
                                         RESPECTIVE FIELD OF ACTIVITY

THE FOURTH INDUSTRIAL USER CRITERION IS: CONTACT WITH COMPLIMENTARY INDUSTRIAL PARTNERS
                                         IRRESPECTIVE OF THEIR NATIONALITY

THE FIFTH INDUSTRIAL USER CRITERION IS : FINANCE

There are some further recommendations how to
treat users. Building the space station should
involve the scientific and industrial users
from the design phase on and not face them with
a 'fait accompli'. Agencies and space industry
should furnish all available information volun-
tarily and not with these strong hesitations.
Public funding should also be made available to
industry at least in the starting phase.

The launcher organizations - mostly government
organizations so far - must learn to be tran-
port service organizations. The customer has
to be the king. User industry should be involved early in the equipment design as well as in
handling and operation decisions. This is a
very important message we must keep in mind
for the space station design.

I think it is very important that we ask users
now to contribute or to participate in the
space station design, the payload system design,
operational design and not decide again everything within the aerospace community and then
show them what we have decided and they may
accept it or not.

Spacelab was developed when we had no user at
all. We had a reference payload which was just
a piece of paper and later on the users showed
up as living organizations and people with own
ideas, suggestions and requirements.

As to the space station, we should begin the
dialogue with the user world very early to
hear what they really want. The user needs a
variety of transport means to have the flexibility in microgravity time, resources, transport price and preparation time. Standard
equipment should be made available and easily
accessible for ground and for on-board purposes.

We must offer more small and inexpensive flight
opportunities that fly frequently and at low
prices. Therefore, INTOSPACE plans to fly the
equivalent of one TEXUS flight per year during
the next five years, starting early '88. That
is about 10 months from now. It is the first
INTOSPACE mission, however by this small missiom
we close and cover the gap between now and the
next possible Spacelab flight as well as
possible flights with the shuttle and with get-
away specials in the early 90's. The more complex missions must be derived from the simpler
missions, this being a very clear experience
we have made.

INTOSPACE proposes a COLUMBUS preparation with
strong industrial involvement. The D-2 Spacelab
mission manifested now for 1991, will have one
third INTOSPACE payloads on board. We suggest
that the German government fly another D-3
mission with about fifty percent industrial
payloads for the COLUMBUS Space station utilization by science and by industry.

Thank you, Mr. Chairman,
Thank you, Ladies and Gentlemen !

JAPANESE CUSTOMER NEEDS FOR SPACE STATION

K. Higuchi[*], I. Iizuka, Y. Fujiwara
National Space Development Agency of Japan
2-4-1,Hamamatsucho, Minato-ku, Tokyo,Japan

Abstract

This paper describes Japan's present utilization requirements as of the end of the preliminary design phase and presents abstracts of the proposed missions, especially in the communications area. The results of mission analysis of an information system reflecting the mission scenarios and the basic concepts of the JEM information network system are then introduced.

1. Japan's Proposed Missions and Scenario

Mission's Classification

Japan has been participating in the Space Station Program which is promoted under international cooperation since the preliminary design phase, with the goal of constructing the Japanese Experiment Module (JEM). Mission requirements for JEM design have been studied by the Space Activities Commission's Ad Hoc Committee, through the space station utilization workshops which were managed by the Science and Technology Agency (STA), the National Space Development Agency of Japan (NASDA), and the Japan Space Utilization Promotion Center (JSUP).

Japan's plans to utilize the Space Station were studied in June 1984 when the Ad Hoc Committee on Space Station Program executed its first series of studies. A total of 165 proposed themes were received from universities, national research institutes, and private companies. The second series of studies on Space Station mission requirements was conducted in October 1985. The number of themes proposed for the utilization, especially in the areas of space environment utilization (material and life sciences) and technology development, doubled to a total of 327 as a result of this study.

These proposed themes from users (which are called Level III) were classified into the following six discipline areas:

1 Scientific observation
2 Earth observation
3 Communications
4 Material processing and production
5 Life sciences
6 Technology development

Fig. 1-1 shows the ratio of themes by area. Considering their similarities and commonalities, these experiments were gathered into 39 groups (which is called Level II). Table 1-1 lists all Level II missions. These were input into NASA's Mission Requirement Data Base (MRDB), which is being jointly developed by the United States, Europe, Canada and Japan. Themes such as missions T-002, E-002, and C-003 which were too general and ambiguous were subdivided and presented as T-002A and T-002B.

Communications Missions

Most of the proposed themes in communications require technologies to be developed and verified in order to realize the Geostationary Orbit Platform and to implement the personal communication systems using them in the future. They are also closely related to the technology development missions such as Large Antenna System Technology (T-002A and T-002B). Their coordination is thus of great importance in developing mission scenarios. The abstracts of communication missions (C-001 to C-005) and some technology development missions (T-002A and T-002B) are presented below.

(1) C-001 RFI (Radio Frequency Interference)
Objective: To provide Space Station design data on RFI due to the electromagnetic transmissions from free flyers and the ground.
Description: Open areas for antennas attached to various places of Space Station. Instrumentation and data storage require a pressurized module.

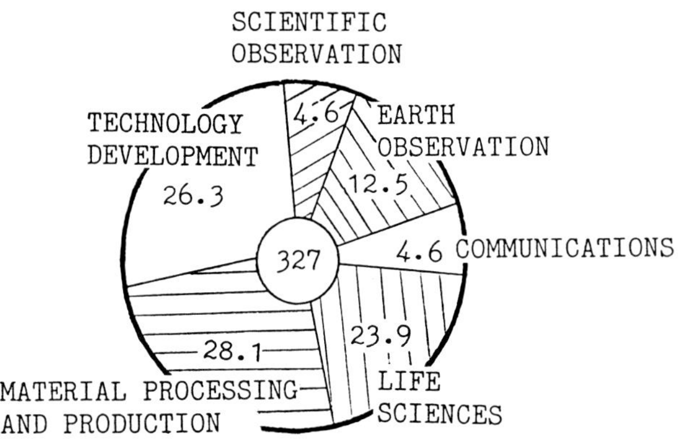

Fig. 1-1 Ratio of proposed themes by area

---

* Manager/Space Experiment Group, NASDA

(2) C-002 Large communication antenna
Objective: To construct a high-accuracy, deployable antenna system and to test its functions.
Description: An antenna (aperture 5m, surface accuracy 0.1mmRMS) will be constructed and tested. Checkout of deployment, verification of the high accuracy surface control system, and antenna pattern measurement will be conducted by utilizing the Space Station facilities.

(3) C-003A ADV communication and data handling system-A
Objective: To conduct high-rate data transmission experiments in the Ka-band between the Space Station and experimental data relay satellite (ETS-VI) to establish the technology of the Ka-band single access high rate of data transmission. Real time data transmission is in the S-band via ETS-VI. Ranging signal relay is in the S-band via ETS-VI.
Description: Perform the low, medium, and high rate data transmission experiments in both the S-band and the Ka-band between the Space Station and ETS-VI by using two antennas installed at the SS upper boom. After the link is established, the following data transmission experiments will be performed:
-Communication functional test
-Antenna tracking functional test
-Data processing functional test
This mission will be performed prior to data relay operation by using the data relay tracking satellite system.

(4) C-003B ADV communication and data handling system-B
Objective: To conduct the high rate data transmision experiments in the millimeter wave band between the space station and experimental data relay satellite ETS-IV. To establish the technology of single access data transmission.
Description: Perform the high rate data transmission experiments (max. 300Mbps) in the Space Station to ETS-VI by using an antenna of T-002A mission and a transmitter on the Space Station.

(5) C-003C ADV communcation and data handling system-C
Objective: To develop application methods through the set up and operation experiment of interstations and interstation network using optical fiber and laser.
Description: Assemble and set up the communication equipment and links. Evaluate computer hardware and software. Measure mechanical and electrical performance such as error rate caused by thermal beam width and pointing inaccuracy. Transmission experiments will be performed between SS and free flyer

(6) C-004 Gravity stabilized deployable antenna test
Objective: To deploy a large mesh reflector stabilized by means of the gravity gradient. To verify the stability of an active tether control system with a scale reflector model on a free flyer.
Description: A satellite system with a gravity stabilized large deployable antenna (40m diameter and 80m focal length) will be mounted on the end of a mast extended from a free flyer. The experiment covers deployment test, reflector surface control test, and the overall satellite system stabilization control.

(7) C-005 TD of large geostationary satellite
Objective: To provide data base for design and operation of large geostationary satellites assembly, checkout, launching and maintenance. Technology development of a Japanese OTV will also be included in this mission.
Description: A large geostationary satellite will be divided into three portions. The satellite will be assembled and tested on SS. It will then be transferred to geostaionary by OTV. The Japanese OTV capability will be verified.

(8) T-002A Large antenna system technology-A
Objective: To aquire assembly and measurement techniques for a large precision antenna reflector used in the high frequency domain. Assembly is to be performed mainly by manipulators. Operational tests using this are also planned.

Table 1-1 Japanese missions

CODE	MISSION (LEVEL II)	CODE	MISSION (LEVEL II)	CODE	MISSION (LEVEL II)
C-001	RFI	L-001	BIOLOGY AND MEDICINE	M-001	MATERIAL SCIENCE EXPERIMENT
C-002	LARGE COMMUNICATION ANTENNA	L-002	SPACE MEDICINE	M-002	SPACE PROCESSING FOR ADVANCED MATERIAL
C-003A	ADV COMM AND DATA HANDLING SYSTEM-A	L-003	CELSS	M-003	COMMERCIAL SPACE PROCESSING TEST
C-003B	ADV COMM AND DATA HANDLING SYSTEM-B	L-004	BIOTECHNOLOGY		
C-003C	ADV COMM AND DATA HANDLING SYSTEM-C			T-001	SPACE ENVIRONMENT TEST
C-004	GRAV STABILD DEPLOYABLE ANT. TEST	S-001	ASTRONOMICAL PLATFORM	T-002A	LARGE ANTENNA SYSTEM TECHNOLOGY-A
C-005	TD OF LARGE GEO SATELLITE	S-002	INFRARED TELESCOPE IN SPACE	T-002B	LARGE ANTENNA SYSTEM TECHNOLOGY-B
		S-003	HIGH ENERGY COSMIC RAYS	T-003	LARGE SPACE STRUCTURE
E-001	LASER RANGING SYSTEM	S-004	COSMIC GAMMA RAY BURSTS	T-004	SPACE ENERGY EXPERIMENT
E-002A	TEST OF SENSOR TECHNOLOGIES-A	S-005	LINE GAMMA DETECTION	T-005A	SPACE ROBOTICS-I
E-002B	TEST OF SENSOR TECHNOLOGIES-B	S-006	X-RAY ASTRONOMY OBSERVATORY	T-005B	SPACE ROBOTICS-II
E-002C	TEST OF SENSOR TECHNOLOGIES-C	S-007	GRAVITATIONAL WAVE DETECTOR	T-005C	SPACE ROBOTICS-III
E-003	OBSERVATIONS OF UPPER ATMOSPHERE	S-008	SPACE VLBI	T-006	PLATFORM SYSTEM TECHNOLOGY
E-004	RTRSOC	S-009	SOLAR ACTIVITY MONITOR	T-007	2D-SOLAR ARRAY MISSION
E-005	DPS FOR EARTH OBSERVATION	S-010	SUBMILLIMETER TELESCOPE	T-008	SOLAR THERMAL GENERATOR
E-006	EARTH OBSERVATION FACILITY	S-011	IMAGING PHOTOMETER FOR AURORA	T-009	LIQUID PROPELLANT HANDLING
				T-010	ADVANCED SPACE POWER SYSTEM

Description: This is a preliminary experiment using small aperture (2m diameter) for the next phase large aperture (10m diameter) antenna assembly experiment (T-002B).

1 Assemble the precision millimeter wave antenna from reflector segments and attach equipment using manipulators controlled by IVA.
2 Measure mechanical parameters and characteristics such as surface accuracy, vibration modes and frequencies, and thermal distortion.
3 Mesure electrical characteristics such as radiation pattern.
4 Perform operational experiments such as weather radar, microwave radiometer, and communication experiments using the assembled antenna.

(9) T-002B Large antenna system technology-B
Objective: Same as T-002A
Description: Perform a large aperature (10m diameter) antenna assembly experiment. The main procedure is the same as in T-002A.

Through the discussion in the workshops, the following technologies have been indicated as the key technologies which should be developed as early as possible using Space Station system:
-Large antenna assembly and handling technology
-Intersatellite optical communication technology (including pointing control system)
-Reliability evaluation of communication subsystems and components
-Maintenance technology of satellite
-Large satellite assembling technology
-Orbital transfer vehicle (OTV) technology

Furthermore, the fundamental technology for free flyers and platforms which are desired for use in many areas and the technology related to the JEM service facility must be cultivated as early as possible.

-Undeveloped fundamental technology related to omnibus design technology (technology for reconstructing the payload on the bus)
-Service technology for the free flyer (repairing facility, orbital servicing vehicle, etc.)

Mission Scenario

Mission scenario was also discussed in the workshops, and its preliminary plan since Initial Operational Capability (IOC) was studied by NASDA. As a result of the workshop's discussion, the step-by-step implementation of the missions was emphasized, that is from fundamental themes to advanced, or easy technology to complicated. Fig. 1-2 is one of the most possible scenarios. Material and life science experiments will be carried out in the pressurized module. Communications experiments will primarily be carried at the exposed area. Fig. 1-3 shows the part of the mission scenario relating to communications. It describes the continuity and the development of these IOC missions.

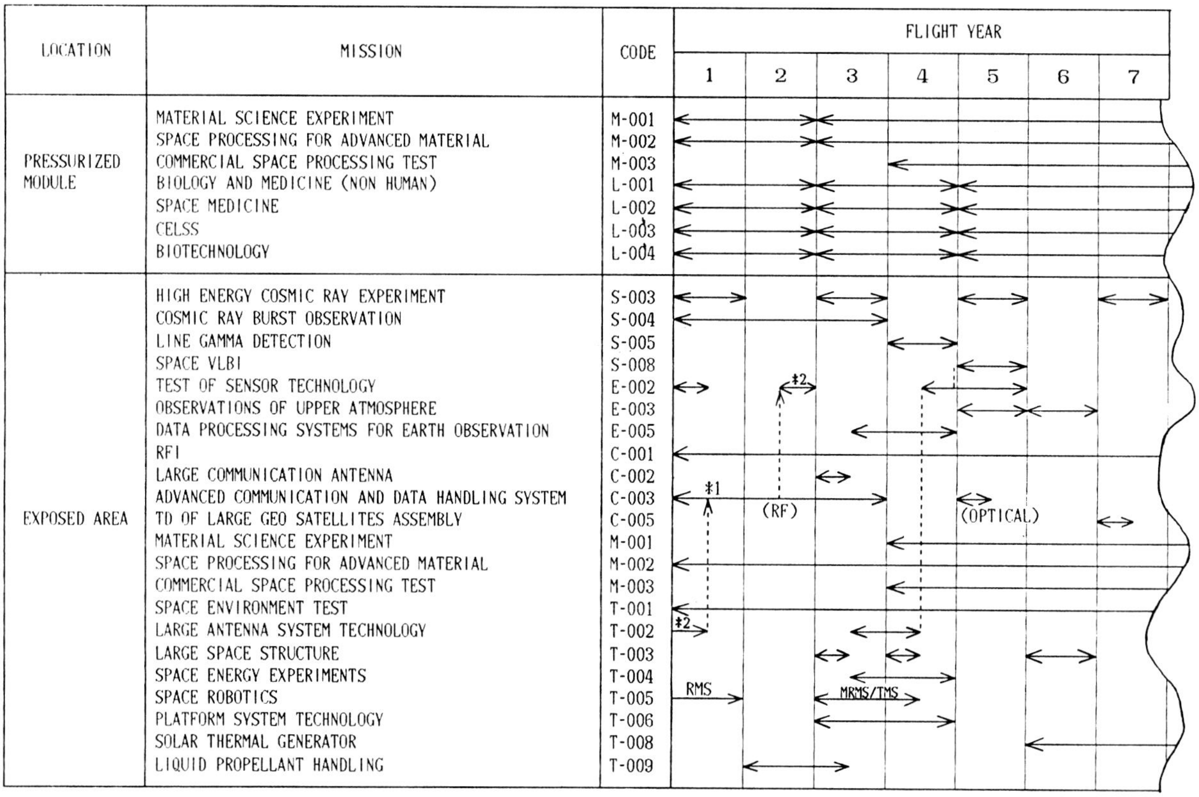

| LOCATION | MISSION | CODE | \multicolumn{7}{c}{FLIGHT YEAR} |
			1	2	3	4	5	6	7
PRESSURIZED MODULE	MATERIAL SCIENCE EXPERIMENT	M-001							
	SPACE PROCESSING FOR ADVANCED MATERIAL	M-002							
	COMMERCIAL SPACE PROCESSING TEST	M-003							
	BIOLOGY AND MEDICINE (NON HUMAN)	L-001							
	SPACE MEDICINE	L-002							
	CELSS	L-003							
	BIOTECHNOLOGY	L-004							
EXPOSED AREA	HIGH ENERGY COSMIC RAY EXPERIMENT	S-003							
	COSMIC RAY BURST OBSERVATION	S-004							
	LINE GAMMA DETECTION	S-005							
	SPACE VLBI	S-008							
	TEST OF SENSOR TECHNOLOGY	E-002		‡2					
	OBSERVATIONS OF UPPER ATMOSPHERE	E-003							
	DATA PROCESSING SYSTEMS FOR EARTH OBSERVATION	E-005							
	RFI	C-001							
	LARGE COMMUNICATION ANTENNA	C-002							
	ADVANCED COMMUNICATION AND DATA HANDLING SYSTEM	C-003	‡1	(RF)			(OPTICAL)		
	TD OF LARGE GEO SATELLITES ASSEMBLY	C-005							
	MATERIAL SCIENCE EXPERIMENT	M-001							
	SPACE PROCESSING FOR ADVANCED MATERIAL	M-002							
	COMMERCIAL SPACE PROCESSING TEST	M-003							
	SPACE ENVIRONMENT TEST	T-001							
	LARGE ANTENNA SYSTEM TECHNOLOGY	T-002	‡2						
	LARGE SPACE STRUCTURE	T-003							
	SPACE ENERGY EXPERIMENTS	T-004							
	SPACE ROBOTICS	T-005	RMS		MRMS/TMS				
	PLATFORM SYSTEM TECHNOLOGY	T-006							
	SOLAR THERMAL GENERATOR	T-008							
	LIQUID PROPELLANT HANDLING	T-009							

‡1: C-003 RELATIONAL ANTENNA WILL BE CONDUCTED AT THE SS UPPER BOOM FOR ETS-VI EXPERIMENT OPERATIONAL LINK.
‡2: PRELIMINARY EXPERIMENT.

Fig. 1-2 Mission scenario of proposed themes

<u>2 Analysis of Mission Requirements
in Data Transmission</u>

<u>Procedure of analysis</u>

JEM design requirements have been analyzed based on the mission scenario previously indicated. The procedure of mission analysis is shown in Fig. 2-1. Typical IOC missions were selected for analysis and studied in detail considering mission operation procedure. Before the analysis, the trade-off study to determine whether the on orbit analysis will be sufficient or not was performed because the effects on the crew skill, crew time, volume of analysis instruments, necessity of quick return, volume of stowage, and impact on the JEM design are extremely significant. Considering the possibility of performing exploratory experiment themes, the study result indicated that on orbit analysis is necessary. The assumptions of the mission analysis were established as follows:

-On orbit analysis   Minimum
-Crew skill   Technician/Scientist
-Volume of analysis   Large
 instruments
-Quick return   Needed

Based on these assumptions, the mission requirement values are now being studied and are being accumulated quarterly. Because of the lack of clarity in mission requirements, the results of the accumulation are not shown in this paper. The reference mission model of each theme will be defined more precisely at a later date.

On communications, the mission requirements for data exchange between JEM and the ground are also being analyzed.

<u>Results</u>

The kinds of data required by the missions are shown below.

-Experiment data   Status data
                   Aquired data
                   Pictures (Static, Dynamic)
-Computer data   Data file
                   Text-graphic
-Voice
-Video

Two types of data, real-time and nonreal-time, were also required.

The amount of data which JEM must handle is now being studied. The quantity of data produced by each experiment is not definite, because the details of the data requirements are not clear. However, the three special missions below will define the capacity of data transmissions.

(1) S-004 Cosmic gamma ray bursts
    Objective: Observation of gamma ray bursts in various energy regions (gamma-ray region 20Kev-5Mev, X-ray region 3Kev-40Kev soft X-ray region 0.2Kev-10Kev, and optical region visible light) to find an optical candidate.
    Requirements: Bursts occur randomly, and the required data differs according to the following conditions.

                   Normal   Burst
    Telescope   20Kbps x 24hr   20Mbps x 100sec
    others   80 bps x 24hr   80Kbps x 100sec daily

(2) T-004 Space energy experiment
    Objective: This mission constitutes an experimental complex contributing to the energy exploitation in space and the assessment of the related environmental impact. Subjects included are
    -Space Plasma Experiment (SPEX),
    -Advanced Propulsion Test (APT),
    -Space Radar (SRADAR),
    -Space Laser (SLASER), and
    -Microwave Energy Transmission Test (METT)
    and are performed in this order.
    Requirements: Real-time is required.
    5Mbps x 2hr x 6cycle
    (36Gbit x 6cycle) daily

(3) E-002A Test of sensor technologies-A
    Objective: To provide technology base of large complex sensors for land, ocean, and atmosphere observation.
    Requirements: 286Kbps x 7.2hr daily

CODE	MISSION	FLIGHT YEAR*					
		1	2	3	4	5	6
T-002A	LARGE ANTENNA SYSTEM TECHNOLOGY-A						
C-003B	ADVANCED COMMUNICATION AND DATA HANDLING SYSTEM-B						
E-002A	TEST OF SENSOR TECHNOLOGIES-A						
T-002B	LARGE ANTENNA SYSTEM TECHNOLOGY-B						
E-002B	TEST OF SENSOR TECHNOLOGIES-B						
S-008A	SPACE VLBI-A						
C-003A	ADVANCED COMMUNICATION AND DATA HANDLING SYSTEM-A						
C-003C	ADVANCED COMMUNICATION AND DATA HANDLING SYSTEM-C						

*FLIGHT YEAR 1 MEANS THE FIRST YEAR

Fig. 1-3 IOC mission scenario in communications area

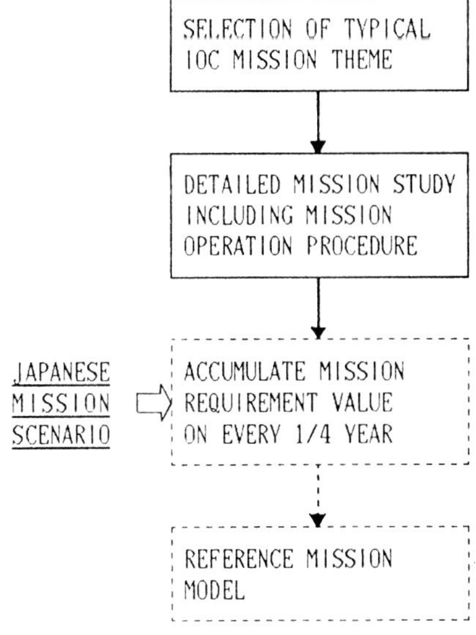

Fig. 2-1 Procedure of mission analysis

To satisfy these requirements, a total of data transmission capacity is now estimated to be about 5Mbps (high-rate, real-time). If this capacity can be achieved, data transmission requirements can be satisfied for almost all themes. If T-004 (space energy experiment) is not considered, the total data transmission capacity is reduced to 1Mbps.

As a result of the analysis, a preliminary interface with the JEM communication subsystem was established, and JEM has been designed to satisfy these requirements. The values are currently being studied in detail. Therefore, there is a possibility that they may be changed. Furthermore, two video channels are required. One of them must be capable of carrying high-quality video pictures. Two voice channels are also required, one for uplink and the other for downlink.

3. JEM Information System (JEMIS)

Overview

To perform the proposed Japanese customer missions described in a Previous chapter, NASDA plans to develop the Japanese Experiment Module (JEM) which will be attached to the Space Station. The module consists of a pressurized module (PM), an exposed facility (EF), and an experiment logistic module (ELM) to satisfy multipurpose mission requirements such as material processing, life science, earth and scientific observations, and communication and technology development experiments. The JEM baseline configuration is shown in Fig.3-1.

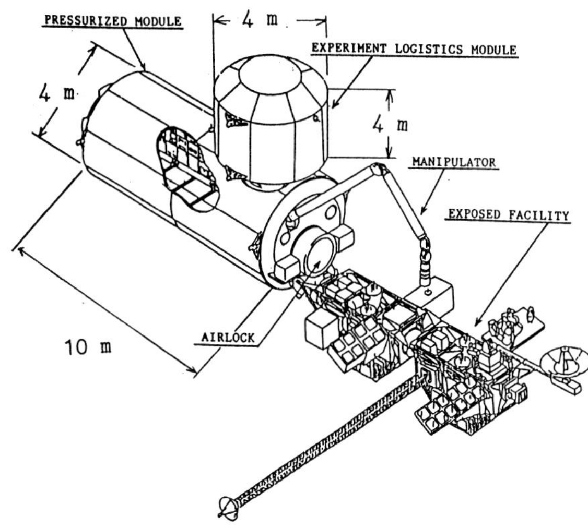

Fig. 3-1 JEM baseline configuration (IOC)

NASDA is planning to develop the JEM Information System (JEMIS) for customers utilizing JEM. JEMIS will provide payload operation support functions and increase the chance of JEM operation while retaining operation flexibility.

JEMIS is an end-to-end communication network that enables continuous JEM operation on orbit from a ground flight operation center in Japan in real time. JEMIS consists of JEM, Engineering Test Satellite-VI (ETS-VI) or Data Relay and Tracking Satellite (DRTS) and the JEM ground operation system. It interfaces with NASA's Space Station Information System (SSIS).

JEM systems and payloads will be operated via JEMIS in coordination with the total Space Station operation plan via SSIS. JEMIS must therefore provide an intimate network interface with SSIS, making the JEMIS-SSIS network the primary communication network for JEM and Space Station operation. Network characteristics will depend on actual TDRSS developments and system capabilities.

The JEMIS ETS-VI communication link will be the secondary communication link because it is an experimental communication link for testing prior to development of DRTSS.
The above JEMIS, SSIS Communication Network concept is shown in Fig.3-2.

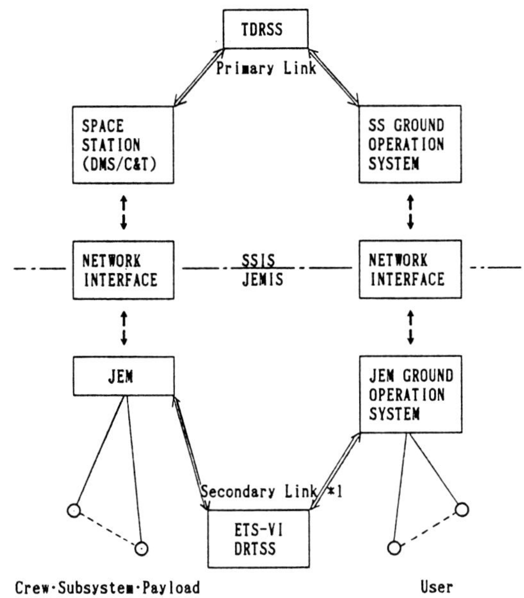

Fig. 3-2 JEMIS,SSIS communication network concept

JEMIS Elements

Major JEMIS elements are listed below.

on orbit elements	· JEM (Bus, payload)
	· Japanese Orbital Service vehicle(JOSV)*
	· Japanese Free Flyer(JFF)*
	· H-II payload*
Orbit-Ground I/F elements	· ETS-VI
	· DRTS*

Ground elements ·Tsukuba JEM Flight
                  operation center
                 ·JSC NASDA Branch (JNB)
                 ·JEM Simulator
                 ·Logistics terminal
                 ·User operations support I/F
                  terminal
                 ·ETS-VI/DRTS ground
                  operation system

        * Growh phase elements

JEMIS Functions

     The major JEMIS functions are described
below.

     a. JEMIS transmits, distributes, processes,
stores, retrieves, and edits information and
data in order to operate and control each of its
elements.

     b. JEMIS can exchange the following user
information and data between orbit and ground :
     Command, telemetry, computer data (operation
and management data, text, graphics,etc.),
voice, video, mission experiment data, and
caution and warning data.

     c. JEMIS handles information and data,
coordinating priority and data volume in
accordance with importance.

     d. To reduce crew and ground operation work
load, JEMIS increases automatic functions such
as equipment status monitoring checking.

     e. JEMIS has the following JEM integrated
control and management operation support
functions :
        Operation planning/scheduling
        Operation sequence control
        Resource Distribution and control

     f. JEMIS is available to install and operate
users' development hardware and software.

     g. JEMIS provides data security to ensure
privacy and protection as required by users.

JEMIS Data Transmission Requirements

     Based on the above Japanese customer needs
and JEM operation concept, NASDA has developed
the following JEMIS information and data
transmission requirements :

     a. JEMIS information and data transmission
         link (see Fig.3-3)

     Primary Link   JEM-SSIS(SS-TDRSS)-JEMIS
                       Ground network
     Secondary Link   JEM-ETS-VI/DRTSS-JEMIS
                       Ground network

     b. The information and data transmission
requirments are shown in Table 3-1.

     c. The critical elements in JEMIS
information and data transmission are ETS-VI and
the international commercial network link
capability, which will restrict JEMIS system
transmission performance.

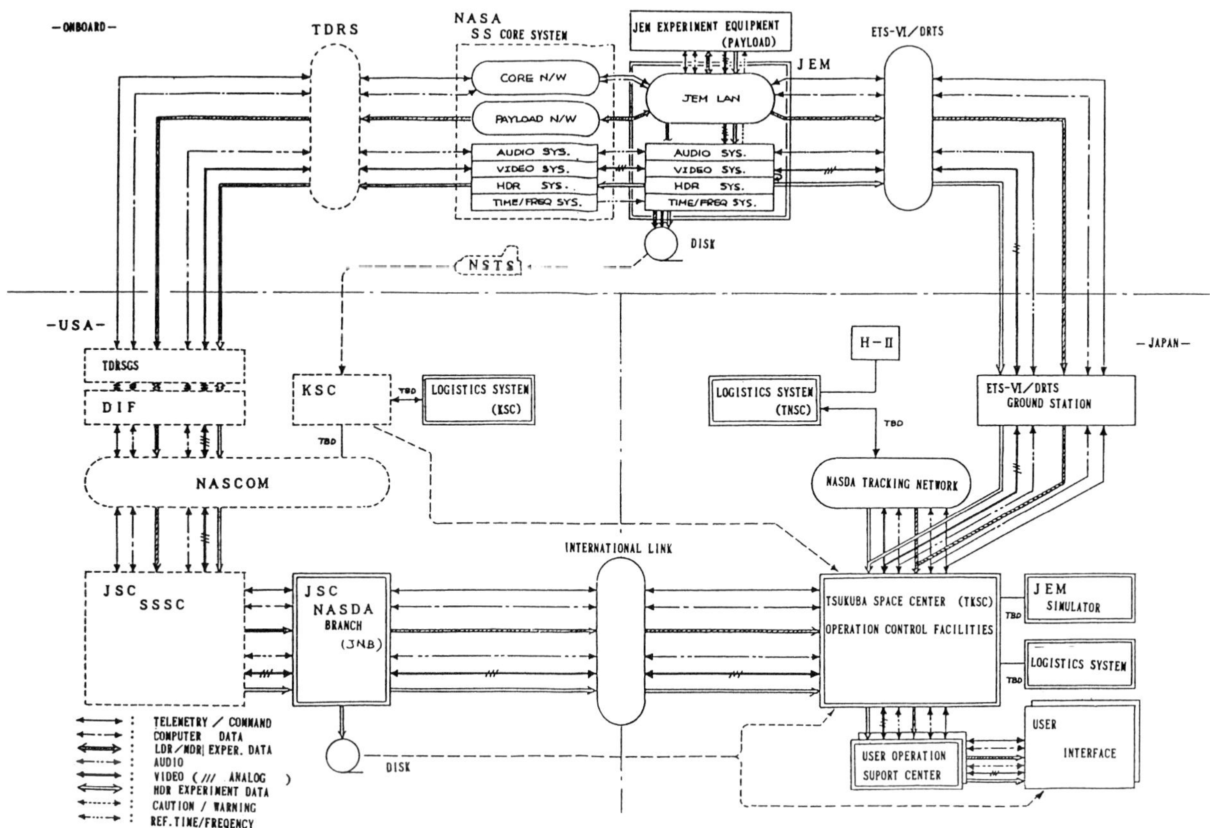

Fig. 3-3 JEMIS data flow

d. ETS-VI/DRTS RF communication link is planned as follows:

	ETS-VI	DRTS
KSA FWD/RTN	1 ch each	4 ch each
SSA FWD/RTN	-	4 ch each
SMA FWD/RTN	1 ch each	-
Feeder Link		
UP/DWN (Ka-Band)	1 ch each	1or2 ch each

The ETS-VI RF link budget currently being studied is shown in Table 3-2 for reference.

Table 3-2   JEM-ETS-VI RF link budget
(Preliminary Draft)

Link/Channel	Data Rate	EIRP	
Inter-Satellite Link	Max.	(dBm)	
KSA FWD (23 GHz)	3 Mbps	68.7	*
KSA RTN (32 GHZ)	23 Mbps	96.2	***
SMA FWD (2.1 GHz)	1.7 Mbps	65.2	**
SMA RTN (2.3 GHz)	1.4 Mbps	76.2	***

*	ETS-VI Antenna	0.8mφ (Non-Deployed)
**	ETS-VI Antenna	Phased Array Antenna
***	JEM/SS Antenna	3.8mφ (Deployed)

4. JEMIS Onboard System

Overview

The JEM system consists of seven subsystems : structure and mechanism, electrical power, communication and control, thermal control, environmental control, experiment support, and manipulator. The communication and control subsystem (C&CS) is considered part of the JEMIS onboard element.

The C&CS provides data communication, processing, recording, and operation control and management functions which support the bus equipment, mission payload operation, and crew-activity.

The C&CS architecture is basically centralized operation management with hierarchical ditributed control so that one crew member can operate and control JEM.

Elements and Functios

The JEM C&CS functionally consists of the following elements :

Computer System
Control & Display System
Optical Network System
High-Rate Data System
Video System
Audio System
Caution & Warning System

Figure 4-1 shows the subsystem architecture.

Table 3-1   JEMIS data handling requirements

NO	DATA TYPE		DATA CONTENTS	DATA RATE	TRANSPORT DELAY MAX	B E R	DATA FLOW
1	COMMAND		OPERATION COMMAND PAYLOAD COMMAND	$\leq$32Kbps	0.1S (ONBOARD) 2~4S(GND→ONBOARD)	ERROR FREE	ONBOARD →ONBOARD GROUND →ONBORAD
2	TELEMETRY		STATUS DATA , SENSOR DATA (ENGINEERING /ANCILLARY DATA)	$\leq$64Kbps	0.1S (ONBOARD) 2~4S(ONBOARD→GND)	ERROR FREE	ONBOARD →ONBOARD ONBOARD →GROUND
3	COMPUTER	DATA FILE	PROGRAM , FILE , DATABASE OPERATION / MANAGEMENT DATA	$\leq$ 1Mbps	MINUTES TO HOURS	ERROR FREE	ONBOARD →ONBOARD ONBOARD →GROUND GROUND →GROUND
4	DATA	TEXT/ GRAPHIC	PROCEDURE MANUAL , DRAWING ELECTRIC MAIL	$\leq$ 1Mbps	MINUTES	$\leq 10^{-5}$	SAME AS ABOVE
5	AUDIO		AUDIO VOICE	64Kbps/CH 3 CH*1	1~ 2 SEC	$\leq 10^{-3}$	SAME AS ABOVE
6	VIDEO		OPERATIONAL VIDEO (CONFERENCE VIDEO,CREW MONITOR OPER./EXPERIMENT MONITOR ETC.)	4.5MHz/CH 2 CH*1	2~ 4 SEC	S/N $\geq$30dB	SAME AS ABOVE
7	EXPER. DATA	LDR	MATERIAL EXPERIMENT DATA ETC LIFE SCIENCE EXPERIMENT DATA	$\leq$ 1Mbps	MINUTES	ERROR FREE ~ 10^{-5}	ONBOARD →ONBOARD ONBOARD →GROUND GROUND →GROUND
8		MDR	IMAGE DATA (STATIC IMAGE) ETC ASTRONOMICAL DATA	$\leq$ 4Mbps	MINUTES	$\leq 10^{-5}$	SAME AS ABOVE
9		HDR	COMPRESSED VIDEO ETC EARTH OBSERVATION DATA	$\leq$100Mbps	MINUTES	$\leq 10^{-5}$	SAME AS ABOVE
10	CAUTION/ WARNING DATA		CAUTION / WARNING EMERGENCY CONTROL COMMAND	TBD	0.1S (ONBOARD) 2~4S(ONBOARD→GND)	ERROR FREE	ONBOARD →ONBOARD ONBORAD →GROUND
11	EMERGENCY AUDIO		EMERGENCY COMMUNICATION AUDIO	4KHz/CH 1CH	0.1S (ONBOARD)	S/N TBD	ONBOARD →ONBOARD (RF LINK/RF FRQ. EVA SYS.COMPATIBLE)

*1:except Ground-Ground requirement

a. The computer system which supports JEM operation management consists of a control unit with high processing capability and high reliability as well as mass memory storage units.

b. The control and display system provides the man-machine interface for centralized monitoring and operation control and consists of input/output equipment.

c. The optical network system, which collects and distributes medium- and low-rate data, consists of the optical LAN and performs highly efficient data transfer and real-time operations. It provides expandability and flexibility through packet swiching. The LAN interconnects to the SS core and payload network through a gateway for clean interface.

d. The high-rate data system collects high-rate experiment data and multiplexes this data using the high-rate data multiplexer before transmission to the ground. This system also has a high-speed data recorder for temporary data storage and replay. This system will directly connect to the TDRS signal processor of the SS core C&T system for transmitting to the ground.

e. The video system collects, distributes, displays, records, and plays back TV image data from JEM, SS, and payloads. This system will interconnect to the SS video system via the analog video bus for data exchange between JEM and SS/ground. This system also has a data compression unit to reduce data transmission bandwidth to the ground.

f. The audio systems has transceivers, headsets, microphones/speakers, and a controller for voice communication to the groud, SS, EVA, STS, etc. Through audio signals will be transmitted to SS and ground through the digital audio bus of the SS audio system in normal operation, an RF transmitter/receiver is provided for equipment faults, breakdown of the physical connection to the adjacent module, or rescue using the experiment logistics module.

g. The caution and warninng system has sensors, displays, speakers, etc, to detect faults which would endanger the crew or impair the total system functions and to notify the crew. In normal operation, the caution and warning signals are transmitted through the optical LAN, but some dedicated lines are provided for transmission of very critical signals.

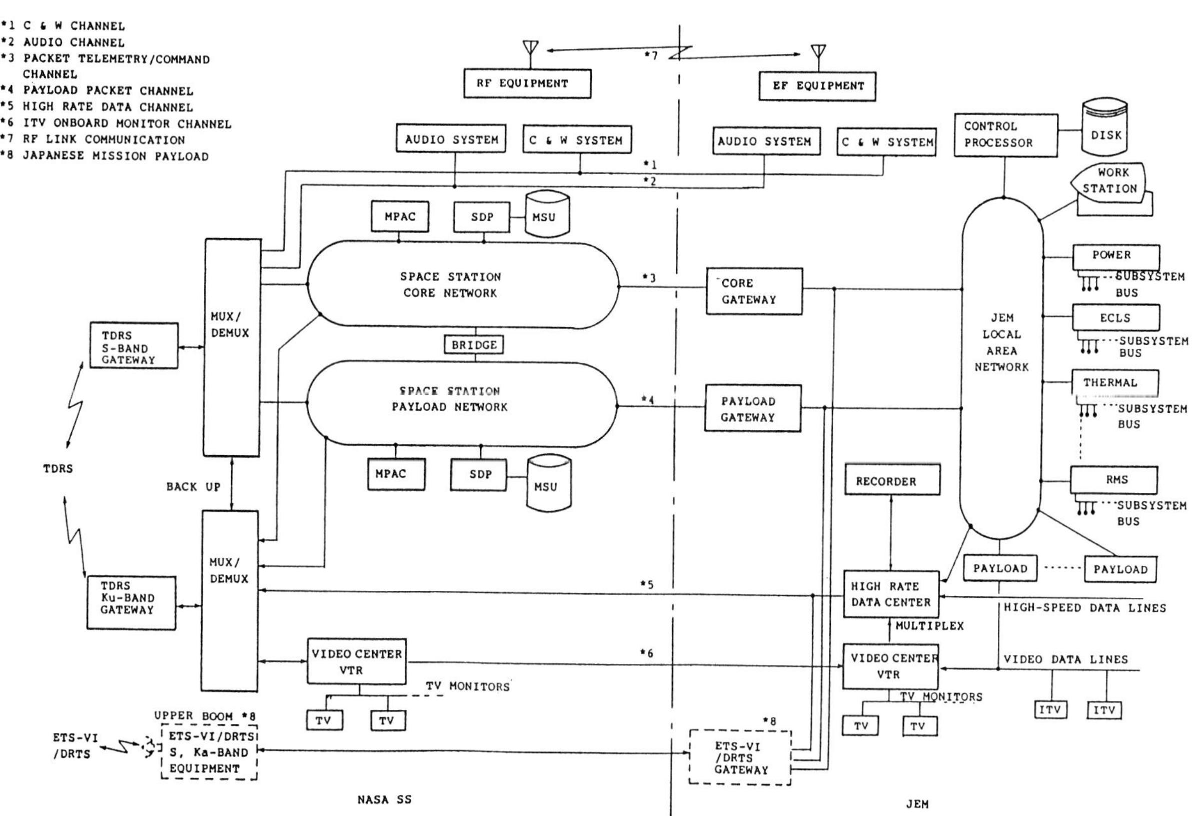

Fig. 4-1   JEM C&CS architecture

<u>Characteristics</u>

The C&CS characteristics are shown in Table 4-1.

<u>Conclusion</u>

In this paper, Japanese customer needs, present results of mission analysis in communications, and the JEM Information System were introduced. Customer needs are becoming more and more high technology oriented and this trend is likely to continue. The results shown here are for the present plan at the end of the preliminary design phase. Based on these, JEM must be constructed to satisfy future needs. Japan looks foward to promoting technical development in the Space Station which will be constructed through international cooperation.

Table 4-1 JEM C&CS Characteristics

```
a. Computer System
     Proccesing Capability        4 MIPS
                                  (Real-Time Processing)
     Main Memory Capacity         1 Mbyte x 6 units
     Secondary Memory Capacity
          Optical Disc Unit       300 Mbyte x 2 units
          Magnetic Bubble Unit     20 Mbyte x 2 units

b. Control & Display System
     Display Capability           Color, Alphanumerics
                                  Graphics
     Input Device                 Keyboard, Trackball, etc.

c. Optical Network System
     Effective Transfer Rate      Max. 32 Mbps
     Transfer Medium              Optical Fiber
     Optical Wave Length          1.3 um Band
     Modulation Method            Optical Intencity Modulation
     Swiching Method              Packet Swiching
     Interface Transfer Rate
     To SS                        Max. 4 Mbps

d. High-Rate Data System
     High-Rate Data Recorder
       Total Storage              4.5 Gbyte/Tape
       Record Speed               Max. 64 Mbps
       Record Time                Min. 10 minutes(at 64 Mbps)
     High-Rate Data Multiplexer
       Output Rate                2 To 100 Mbps
       Input Rate                 2 To 64 Mbps(Synchronized Data)
                                  Max. 100 Mbps
                                        (un-synchronized Data)

e. Video System
     Signal Type                  NTSC
     Monitor TV                   9 inch x 2 units
     VTR Record Time              60 minutes
     Data Compression Ratio       Less than 1/2

f. Audio System
     Channel                      3 channel(Duplex)
     Digital                      64 Kbps/channel

g. Caution & Waiting System
     Warning Speeker Output Power  Max. 5w
```

SCIENTIFIC CUSTOMER NEEDS - EARTH SENSING

Francis P. Bretherton
National Center for Atmospheric Research
Boulder, Colorado

<u>Abstract</u>

Earth Sensing requirements are characterized by very high downlink data rates from the polar platforms. These are driven by long-term needs to document changes in our global environment because of human activities. Limited uplink capabilities for commanding instrument operation are also needed.

My talk today is about observing the Earth from space. It is often overlooked than an integral part of the space station program is, in fact, the long-term platforms in polar orbit. Of course, there is little direct connection between the physical space station and polar orbit in the IOC era. Nevertheless, there is a great deal of commonality in terms of subsystem components. Also, as we will see, some of the drivers on the information system are, in fact, in the polar orbital component.

One of the earliest applications of space technology was looking at the Earth. The weather satellites, the LANDSAT Program, and more recently, the SPOT Program of France, and the Japanese Earth Resources Satellite Program are all clear examples of the importance and ongoing applications of remote sensing. However, this area has a somewhat different emphasis in its information, data and information system requirements from that of the manned space station proper. Elements in common are major emphasis on long-term commitments, on continuity, stability, data access, and also on the connectivity of the system in terms of the ability of user groups to work together and where appropriate, to command operations on the spacecraft. However, there is a major need for distribution of large amounts of data in real-time or near real-time which is, perhaps, rather different from that which is perceived for the manned spacecraft itself. Characteristically, this is an area with really high data transmission rates. When the polar platforms are in full operation in the late 1990s, the projected data rates are around 5×10^{12} bits/day, i.e., a sustained average of 50 M bits/s. The largest archive of any space data at this time is probably that of the LANDSAT dataset which is something of the order of 100×10^{12} bits, or 100,000 nine track tapes. That data volume will be produced every three weeks. These data rates are at least an order of magnitude greater than any we handle today, a fact that few of us have begun to comprehend.

A secondary space station requirement, which is very important for these polar platforms in the long run, is the capability for remote servicing and telerobotics. Though clearly a post IOC activity, the integration of this Earth sensing program with the manned activities which during IOC are confined to tropical orbit will be one of the major drivers upon further space station development.

We now consider some of the forces giving urgency to these activities in remote sensing, from the point of view of a research scientist. A theme which is receiving more and more international attention is the impact of human activities on our global environment, exemplified by the steady increase of carbon dioxide in the atmosphere due to the burning of fossil fuel, of oil and natural gas and coal by all the nations of the world, a trend which we see no signs of decreasing. Also increasing is methane, or marsh gas. In this case the reason for the observed increase is much less clear. Methane comes out of wet environments with strong biological activity. Respectable theories range from the impact of the green revolution on increasing productivity of rice paddies around the world. Another theory is that, in fact, it is not a changing source of methane but it, in fact, a changing mechanism of destruction of methane due to industrial air pollution. These gases are important because they act like a greenhouse. In other words, though they allow radiation from the sun to reach the surface of the Earth unhindered, heat radiated back to outer space tends to be retained in the atmosphere. Thus greenhouse gases act like a blanket and warm things up. Many computer models show that over the next fifty years or so, the climate of the world should change due to this effect by an amount comparable to the change that has taken place since the last Ice Age, i.e., global warming of $2\text{-}3^{\circ}C$ with substantially larger values in high latitudes. Though this change may seem small, it implies in fact a fundamental change in the living environment of all the people of this world at a rate unprecedented in human experience. The details of this warming and associated changes in rainfall are not known, in particular, from one country to another and many of the underlying cause-effect relationships. However citizens and scientists both in the United States and in other countries are becoming more and more conscious of the need for an urgent program of global observation. Though by some standards these changes are slow, obtaining the knowledge base that is necessary for our children and grandchildren to react intelligently will take all the time that is still available. This requirement is likely to be a major driver for Earth information systems for the next many decades.

This observation program involves extensive measurement and imaging of the atmosphere, oceans, ice and earth surface, much of it on an ongoing basis. New instrumentation will include high resolution infrared spectrometers and synthetic aperture radars, as well as integrated systems involving a variety of lower data rate systems. A sophisticated command capability will control the operation of these instruments. In conjunction with in-situ measurements these data will be used to develop and test predictive models of the entire Earth system. Though much of the image data will initially be interpreted by

human analysts on a case by case basis, the
trend will be towards automated processing of
large data sets into products summarizing the
global distributions of key variables.

Further details are available in the reports
of the Earth System Sciences Committee: "Earth
System Science, An Overview," which was publish-
ed in 1986, and "Earth System Science, A Closer
View," which is in preparation, and of the
Working Group on the "Earth Observing System:
Science and Mission Requirements."

SCIENTIFIC USER REQUIREMENTS FOR
MICROGRAVITY RESEARCH (EUROPEAN ASPECTS)

I. Egry, G. Otto and B. Feuerbacher

German Aerospace Research Establishment (DFVLR)
5000 Köln 90, F.R. Germany

Abstract

In order to be prepared for the capabilities of the Space Station ESA has initiated several acitivities to analyse future user requirements. Model payloads from the various disciplines serve as a requirement envelope for the system designers. An important aspect for future experimentation is the principle of telescience. Past experience with the German D1 mission has demonstrated the benefits of this principle and resulted in an increased scientific output. Additional support to the experimenter will be provided by the installation of User Centers realizing that the number of experiments to be performed during the Space Station era will be far beyond the present level.

Introduction

In 1985 the European member states of ESA decided, following President Reagan's invitation, to participate in the International Space Station Project and initiated the Columbus programme. This presently consists of a polar platform, mainly for earth-sensing, one pressurized laboratory module attached to the Space Station, a resource module and a second pressurized module, to be used in combination with the resource module as a Man-Tended Free-Flyer (MTFF). The pressurized modules are mainly used for microgravity research. In addition, Europe is developing Ariane 5 as a launcher capable of carrying Hermes, the European space shuttle.

In order to be prepared for the capabilities of this system, ESA has set up a Columbus Preparatory Programme (CPP) and is planning to spend some extra effort in the Columbus Utilisation Preparation Programme (CUPP).

In the framework of these activities, DFVLR has performed, on behalf of ESA, a series of studies entitled "European Users' Requirements Analysis" [1]. These studies were monitored and supervised by the Space Station User Panel (SSUP). This panel brings together 10 scientists of different disciplines to represent the European scientific user community in the Columbus programme.

Model Payloads

The main outcome of these studies are so-called model payloads that summarize user requirements of a specific discipline and a specific flight element. They represent a requirement envelope for system designers. For the microgravity sciences, namely materials sciences and life sciences, the following model payloads have been defined [1]:

LIF 111   General Purpose Life Sciences Research Laboratory for the Attached Pressurized Module (APM)

LIF 141   (Radiation-) Biology and Bioprocessing Laboratory for the MTFF

LIF 311   Exo- and Radiation Biology Payload, unpressurized, attached to S/S or coorbiting

LIF 313   Exo- and Radiation Biology Payload unpressurized on the Polar Platform

MAT 110   Materials Sciences Research Laboratory for the Attached Pressurized Module (APM)

MAT 130   Materials and Bioprocessing Payload, unpressurized, attached to S/S or coorbiting

MAT 140   Semi-Automated Materials Sciences Laboratory for the MTFF

As is evident from this list, the user community in Europe has requirements for both, manned and unmanned systems.

Whereas the first German Spacelab mission, D1, has provided some experience in operating and utilizing a manned mission, it is hoped that the first EURECA (European Retrievable Carrier) mission, to be launched 1991, will play the same role for automated microgravity missions.

Information Systems

In order to make most effective use of the payloads in orbit, an efficient information system is necessary. This system not only comprises the communication links during missions - those will be addressed in greater detail later - but it contains also elements that are in use before and after the mission. These are essentially data bases containing information on planned or performed missions. Within the Columbus project, ESA is setting up a Management and Technical Information System (MATIS). However, this system is tailored for industrial contractors and less adapted to scientific user needs. A user/payload related data base has been established within the above-mentioned study under the name Columbus Payload Data Base (CPDB). Presently, it contains the detailed requirements of the model payloads. In a future step, the model payloads will be substituted by real payloads. It is implemented as a relational data base so users of the CPDB can extract information according to their specific interest, i.e. accommodation planners as well as potential investigators will find the details they need.

Similarly, ESA plans to install a data base containing the results of microgravity experiments in a systematic way allowing for easy information retrieval and evaluation.

Telescience: Definition

Of course, the key issue of the optimum scientific utilization of the payloads in orbit is the performance of the experiment itself and the interaction with the payload. For the interactive operation of space-borne experiments, the concept of telescience was introduced by P. Banks [2]. The basic idea is the amalgamation of teleoperation and telepresence. Telescience satisfies a genuine scientific user requirement, namely to perform experiments in space just like in a terrestrial laboratory as illustrated in Fig. 1.

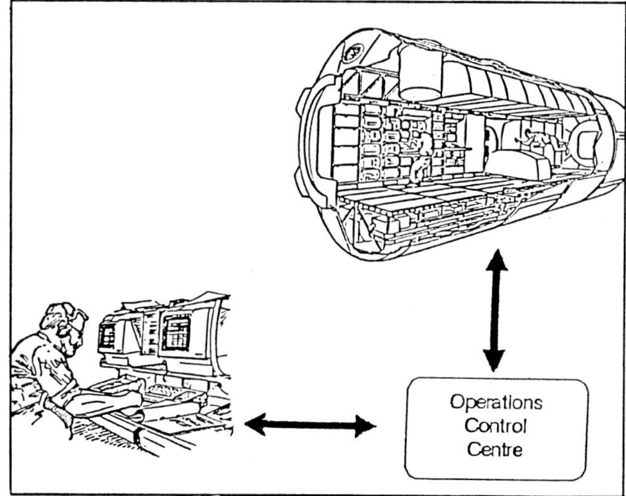

Fig. 1: Telescience scenario

The user himself should be able to take care of his experiment to the largest possible extent through an Operations Control Center. The crew should be used only where it cannot be replaced, and routine operations should be done automatically. This will leave sufficient flexibility to the crew for scientific interactions and analysis, for unplanned trouble-shooting activities, and for a stand-by capability.

ESA has taken up this idea and installed a telescience user team under the leadership of A. Balogh. this team agreed on the following definition [3]:

"Telescience is the mode of interactive payload operations whereby investigations on the Space Station and Columbus elements are operated under the effective and direct control of the investigator teams."

Please note that this definition is not restricted to Columbus elements and is equally applicable to manned and unmanned operations. Of course, the implementation of this idea will be different for the two cases. In this context it is worthwhile mentioning that telescience has a long and successful history in the area of space sciences where astronomy missions like, e.g. the International Ultraviolet Explorer (IUE) are operated this way. However, for microgravity payloads telescience is still a new concept. ESA therefore has installed a telescience engineering team that provides the interface between the telescience users and the project engineers. Technically speaking, the implementation of telescience implies, in addition to transmission of digital data, voice and video links in near-real-time. Depending on the required quality of the video, the data stream may range from 64 kbps up to 88 Mbps. The time delay is determined mainly by the number of nodes through which the signal has to pass, and, to a smaller extent, by the distance which the signal has to travel. Roughly speaking, a time delay of 5 - 10 seconds cannot be avoided.

Telescience: Experience from D1-Mission

Preliminary experience on telescience operations has been obtained during the D1 mission in the fall of 1985 [4]. The objectives of the German D1 mission were mainly directed towards the investigation of basic phenomena in materials science, physical chemistry, biology and medicine. A prerequesite of most of the experiments performed was their microgravity relevance, with exception of two navigation experiments.

The operation of the D1 payload during the mission was carried out in two shifts both on board and on ground. The Payload Operation Control Center (POCC) of DFVLR was located at Oberpfaffenhofen near Munich, the Mission Control Center (MCC) for the Shuttle System at Houston. Figure 2 shows the communication links between the

shuttle-spacelab-system and the two control centers. A user control room for
interfacing with the POCC was also co-
located in Oberpfaffenhofen.

Due to the availability of only a single
Tracking and Data Relay Satellite, direct
communication and real time data acquisition between Spacelab and the POCC was
limited to about 40 % of the mission experiment operation time (157 hrs). During
the mission, real time television transmission turned out crucial for many of the
interactive experiments and therefore the
planned duration (19 %) was increased to
22 %. However, only 2/3 of that was relayed to the POCC in Oberpfaffenhofen.

The total payload consisted of eight payload elements [5].

o The Werkstofflabor (WL) was designed for
  materials science experiments and included six experiment facilities in a
  double rack.

o The payload element MEDEA (MD) was composed of three autonomous experiment
  facilities, a gradient furnace with
  quenching device, a monoellipsoid mirror
  heating facility and a high-precision
  thermostat.

o Another payload element for fluid and
  materials sciences was the Prozeßkammer
  (process chamber PK) consisting of three
  facilities.

o Biological research was performed in the
  payload elements Biorack (BR) and Bio-
  Wissenschaften (BW), while the Vestibu-
  lar Sled (VS) was devoted to medical
  research.

o Outside the module, two mainly automatic
  experiments were located in the payload
  bay, the Material Experiments Assembly
  (MEA) of NASA and Navex (NX), an instrument dedicated to navigation investigations.

Table 1 gives a summary of payload operations during the D1 mission. The first
line indicates the number of experiments
carried out in each payload element. The
"work horse" of the mission was the WL
double rack with 30 experiments, followed
by the Biorack with 14 experiments. The
following three lines in the table illustrate to which extent the handling of
hardware was limited to loading and un-
loading of samples and where additional
handling and experiment reconfiguration
from the pure manipulative viewpoint was
required. It is interesting to note that
the crew intensive WL payload element of
560 kg mass had a much higher "experiment
density" than the fully automated MEA element of 1000 kg mass.

Table 1 also illustrates the extent to
which experiment processes were controlled
by software only, by the astronaut or by
the astronaut supported by the principle
investigator. The latter situation is characteristic of interactive telescience
operations. Therefore, the shaded area indicates the degree to which telescience
was used during D1, namely about 25 %.
Finally, the table indicates the efficiency of crew repair work and the success of
experiment operation during the D1 mission. With respect to the interpretation
of these numbers, it may be concluded that
due to the high amount of interaction between flight and ground crew, the operational success of the mission was outstanding.

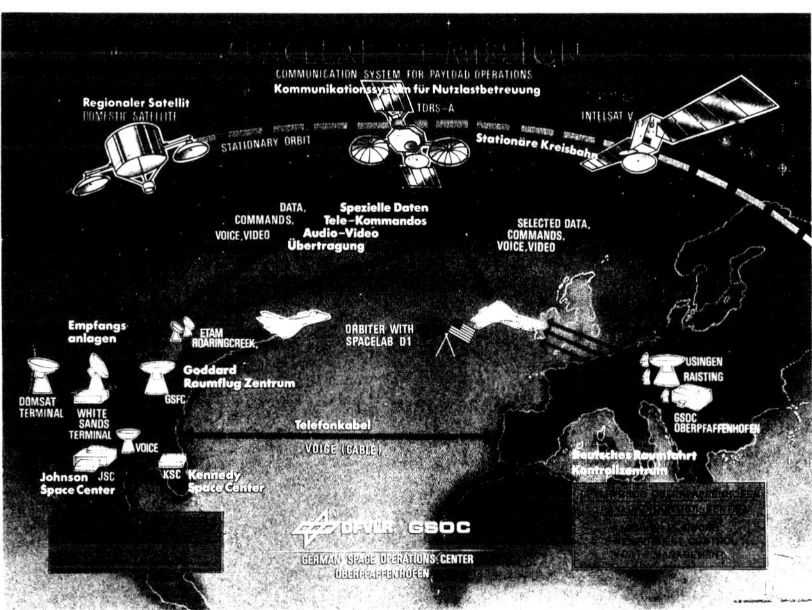

Fig. 2: Representation of the communication system necessary during the D1 mission

		Payload Element								Sum
		W L	M D	P K	B R	B W	SLED	N X	M E A	
Number of experiments		30	7	6	14	7	10	2	5	81
crew handling	automatic	1						2	5	8
	samples only	15	4		5	1				25
	reconfigure	14	3	6	9	6	10			48
experiment control	software	18		3	5	1		2	5	34
	crew	4	4	3	7	4	5			27
	crew + PI	8	3		2	2	5			20
successful repair		4	7					1		12
experiment success	0	1			1	1				3
	< 100 %	4			2	2			3	11
	100 %	20	5	2	11	7	8	2	2	57
	> 100 %	5	2	1			2			10

Table 1. D1 crew - payload interactions

Telescience - Plans for EURECA

As is evident from the preceeding chapter, the D1 mission offered an opportunity to test the telescience concept in a manned environment. For fully automated missions, EURECA will be the first possibility for testing telescience under such circumstances (Sounding rocket flights have too short mission durations for a sensible application of telescience). The core payload of the first EURECA mission consists of the following facilities:

- Automatic Mirror Furnace (AMF)
- Solution Growth Facility (SGF)
- Protein Crystallization Facility (PCF)
- Multi Furnace Assembly (MFA)
- High Precision Thermostat (HPT)
- Exobiological Radiation Assembly (ERA)

The PCF is equipped with a video camera in order to monitor the growth of the protein crystals. The pictures are downlinked to the principle investigator who may, based on this information, take corrective measures if necessary. However, due to limited ground contact times, the interactive cycle times are in the order of the orbital period. Nevertheless, this will be a demonstration of telescience for a fully automated payload.

Telescience: Plans for Columbus

The novel features of interactive payload operations in the context of the Columbus program stimulated ESA to establish a telescience testbed program. The ojective is to use a set of pilot investigatons to simulate interactive operations on the ground with a view to study the requirements and operational procedures needed for the planning of payload operations on Columbus.

The selected pilot investigations are in the field of human physiology, fluid and materials sciences and in plant biology. A summary of the various topics proposed by the corresponding Telescience User Team are given in Table 2. Additional information can be found in Ref. 3.

Table 2: Pilot investigations proposed by the Telescience User Team.

o Human physiology (contributed by N. Foldager): the influence of microgravity on the cardiovascular adaptation to dynamic leg exercise in man

o Fluid science (contributed by R. Monti/ I. Martinez): Fluid dynamics experiments on Marangoni flows

o Material sciences (contributed by H.P. Schmidt): optical observation of solidification of undercooled metallic drops

o Botanical sciences (contributed by D. Heathdote): determination of the gravitropic responses of seedlings to gravitational stimuli of controlled duration and magnitude

o Medical diagnosis (contributed by N. Foldager): investigation of telescience techniques for tele-diagnosis

Compared to Spacelab operations, payload crew time will be a very limited resource on the Space Station. Therefore, the concept of crew utilization has to be carefully balanced between a Spacelab-type operation and the apparent limitations on crew availability. This means that crew should be used only where it cannot be replaced, and routine operations should be done automatically wherever possible. In addition, the user himself should be allowed to take care of his experiment to the largest possible extent.

User Support

The utilization of the Space Station will be different from past space flight activities in essential points. Preparation, actual operation and evaluation of experiments will no longer be sequential but in parallel. Practically unlimited experiment time in space will allow iterative scientific investigations. After a quick and preliminary evaluation of results, the experiment procedure can be optimized with the help of engineering ground facility data and then repeated in space with modified parameters [6].

At DFVLR in Köln, Germany, three institutes, the Institute for Aerospace Medicine, the Institute for Space Simulation and the Institute for Materials Research, participate in the formation of a Microgravity User Support Center (MUSC) for the Space Station. It will support space users from university and industry from the initial idea to the qualified space experiment. Operation and control of experiments during flight can be interactive according to the telescience principle. Necessary support will be given for characterization of samples, analysis and data evaluation during all phases of the experiment.

The following installations will be available for the user:

o Engineering models of multiuser facilities for materials and life sciences which are installed at the respective institutes

o A control room for monitoring and control of experiments during space flights

o A microgravity information center and a data base for experiment and facility data

o A scientific and technical infrastructure for the support of the investigator and the operation of the engineering models

The support center philosophy is based on scientific cooperation between external users and scientists of the User Center who can share their knowledge from earlier microgravity experiments.

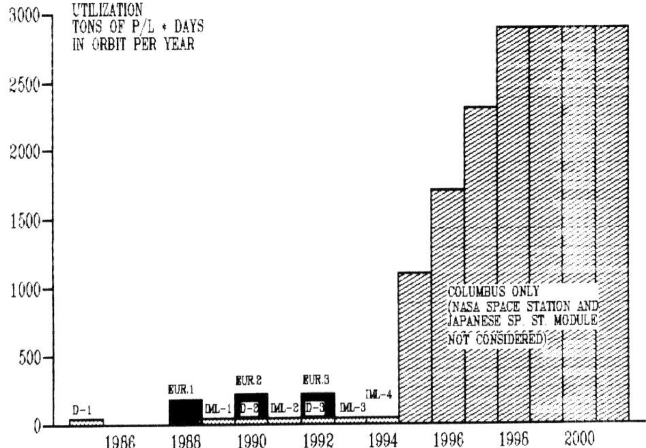

Fig. 3: Utilization potential of Columbus. Comparison of the utilization of Spacelab, EURECA and Columbus by microgravity disciplines.

Outlook

In the preceeding chapters we have discussed the user requirements on the Columbus elements and their implications especially with regard to telescience. We have also touched upon the question of information systems, e.g. data bases, necessary to prepare and evaluate microgravity experiments. However, a discussion of the utilization aspects of the Space Station era would be incomplete without recalling the dimensions of the whole enterprise. Compared to the present Spacelab missions, the Space Station offers a tremendous utilization potential. Defining this potential as payload mass times mission duration, ESA has produced the diagram shown in Figure 3. It clearly demonstrates the dramatic increase in utilization potential that Columbus offers compared to Spacelab or EURECA flights. In order to fully exploit this opportunity, a huge ground infrastructure will be necessary, involving mission control centers, user support centers and user operation centers, all interconnected and constituting a network. Last but not least, also the user community will have to grow accordingly. It has been estimated that Columbus will keep up to 1000 scientists busy preparing, performing and evaluating their experiments. This is the real challenge of the Space Station era, and we can only express our hope that instead of discussing scientific user requirements we will not have to consider requirements for scientific users.

<u>References</u>

[1] European User Requirements Analysis
(WP 1000), ESA Contract No. 6610/85/F
(1986)

[2] P. Banks: "Telescience" Presentation
to the SSUP, Stockholm 1985

[3] A. Balogh "Interactive Payload Operations on Columbus", Report of the
Telescience User Team, Phase 2, February 1987

[4] H.P. Schmidt, B. Feuerbacher and E.
Messerschmid, "Telescience – A Concept
for Scientific Experiment Operations
in Space" IAF Paper 86-25 (1986)

[5] Scientific Goals of the German Spacelab Mission D1, WPF Secretariat,
DFVLR, Köln 1985

[6] H.P. Schmidt and K. Wittmann, "Microgravity User Support Centre for
EURECA-1", Z. Flugwissenschaften und
Weltraumforschung, Vol. 10, p. 6-12
(1986)

SCIENTIFIC CUSTOMER NEEDS - NASA USER

David C. Black*
NASA Headquarters
Washington, D.C.

Abstract

The Space Station era will be one of significant new challenges for NASA and its international partners in the Space Station program. The manned base of the Space Station, along with its unmanned platforms, bring capabilities to a wide range of user communities, capabilities that can only be realized fully if adequate attention is given to the information systems developed as part of the program. This paper outlines some of the major needs of the scientific community with regard to this information system.

Introductory Remarks

The scientific user of Space Station places several requirements on the information system associated with the Station. These requirements involve high peak data rates from the polar platform, extensive slow and rapid scan video, and two-way voice communication between the crew on the manned base and scientists on the ground. A related aspect of the information system involves the dissemination or distribution of data to the user. Much of the data from Spacelab and from NASA scientific satellites (e.g., IRAS) was or is being distributed in the form of magnetic tapes to scientific users. While tapes serve a useful archival function clearly they should not be the mode of data distribution during the Space Station era. In fact, one of the major challenges to information systems in the 1990's will be to provide not only rapid dissemination of data from a space system to a ground-based user, but also to permit that user to interact in a real-time command/control sense with the Space system. This type of interaction has been termed "telescience" by

NASA's Task Force on Scientific Uses of Space Station (TFSUSS).

The detailed quantitative requirements of scientific users are well documented elsewhere in this volume (e.g., paper by Bretherton and by Feuerbacher). I will concentrate here on more general issues and needs associated with scientific users.

A Need For Interactive Testbeds

A hallmark of scientific data from past space programs is that only a small fraction of the data that was taken as part of a given program has been reduced or examined in detail. While some programs have been good in this regard, a generally accepted average figure is that less than five percent of all data taken in the space program has been analyzed at a level leading to publishable results. In some cases the unanalyzed data is lost for all time because calibration data and even adequate knowledge of data format is unavailable. In other cases the nature of the data distribution process has been inefficient and slow.

In view of the rising cost per bit of data from space systems it is essential that information systems in the Space Station era place data in the hands of the users as rapidly as possible and in a form that facilitates rapid analysis. This process takes interactive participation in the design and development phases by the designers of the information systems and the users of those systems. This interaction is most effectively done through testbeds which allow users to check out design concepts, offer suggested changes, and learn how best to utilize the final information system. The key ingredient here is the _interactive_ nature of this process. Users cannot and should not simply pass requirements to the designers of information systems and then await a finished product. Nor can the designers presume that they can, in isolation from the users, make critical decisions regarding the nature of the information systems of the Space Station era.

---

*Chief Scientist, Office of Space Station

One consequence of the diversity in information systems is that there must be an equally diverse set of testbeds.

One can see a need for video testbeds, interactive voice/video testbeds, and instrument protocol testbeds just to name a few. One of the more demanding tasks for managers of information systems will be to assure comprehensive involvement by the user communities in both the identification and operation of information system testbeds. Further, given the relatively long lead time for development of information systems particularly where these needs to be extensive iteration between the users and designers of those systems, it is important to start soon on testbeds for the information systems of the Space Station era.

<u>The Need for "Telescience"</u>

The Space Station era will be one in which there will be a great "mix" of research activities in space. There will be major unmanned free flyers such as the Hubble Space Telescope, the Gamma Ray Observatory, the Earth Observing System, and the Advanced X-Ray Astrophysics Facility. These will be operating in parallel with research on the manned base of the Space Station involving payloads such as the Astrometric Telescope Facility attached to the exterior truss structure along with material and life science experiments in the pressurized modules.

An important characteristic of the information systems of this era will be their ability to permit effective interaction between ground-based users and their space-based equipment. The nature of this interaction depends on the user. Scientists working with free-flyers typically need to be able to rapidly obtain data, compare that data with information available in existing, often extensive, data bases, and then command their space-based instrument(s) to continue or alter its operation. The closest approximation to this mode of operation is that of the International Ultraviolet Explorer Satellite.

Perhaps the major challenge will be that associated with research on the manned base. One reason for this challenge is that the research on the manned base is <u>experimental</u> as contrasted with the <u>observational</u> research typical of unmanned research.

The Space Station can be viewed as a laboratory in space. Although it is only a few hundred miles distant from the ground it is as remote as research laboratories in the Antarctic. One lesson that has been learned from years of research activity in remote sites such as the Antarctic is that the quality of the research is significantly improved if a trained researcher is on-site. Given the obvious transportation limitations of the next one to two decades, it is not realistic to suppose that most scientists doing space experiments will be able to be present at the Space Station. We must find a way to bring, in effect, the space environment to structures on the ground. That is the essence of telescience.

Viewed from the prospective of information systems, the telescience concept is particularly challenging. On one hand, it requires flexibility in terms of interactive exchange between ground-based scientists and their space-based instruments. On another, it requires access to and manipulation of extensive data bases and perhaps theoretical models on time scales near real-time. The latter could require networking of scientists at various locations within a given country or perhaps even globally.

The telescience concept also places requirements on information systems in terms of video and voice exchange between space and ground. The details of video requirements is just beginning to be understood and much work remains to be done. Many of the experimental activities currently envisioned in the Space Station era require still frame video data, although there are some which require both high spatial and modest temporal resolution. As scientists come to realize the full potential of telescience, it is expected that additional and far more demanding

requirements will be placed on the
visual component of space information systems.

<u>Concluding Remarks</u>

I have summarized here a very limited
science user perspective of needs that
they will have for information systems
in the Space Station era. Two major,
if somewhat general, needs appear
prominent. First, the users of
information systems and the
designers/builders of such
systems must work together
throughout all phases of the
design and development of information systems. This interaction
can be best accomplished through
jointly conceived and operated
testbeds. Second, because of the
limited access of personnel to
space it will be essential that
information systems of the Space
Station era have characteristics
which permit, as much as is
possible, ground-based scientists
to conduct their research using
space-based equipment as though
the scientist and equipment were
co-located. This concept of
"telescience" places special
demands on information systems
ranging from data/command flow
between ground and space, to
potentially extensive ground
manipulation of data sets and
user interaction.

In many ways the information
systems of the Space Station era
are pivotal to the quality and
the quantity of scientific space
research in that era. If the
implemented information systems
have the capability that many of
us hope they will, they will
enable entirely new ways of
conducting space research and
they will also permit the active
involvement in space research of
far more people than is possible
using current information systems.

PROCESS CONTROL AND DATA ACQUISITION FOR COMMERCIAL
MATERIALS PROCESSING IN SPACE

Earl L. Cook
Space Research and Applications Laboratory
3M Company
St. Paul, Minnesota 55144

Abstract

Commercial, materials processing in
space is discussed with emphasis on process control and data acquisition requirements. The research aspect of materials
processing requires a capable, flexible,
adaptable, general data system that can
support process control, data acquisition,
and real-time interaction, with significant mass memory. In-space manufacture is
more autonomous, with the apparatus providing most of the process control and
data acquisition requirements internally
and requiring less from the general data
system. Some specific materials processing experiments aboard the Shuttle are reviewed, and the support currently available on the Shuttle examined. A prototype
system recently developed for multiple
secondary payloads on the Shuttle is described. This system, known as the Payload Support Network (PSN), may serve as a
model for Space Station data systems support for materials processing.

Introduction

One of the primary objectives of the
Space Station is to provide a national resource for research and technology development, equipped with the facilities and
capabilities to assure that the station is
an effective and beneficial tool for the
advancement of U.S technology. One of the
disciplines to be emphasized on Space Station is materials processing. This discipline is one of the more commercially attractive and is one of significant interest to American industry.

The success of the Space Station in the
arena of materials processing requires
careful consideration of the facilities
and supporting services provided on the
Station and the way they are configured
and managed. Indeed, considerable effort
is currently being expended to assure that
materials processing will be adequately
supported in the U.S. Microgravity Materials Processing Facility (MMPF) and that
the requirements of all potential users
will be met. However, there is a danger
in an all things to all people approach;
the final product, especially with regard
to the information and data management
system, may be so general that it serves
few adequately and most less than adequately. It may in fact be so complex and
cumbersome as to be counter-productive,
actually impeding and discouraging some.

This paper is addressed to the concerns
of commercial, materials processing interests and those aspects of information and
data management systems provided on the
Space Station that are important to materials processing. However, much of what
is discussed here may be equally applicable to other types of Space Station payloads, particularly smaller, secondary
payloads of an experimental variety.

The perspective presented here derives
from the experience gained from materials
processing experiments flown aboard the
Shuttle. 3M has been a participant in
space research for over two years and has
flown three mid deck experiments. Moreover, 3M is committed to long-range space
research and has demonstrated that commitment by entering into a 10 year Joint Endeavor Agreement (JEA) with NASA A large
number of materials processing experiments
are planned for the period covered by the
JEA.

In support of the this long-term effort, 3M has developed a generic system
for the simultaneous control and operation
of multiple experiments connected in a
network. This system is independent of
existing Shuttle avionics, and provides
sophisticated process control and data acquisition support for a large variety of
payloads while providing a capable, central interface between the payloads and
the operating crew members. The proposed
system, known as the Payload Support Network (PSN), has been well-received by potential users, and should be of significant advantage to the space experimentation community generally when fully
implemented.

What is important about this payload
control system (PSN) to the subject of
this paper is that it may serve as a model
for information and data management systems support required for effective materials processing on the Space Station.

Research, Development and Manufacture in Space

The familiar product cycle of research,
development and manufacture is particularly distinctive in materials-oriented,
product development endeavors on earth; it
is even more so when used in space. For
materials processing in space, each stage
of this cycle has its own distinct, special concerns that must be met and adequately supported if the effort is to be
successful.

Data Systems

In the above, the data systems were not very definitive and the requirements outlined were rather general. It is appropriate at this point to discuss some of the specifics and put them into the proper perspective.

Materials processing has some data system requirements that are unique to materials processing users and not of general interest to other station users. The questions are: Should the Space Station data system attempt to provide these discipline-unique services or only supply those services of general interest? If only the more general services are supplied, how will the distinction between the two be made and where will the line be drawn?

Most of the problems can be minimized by structuring the data system as a hierarchical, distributed system with well-defined interfaces between the various hierarchical levels. In this way, specialized services can be integrated into the total system when and where needed in a manner that minimally impacts the total system.

For example, a material processing user may require some unique process control and data acquisition functions that would be of little or no interest to other users. The same user may also require an display interface with the operators of the experiment so data acquired via the unique hardware can be displayed. This display interface function is, of course, not at all unique to the materials processing user, but of interest to most users. Obviously, the display interface function should be supported at a total system level while the user-unique functions supplied by the user.

This hierarchical structure requires carefully designed and implemented interfaces so that the unique functionalities can be easily integrated into the total system in a way that does not adversely impact other users or services in the system.

The hierarchical structure is quite compatible with the distributed system concept and would allow individual users to provide as much computational capability as needed for their experiment without running the risk of overtaxing the total system capability.

Materials Processing Aboard the Shuttle

During the early days of the space program, materials processing interests were not really significant in most missions and usually were given attention only if there were no other "important" activities to concern the crew. Materials processing experiments were indeed considered "secondary" payloads and afforded no system-level support. There were real reasons for this posture; the operation of the spacecraft and the objectives of the mission were demanding and the crew had little time for "secondary" experiments.

With the Space Shuttle, more opportunities for secondary payloads, including materials processing experiments, were available. However, support for the operation of them was not included in any significant way.

Currently, the Shuttle offers secondary payloads essentially two options as far as process control and data acquisition are concerned: 1) total autonomy where the payload is completely and totally controlled through a self-contained control system, and 2) limited process control and data acquisition through the Shuttle's General Purpose Computers (GPC).

Both options are less than satisfactory. The major disadvantages of the first option are: 1) the interface between the payload and the operating crew member and 2) the capability to communicate with the payload from the ground. The interface disadvantage creates on-orbit operational problems in that each autonomous secondary payload has its own unique crew interface each requiring special training and familiarization by the crew. The major disadvantages of the second are: 1) the GPCs offer rather limited control and data acquisition capabilities and 2) the interface between the GPCs and the payload must be specified some 12 to 18 months before the flight.

3M's first experiment (DMOS, STS 51-A) was performed in the mid deck of the Shuttle and was totally autonomous with the internal control electronics completely responsible for the experiment's operation; the interface between the experiment and the crew consisted of an ON-OFF switch and a power indicator. This proved to be acceptable but only because the experiment operated in a totally nominal fashion. Had the experiment not operated properly, it might have been a complete failure. The absence of an interface did not allow any indication of the progress or health of the experiment or permit any corrective interaction had there been a problem. This situation resulted from the fact that there is no support available for secondary payloads in the mid deck. If the experiment could have been interfaced with the Shuttle's GPCs, the time required to implement the interface would have caused a delay of a year or more in the flight of the experiment.

In the second (PVTOS, STS 51-I) and third (DMOS/2, STS 61-B) 3M experiments, a small keyboard/display unit was incorporated into the experiments. The experiments were designed to operate quite autonomously with little interaction with the crew required provided no off-nominal

The materials research phase is primarily concerned with fully understanding the materials under study, carefully characterizing their properties, and optimizing those that are important in the application of the materials to a specific product.

The development phase is primarily concerned with the scaling up the materials processing techniques developed in the research phase and exploring ways to process commercially significant quantities of the materials in a economical and realistic way.

The manufacture phase is that phase where the materials are routinely produced for very specific applications.

Research

The research phase of this cycle is perhaps the most critical and demanding as far as in-space facilities and capabilities are concerned. Two of the more distinguishing characteristics are that materials research is an iterative endeavor, and that it requires knowledgeable experts to interpret the progress and results of the experiments.

The iterative aspect is simply inherent to the research process; research is a step-by-step process. Most often, one experiment must be performed, analyzed, and interpreted before the next experiment can be fully specified, designed, and implemented. Any facilities involved in the support of the research effort must be flexible, adaptable, and capable of quick alterations so as not to impede the iterative character of research.

The knowledgeable expert aspect of research is equally important. Experiments often produce unexpected results in that the phenomena that are expected to be dominant are either not observed or are not as dominate as expected, and other, unexpected phenomena dominate instead. These unexpected phenomena may be very important to the research process by providing new insight into the process being investigated; every effort should be made learn as much as possible about them. This may mean that the experiment should be altered when the unexpected phenomena are recognized. Of course, the recognition, interpretation, and alteration of the experiment requires a knowledgeable expert; the ideal situation would be to have the expert operate the experiment either hands-on or remotely. Alternatively, the experiment could be supported in such a way that the expert could easily access, from the ground, information about the experiment in real-time and use that information to recommend alternative experimental procedures to the operators of the experiment.

As far as data systems are concerned, the requirements center on adaptability, interactability, and capacity. In other words:

- the data systems must be flexible and capable of relative quick reprogramming and restructuring as the experiments change;

- the data systems must be capable of interfacing the experiment, the operators, and the knowledgeable experts, displaying sufficient information to allow real-time control and alteration of the experiment; and

- the data systems must be capable of acquiring and storing relatively large amounts of data for post-experiment analysis.

Development

The development phase of the process is not quite as demanding as the research phase. Most often, the objectives are rather more clearly defined with the apparatus and procedures to meet them more amenable to methodical, predictable exploration, longer range planning, and a more typical engineering approach. By the time this phase is achieved, the problem should be rather well-defined, with the goal being to implement a solution to the problem rather than discovering one as is the case in the research phase.

Although lessened, this phase does have some special requirements as far as data system support is concerned. Basically, the requirements are not so much for interaction and real-time response as they are for extended capabilities and capacity. With these, large amounts of data can be acquired and processed for a complete analysis of the process under development. Of course, the goal of this phase and the objective of the data acquired are the optimization of the processes in preparation for possible in-space manufacture. There is still a need for a rather capable information and data management system to support the activities of this phase.

Manufacture

The manufacture phase of the materials processing may be the least demanding of the three phases. By the time this phase of the cycle is achieved, the process and the parameters governing it should be sufficiently well-known and understood that the manufacture system should be rather independent of external systems and capable of autonomous operation. In other words, this phase should be characterized by a well-developed manufacturing facility that can be routinely operated, producing product with a minimum of intervention and real-time changes like those that characterize the research and development phases of the cycle.

conditions occurred in the experiments.
If something unusual was detected, however, the crew could control the experiment via the keyboard/display unit and
thoroughly test it. This capability
proved to be quite important in the PVTOS
experiment. Early in the on-orbit operation of the experiment, an off-nominal
condition was detected by the control
electronics and reported to the crew. In
consultation with experiment developers on
the ground, the problem was diagnosed via
tests executed from the experiment's key-
board/display unit and a scheme was de-
vised to circumvent the problem thereby
saving the experiment.

Payload Support Network

Based on the experience gained from
flying three materials processing experiments aboard the shuttle, 3M has designed
a payload support system that significantly improves support for secondary payloads flown on the Shuttle. This system,
known as the Payload Support Network (PSN)
permits the simultaneous control of and
significant operational interaction with
multiple payloads connected in a network
in a way that does not impose any severe
implementational or operational demands on
Orbiter resources.

Although the PSN system has not yet
flown, it has been prototyped and its operation simulated in the laboratory. The
results of the simulations indicate that
the system, when implemented, will create
a very adequate and attractive environment
for secondary payloads on the Shuttle.

From the materials processing user's
point of view, the PSN system is a good
model for a reasonable process control and
data acquisition environment for Space
Station.

System Overview

The PSN system consists of three major
subsystems:

- a shared interface between the payloads in the network and the crew,

- dedicated experiment processors for
  each of the payloads in the system,

- interconnecting network that supports
  communication between the nodes in the
  network.

and can be characterized by the following
attributes:

- complete payload control system,

- multiple payloads in either the mid
  deck and/or cargo bay,

- distributed system,

- familiar, central crew interface on
  aft flight deck,

- large mass memory systems,

- command uplink and data downlink,

- flexible and adaptable.

The basic system configuration is shown
schematically below. The left-hand drawing shows the basic topology of the network and the right-hand drawing shows how
the system might be configured to support
experiments in both the cargo bay and mid
deck of the Orbiter.

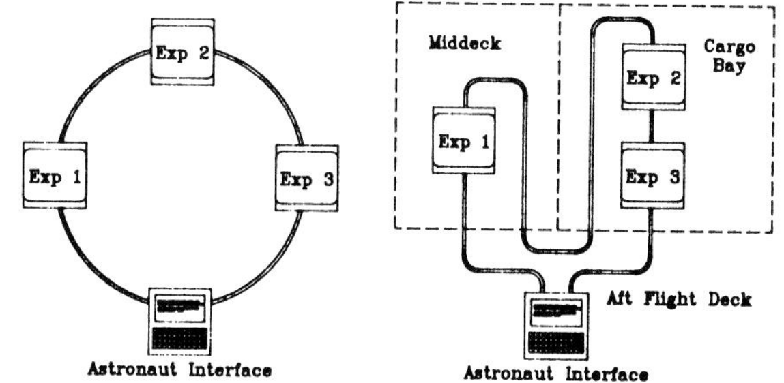

Crew Interface

The crew interface for the PSN system,
which could reside on the aft flight deck
or any other convenient place in the Orbiter, basically functions as a common in-
put/output device for the network. Any
experiment can display information to or
request information from the crew via the
interface as appropriate to the operation
of the experiment. Since the interface
must be capable of handling the in-
put/output requirements of all payloads in
the network concurrently, the actual functionality of the interface is rather complicated and the device itself must be
rather capable.

The interface is implemented with a
GRiD Systems 1137 computer. This computer
is essentially an IBM PC compatible using
MS DOS as the operating system. The GRiD
is very small and rugged; it has been
flight qualified and routinely flies
aboard the Orbiter. For Shuttle applications, the GRiD is known as the Shuttle
Portable Onboard Computer (SPOC).

The PSN system uses the concept of
'windowing' to implement the concurrent
use of the central crew interface. In
this scheme, each payload is assigned a
one or more windows and can use those windows for information display in any way
that is appropriate to the operation of
the payload. The crew interface is configured to support these windows and provides a method for the crew to quickly and
efficiently move from window to window
and, effectively, from payload to payload.

The PSN system supports both text and
graphic windows. Text windows are used to
display messages and numeric data from the
payload while graphic windows are used to
display information graphically. The man-

ner in which these windows are used is left up to the payload developers, there are relatively few system limitations.

From a payload developer's point of view, the crew interface is a standard input/output device and can be used without concern about how other payloads may be using the interface. Moreover, the crew interface is quite standard in that it supports most of the text and graphic functions found on the more familiar terminals.

The crew interface is more than an input/output device, however. It is also responsible for management and health of the network. Additionally, some system-wide functions like time keeping are also implemented in the crew interface.

Dedicated Experiment Processors

The PSN system is a distributed system where each payload provides its own computational needs with dedicated processors rather than deriving them from a central system. The Dedicated Experiment Processors (DEP) are intended to be the basic computational building block for the PSN system.

The Dedicated Experiment Processors are designed and implemented with great importance placed on maximizing the functionality of the processors so that the requirements of sophisticated process control and data acquisition will be adequately satisfied. All peripheral functions like A/D and D/A conversion, digital input and output, etc., required for experiment interaction are fully supported. Large amounts of mass memory (typically 100 megabytes) are available to allow significant amounts of data to be acquired on-orbit and returned for post-flight analysis.

The Dedicated Experiment Processors are based on the Motorola 68010 micro processor and configured on the VME bus. The mass memory is implemented using 20 megabyte hard disk systems and 40 megabyte magnetic tape cartridge systems. Cost effectiveness and reliability have been emphasized in the implementation of the processors.

The operating system is a real-time, multi-tasking operating system that supports high-level languages like PASCAL, ADA, C, etc. providing the experiment developer with a familiar programming environment for the development of software for the experiment. The software required for exchanges over the network have been embedded into the operating system; network transactions are totally transparent to the programmer.

The Dedicated Experiment Processors are housed in hermetic containers that are equivalent to one mid deck locker and can be installed in one mid deck locker space for experiments operated in the mid deck

of the Orbiter. The containers are actively cooled and are capable dissipating 75 Watts.

Slight modifications to the containers allow them to be used in the cargo bay also.

Network

The network connecting the experiments and the crew interface is implemented in a way that minimally impacts the orbiter resources. In particular, the PSN system can be installed using twisted pair cables of the type that are currently installed and available on the Orbiter. The communication protocol for the network is standard RS232 at 19.2 kilobaud.

The network is structured as a token-passing, ring network. This structure is particularly good in that the number of network interconnections is kept to a minimum and the management of the network is quite positive. The token passing scheme results is a well-behaved network that does not suffer from the packet-collision problems that other network structure exhibit. Basically, token-passing is based on collision avoidance rather than collision detection.

Information is communicated from node to node via packets. Packet integrity is monitored using cyclical redundancy checks.

The GRiD computer performs some network management functions like, for example, regulating which nodes are allowed to communicate with each other and performing corrective actions in cases where a packet becomes corrupted for one reason or the other.

System Support Functions

The PSN system also supports a number of functions that are available to all payloads in the network. One important function is a link, through standard Orbiter interfaces, between the Orbiter and the ground and supports both command uplink and data downlink.

There are some safety and philosophical considerations associated with command uplink that are not yet resolved. Most importantly: Can commands be sent from the ground to a payload independent of crew interaction without compromising the safety of the payload or the Orbiter? Secondarily: If a payload is operated totally from the ground, what is the need of a crew? The current orientation of the PSN system is to have the crew be the primary operators of all payloads and employ the uplink only when absolutely necessary.

The downlink function is less philosophically objectionable than uplink and can be of great advantage to payload de-

velopers on the ground. It allows them to
monitor the progress and status of their
payloads and brings the 'expert' aspect to
payloads like materials processing where
real-time evaluation of the experiment can
be of great benefit.

     The PSN system also provides the capa-
bility for storing fairly reasonable
amounts of data at the crew interface.
This capability permits an on-line 'help'
system for the operation of the payloads
that can be updated from the ground so
that the latest information concerning the
payloads are always available to the crew.
It also allows the operating crew members
to conveniently record observations
concerning the operation of the payloads.
This capability is known as the PSN
'Notebook' facility.

Technology 1: Communications and Data Systems
Co-Chairmen: John L. McLucas and Michiro Kusanagi

SYSTEM AND TECHNOLOGIES OPTIONS
FOR THE EUROPEAN DATA RELAY SATELLITE

Claudio Soprano*, Mario Lopriore**
ESTEC, Noordwijk, the Netherlands

Abstract

System and technology options for the European
Data Relay Satellite are outlined, as emerging by
currently on going studies; key issues are also
addressed.

Introduction

Several options have been studied in the last
years for a European Data Relay System in a number
of contracts. Presently the European Space Agency
is entering in a competitive phase for the detailed
definition of the system comprising space and earth
segments and user terminal.

In parallel a number of technologies and techniques are being investigated under separate funding budgets. For example in the P.S.D.E. programme
a comprehensive development of all elements of an
optical Inter Satellite Link has been initiated.
This is to be considered as a possible, long term,
alternative to the IOL inter orbit links at 26 GHz
that constitutes the current baseline.

For a much closer time scale a laboratory
simulator of the TDRSS link(s) is being manufactu-
red and integrated. This "TDRSS" laboratory breadboard includes a user terminal, a TDRSS equivalent
transponder and an earth station simulator. It will
be used for evaluation purposes should the Columbus
programme operate initially via TDRSS.

Other techniques are being studied to evaluate
fully the implications of a multiple access scheme
in S-band together with a decentralized access system approach.

In this paper the transmission techniques and
the RF technologies that are being investigated in
connection with the DRS programme will be discussed
and the current status outlined.

Baseline System Definitions

The European Data Relay Satellite (D.R.S.)
System includes the following elements:

-   the "space segment", that is the relaying pay-
    loads carried by two active satellites on geo-
    stationary earth orbit (GEO);
-   the communication terminals of the "user
    satellites" flying on low earth orbit (LEO);
-    the "earth segment" consisting of several
    earth stations, which receive and transmit the
    data, one center which controls the system and
    one or more stations to monitor its performan-
    ce.

These elements are connected by feeder links,
which link the earth segment to the space segment,
and (IOL) at S-band and K-band, which link the user
satellites to the space segment. Signals transmitted from the earth stations to the user satellites
via DRS are said to travel in the "forward" direction, those transmitted in the opposite direction
travel in the "return" direction.

The services provided by the system to its
users include, the support to transmission, to and
from LEO vehicles, of telecommand, ranging, house
keeping telemetry, and, on both the forward and the
return direction, the transmission of data; besides, accurate localisation may also be provided at
S-band.

Though several options are presently under
consideration, a baseline configuration of the Data
Rely Satellite Payload includes four independently
steerable reflector antennas supporting four independent inter orbit communication links, each one
on both the forward and the return directions, two
at K-band and two at S-band.

The reflector antennae are pointable across a
20° field of view, compatible with up to 1000 km
altitude low earth orbits; pointing is achieved by
open lop control at S-band, while, at K-band, a
radiofrequency tracking closed loop control is implemented on board the DRS payload.

The initial acquisition time, at K-band, may
be shortened by a switchable configuration of the
feeder elements producing a wider main beam.

The support, via S-band of critical link
acquisition, as required by specific LEO users, has
also been studied.

The baseline G/T figures, as provided to
support the Return Links are 23.7 dB/dB°K and 4.2
dB/dB°K, respectively at 26 GHz and at S-band, at
boresight.

Satellite platforms under consideration are
the ones based upon existing European technology
such as the Eurostar, the Spacebus and possibly the
Olympus platforms, depending upon the eventual
mission requirements and system configuration.

Table 1 shows the frequency bands presently
being considered for the various communication
links supported by the system.

---

* System Engineer in the Communications Satellites Department
** Head, RF Systems Division

TYPE OF LINK	FREQUENCY BAND

FEEDER LINKS
(Earth Station - DRS)
  Up-links
  Down-links
	
	27.5 - 30
	18.8 - 20.2

INTER-ORBIT LINKS (DRS-User Satellite) . Forward links (DRS - User Satellite)	 (a) 2.025 - 2.110 (b) 25.25 - 27.5
. Return links (User to DRS)	(a) 2.2 - 2.29 (b) 25.25 - 27.5

Table 1 - Frequency Utilisation plan

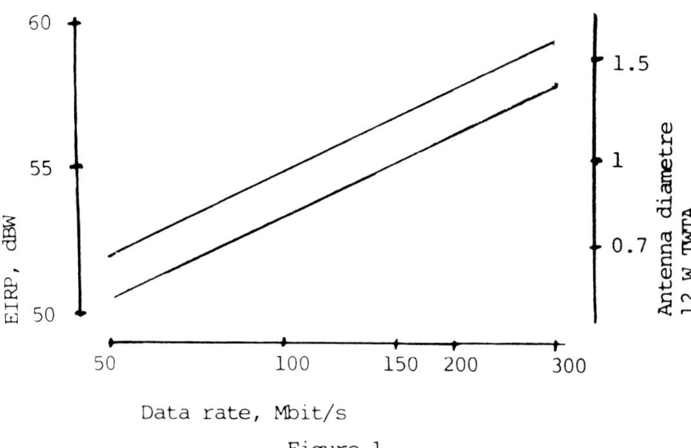

Figure 1

The 20/30 GHz frequency bands have been base-lined for the feeder links because of the severe problems, of institutional nature and of tight frequency coordination with other services, which would have to be faced on other frequency bands.

For the inter orbit communication links at K-band the 26 GHz frequency band (25.25 GHz/27.50 GHz), worldwide assigned to satellite data relay services on a secondary basis, has been baselined* to allow a reduced antenna size on both the D.R.S. payload and the user satellite.

The transmission capacities for the various communication links are shown in Table 2.

The transmission of the data at 500 Mbit/s, as baselined at K-band is achieved by means of parallel transmission through frequency division multiplex channels at lower data rates; besides, forward error correction by means of the standard convolutional code of constraint length 7 and rate 1/2 is baselined for all the data transmitted in continuous mode.

	FORWARD LINK
S-band	up to 300 Kbit/sec
26 GHz band	up to 25 Mbit/sec
	RETURN LINK
S-band	up to 5 Mbit/sec
26 GHz band	up to 500 Mbit/sec

Table 2 - Projected transmission capacity

The number of channels and their capacity has not yet been finalised; however, fig. 1 shows the projected performances of the inter orbit Return Link in terms of L.E.O. user spacecraft E.I.R.P. versus data rate (information rate) per channel.

A 10exp(-5) B.E.R. has been assumed and two conditions are shown for balanced and unbalanced links (dotted lines).

The R.F. (radio frequency) power at the L.E.O. user terminal may be generated either by separated TWTAs (travelling wave tube amplifier) frequency multiplexed at the antenna input, or by a single TWTA providing the requested RF power level with linearity under multicarrier operation; the trade-off between the two configurations has to be based upon the specific L.E.O. user spacecraft constraints.

The D.R.S. system is presently conceived as having a decentralised architecture: direct communication links are foreseen from the D.R.S. Payload to earth stations located at the premises of the primary users of the data.

The advantaged of a decentralized configuration are: i) no need for further wide band communication infrastructure for the (raw) data distribution to the end users (ii) multiple reception of the data by many earth stations provides an intrinsic redundancy to the system.

Some capability of direct access to the Forward Link is also required because of the specific interest of some major potential users.

The decentralised architecture is also driven by the need of real time distribution of voice and video data originating from manned space vehicles.

The requirement of near real time distribution of the data to non-European users of the system may lead to the consideration of extending the coverage of the feeder down link to U.S. and Japan via dedicated transponder and antenna beam.

The feeder link antenna has to provide a wide European coverage. This and above requirements of feeder links to US/Japan severely constraints the selection of viable positions.

The issue of the actual methods to implement the frequency and phase coherency of the communication links in a decentralised configuration is still under study, the relevant options are discussed further on.

---

* Some consideration is still being given to other frequency bands, including the 60 GHz, the 23/32 GHz and the 14/15 GHz.

Key System issues at K-band

The feeder links, at 20/30 GHz, with the margins by them requested to provide specified service availability against the fadings due to atmospheric propagation, set the performance of the data transmission system at K-band in terms of channel capacity and achievable coverage.

Fig. 2 shows the margins to be accounted for in the Return Link budgets versus orbital positions for given service outage time as a percentage of one year. The curves are based upon propagation model for Europe including the effects of gaseous absorption and clouds; potential locations for receive earth stations are considered.

The impact of the orbital positions on the coverage of the inter orbit links is shown in fig. 3 for D.R.S. West and DRS East. As an example, extreme orbital positions are considered at 50°W and 59°E with a 1000 km LEO altitude.

A simplified link budget is shown in fig. 4 which gives the requirements of satellite EIRP for the downlink versus the G/T of the earth station for various data rates. Equal noise apportionment is assumed for inter orbit link and down link with 7 dB margin for athmospheric propagation and 10^{-5} B.E.R.

The European coverage area in satellite coordinates as visible from various positions on geostationary earth orbit is shown in fig. 5.

The above is hoped to have cast some light on the intricacy of the trade-offs requried to define a viable system configuration.

A detailed analysis of possible configurations of the antenna supporting the feeder link is clearly of paramount importance; the required coverage may be achievable with a continguous beam pattern.

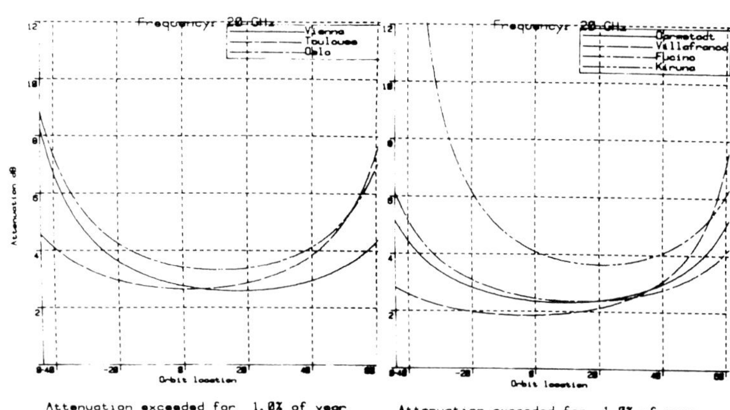

Figure 2

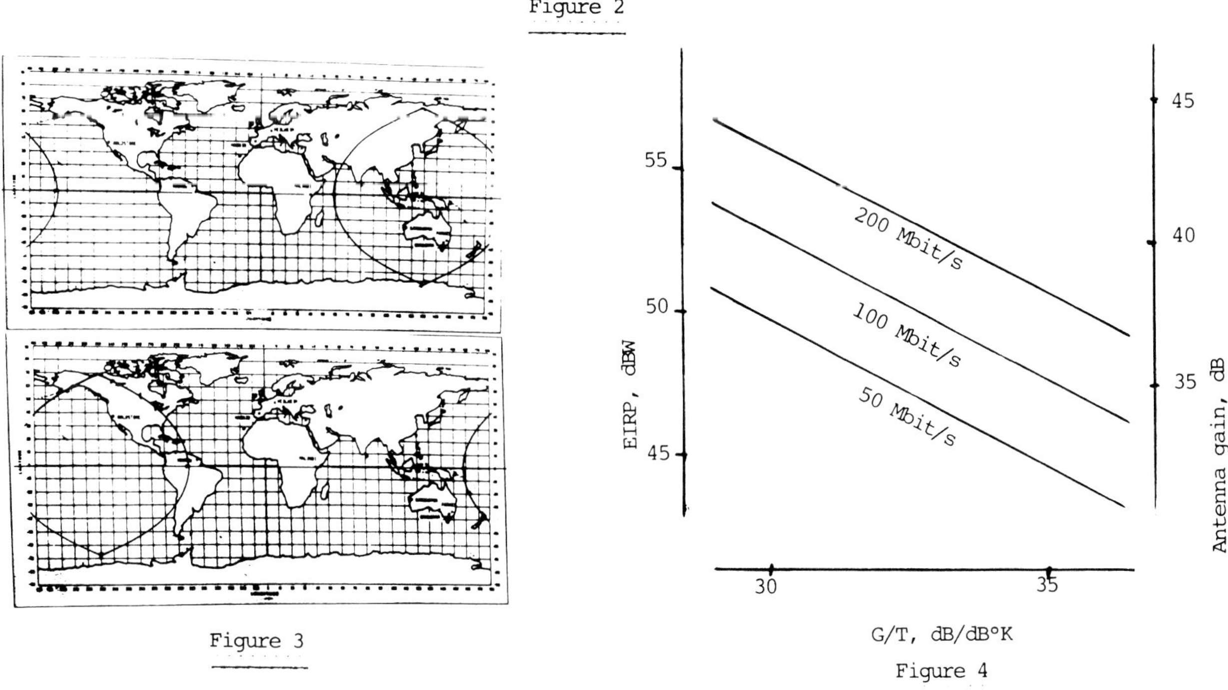

Figure 3

Figure 4

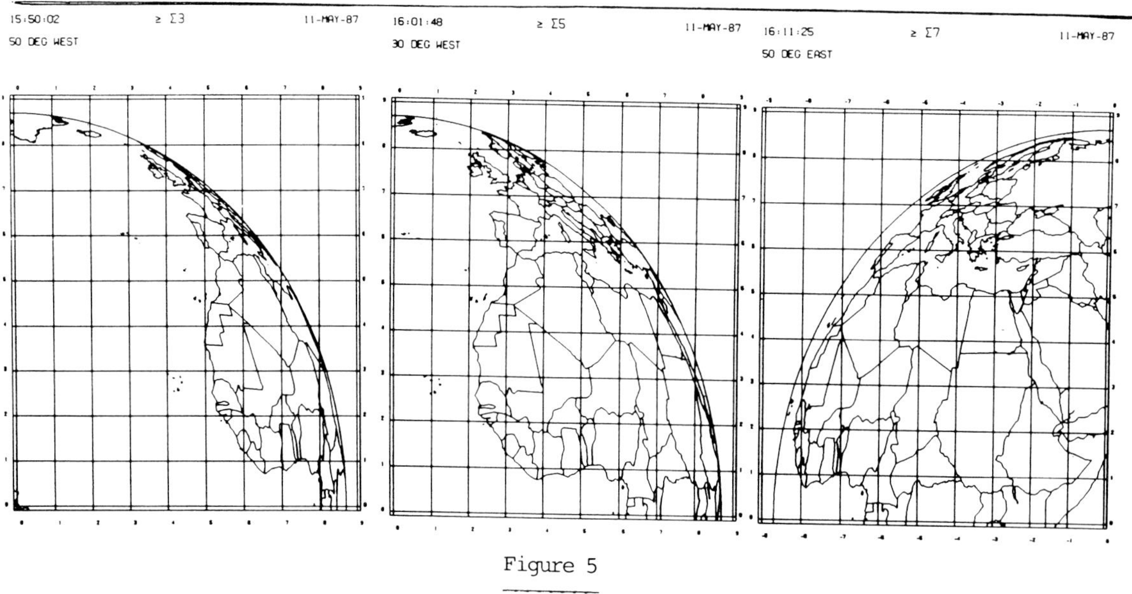

Figure 5

System and technology issues at S-band

The peculiar aspect of the S-band is to be widely used for the space operation service while only being assigned to it on a secondary basis; the special case of space operation service supported by space to space communication links is furtherly restricted.

The co-existance of several services has led to a situation characterised by a R.F.I. (radio frequency interference) environment, also of military nature, and by restrictions to the allowable power flux density at the surface of the earth as generated by space-born transmitters; besides, inter orbit communication links established with a L.E.O. vehicle equipped with a low directivity antennas may experience severe multipath.

Such a situation is believed to have led NASA to define, for its TDRSS, a transmission system including low rate data encoding, interleaving and spread spectrum modulation; the same situation and the actual need of co-existance with the T.D.R.S.S, has led ESA to adopt the same transmissionsystem as an initial reference, to which some modification could be proposed.

Co-existance with Other Systems

The modest frequency resources available at S-band and the comparatively low antenna gains which there may be achievable, lead to the consideration of potential problems of radio frequency interference between data relay satellite systems; these problems may be controlled by coordination of frequency, polarisation, orbital phasing of L.E.O. vehicle, orbital separation of the relaying payloads on G.E.O., levels of radiofrequency radiated power.

The constraints on the phasing of spacecraft on low earth orbit may be appreciated by considering the interorbit link coverage as provided by the DRS S-band antenna pattern extended to include the first side lobe.

It may be seen that moving the boresight of the antenna from the subsatellite point, a significant part of the available low earth orbit coverage is taken by that single access at its operational frequency, the worst case, about 43% arising for a 4° depointing. A similar situation, though with different constraints, may exist for the TDRSS.The first side lobe, being at not less than 20 dB below peak gain, plays a relevant role in the build up of the R.F.I.

Considering then the antenna gain requirements versus the data rate for a LEO spacecraft operating in the single access Return Link as presently baselined, and the angular deviations from the boresight corresponding to the peak of its first sidelobe, it may be seen that an important orbital separation may be required between two data relay payload, operating at the same frequency, in order to achieve a significant C/I (carrier to interference ratio).

The assumption of operation at single R.F. carrier per single access is indeed constraining the overall capacity at S-band.

Considering a 20 MHz gross channel bandwidth for each one of the single access, as reported for the TDRSS, only 8 users would be serviced over a wide orbital arc on a full non-interference basis.

The above issues have been developed only into a very preliminary stage; quite an important amount of work is requested for a complete assessment of the R.F.I. scenario. However, as a first step, it is proposed to increase the overall capacity at S-band by making use of the signal isolation properties provided by the spread spectrum modulation already used, to other purposed, in the TDRSS Single Access.

Besides, an increased chip rate, for instance up to 12 Mchip/s, as used by the TDRSS for support to the Space Transportation System, though requiring proper output filtering at the LEO user transmitter, would increase the signal isolation for data at rate above 300 kbit/s.

For what is concerned with the investigation of the effects of pulsed radio interference on the Return Link, in addition to classical approaches such as data interleaving and low rate (1/3 or 1/4) inner channel convolutional coding, advanced solutions are also being evaluated based upon erasure decoding supported by specific processing at earth stations and on board the DRS payload; besides, the usefullnes of wideband spread spectrum is also receiving attention.

Configuration options at S-band

In parallel to the baseline, the feasibility is being evaluated of configurations leading to high performance single and multiple access schemes including, respectively, large reflector antennas or directly radiating phased array antennas.

The latter configuration would simultaneously support several Return Links providing to them a goal G/T up to 4 dB/db°K. Such a solution would work in FDMA (frequency division multiple access) and CDMA (code division multiple access) may be provided as an operational mode of each one of the FDMA accesses, depending upon LEO user needs and ground processing capabilities. Data rates exceeding 1 Mbit/s are foreseen.

Receive and transmit beams are synthesized on board the payload.

For the return link beam forming two solutions are being evaluated, one in which a fixed multibeam antenna pattern is generated and one in which complex electrical weighting is applied to the antenna elements to generate a scanning beam.

In the former scheme a switch matrix may be used to scan the coverage area by selecting for transmission to the down link only the beams in which a LEO user is appearing.

The latter scheme generates a number of electronically scanned beams which may follow a LEO spacecraft across the coverage area; besides the scheme leads itself to be driven by the control station to generate pattern nulls to reject potential sources of RFI.

The capacity of the system is under investigation but it is anticipated that possibly two simultaneous Forward Links with up to eight simultaneous Return Links could be supported.

A specific problem arising by the decentralised architecture of the feeder link is the provision of frequency and phase coherency to all the users for range/range-rate measurements and for Doppler compensation to support the initial acquisition of the inter orbit link.

Frequency and phase coherency may also be requested by full compatibility with the TDRSS transmission standards.

One of the considered solutions includes the generation of reference frequency at the control centre which is also responsible for link acquisition and Doppler compensation; user's earth stations accessing the system with radio frequency carriers at 30 GHz would be demodulated on board and remodulated on a radiofrequency carrier controlled by the control earth station.

Other solutions including the generation of pilot tones at each earth station accessing the system, or the distribution, via the DRS payload, of the stable reference frequency, are also being evaluated.

The technology issues at Ku/Ka bands

Two main activities have been initiated. One is addressing exclusively the Ka antenna design and the corresponding deployment/pointing/feeding problems due to its location on top of the longs masts both on the Polar Platform and on the Resources Module. The other activity is the development of a dual frequency terminal operating at Ku and Ka and the relevant channel simulators for the transmission via TDRSS, respectively DRS.

For the LEO user Ka antenna two options have emerged. In the first one the antenna on top of the mast is fed by a guided wave beam system, all the signal conditioning and RF amplification being implemented on the spacecraft proper. In the second one the transmit and receive front-ends are located in a compartment behind the antenna dish, on top of the mast, and are connected via a low-level IF cable to the signal conditioning electronics on the spacecraft body. The concept studied, is best suited to the small dish (70 cm) retained at the time but a centered solution, similar to the one adopted in the earth stations should be better matching the larger dishes (160 cm) considered at present.

Extensive investigations have been carried out to assess the acquisition time as a function of the LEO antenna size for various assumptions on attitude and orbit indetermination of the LEO, deformations, sweep rate of LEO and GEO antenna beams etc. The result differs according to the acquisition concept adapted (cross-search or wide-beam approach) but in both cases is not much depending on the LEO antenna diameter.

For what concerns the dual frequency user terminal and Link simulator the antenna concept studied is a dual frequency Ku/Ka (possibly Ku/Ka/S) concept that would be able to operate with either TDRSS or DRS. The study is still in an initial phase. For the transponder, because of the requirements on communication link acquisition time, the (relatively!) favourable Forward link acquisition budget and the hardware complications, a coherent 26 GHz transponder is not foreseen.

The breadboarding of a complete Ku band hardware simulator for the TDRSS link has been initiated and is currently envisaged to expand it to the DRS link(s). The DRS link simulator will be initiated, in hardware, as soon as the link transmission study will have been completed. A high speed modem (up to 300 Mb/s) and a high speed encoder and decoder are the Key new development of this work.

The key difference with regard to the TDRSS link is the fact that the DRS link (LEO to GEO) will have to be channelized. The issues of this channelization are affecting the hardware configuration on the LEO terminal. The options for the maximum foreseen capacity (500 Mb/s) are either four channels i.e. four TWTA plus redundancy or a higher power TWTA operating in back-off and using a lineariser.

This second choice is only slightly penalised in RF power but quite advantageous in the hardware arrangement should the solution with all RF front-ends on top of the mast be retained.

The breadboarding of a complete K-band hardware simulator has been initiated including both the TDRSS and the DRS systems, with high speed modems (300 Mbit/s as foreseen by the TDRSS) and high speed Forward Error Correction encoder and decoder.

JAPANESE DATA RELAY SATELLITE SYSTEM

M.IKEUCHI,T.TANAKA,M.KAJII,H.AWAZAWA,T.DOURA,Y.TSUJINO

National Space Development Agency of Japan

Abstract

This paper describes one possible Japanese Data Relay Satellite System concept in the Space Station Era. Data Relay and Tracking Satellite System(DRTSS) will provide S-band and Ka-band communication and tracking service for orbiting spacecrafts with data rate up to 300Mbps.

Mission objectives and analysis are discussed, and the related experimental ETS-VI program is also presented.

1. Introduction

When the Space Station Era comes,various kinds of spacecrafts, including manned-spacecrafts, will fly in low earth orbit. In order to operate these spacecrafts, it is insufficient to control directly by only earth stations. So Japan should have its own data relay satellite system to promote Japanese space activities.

On the other hand, international cooperation is indisdensable.

Under these conditions, we(NASDA) are conducting Pre-phase A study on Japanese DRTSS. And this paper describes the intermediate results.

2. Mission Requirements

Mission Objectives of DRTSS are as follows:

(i) Transmission of earth observation data which can't be stored by onboard tape recorders

(ii) Real-time and dual-direction transmission for telepresence such as TV data and space experiment data

(iii) Simultaneous and real-time operation of Multi-spacecrafts for rendezvous-docking

(iv) Tracking & control of spacecrafts including launch phase

Within the future programs foreseen for the next 15 years, potential candidates as DRTSS users are

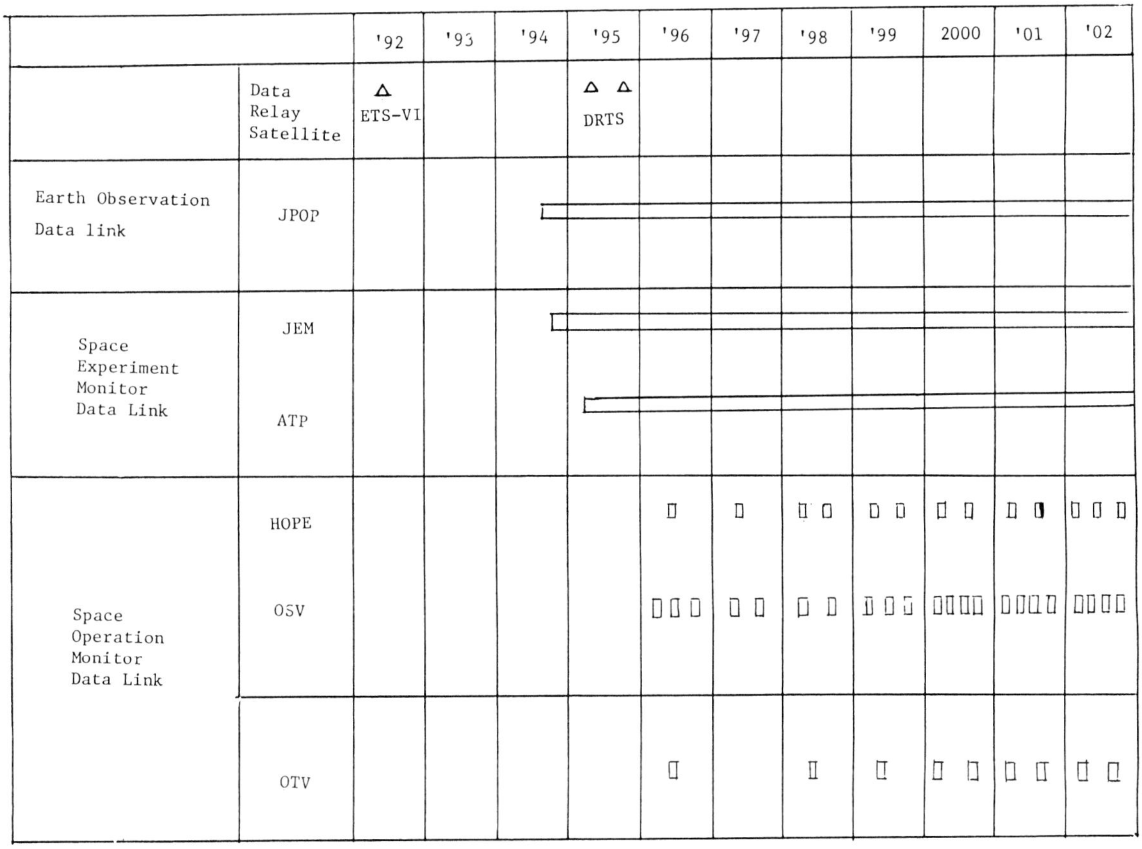

		'92	'93	'94	'95	'96	'97	'98	'99	2000	'01	'02
	Data Relay Satellite	△ ETS-VI			△ △ DRTS							
Earth Observation Data link	JPOP			┌────	────	────	────	────	────	────	────	────
Space Experiment Monitor Data Link	JEM			┌────	────	────	────	────	────	────	────	────
	ATP				┌────	────	────	────	────	────	────	────
Space Operation Monitor Data Link	HOPE					□	□	□ □	□ □	□ □	□ ▮	□ □ □
	OSV					□ □ □	□ □	□ □	□ □ □	□ □ □ □	□ □ □ □	□ □ □ □
	OTV					□		□	□	□ □	□ □	□ □

Fig. 1 Operation Plan

Table 1 Mission Requirements

Spacecraft	Phase		Contents of data	Forward / Return	Data Rate / O.D.	Support Period	Link	Remarks
JEM	Early Phase	T & C	Telemetry	R	64 Kbps		S-band	
			Command	F	30 Kbps		〃	
			Experiment & Video data	R	23〜33 Mbps	100 % in-view coverage	K-band	TV data included TLM data included
			Experiment & Control data	F	1 〜8 Mbps	〃	〃	
	Middle Phase		Experiment & Video data	R	100 Mbps	〃	〃	TV data included TLM data included
			Experiment & Control data	F	30 Mbps	〃	〃	
All Spacecrafts (except HOPE, JEM)	Launch Phase	T T & C	Telemetry	R	Max 2 Kbps	〃	S-band	
			Command	F	Max 1 Kbps	〃	S-band	
			Range & Range Rate	R & F	1 km	Several hours / day	S-band	GPS used. 1km／5 min. at Rendezvous
JPOP	Operational Phase	T T & C	Telemetry	R	1024 bps		S-band K-band	
			Command	F	512 bps	Several hours / day	〃	
			Range & Range Rate	R & F	30 m	〃	〃	GPS used. 1km／5 min. at Rendezvous
			High data rate image data	R	300 Mbps	〃	K-band	SAR Visible & IR sensor. TLM data included
			Medium data rate image data	R	10 Mbps	100 % in-view coverage	S-band K-band	TLM data included
ATP	Operational Phase	T T & C	Telemetry	R	4 Kbps		〃	
			Command	F	8 Kbps	Several hours / day	〃	
			Range & Range Rate	F & R	1 km	〃	〃	GPS used. 1km／5 min. at Rendezvous
			Experiment & Video data	R	50〜70 Mbps	100 % in-view coverage	K-band	Mainly TV data TLM data included
HOPE	All Phase	T T & C	Telemetry	R	100 Kbps		S-band K-band	
			Command	F	8 Kbps	Several hours / day	〃	
			Range & Range Rate	F & R	1 km	〃	〃	GPS used. 1km／5 min. at Rendezvous
	Operational Phase		Non operation	R	〜10 Mbps	100 % in-view coverage	S-band K-band	TLM data included
			Operation	R	70〜100 Mbps	〃 at experiment	K-band	TLM data included
OSV	Operational Phase	T T & C	Telemetry	F	10 Kbps	Several hours / day	S-band K-band	
			Command	R	1 Kbps	〃	〃	
			Range & Range Rate	F & R	1 km	〃	〃	GPS used. 1km／5 min. at Rendezvous
			Mission data	R	50〜70 Mbps		K-band	Mainly TV data
OTV	Operational Phase	T T & C	Telemetry	R	1 Kbps	Several hours / day	S-band	
			Command	F	1 Kbps	〃	〃	
			Range & Range Rate	F & R	1 km	〃	〃	
H-Ⅱ Rocket	Lanch Phase	T & C	Telemetry	R	130 Kbps	100 % in-view coverage	〃	Impossible for DRTSS
			Command	F	several Kbps	〃	〃	

Space Station(Japanese Experiment Module(JEM)),
Japanese Polar Platform(JPOP),Applications Technolo-
gy Platform(ATP),H-II Orbiting Plane(HOPE), Orbital
Service Vehicle(OSV), and Orbital Transfer Vehicle
(OTV).

These requirements are shown in Table 1 which include data rates with expected duty cycle. The
operation plan of each program is shown in Fig.1.

3. DRTSS Concept

In the mid 1990's, DRTSS will become operational.
The DRTSS configuration consists of two DRTSs in
geosynchronous orbit and associated ground system
mainly in Japan. The system configuration is illustrated in Fig.2.

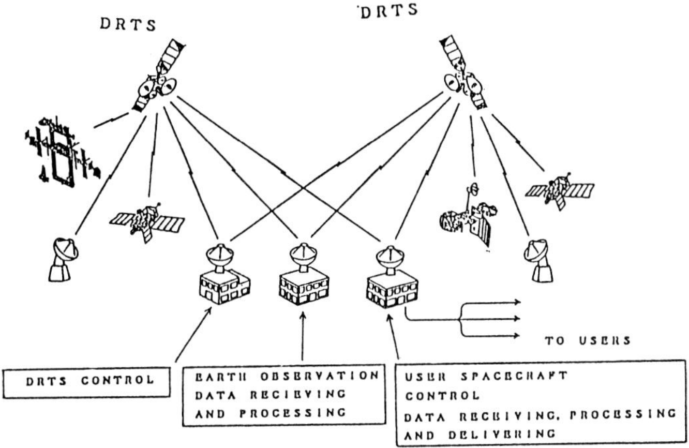

Fig. 2 Configuration of DRTSS

The two DRTSs will be placed approximately 100°
apart at 90° East and 170° West. Coverage for low
altitude spacecrafts is shown in Fig.3.

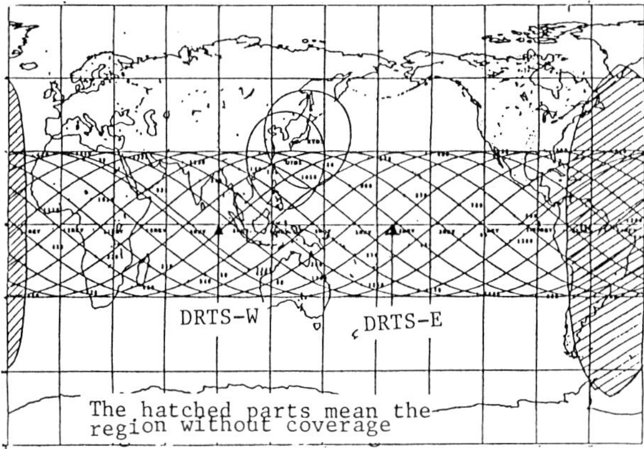

Fig. 3 Coverage for low altitude spacecrafts

4. Telecommunication System

Mission requirements defines the telecommunication
performance of DRTSS between the radio frequency
(RF) interface with user spacecrafts and the interface with earth stations.

<u>Preferred Frequency Band</u>

S-band and Ka-band are used as an intersatellite
link, and Ka-band is used as a feeder link.

<u>DRTS---User Spacecraft</u>

(1) Ka-band

NASDA chooses 23/32 GHz band for high data
rate intersatellite links because of following
reasons :

(i) 23/32 GHz are allocated as primary service
for intersatellite service

(ii) There are wide usable bandwidths for
inter satellite service

(iv) No harmful interference is considered to
occur

23GHz is assigned as DRTS to User Spacecraft(S/C)
(forward link) and 32GHz is assigned as User S/C
to DRTS(return link).

(2) S-band

S-band is mainly used for TT&C. And TT&C is most
important for spacecraft health check and life
support. So S-band frequency of DRTSS is designed
compatible with Tracking and Data Relay Satellite
System(TDRSS) of NASA and Data Relay Satellite(DRS)
of ESA.

<u>DRTS---Earth Station</u>

NASDA chooses 20/30 GHz for feeder link because
of following reasons:

(i) This band is the same frequency as that of
the related experimental ETS-VI program,
which will be described in Section 6

(ii) These bands(19.7-21.2GHz,29.5-30GHz) are
allocated as primary service only for fixed
satellite service

(iii) No harmful ineterference is considered to
occur

(iv) The effect of rain attenuation can be
avoided by operation
(Rain attenuation at Ka-band is about 10dB
worse than that at Ku-band)

The frequency plans for the forward,return, and
feeder link are shown in Fig.4.

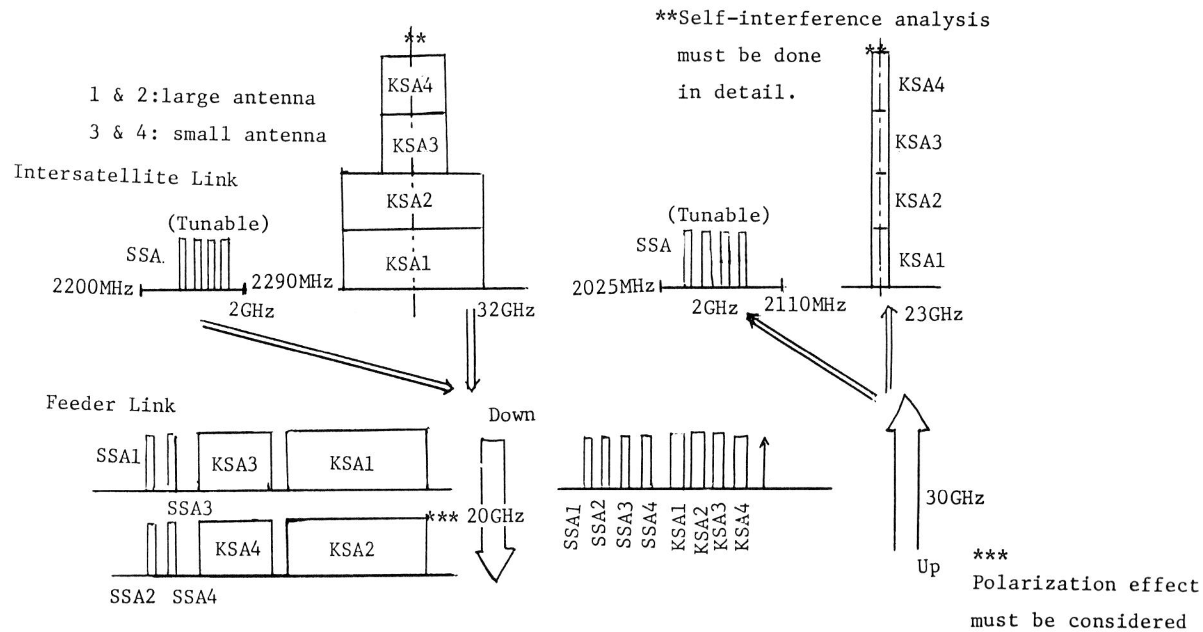

Fig.4 DRTSS Frequency Plan

Table 2 DRTSS Link Service

(1) F o r w a r d   L i n k

	S S A		K S A	
	Small Antenna	Large Antenna	Small Antenna	Large Antenna
Frequency	2025~2110MHz (tunable)	2025~2110MHz (tunable)	23~23.55GHz	23~23.55GHz
Field of View	±10°	±50°	±10°	±50°
Bandwidth	T.B.D	20MHz	30MHz	30MHz
Number of Channel	T.B.D	2	2	2
Kind of Antenna	2 mφ	5 mφ	2 mφ	5 mφ
Antenna Gain	T.B.D	36dB	50dB	58dB
Transmitter Power	T.B.D	13.5dBW	4.7dBW	4.7dBW
Data Rate	T.B.D	~5 Mbps	~5 Mbps	~30Mbps

(2) R e t u r n   L i n k

	S S A		K S A	
	Small Antenna	Large Antenna	Small Antenna	Large Antenna
Frequency	2200~2290MHz (tunable)	2200~2290MHz (tunable)	32.3~33GHz	32.3~33GHz
Field of View	Same as forward link			
Bandwidth	T.B.D	10MHz	100MHz	300MHz
Number of Channel	T.B.D	2	2	2
Kind of Antenna	Same as forward link			
Antenna Gain	T.B.D	35.8dB	52dB	61dB
System Noise Temperature	T.B.D	450K	900K	900K
Data Rate	T.B.D	~10Mbps	~100Mbps	~300Mbps

Link Analysis

From mission requirements and some design constraints, the transmitting powers and antenna size of DRTS and User S/C are evaluated.

S-band

As there are not so many users as NASA, S-band Multiple Access(SMA) is considered unnecessary in Japan. Only S-band Single Access(SSA) service is adopted now.

SSA link must be considered within Power Flux Density(PFD) restraint because the altitude of user spacecrafts is rather low. In case of low data rate, energy dispersal techniques are necessary.

Communication characteristics of S-band is shown in Table 2.

Ka-band

For Ka-band communication, spacecraft stability and pointing accuracy, antenna size limitation, transmitting power capability, and acquisition method would be critical.

Communication chracteristics of Ka-band is shown in Table 2.

Tracking System

In the Space Station Era, requirements of orbit determination is very severe.

In DRTSS, orbit determination of both DRTS and User S/C must be considered.

Orbit Determination of DRTS

When determining the position of the low orbit spacecrafts, the movement of DRTS must be taken into account. So it is necessary to get precise information on location of DRTS.

For precise orbit determination of DRTS, some types of ranging operation are considered, one is Bitaleration Ranging.

Precision of measurement at Ka-band is better than that at S-band.

Orbit Determination of User S/C

DRTSS is capable of providing one-way doppler measurement and two-way range and dopplermeasurement. Doppler measurement at Ka-band is more precise than that at S-band. And the coherent frequency can be used.

5. DRTSS

DRTS

DRTS(Fig.5) is a large spacecraft, being launched by H-II Vehicle, and weighing about 2 tons in geosynchronous orbit. The bus module is basically the same as ETS-VI, three-axis stabilized configuration with sun-oriented solar panel and normally pointed at the nadir. The telecommunication equipments are changed to implement DRTSS mission.

DRTS has four antennas for intersatellite communications;two are small,two are large.

DRTS also has two steerable antennas for a feeder link, which can be directed to several places.

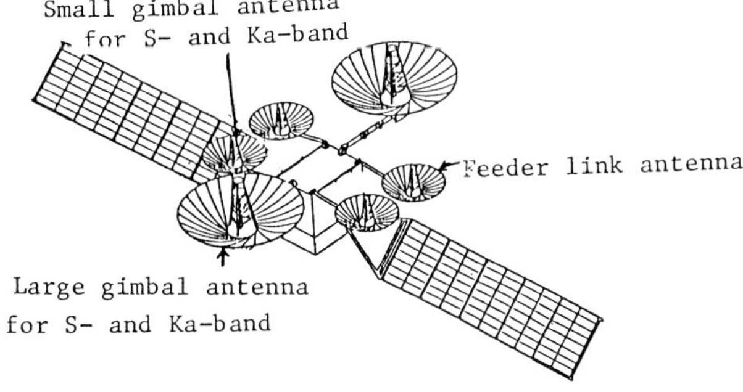

Fig.5 Configuration of DRTS

User S/C Communication System

User S/Cs are divided into three types by transmission data rate and acquisition method.

The first is only S-band users whose system is rather simple. Only OTV belongs to this type.

The second is only Ka-band users whose antenna is not large. This type can use only small antennas of DRTS and medium data rate can be transmitted.

The third is S-band and Ka-band users. Most User S/Cs belong to this type. Usually S-band is used for launch phase,and tracking of operational phase.

Ground System

DRTSS has several earth stations and each station can access User S/C via DRTS.

DRTS Operation Control Center controls DRTS and allocation of intersatellite antennas and feeder antennas.

Conceptual operation of DRTSS is shown in Fig.6.

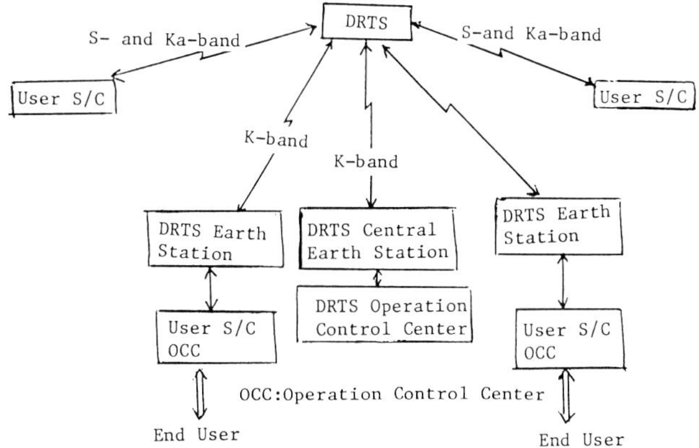

Fig. 6 Configuration of DRTSS Operation

6. Preliminary Experiment

Intersatellite communication experiments are being conducted with ETS-VI, which is being launched to geosynchronous orbit by H-II Vehicle in 1992.

Mission Objectives

Mission objectives of ETS-VI intersatellite communication experiments are as follows:

  (i) Establishment of technologies for DRTSS

    a) Key technologies

      23/32GHz communication

      Antenna pointing mechanism

    b) System technologies

      Intersatellite communication

      Acquisition of User S/C

      Tracking & control of several User S/C

  (ii) JEM---ETS-VI data transmission

Concept of ETS-VI Intersatellite Communication

The system configuration of ETS-VI intersatellite communication experiments is shown in Fig.7.

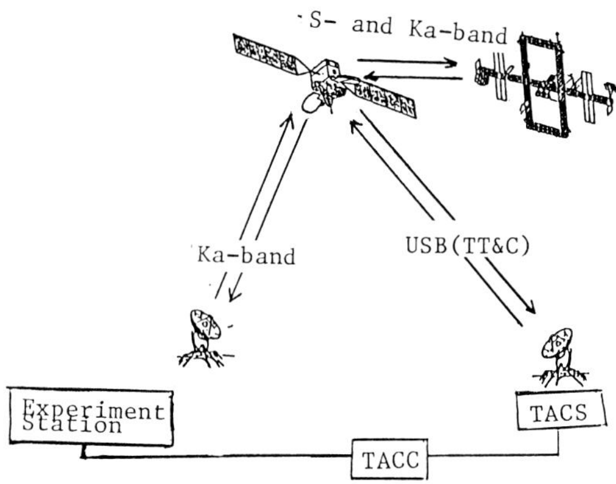

Fig. 7 Configuration of ETS-VI intersatellite

communication experiments

TACS:Tracking &Control Station
TACC:Tracking & Control Center

ETS-VI will be placed approximately 140° East.
Coverage for low altitude spacecrafts is shown in
Fig.8.

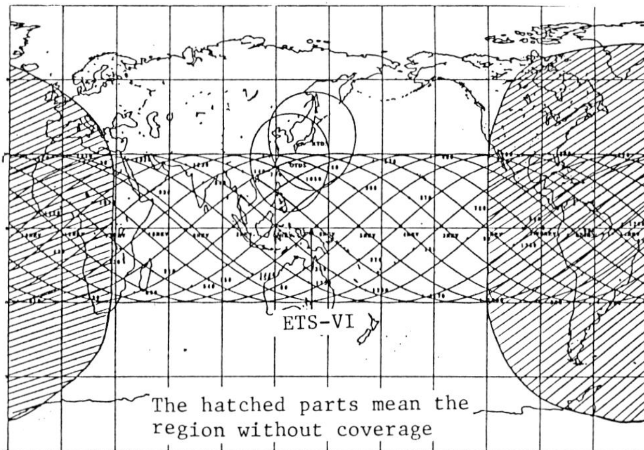

Fig.8 Coverage for low altitude spacecrafts

ETS-VI

Configuration of ETS-VI is shown in Fig.9.
Basic design of ETS-VI has been started since this
Apirl.

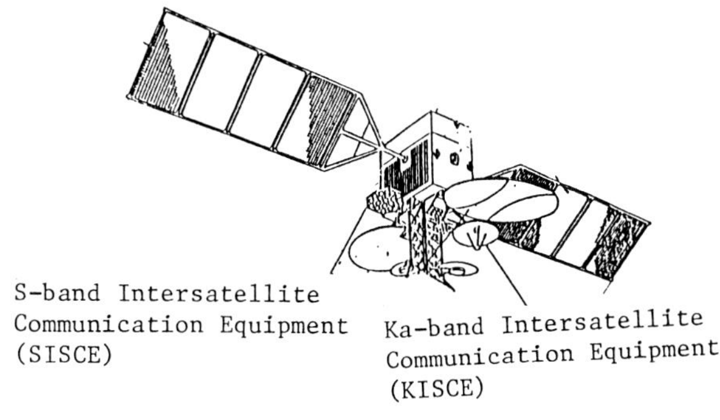

Fig. 9 Configuration of ETS-VI

The characteristics of ETS-VI is shown in Table 3.

Table 3   Characteristics of ETS-VI

I t e m	C h a r a c t e r i s t i c s	
Launch Vehicle	H－Ⅱ Vehicle	
Launch Date	Summer 1992 (Approved)	
Stationary Position	140˚ East (Tentative)	
Life	10 years	
Total Weight	3.750 kg in GTO	
Payload Weight	660 kg	
Payload Power	1.6 kW	
Orbit Control	North-South ± 0.1˚ West-East ±0.1˚	
Attitude Control	Roll = ±0.05˚	
	Pitch = ±0.05˚	
	Yaw = ±0.15˚	

Intersatellite Communication Equipments on ETS-VI

Experiments of both S-band InterSatellite Commu-
nication Equipment(SISCE) and Ka-band InterSatellite
Communication Equipment(KISCE) are being conducted.
These equipments are under Engineering Model (EM).
Each characteristics of communication equipments
is shown in Table 4.

Table 4   ETS-VI Link Capability
(1) F o r w a r d   L i n k

	SISCE	KISCE
Frequency	2106.4MHz (Tentative)	1 band between 23 and 23.55GHz
Field of View	±10˚	±10˚
Band width	6 MHz (and 10MHz)	30MHz
Number of Channel	1	1
Kind of Antenna	19 Poles Phased Array	0.8 m∅
Antenna Gain	26.8dB	36.9dB
Transmitter Power	～12.8dBW	4.8dBW
Data Rate	～ 2 Mbps	～ 3 Mbps

(2) R e t u r n   L i n k

	SISCE	KISCE
Frequency	2287.5MHz (Tentative)	1 band between 32.3 and 33GHz
Field of View	Same as forward link	
Band width	6MHz (and 10MHz)	300MHz
Number of Channel	2	1
Antenna Gain	27.5dB	41.9dB
System Noise Temperature	293K	1800K
Data Rate	～ 2 Mbps	～23Mbps

Development schedule on DRTSS and ETS-VI is shown in Table 5.

Development of ETS-VI is approved, and development of DRTSS is expected to be approved this year.

We have mentioned the intermediate results of Pre-phase A study on DRTSS.

In march '88, the Pre-phase A study on DRTSS is expected to result in the definition of DRTSS three segments; DRTS, User S/C, and ground system.

Table 5　Development Schedule on DRTS and ETS-VI

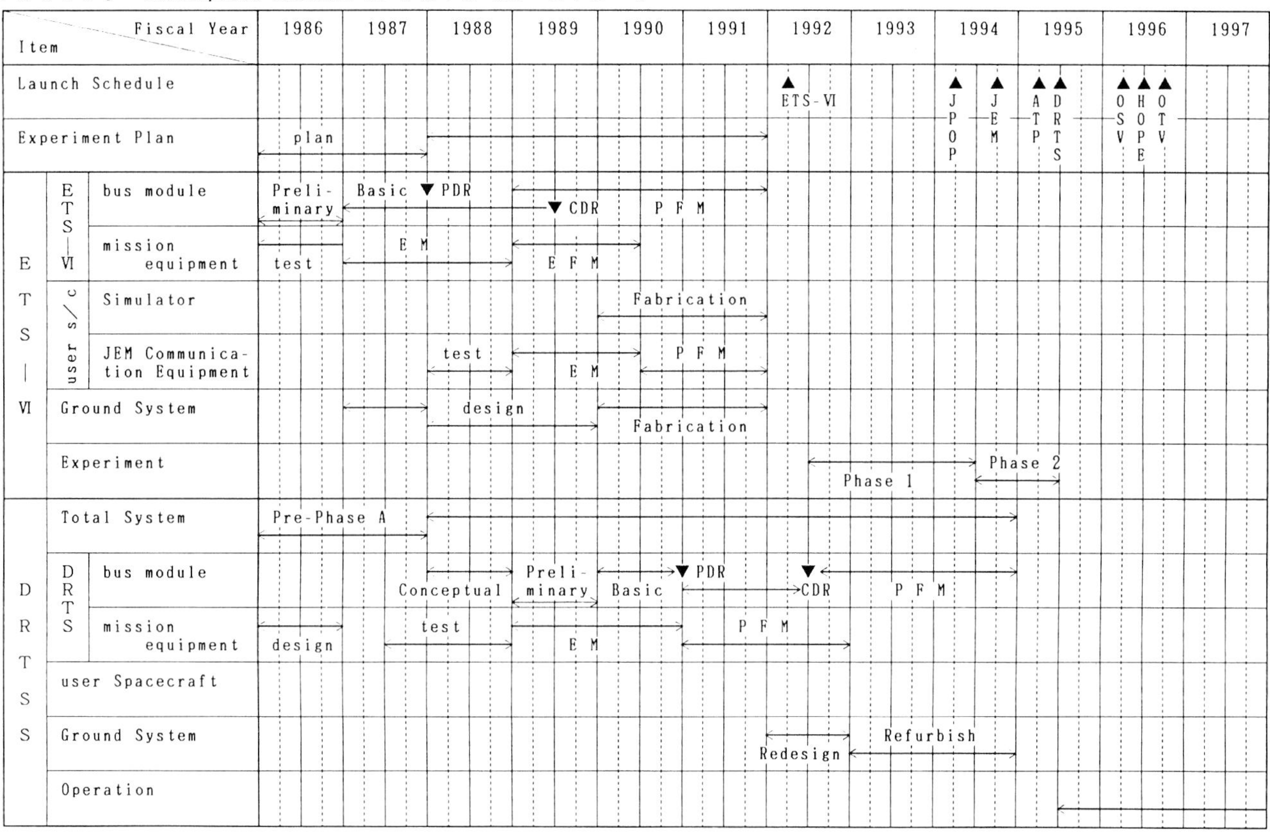

Item			1986	1987	1988	1989	1990	1991	1992	1993	1994	1995	1996	1997
Launch Schedule									▲ ETS-VI		▲ JPOP　▲ JEM	▲ ATPS　▲ DRTS	▲ OSV　▲ HOPE　▲ OTV	
Experiment Plan			plan											
ETS-VI	ETS	bus module	Preliminary	Basic　▼ PDR		▼ CDR	P F M							
		mission equipment	test	E M		E F M								
	user s/c	Simulator					Fabrication							
		JEM Communication Equipment			test	E M	P F M							
		Ground System			design		Fabrication							
		Experiment							Phase 1			Phase 2		
DRTSS		Total System	Pre-Phase A											
	DRTS	bus module			Conceptual	Preliminary	Basic	▼ PDR	▼ CDR	P F M				
		mission equipment	design		test	E M		P F M						
		user Spacecraft												
		Ground System							Redesign	Refurbish				
		Operation												

TRACKING AND DATA RELAY SATELLITE SYSTEM (TDRSS) AND THE FUTURE

Robert E. Spearing and Robert D. Godfrey
Mission Operations and Data Systems Directorate
NASA-Goddard Space Flight Center
Greenbelt, Maryland

Abstract

Prior to the 1980's, NASA employed numerous ground-based tracking and data communications facilities to provide Telemetry, Tracking, and Command (TT&C) support to spaceflight missions. This support was limited to brief contact periods during each satellite's orbit of the Earth, restricting the types and degree of telescience and teleoperation that could be performed. Introduction of the Tracking and Data Relay Satellite System (TDRSS) as the primary space mission support system represented a radical change in the concept of space mission operations; to a large extent eliminating the severe orbital coverage constraints inherent in the ground-based station configuration. This new system architecture was designated the NASA Space Network. The implemented TDRSS concept was intended to provide service for a 10-year period, beginning in 1980.

TDRS-1 was launched in April 1983. Experience derived from 4 years of TDRSS operations, new space mission operational concepts including those associated with Space Station, evolving Shuttle support requirements, and the prospect of international data relay systems have modified and expanded the initial TDRSS concepts; resulting in a need to revise and upgrade Space Network capabilities. Starting with the existing TDRSS as a reference, this paper discusses planned changes in the Space Network which will be applicable in the early Space Station operational era, as well as potential capabilities for the follow-on Space Network.

Introduction

The primary function of the NASA Space Tracking and Data Acquisition Network is to provide a data and information connection between a spaceflight mission and the ground-based controllers and experimenters associated with that mission in a manner that is efficient, reliable, and as transparent as possible. The Network consists of equipment for transmitting information to and receiving information from the space mission. Planning and scheduling systems are employed to assure that this data and information transfer achieves the objectives of the space mission.

The function of a space mission support Network is illustrated generically in Figure 1.

The Network must support a variety of space missions and, hence, must meet a variety of support requirements. The resulting Network implementation should be totally transparent to the mission (transparent meaning that the Network should not delay or alter the information transfer or impose any constraints on the transfer).

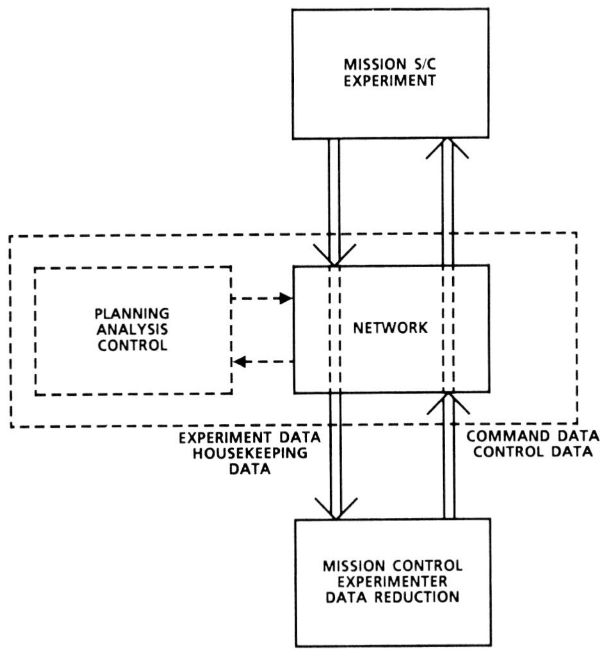

Fig. 1 Space Mission Support Network - Function.

Such an ideal Network is never achieved since financial, physical, and technological constraints must be considered. Any realizable Network will be a compromise between mission requirements, which the Network must attempt to satisfy, and "real world" constraints. The implemented configuration will be the result of an iterative process, whereby mission requirements (the primary driver) and Network constraints are considered together, resulting in a modified set of mission requirements which can be satisfied by a realizable Network. The process must be continuous as mission objectives, technology, and support concepts change. Thus, the resulting Network evolves with time, generally in an incremental manner due to the imposed constraints. Of course, user commitments must be met during evolutionary transitions, necessitating the maintenance of downward compatible services as enhancements are made.

An illustration of the evolving support requirements placed on the NASA Network since the beginning of the Space Age is shown in Figure 2. Figure 2 illustrates the magnitude of the support requirements, in terms of the total amount of data transmitted on a per-day basis, for the composite mission model. Representative single mission requirements are illustrated in Table 1.

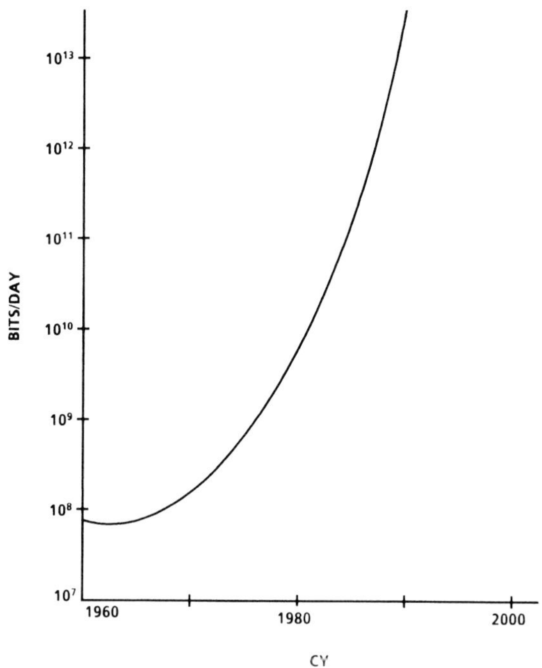

Fig. 2 Support Requirements Growth In Bits/Days.

The TDRSS Network

In the early-1970's, it became apparent that the existing Network (as shown in Figure 3) and evolving support concepts could not satisfy the changing mission/experiment requirements in the 1980 time frame. The primary mission/experimenter requirements were, again, increased data bandwidth, real-time experiment and control data reception at the control centers, real-time mission/experiment control, and experiment/experimenter interaction. None of these support objectives could be accommodated by a Network consisting of remotely-located ground facilities for both cost and physical reasons. The early 1970's concept of the 1980's mission support requirements is shown in Table 1.

Table 1 TDRSS Planning Mission Model, July 1975 Update.

MISSION NAME OR CLASS		CMD	RT TLM	MAX OR DUMP TLM
SHUTTLE (NOTE 1)				
SPACELAB		1925	256 S (KU)	50,000 (KU)
HEAO C2		1	64 S	128S
FOLLOW-ON		1	2 S 12 S	12 S 212 S
NIMBUS G		<1	4 800	1600 800
AIRSAT		1	12 1200	30 M –
SEASATS A		1		
SEASATS B		1	48 S to 1200 S	
GRAVSAT A, B			1	
MINI LAGEOS			LASER	
GRE			4 S	
LST		1	1.6 S 1000	51 S 1000
RADBUD A		1	1	–
RADBUD B		1	1	–
RADBUD C		1	1	–
SMM A			4.2 S	128 S
SMM B		1	6.4 S	128 S
FOLLOW-ON (CLASS)			6.4 S	128 S
TIROS O			20 S 2050 S	
AEM-E, -F, -G, -H, ETC.		≤1	1.2 S TO 40 S	245 S TO 256 S
COSM BKG A			41	4
COSM BKG B			41	4
BESS, A, B, C, D				1.6 9.6 M
EARTH RESOURCES EOS, ERS, (LANDSAT) (DOI, NASA)	1	1	12 S 1200 S	300,000 (KU)
EARTH RESOURCES EOS, ERS, (LANDSAT) (DOI, NASA)	2	1	12 S 1200 S	300,000 (KU)
EARTH RESOURCES EOS, ERS, (LANDSAT) (DOI, NASA)	LOW INCL	1	12 S 1200 S	300,000 (KU)
EXT XRAY			57 S	1 MS
GRAV PRB				2.3 S
MAGSAT			5 S	
UPPER AE			9.8 S	130 S
INTERNATIONAL EXOSAT			10 S 100 S	
INTERNATIONAL 1			2 S	40 S
INTERNATIONAL 2			2 S	40 S
ADVANCED LEO SCIENCE 1			5 S TO 40 S	128 S
ADVANCED LEO SCIENCE 2			5 S TO 40 S	128 S
ENVIRONMENTAL MONITOR 1			20 S 560 S	
ENVIRONMENTAL MONITOR 2			20 S 560 S	
NOTE				S = S BAND KU = KU BAND
1. SHUTTLE AND SPACELAB SUPPORTED AS ONE MISSION.				

To satisfy data transfer and mission operational requirements, the concept of an orbiting communications relay system was reexamined. The initial concept was proposed in the mid-1960's to provide continuous support to the Apollo spacecraft in earth orbit and was later investigated as a potential approach for support of low-earth orbiting missions in the late 1960's. The initial concepts were not implemented due to cost, technology, and a lack of adequate requirements at that time. Network planning activities in the early 1970's, however, showed that the technology would be available for mission support in the 1980's time frame, that there was no other practical method of supporting the mission model requirements, and that such a system could be cost effective. These planning activities resulted in the development of the TDRSS as it is currently configured. This system is illustrated in Figure 4.

The operational and support philosophy of today's TDRSS configuration will provide two spacecraft located 130 degrees apart, with both visible to a single ground terminal located at White Sands, New Mexico. This system is designed to provide a minimum of 85-percent visibility to low-earth orbiting missions. Full visibility for all missions was not provided due to a lack of definitive requirements and cost considerations. A total of six spacecraft are available to provide for 10 years of service. The use of a ground station in the continental United States provides the capability to relay all command and telemetry data in real time without storage within the Network, thereby providing the mission projects with rapid data return and interactive control capability.

Referring back to Table 1, it can be seen that the model consists of a large percentage of missions requiring data rates (telemetry) of 50 kilobits per second (kbps) or less; a much smaller percentage requires higher rates. These requirements translated into a TDRSS configuration that provides dedicated telemetry coverage to the low data rate users, via the TDRSS Multiple Access (MA) system, with a time-shared command link (20 telemetry and two command links for the TDRSS) and time-shared high-gain Single Access (SA) systems for support of the smaller number of high data rate missions. These SA-supported missions operate at S-band (medium data rates) and Ku-band (high data rates) with data rates ranging from a few hundred kilobits to a few hundred megabits. Each SA system provides both telemetry and command simultaneously.

Fig. 3 Ground Station Locations - 1970.

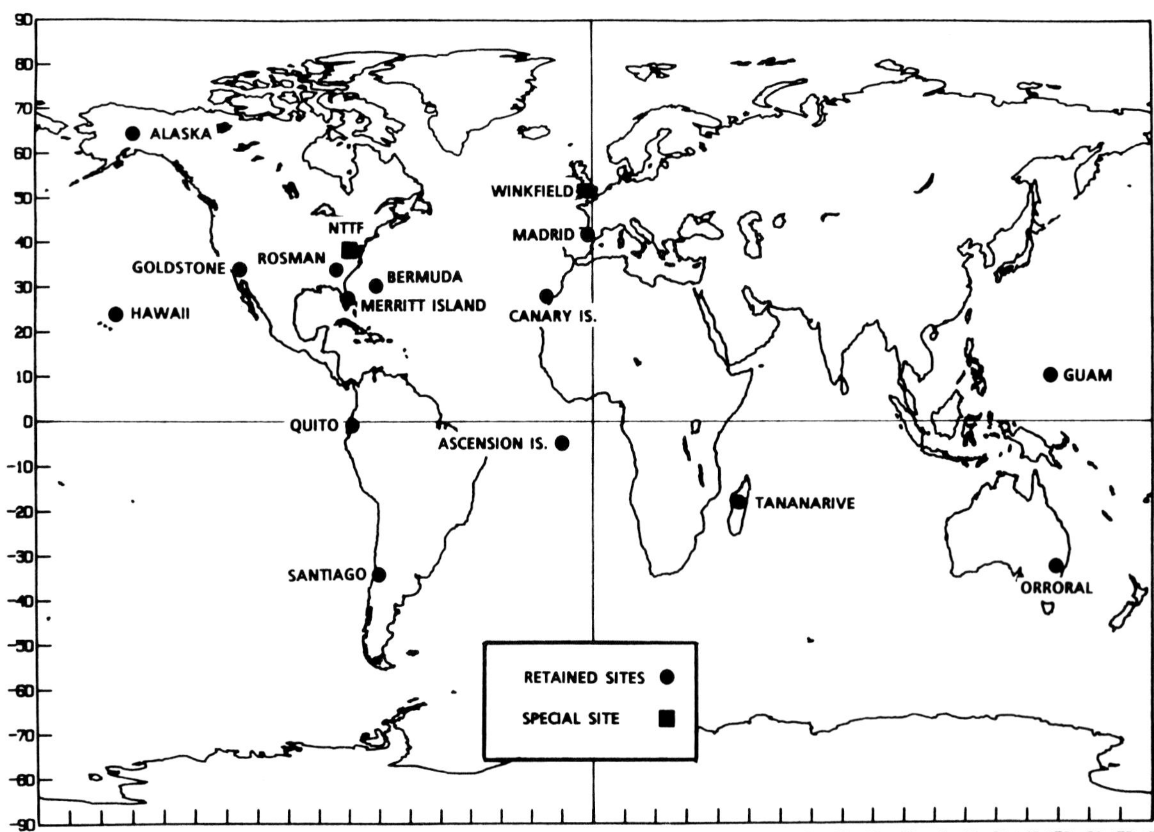

The current configuration provides a total of four SA services for the operational system. Tracking services by range and range rate determinations are also provided at S-band, as with the earlier Ground Network.

As originally planned, the TDRSS was to be operational in 1980 and maintained in the initial configuration for the 10-year period ending in 1990. The first TDRS became operational in 1983 with the launch of the F1 spacecraft. The F2 spacecraft was lost with the Challenger Shuttle in 1986. It is currently anticipated that a full TDRS System will be operational in 1989. Experiences to date with the F1 spacecraft have shown that the system will perform as planned, although some problems have been experienced which impact the operational efficiency and adaptability of the configuration. Also, due to launch delays (TDRS, as well as other planned missions), the currently-completed TDRS spacecraft will be supporting a significantly different mission model than that expected.

The Early Space Station Era

The most significant change in mission requirements that must be supported by the current TDRSS configuration is the planned implementation of the Space Station in the mid-1990's. The Space Station, along with its associated platforms, supply vehicles, and tugs, shifts the Network support mission model from one dominated by the lower data rate, MA-type user missions to one which requires multiple, high-gain communications links. Video data transmission is expected to become an increasingly important driver for duplex communications. The possible operation of all the various Space Station elements at a common communications frequency (current discussion includes Ku- and Ka-band operation) is another important consideration. Other potentially impacting factors include the remote operation of the tug vehicles (OMV, etc.) from the ground and possibly from the Space Station itself. This control mode would require a real-time, inter-

Fig. 4 Current Space Network.

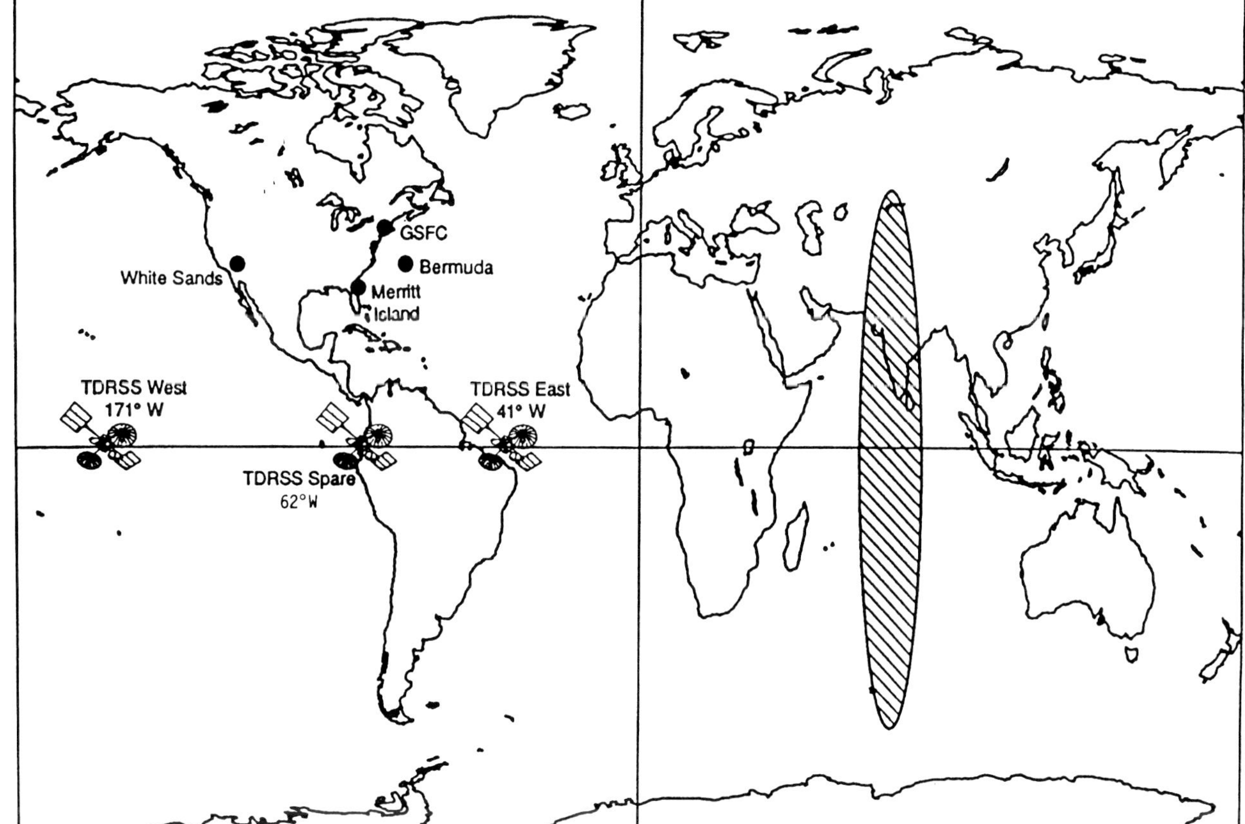

active, dedicated command and control relay. These requirements were not previously a significant factor in determining the Network support configuration.

It is expected that the Space Station and station-associated requirements will build up slowly over a period of time. For the early Space Station era, the primary objective of the Network configuration changes is to provide additional SA services. The early Space Station program has been constrained to be compatible with existing TDRSS capabilities. Since some of the existing TDRSS spacecraft are expected to be operational in this time frame, the needed SA services will be provided by placing additional spacecraft in orbit to form a four-operational satellite constellation. A Second TDRSS Ground Terminal (STGT) will be available to expand existing operational capabilities and to control and support the additional spacecraft. Also, the STGT will provide a redundant capability in the event of a major ground terminal failure. To assure availability of services in this time frame, up to two more TDRS spacecraft (F7 and F8) are being procured.

During the early Space Station era, the Network will retain the concepts embodied in the TDRSS with relatively minor changes in systems and operational techniques. Planning and evaluation activities related to the implementation of significant new support and operations concepts, and hardware configurations for the mature Space Station, are currently in process and will be discussed in the following sections. It is important to note that, although the primary discussion has centered on the Space Station and new associated requirement's, the Network will continue to support a number of other missions, such as the Hubble Space Telescope (which will be launched in the pre-Space Station era). The Network must support the requirements of the composite scenario as efficiently as possible within the specified constraints. As a result, the Network must make a smooth transition from one configuration to the next.

The Space Station Era Space Network

The requirements imposed on the Network by a fully-configured Space Station will affect a significant change in Network support systems and operational concepts. Since many of the support requirements are still in the concept and planning stage, a specific Network configuration cannot be defined. However, studies are in process to develop the "follow-on" Space Network concept.

The remainder of this paper will concentrate on the possible configuration options which will be applicable during the first 10 years of Space Station operations. This era is represented by the F9 through F12 TDRS relay vehicles.

Another factor which will have an impact on the Space Network configuration involves the planned international nature of the Space Station and associated elements. These plans have generated a desire and stated intent on the part of the European Space Agency (ESA) and the National Space Development Agency of Japan (NASDA) to implement data relay systems to support their own spaceflight objectives. If the ESA and NASDA systems are implemented, there exists the possibility of providing joint telecommunications support at common frequencies. At the same time, there exists the possibility of generating severe interference. This problem must be considered and coordinated in any advanced Space Network design, both from a systems design point of view and an operations point of view. The possibility of these additional relay systems may also impact the types of signal structures employed.

Some areas of potential improvements to the Space Network, now in the investigation stage, are listed as follows, along with a discussion of the reason for the change from the existing TDRSS configuration.

o Increase the number of SA systems - possible implementation of a high-gain multiple access system. As a minimum, the implementation of additional SA systems for support of the Space Station and associated elements.

o Maintain S- and Ku-band operation - possible additional system operating at a higher frequency for international compatibility and bandwidth.

o Provision of communications service between two (or more) orbiting user space elements; e.g., OMV and Space Station.

In general, these improvements are oriented toward providing increased SA capability, higher data rates for the forward link (command and interactive information), the return link (telemetry or interactive information to the ground), dedicated support systems, total system interference reduction, and provision for full orbital support. Along with technical improvement, is the development of concepts to improve the operational response of the Network and its transparency to the mission and mission/experiment controllers.

These changes will impact both the Space Network space components, as well as its ground components. Included in the ground components are the ground-based data distribution system (NASCOM) and the network control systems (NCC), as well as the mission and experimenter control centers (POCC). Special consideration is being given to improving the user interface to the Space Network. Network. Changes are planned in the areas of mission planning, mission support, fault isolation, and scheduling. These functions must be made as efficient and automated as possible, within the limits of current or projected technology.

The continued existence of many non-Space Station associated missions must be considered. These missions will require the continuation of existing support systems, at least for the communications interface presented by the new Space Network. It is, therefore, expected that the Network will maintain the existing communications interface along with any Space Station-unique systems that may be required.

Possible Advanced Space Network Concepts

The previous paragraphs discussed technical improvements and other changes in the Space Network that may be implemented to satisfy the requirements of the Space Station era. Other changes, which cannot be defined at this time, are possible if all of the Space Station operational and experiment "requirements" and desires are considered. Some of these improvements may be implemented during the first 10 years of Space Station operations, while others may be deemed undesirable or implemented at a later date. These possible configuration changes are listed as follows, along with rationale for the change:

o Implementation of bandwidth efficient modulation systems such as 8 PSK, TFM, or other systems to permit greater data transport over existing channels.

o Development and implementation of more efficient coding systems to reduce the missions communications terminal requirements or to improve overall performance.

o Space-to-space link operation at higher frequencies, such as 60 GHz, or optical frequency bands.

o Implementation of a frequency division multiplex configuration for all SA operation (Ku-band, Ka-band, etc.) for both the forward and return links, thereby allowing a single SA antenna to simultaneously service several users within the antenna beamwidth.

o Implementation of multiple channel forward links to more efficiently service the Space Station and its associated components.

o Development and implementation of more effective interference rejection schemes to more reliably support multiple users at the same frequency in the same spatial location.

o Implementation of direct duplex spotcasting capability to facilitate the transmission of data to the user's mission data facility and to aid in interactive control.

o Support extending to and beyond geosynchronous altitudes (for OTV support).

o Addition of ultra-high performance communication payloads for experimental purposes related to improvement of future Network communications services.

Implementation of any of these capabilities will depend upon planning and requirements development activities currently in process, as well as the various technology evaluation studies currently underway.

Summary

In the foreseeable future, NASA Network for support of low-earth orbiting missions will continue to be based on the use of synchronous satellites relaying data to one or more ground facilities or directly to the user facilities, as may be appropriate. The implemented system may operate in cooperation with other relay systems (at least at a common cross-support frequency).

The current concept for the NASA Network through the end of the first decade of the 21st century provides for near-continuous, transparent communications coverage to all space missions. The implemented system will support the widest possible range of requirements given the usual constraints of funding and contemporary technology.

An examination of the evolution of the NASA Network shows a continuous increase in ability to meet the requirements for support of the various missions while, at the same time, minimizing Network-induced constraints on mission planning and operations. The early Network imposed a significant constraint on mission planning and operations; the TDRSS reduced these constraints. The Network of the Space Station era will reduce these impacts much more.

Richard B. Miller
Manager, Information Systems Program Office
Jet Propulsion Laboratory/California Institute of Technology
Pasadena, California 91109
and
David A. Nichols
Assistant Manager, Information Systems Program Office
Jet Propulsion Laboratory/California Institute of Technology
Pasadena, California 91109

Abstract

NASA is facing data system challenges in the 1990's which include more than an order of magnitude increase in data volume and rate. Characteristics of the two highest rate instruments, the HIgh Resolution Imaging Spectrometer (HIRIS) and the Synthetic Aperture Radar (SAR), are described. The system description from the instrument on-board a future space platform through to the analysis and processing by the end user is described as currently conceived. The technological challenges inherent in the system and the management issues associated with implementing an operating system are also discussed.

1.0   Introduction

NASA is currently planning earth orbiting missions for the mid-1990's which will result in an increase of at least an order of magnitude in the total volume and average data rate as compared to existing missions. This is due both to the nature of the scientific questions facing mankind and to the evolution of the capability of individual instruments and associated space-based servicing concepts. In addition, Space Station platforms will present a new environment in which we move from traditional satellite observational periods of days, months and years to decades. The challenge of understanding global change or the earth as a system requires consistent observations over extended time periods with a variety of sensors to measure complex environmental features and processes.

Two instruments which have the potential to be major contributors to understanding global change are the HIgh Resolution Imaging Spectrometer (HIRIS) and the Synthetic Aperture Radar (SAR), to be flown as research facility instruments on the NASA Earth Observing System (Eos) polar orbiting platforms. These instruments are capable, individually, of producing data rates in the hundreds of megabits/second, with current science scenarios placing the sustained, combined rate at 100 megabits/second. The next lower data rate instrument planned for Eos is the MOderate Resolution Imaging Spectrometer (MODIS),[1] expected to produce 10 to 15 megabits/second. The Goddard Space Flight Center has estimated that the aggregate average data rate for all Space Station-associated payloads in earth orbit, excluding the HIRIS and SAR instruments, will be on the order of 20 megabits/second. The HIRIS and SAR are the major concerns when addressing the data rate challenges of the 1990's. This paper will focus on HIRIS and SAR specifically, rather than other Eos or Space Station payloads.

While initial data system planning is progressing based on preliminary scientific requirements, much work remains in understanding the scenarios for the scientific utilization of these instruments--how the instruments will actually be utilized in such areas as spatial and spectral sampling, effective ground resolution, coverage, and in conjunction with other instruments. These scientific scenarios, which will determine the instrument duty cycle and data resource requirements, will be first-order factors in the cost of data systems to service these instruments.

Section 2 of this report will provide a description of the HIRIS and SAR instruments and will explain the source of such high data rates. Section 3 will present a functional-level description of the end-to-end data flow for HIRIS and SAR. The following Section will then focus on major technological challenges that must be met in achieving an implementation of the system. Finally, Section 5 will discuss some of the management issues associated with high rate, high volume data.

2.0   Description of the HIRIS and SAR Instruments

HIRIS

The HIgh Resolution Imaging Spectrometer (HIRIS)[2] instrument is a pushbroom imaging spectrometer which uses area array detectors to provide simultaneous, and inherently registered imaging in 192 spectral bands from 0.4 to 2.5μm. The instrument functional parameters are summarized in Figure 2.1.

HIRIS is intended for use on the Eos polar platform with a current design altitude of 824 km. Because of the high orbital altitude, which in turn requires a large optical system, the size and mass of the instrument has been a major factor in the design. The instantaneous field-of-view (IFOV) of 30 meters provides the same spatial resolution as the Landsat Thematic Mapper (TM). Crosstrack and downtrack pointing is provided for target acquisition, multiple-look angle imaging and atmospheric correction. The downtrack pointing capability may also be used to provide image motion compensation, which results in a signal-to-noise improvement at the expense of continuous downtrack coverage.

The spectrometer is separated into a visible and near infrared (VNIR) and short wavelength infrared (SWIR) section. The focal plane assembly houses the VNIR and SWIR detector, the spectrometer entrance slit, and a field-flattening mirror. The SWIR focal plane is a mosaic of three

128 x 256 hybrid arrays, arranged with a butting gap of four pixels between arrays. This results in a total format of 128 x 776. Each hybrid array is comprised of a HgCdTe detector and a silicon swath-array multiplexer. The VNIR focal plane is a silicon charge-coupled device (CCD) which is designed to match the format of the SWIR mosaic. The pixel size for both detectors is 52μm.

HIRIS Functional Parameters

Design Altitude	824 km
IFOV	30 m
Swath Width	23.3 km
Spectral Coverage	0.4 - 2.5μm, 192 channels
Spectral Sample Interval	
VNIR, 0.4-1μm	9.4 nm
SWIR, 1-2.5μm	11.7 nm
Pointing	
Down-track	$+60^0$ / -30^0
Cross-track	$+20^0$ / -20^0
Encoding bits/pixel	12 or 8
Maximum Internal Rate megabits/second	393
Image Motion Compensation (IMC)	Gain States of 1 (off), 2, 4 and 8

Figure 2.1

To implement the down-track pointing requirement, a mirror is located in front of the foreoptics at a nominal angle of 45^0 to the axis. This mirror can be rotated from this position 30^0 toward the foreoptics, and 15^0 away. This provides the required $+60^0$ / -30^0 object space pointing. The cross-track pointing is implemented by rotating the entire optical assembly, including the down-track mirror.

In order to manage the high data rate of the instrument, a variety of edit modes are provided. These modes fall into three basic categories: spectral selection, spatial averaging, and encoding selection. Since no on-board gain and offset correction is performed, there will be some loss of calibration if an averaging mode is used which is not specifically calibrated. For this reason, the number of averaging modes is limited.

SAR

The Eos SAR[3] will be a multi-polarization, multi-frequency synthetic aperture radar (SAR) with variable imaging geometry. Nominally, this SAR will be a three frequency (L-, C- and X-band) system with quad-polarization available for the L- and C-bands and dual polarization for the X-band system. It will be capable of acquiring multi-incidence angle data using electronic beam steering and other imaging geometries by mechanically pitching, yawing and rolling the antenna.

In the quad polarization mode, the H and V L- and C-band transmitters operate on alternate transmitter pulses. The H and V receivers both receive all return pulses half of which are horizontally transmitted pulses and half of which are vertically transmitted pulses. This procedure requires twice the pulse repetition frequency (PRF) of a single polarization mode. Thus, a multi-polarization radar requires four times the data rate as a single polarization radar for the same swath and resolution.

Each of the 10 channels is capable of producing over 100 megabits/second (6 bits/sample, 45^0 incidence angle, 100 km swath). The channel data must be fed into one or more of six parallel Digital Data Handling (DDH) subsystems. Each DDH is capable of a maximum of 45 megabits/second. Therefore, the peak rate of the instrument is constrained to 270 megabits/second. Data acquisitions are made to fit within this constraint by selection of quantization level, channel selection, resolution, swath width, and incidence angle. The SAR is expected to be used in a mapping mode approximately 20% of the time, acquiring two simultaneous channels at 100 km swath width. In a typical research mode, (5% of the time) all 10 channels would be operated at reduced data rates so as to meet the 270 megabits/second limit. The SAR science scenarios at this time indicate the need for a sustained average data rate of 55 megabits/second.

Eos SAR (IOC)

Frequency/Polarization	
C Band	HH, VV, HV, VH
L Band	HH, VV, HV, VH
X Band	HH, VV
Incidence Angle	15^0 - 60^0
Swath Width	30 - 200 km
Azimuth Resolution	30 M (4-look)
Range Resolution	47 - 15 M
Peak Data Rate	270 megabits/second

Figure 2.2

3.0 Strawman High Rate Data System Description

The end-to-end data system view presented here is not intended to show the current design but rather to show how the high rate instruments would operate in a design representative of typical current concepts. The end-to-end system view includes the flow of data from the output of the sensor through the various flight components, communications to the ground, the successive stages of ground data capture, processing and archive, and its use by individual investigators, culminating in scientific results. The view of the end-to-end system in the opposite direction is referred to as the command system. This system involves individual observation planning by each investigator coordinated into a consistent sequence of events and commands, checked against conflicts with other spacecraft subsystems and payloads and then relayed as instructions to the payload at the appropriate time. The command traffic rate and volume for these high rate instruments is basically the same as low rate instruments. Therefore the command, or "uplink," end-to-end system will be described only insofar as it relates to providing quick-look versions of the high rate data for purposes of observation planning.

On-Board Processing

At least two phases of on-board processing associated with the high rate instruments are anticipated. Figures 3.1 and 3.2 show how this evolution might occur for the HIRIS instrument. Figure 3.1 shows a baseline configuration for the instrument at IOC (Initial Operations Configuration). In this case, very little on-board processing would take place and the principal control on the data rate would be selection of spectral channels, spatial averaging and quantization level. This is represented on the drawing by the command system interface with an uplink commandable editor.

One scenario under study by the Imaging Spectrometer Science Advisory Group (ISSAG) is to limit the total number of channels which can be acquired at any given time to 1/3 of the total number of channels. This would reduce the peak data rate of 393 megabits/second to 131 megabits/second. This limitation, however, presupposes that the same 64 channels would satisfy requirements for coverage by all disciplines desiring data to be acquired concurrently. The spectra of interest for vegetation mapping has essentially no overlap with the spectra of interest for surface mineral identification. However, geologic observations do not have synoptic dependence; therefore, it may be acceptable for the vegetation and the geologic observations of a given area to be acquired at different times. Other classes of observations may be best done if the observations are simultaneous, driving up the requirement for concurrent channel availability. Until the technology can catch up, however, selecting a limited set of spectral channels, reduction of spatial resolution or extent, and reduction of spectral accuracy, appear to be the only methods available to reduce the volume of instrument data output.

Figure 3.2 depicts a later generation on-board data system which would provide several techniques to reduce the downlink rate and to optimize the information flow within a given bandwidth resource. The two approaches are data reduction, by either data selection or data compression, and information extraction.

Beyond the simple pre-selection of a set of spectral bands by up-link commands to reduce the amount of data transmitted, one may consider autonomous editing where the scene context is used to automatically select the most appropriate subset of spectral bands for transmission. The challenge in this case is to develop algorithms capable of accurately recognizing a sufficiently broad set of classes such that the science utility of the data is not compromised. In one oft-cited example, data acquisition could be terminated when viewing cloud-covered scenes.

In addition to editing as a means of reducing transmitted data rate and volume, two classes of data compression approaches seem promising: 1) distortion-free source coding, which for HIRIS data would yield, at most, a factor of two to three in compression; and 2) two- and three-dimensional source coding with distortion measures linked to the science utilization of the data. A number of techniques and algorithms exist which could potentially be adapted to noiseless compression. The most promising approach to two- and three-dimensional source coding (with distortion) is the use of vector quantization, which when applied in a one-dimensional fashion to spectral data yields a coding algorithm closely analogous to vector quantization of speech data. Initial evaluation of the vector quantization approach indicates that compression factors of 1:50 to 1:100 may be achievable while still producing scientifically useful data.

On-board information extraction would be another method of greatly reducing the total volume of data that must flow through the end-to-end system. For example, certain groups of imaging spectrometer data users may be more interested in obtaining accurate classified maps than in acquiring detailed data sets which must then be processed to extract the desired information. For geologic exploration, it might be possible to implement on-board the algorithms for doing mineral spectra matching, sending down only the classification and position information.

In general, information extraction (or data interpretation) operations on imaging spectrometer data require sophisticated pattern recognition approaches which must rely on extensive knowledge bases tailored to specific science disciplines. Accurate interpretation of remotely sensed surface reflectance spectra also requires pre-processing for removal of the solar spectral irradiance and a first-order correction for atmospheric scattering effects. When performing geological mineral identification, the total number of multiply/add operations per image pixel for accurate interpretation is of the order of $20L^3NlogN$, where L is the number of spectral channels and N is the number of stored prototype spectra. Building such an intelligent instrument will obviously require a great deal of on-board computing resources.

The SAR instrument, described in Section 2.0, has ten channels which can be independently operated. In the early phase of the Eos SAR mission, data rate can be controlled by selecting channels, swath width, incidence angle and resolution. The later version of the SAR flight system may include a processor to convert the raw radar return data into images. On-board image formation would allow on-board multi-look averaging, the saving of transmission of multiple images for multi-look averaging (typically two to five). A flight SAR processor that can operate at 50 megabits/second, however, is a significant technological challenge.

HIGH RATE ON-BOARD DATA SYSTEM

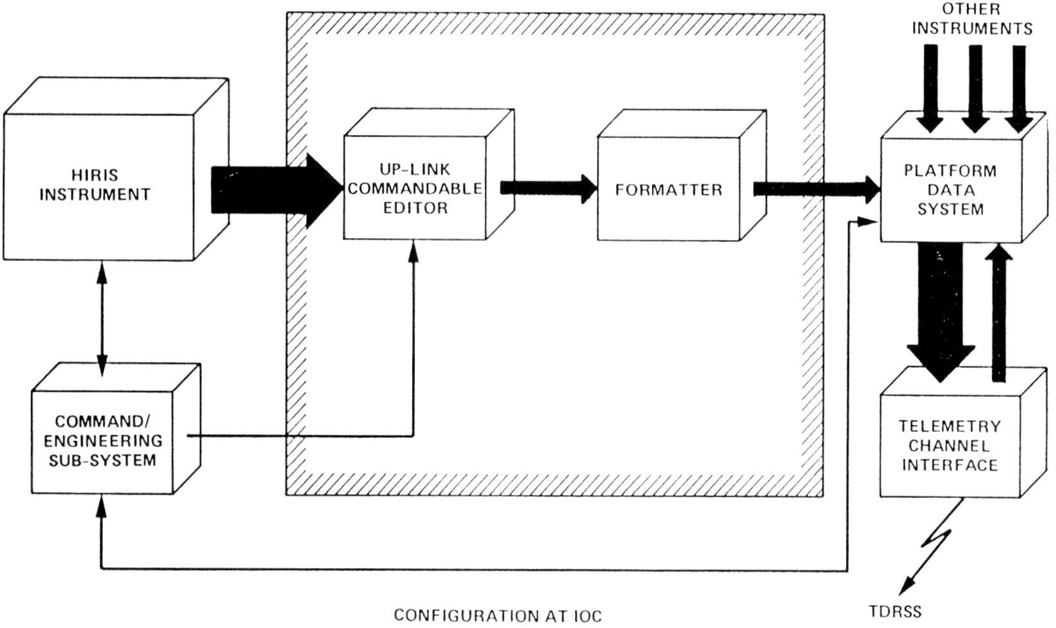

Figure 3.1

HIGH RATE ON-BOARD DATA SYSTEM

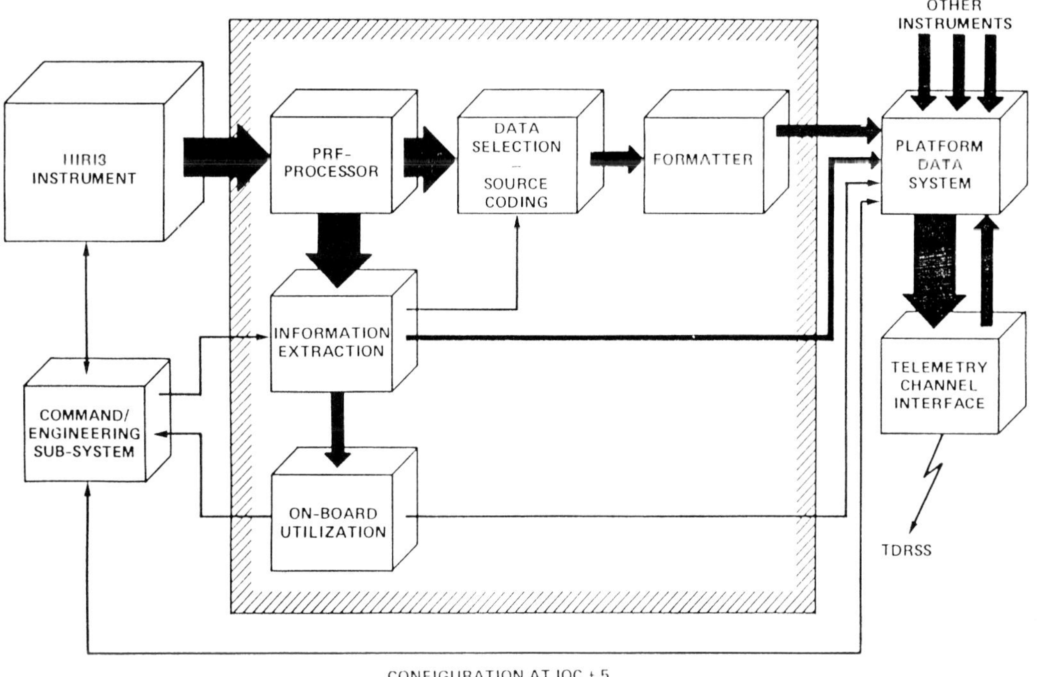

Figure 3.2

Once in the image domain, an additional compression factor of on the order of 2-3 times could be achieved. The combination of multi-look averaging and data compression would reduce the data rate by a factor of 10 to 15. A third stage of flight capability might implement information extraction techniques analogous to those described for the HIRIS instrument. At these data rates, even doing conventional data compression for a factor of two to five is a serious undertaking.

Figure 3.3 shows additional aspects of the flight data system. The upper part of the diagram shows the data system associated with the instrument itself. The middle part depicts data system elements associated with the carrier or platform. The lower portion of the diagram represents the data system ground segment.

For NASA data, the assumption to date has been that all data must be received on the ground through the Tracking Data Relay Satellite System (TDRSS). TDRSS has a peak data rate capability of 300 megabits/second but with conflicting demands on its utilization. The Eos Polar Orbiting Platforms have been allocated 1/3 utilization of this 300 megabits/second transmission resource. As a consequence, a policy has been established wherein instrument designs must assume TDRSS availability for moving data from the platform to the ground will be asynchronous with the observation times. For this reason alone, a mass storage subsystem must be available on the platform to buffer data up to the aggregate of the instrument peak data rates (in excess of 300 megabits/second) and with a total capacity sufficient for 2 orbits of data acquisition (currently scoped at a terabyte of data storage).

Figure 3.3 shows two mass storage devices: one located in the instrument data system and one located in the platform data system. The concept is that some form of mass storage device with random access would be necessary in later versions of the instrument data systems to enable sophisticated forms of information extraction and data reduction to be carried out on board. The mass storage device in the platform is considered the bulk memory device used for data buffering in achieving scheduling independence between observation times and TDRSS availability. This device in the mid 1990's will most likely still be a serial access tape recorder.

The current baseline Eos platform data system design has a single high speed network for frame transfer that interfaces with the RF subsystems. A lower rate local area network is shown for the flow of command traffic and low rate data. The rate for this local area network is expected to be 50 megabits/second, meeting the requirements of all payloads except for HIRIS and SAR. This particular configuration shows all "low" rate downlink data being bridged to a single high rate

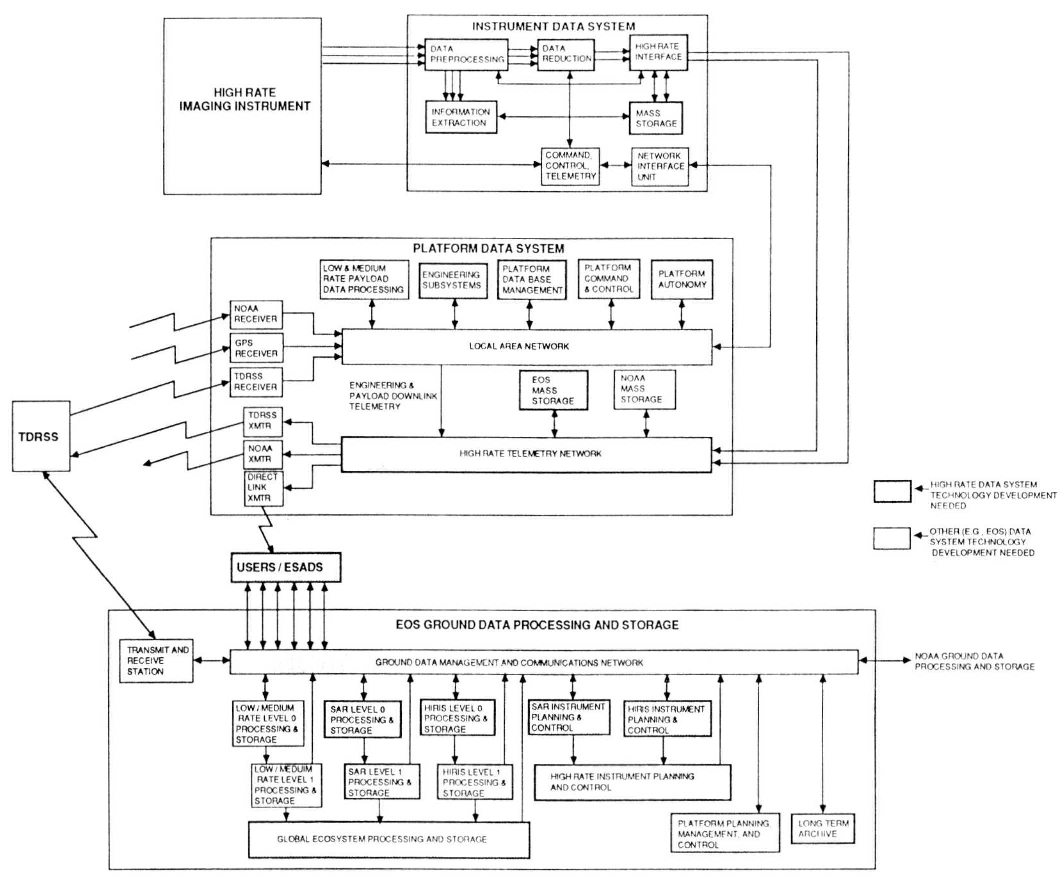

Figure 3.3

network. The high rate network is also shown as the only access for the platform mass storage device. Another approach favored by some designers is to carry the separation of "low" rate network and the high rate channel even further. In this design, the RF subsystem would have two interfaces, one to the low rate bus and the other to the special high rate channel. Likewise, the mass storage device would have a dual interface: one to the low rate bus and the other to the high rate channel. It should be kept in mind that the entire flight data system design is very preliminary and in some cases anticipates available technology. There are many trade-offs yet to be made before a design is finalized.

Packet telemetry[4] is an important aspect of the information systems architectures being proposed for the Space Station and associated platforms. With this concept, individual instruments produce source packets which are then packaged into transfer frames for transmission to the ground. Transfer frames would be fixed length, whereas the source packets would be variable length. There are two means of implementing individual channels for SAR or HIRIS in a packetized system. One method is to have uniquely identified source packets wherein each spectral band of the HIRIS or each frequency/polarization channel of the SAR looks effectively like a different source instrument. This is implementing virtual channels using source packets. A second method is virtual channel identification at the transfer frame level.

An important feature of the packet design is that with a fixed length transfer frame, frame synchronization can be easily implemented with foreseeable technology at the necessary data rates.[5] The source packets (of variable length) start and end points are identified by length information in the headers eliminating the need to have a second layer (at the packet level) of sophisticated frame synchronization hardware.

<u>Ground Processing</u>

As described above for NASA data, the Tracking and Data Relay Satellite System (TDRSS) will be the means of transferring data from space to ground. Once TDRSS coverage is available, data from the high rate instrument will be played back from the mass storage device on the platform. The data will flow through TDRSS inside transfer frames and will be received by a TDRSS ground station. A first step of ground processing is decoding whatever channel coding has been applied to decrease the error rate over the RF link. For 300 megabits/second data, the channel may be uncoded. If this is true, then special codes must be applied within the transfer frame to error-protect the length and ID information necessary for locating source packets. The next step is a frame synchronization process (which may or may not be associated with the decoding process) resulting in identified and located transfer frames. At this point, virtual channels and/or source packets may be separated and individually routed. This implies that it is only up to this point in the ground processing that the system must be able to operate at the peak rate of 300 megabits/second. The packet concept allows the use of parallelism in the remaining processes appropriate to whatever technology and budget

allows. For example, the SAR instrument could operate with as many as six 45 megabits/second capacity processors or three 90 megabits/second processors. Likewise, the HIRIS could utilize whatever level of parallelism desired, conceivably even to the limit of one set of processors per spectral channel (192).

The first step of processing after packets are extracted is termed level zero processing. This step consists of putting the data in time order, eliminating any duplicate data, and recording the data to some permanent form. Level zero processing simply removes the artifacts of the intervening data system, adding no real scientific value to the data. Two independent studies, one done at Goddard Space Flight Center and one at the Jet Propulsion Laboratory, roughly estimated a cost of 50 million dollars to implement a level zero processor to handle high rate data at a peak rate of 300 megabits/second and a sustained throughput of 100 megabits/second. This is a surprisingly high cost considering the relative simplicity of the level zero processing. Clearly, further work needs to be done to reduce the potential costs of level zero processing. It may be possible to limit certain operational modes to simplify the level zero processing to reduce cost. It may also be fruitful to pursue custom processing hardware which might do the job less expensively. In any case, such a frightening cost estimate for this seemingly trivial level of data processing points out the importance of critically examining the rest of the system for high data rate implications.

In the case of the SAR, the next step of processing is to convert the raw radar return data (assuming this has not been done on-board) to an image. The technology necessary to do this at the rates required was demonstrated in the summer of '86 in a culmination of a six-year NASA (Office of Aeronautics and Space Technology) funded project to develop an Advanced Digital SAR Processor (ADSP). This pipeline processor has demonstrated sustained range and azimuth correlation at over 100 megabits/second. In reaching this throughput, the ADSP performs at a rate of 6×10^9 floating point operations per second. A single one of these devices would be adequate for doing the ground correlation of all the SAR data in the Eos era, since the SAR instrument is envisioned to have a duty cycle of less than 20%. Rate buffering may be used in front of the SAR processor to keep the rate at or below 100 megabits/second.

The SAR data converted to the image domain is termed a level 1 product. The SAR and HIRIS image data corrected for geometric distortions would be termed level 1.5 or 2.0 data. It is the level 1.5 or 2 data products that are the lowest level of data of interest to investigators and it is the data at this level that should be placed into a permanent archive system. The traditional approach to data distribution has been delivery of digital and hardcopy products many months after data acquisition. With the duration and volume of data associated with Space Station and Eos payloads, NASA cannot afford a system in which data accumulate and are then distributed at some much reduced rate. A reasonable objective is for all catalog entries to be made and the initial data delivered both to the archive system and the active investigators within 24 to 48 hours; on a continuous basis with no backlog. Accomplishing this turnaround would reduce the need for expensive on-line buffering. Quick look image data, on the other hand, would be provided within hours so that engineers can verify the health and safety of the instrument as well as operational modes. The quick-look data would also assist the investigator in determining pointing and cloud cover characteristics.

The delivery of data to an investigator will be of two types: a subset of the data transmitted by electronic means, and a complete set of the data due a particular investigator delivered on either magnetic or optical media. We will do a little soothsaying at this point and project that by the mid-1990's the medium of choice will be the 5-1/4 inch format WORMS (Write Once Read Many optical platters).

Scientific questions associated with global habitability or earth as a system along with the sheer volume of data involved leads to the need for routine systematic processing downstream from the fundamental level 1.5 or 2 data products. Systematic processing is meant to imply processing done on a routine basis as a production job. For many NASA missions, processing has ended with the delivery of level 1, or in even some cases, level zero products to investigators and with later delivery into an archive. This processing would be provided as a service for those cases in which it makes sense to be performed at a centralized facility and in cases where individual investigators are unable to provide their own computational resources. At a minimum, such additional operations as calibration, co-referencing, geometric registration and rectification, mosaicking and atmospheric correction will be desired. In fact, with the Eos mission, the need for systematic processing services may extend to the inversion algorithms themselves to support certain types of investigators.

Figure 3.4 depicts one scenario for a mature Eos program that involves systematic processing. In this case, the processing of calibration, geocoding, mosaicking, and spatial sub-setting is necessary to inter-compare the high resolution high rate instrument data with the lower resolution global coverage data. The processing becomes even more difficult when the data are not acquired from the same spacecraft platform. It is very important to initiate a concerted effort of the scientific community to look at the real operational scenarios and the potential requirements for systematic science processing so that the right technological areas can be pursued for these high rate instruments.

Telescience

The last data systems area to be discussed is uplink command and control. The concept for the 1990's is for the investigator to be able to operate from his home institution interactively with the payload. This has sometimes been termed "telescience." It is possible to conceive of operational scenarios where this would make sense even for facility class instruments of the scale of HIRIS and SAR. For example, an investigator may have selected a set of spectral channels for

EOS SYSTEMATIC GROUND SCIENCE DATA PROCESSING

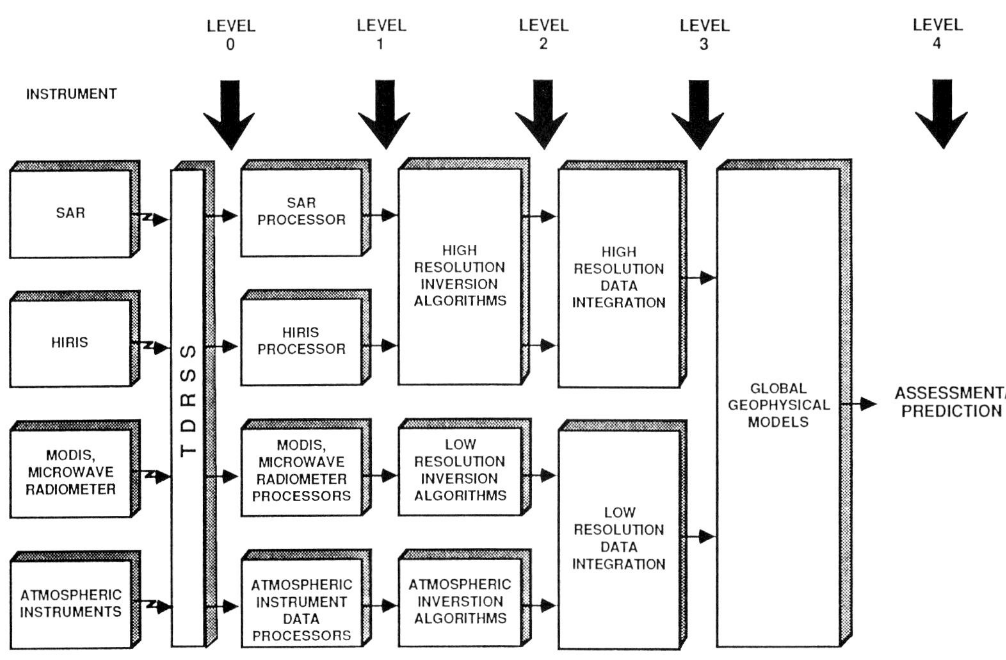

Figure 3.4

the HIRIS instrument which, with the right set of
tools in a quick-look sense, he could evaluate
whether that was the right choice in time to
interact with the next scheduled observations over
the region of interest. For the SAR, it might be
desirable (but yet to be proved feasible) to
interact even in real time since sometimes the
optimal observational parameters are scene
dependent. To make such near real time (say $\leq$ 24
hours) commanding possible would require providing
the "quick-look" data products, which were touched
on earlier, to the investigators. The remaining
aspects of operations and telescience are
envisioned to be common with the lower rate
instruments, and therefore, are not the subject of
this paper.

4.0 Technology Drivers

Although the HIRIS and SAR instruments are
technological challenges in themselves, at least
we have proceeded far enough into the details of
the designs that we know how to build them and
know the areas of risk. In fact, we have flown or
are flying prototype instruments of those
conceived for the 1990s. Technological
advancements, though, are needed in a variety of
data systems areas, some of which are directly
attributable to the high rate data streams, such
as on-board recording, and some indirectly, such
as automated planning and scheduling. The
enabling data system technologies for Eos relate
primarily to accommodation of the peak instrument
data rates. Most of the data systems technology
requirements are enhancing in the sense that they
will either significantly reduce mission costs,
make more efficient use of the instruments, or
provide for increased scientific productivity.

This section will review the current
state-of-the-practice of some major data systems
elements, both ground and flight, and suggest
where we need to improve for the next generation
of NASA missions--for which Eos provides a good
model. A summary of this roadmap is provided in
Figure 4.1.

Flight Data Systems

General-purpose flight computers will be
needed for future missions with performance of 10-
25 MIPS and above. These speeds are needed to
support applications such as telerobotics control,
active structures control, on-board information
extraction, payload autonomy, and simply
controlling devices necessary to support extremely
high data rates. Processors currently being used
are in the 1 MIP, and below, range. For Eos,
advanced general purpose processor requirements
are expected in the spacecraft data system to
support the payload control executive, as a data
storage controller and as a LAN controller.

There are generally two classes of processors
needed for flight applications. One class is used
for non-critical functions in non-severe
environments. Low-earth, equatorial orbits,
combined with the relatively non-critical
functions of payload experiments means that
commercial and Mil-Spec processor parts can
typically be used. Another class of applications,
however, as exemplified by Earth polar orbiting
and planetary missions, must be concerned with

radiation and Single Event Upset (SEU)
sensitivity. Meeting the radiation and SEU
requirements eliminates all current ground,
commercial processors from these applications and
makes fault-tolerant architectures for future
computing systems extremely important.

Current, and anticipated, severe-environment
flight computers are all modest performers.
Galileo, for its most demanding application, uses
a Sandia Labs version of the AMD2901 bit-slice
processor to emulate the ATAC-16. It is rated at
.6 MIPS (Million Instructions Per Second). Mars
Observer is planning to use a Harris Corp. flight
qualified 80C86 which performs at approximately .5
MIPS. The Space Shuttle computers are IBM
derivations of the 360/370 architecture, built to
Mil-Spec, which perform at less than 1 MIPS.
However, these computers would have to be replaced
or upgraded, were the shuttle to be launched into
polar orbit. Gaining the performance increases
necessary for the next generation of flight
computers is not likely to come from speed
increases of individual microprocessors. Rather
it will come from concurrent architectures which
take advantage of the parallelism not only for
speed but also for fault tolerance.

The development of increasingly autonomous
spacecraft and the addition of substantial
intelligence to on-board systems will require the
use of flight-qualified fast random access memory
similar to that used in ground-based systems. The
Galileo spacecraft initially used 1K static RAMs
but the Command Data System is now being upgraded
to 4K static RAMs. This density of memory will
obviously not allow the deployment of large
random-access memory sub-systems in the millions
of bytes because of power and weight constraints.
Progress in this arena is slow. However, the
early 1990's should see a flight qualified 64K
static RAM. With the 64K RAMS and beyond, we can
then begin to think about building memory
sub-systems for polar orbiting and planetary
spacecraft capable of supporting the on-board
intelligence currently being planned.

The other common form of flight data storage
is large mass memory used for on-board buffering
and rate smoothing. Spacecraft data recorders are
available today capable of storing 10^9 bits. The
Eos spacecraft have a requirement to store two
orbits of data (estimated at 10^{12} bits) with a
peak record rate of greater than 300
megabits/second. As we continue to gain better
understanding of the Eos science scenarios, it is
the flight data buffer which appears to be the
most critical resource, requiring, if not
technological development, at least significant
advances over current practice. The serial nature
of the tape recorders typically used for such
applications also presents problems downstream in
that the data must then be reversed to be put into
time order. For Eos data rates, this constitutes
a substantial expense which could be avoided if
random access flight data buffers were to be
developed.

On-board data compression is a technique
which has been used successfully on several
spacecraft in order to make maximum utility of
communication link bandwidth. However, data
compression requires a non-negligible processing
resource. Galileo contains a data compressor,

NASA HIGH-RATE DATA SYSTEMS TECHNOLOGY
PRESENT AND FUTURE

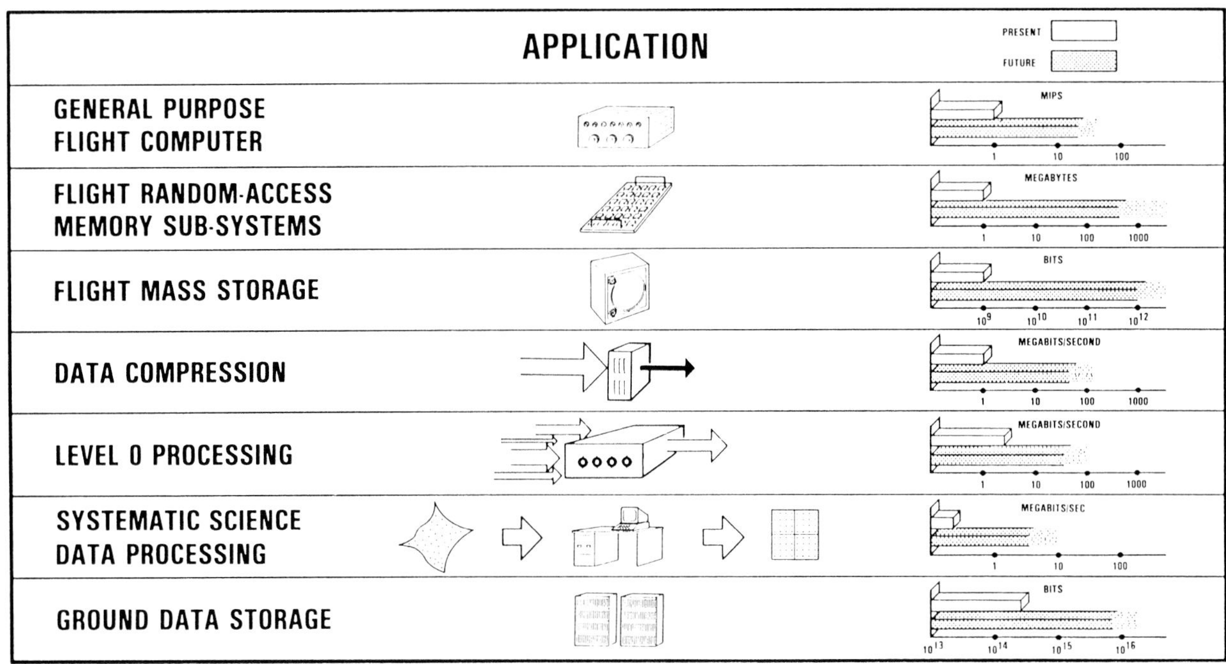

Figure 4.1

which consumes a full PC board, capable of 3:1 compression at a 1 megabit/second input rate. Although the ability to do noiseless compression is likely to remain between 3:1 and 2:1, depending upon the data, the need exists for much higher performance compression sub-systems, capable of supporting a 50 megabits/second data stream. Other data compression applications may arise in which compression-induced noise is traded-off against data volume factors such as spatial and spectral coverage. This will give rise to the need for subsystems capable of performing a diverse range of data compression algorithms at extremely high rates.

Ground Data Systems Technology

As discussed earlier, level zero processing involves essentially undoing whatever the flight data system and communication system has done to the data such that it appears as if the data came directly form the instrument. This operation involves primarily removing any data overlaps and placing the data in time-order. Current practice is to perform this function with large general-purpose computers and extensive random access mass storage. Level zero processing performance for the Landsat Thematic Mapper is about 4 megabits/second. The combined Eos data stream will require significantly higher data rates.

It appears as if level zero processing for Eos data rates is certainly feasible using conventional approaches. However, there is a great deal of incentive to do everything possible to minimize the cost of this processing step. The challenge is to increase our capability by an order of magnitude while decreasing costs by an order of magnitude. Several approaches are being pursued in this arena, including the development of specialized and custom hardware, integrating level zero processing with other steps to minimize data handling overhead, minimizing the computational requirements by restricting certain operational modes for the high rate instruments, and finally, attempting to eliminate the step altogether by not allowing the artifacts to be introduced by the flight data system from the beginning.

Once level zero processing has been accomplished, the real ground computational requirements begin. Systematically processing the scientific data requires enormous amounts of computational resources for operations such as calibration, atmospheric correction, geometric processing, and information extraction. In the Eos era, as now, these operations are expected to be performed both as a service by a central facility, and in a custom fashion at investigator facilities. With current workstation and mini-computer systems, it is not reasonable to expect much more than a sustained rate of less than .5 megabits/second for even the most simple systematic processing. Off-the-shelf concurrent computers and supercomputers have turned out to be less than ideal for these tasks because of their relatively slow I/O. An order of magnitude increase in throughput performance is needed for

centralized facilities performing such processing
on a routine, sustained basis. Because of the
complexity of this processing and the fact that
much of it is human-intensive, increases in
processor and computer system speed alone will not
suffice to meet this challenge.

Ultimately the goal is to move selected
science data processing functions to the
spacecraft itself. On-board information
extraction has the potential of reducing the data
rates by orders of magnitude for imaging sensors.
One aspect of science data processing peculiar to
imaging radars is that of image correlation. This
is a necessary step for practically all Synthetic
Aperture Radar data for it to be of any use. It
is currently possible to perform this operation
with the Advanced Digital SAR Processor, a custom
system, at throughput rates of 100 megabits/second
on the ground. Reasons for wanting to put this
function on-board the spacecraft are that it would
allow data reduction by multi-look averaging, data
compression, or information extraction. On-board
image formation would also open up the possibility
of direct downlink to receiving stations which
would not need expensive correlation equipment.

Accommodating high date rates for extended
periods of time, particularly without data
reduction strategies, ultimately leads to very
high data volumes which must either be archived
and managed or discarded. The latter option is
very seldom considered appropriate. The EROS Data
Center, which maintains an archive of much of the
Landsat MSS data, holds approximately 3^{14} bits.
The NOAA NESDIS holding of environmental satellite
data is estimated at approximately 51^{4} bits. It
has taken many years to acquire these data from
many space missions. If Eos averages only 10
megabits/second (substantially less than current
estimates) and archives all data, the archive
would be 5^{15} bits at the end of its estimated 15
year lifetime. Again, we are faced with an order
of magnitude increase in capability above current
practice which we must achieve in order to be
responsive to the needs of NASA's next-generation
of missions.

Assuming optical recording technology will
automatically rise to the occasion in supporting
future archives overlooks the issues associated
with the management of such a library, regardless
of whether or not it can be recorded. Given
current optical disk recording densities, 1.5
million 12 inch optical disk platters would be
required to hold the Eos archive as sized above.
Simply managing and making available the knowledge
about the location and content of such holdings is
indeed a challenge.

5.0 Management Issues

One overriding management issue is the
development strategy for bringing the Eos data
system into existence. The combination of an
order of magnitude increase in data rate and the
long term, evolutionary nature of the mission
suggest that traditional approaches to
requirements definition, specification and
implementation will be inadequate. The report of
the Eos Data Panel states "... the operation of an
Eos Data and Information System will create
management problems of a magnitude that cannot

even be fully appreciated at this time by either
NASA management or the scientific research
community who must cope with these data in their
research."[6]

The first question is how the
responsibilities for the design, development and
implementation of the end-to-end system which has
to be able to support the high rate instruments is
spread across the NASA organization. Figure 5.1
portrays the end-to-end system as a set of serial
functions and highlights the most significant
interface points between major system elements.
The letters S, E and T across the top of the page
are symbols for the major NASA organizational
elements. E stands for the Office of Space
Science and Applications (OSSA) who is responsible
for the scientific and application missions of
NASA. The certain responsibility for OSSA is to
provide the two ends of the system: the payload
of instruments in the flight element and on the
other end, the ground mission science data
systems, science data processing systems, support
of individual investigators and archival systems.
Code S stands for the Office of Space Station
(OSS) who in current NASA planning will provide
some of the orbital platforms. Next comes the
Office of Space Operations (OSO) which is
responsible for the tracking systems of NASA
(including the TDRSS) the ground operational
communications functions of NASCOM, and the non-
operational communications of the Program Support
Communication Network (PSCN), and for certain
types of data processing, control centers, and
navigation functions when those functions occur at
the Goddard Space Flight Center. Not shown is the
technology development supporting role of Code R,
the Office of Aeronautics and Space Technology.
The area of the end-to-end system where the
responsibilities are less obvious is in the first
stages of data processing and in the operations
roles and responsibilities.

Current concepts for the platforms divides
the on-board system into two categories: the
spacecraft or platform data systems necessary for
the operation and control of the physical platform
itself, and the carrier data system which is there
to support the payload of instruments. There is a
corresponding potential division of responsibility
on the ground for the operational control of the
platform and its payloads. First is a platform
support center, or operations control center,
responsible for the monitoring and control of the
physical platform. Next is a carrier control
center (which may or may not be co-located with
the platform control center) responsible for the
overall coordination of the flight data system
support to the payloads. The third level of
responsibility is the actual control of individual
instruments. Current concepts have the platform
and carrier control functions residing at the
Goddard Space Flight Center with these facilities
designed, implemented and operated by OSO. The
instrument control center responsibilities are
envisioned as being geographically distributed.
For example, the HIRIS and SAR which are
envisioned as JPL-provided instruments would be
operated from an instrument control center located
at JPL in Pasadena. The instrument control center
located at Goddard may be provided by OSO whereas
those not located at Goddard would have to be
provided by OSSA. There is an interest at GSFC
and OSO to expand their multimission facilities to

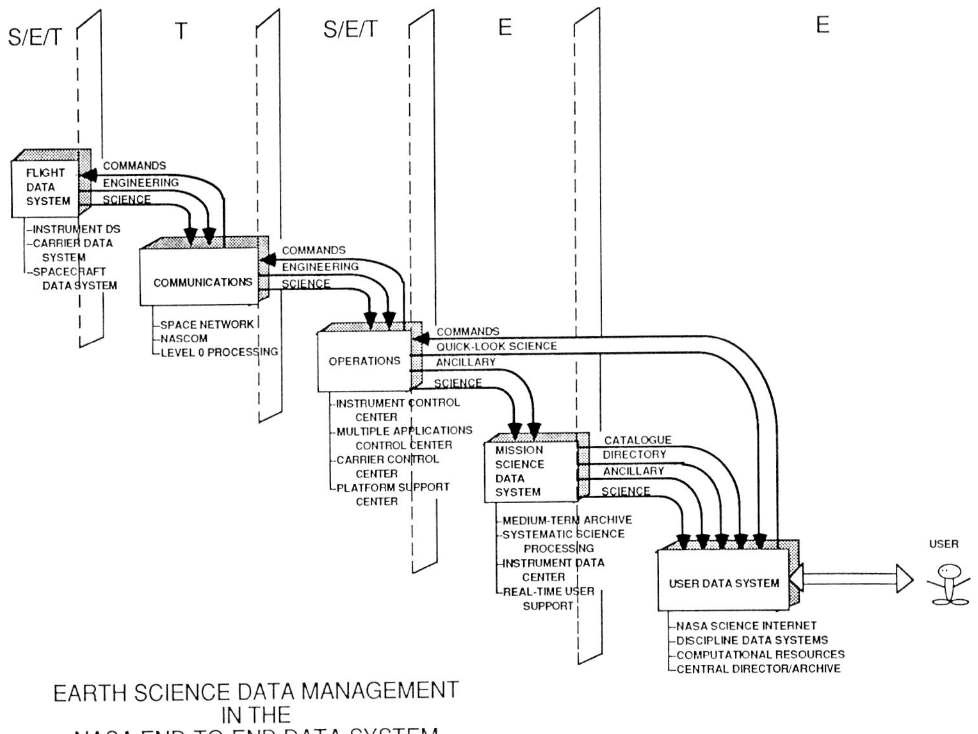

EARTH SCIENCE DATA MANAGEMENT
IN THE
NASA END-TO-END DATA SYSTEM

Figure 5.1

accommodate as wide a set of operations and requirements as possible. This may result in some combination of the platform, carrier, and instrument control functions taking place in an expanded multimission operations control center and associated facilities.

A natural question is who is responsible for the end-to-end design. Historically in NASA for free-flying spacecraft the responsibility of integrating the end-to-end information has fallen upon the individual missions. By analogy, the responsibilities in the era of Space Station and platforms would fall on OSS for the platform systems and would fall on OSSA for the payloads. Unfortunately, with few exceptions, another historical truth is that prior to project approval funds tend to be rather limited and projects tend to focus on instrument and specific spacecraft hardware and defer dealing with the information system necessary to support the mission until rather late in the development process. Fortunately, the scope and complexity of the new information challenges in the Space Station era have been recognized and NASA has been making efforts to organize design activities and teams to deal with the end-to-end information systems. It can be debated, however, as to whether the organizational elements yet established and the resources being applied are really adequate to address the issues inherent in the high rate systems with enough time left to solve the technological problems that remain to be identified.

An example of both the management and technological challenges is the issue which has already been described of the level zero processing. OSO has traditionally provided the level zero processing for near-earth missions. Studies completed at Goddard show that the majority of payloads add to an aggregate average data rate of less than 20 megabits/second with the exception of HIRIS and SAR. Because of this study, OSO has set a tentative proposed policy that they will provide level zero processing only up to an aggregate of 20 megabits/second. As we have already pointed out, in two tentative studies, the cost for level zero processing at an aggregate rate at something on the order of 100 megabits/second looks to be very high. The OSO policy, if upheld, would force OSSA to build the capability to handle the level zero processing for the HIRIS and SAR instruments. Goddard argues that that would not be a cost effective approach, that the incremental to go up to the 100 megabits/second from 20 megabits/second would be much smaller than the cost of building a completely separate facility. Some of us at JPL argue that before we make any such decision we ought to see if there are some completely different and novel technical approaches to the level zero processing at high rate which might significantly lower its cost. In addition, there is the open question of what are the actual scientific scenarios for the operational use of the HIRIS and SAR instruments and the corresponding aggregate real average data rates for these instruments.

The operational scientific scenarios for
these instruments are not well understood and can
be tremendous swingers in the actual requirements
and corresponding cost of the system to support
the instruments. For example, early estimates of
the utilization of the HIRIS instrument assumed
about a 15% duty cycle at 160 megabits/second.
This results in an average data rate of 24
megabits/second. A recent look by the Imaging
Spectrometer Science Advisory Group arrives at an
aggregate average rate of only 3 megabits/second
by assuming that there would be less than 100
investigators and that each investigator would be
content with just one or two scenes per day. It
is this author's opinion that the correct number
for HIRIS probably lies between these two
estimates. For sure there will continue to be a
category of scientific investigation which
consists of an investigator examining a handful of
scenes over a very extended period of time.
However, to really successfully tackle the
scientific questions in global change and
understanding the earth as a system will require
the kind of systematic science processing of large
volumes of data that was discussed earlier in this
paper. When the scientific scenarios and
resulting systematic processing become understood,
then we can expect to see the requirements for the
volume of the HIRIS data to grow again. It
appears that the SAR using community is beginning
to step up to these questions.

There is a three-stage process which must go
on. First is getting the right group of
scientists together to address the scientific
operational scenarios for the utilization of the
SAR and HIRIS instruments, particularly for
addressing the issues of global change and earth
as a system. From those scenarios the right
combination of NASA, industry, and university
engineering talents must provide tentative designs
for the end-to-end system. From those designs,
critical missing technological components must be
identified. This three-step process must be
iterated so that a set of technological needs
achievable by the time of flight in the mid-1990's
match the operational scenarios. This process is
at an embryonic state at NASA, but at least it has
started.

There are significant management,
technological, and scientific challenges to be
faced in preparing for attacking the global change
and earth as a system questions with the earth
observing systems of the 1990's. There are three
possible scenarios for the future of the HIRIS and
SAR class of instruments: (1) although the
instruments are affordable, the associated data
systems are viewed as so expensive that we never
fly the instruments; (2) we fly the instruments
but with a data system that results in grossly
under-utilizing the instrument capabilities and we
only dabble at solving the earth as a system
scientific challenges; or preferably (3) we do
enough advanced work to find the right compromise
between the scientific objectives and scenarios
and solve the right set of technological problems
to provide an affordable, highly capable system
that gives us a real shot at successfully
addressing the global scientific questions of
interest to the long-term survival of mankind.

Efforts have been initiated on several fronts
which we feel, if they are given enough attention
and resources, can lead us in the direction of the
third scenario.

References

1.   "MODIS, Moderate-Resolution Imaging
     Spectrometer," Instrument Panel Report, 1986,
     Volume IIb.

2.   "HIRIS and SISEX, High Resolution Imaging
     Spectrometry: Science Opportunities for the
     1990s," Instrument Panel Report, Volume IIc,
     in press.

3.   "Spaceborne Imaging Radar for the Earth
     Observing System," Eos SAR Panel Report,
     March, 1986.

4.   "Recommendations for Space Data System
     Standards: Packet Telemetry," Consultative
     Committee for Space Data Systems, May, 1984.

5.   Hockensmith, Richard P., et. al., "NASA Data
     Link Module (NOLM) Interfaceability Study,"
     NASA Technical Memorandum 87793, July 1986.

6.   "Eos Data and Information System," Data Panel
     Report, Volume IIa, 1986, NASA Technical
     Memorandum 87777.

Acknowledgement

The research described in this paper was
carried out by the Jet Propulsion Laboratory,
California Institute of Technology, under a
contract with the National Aeronautics and Space
Administration.

DATA STORAGE SYSTEMS TECHNOLOGY
FOR THE SPACE STATION ERA

John Dalton, Fred McCaleb, John Sos, James Chesney,
David Howell/NASA, Goddard Space Flight Center, Greenbelt, MD;
Linda Kempster, Michael Healy/Computer Technology Assoc., Lanham, MD

ABSTRACT

NASA's Space Station and the associated Polar
Orbiting Platforms, due to be operational in the
early 1990's, will provide scientists, industry,
and the United States Government the vehicle to
deploy an unprecedented number of data producing
experiments and operational devices. By the
fourth year of the mission, peak down-link data
rates are expected to be in the 500 megabit per
second range and the daily volume of user data
could reach 2.4 terabytes. Such startling
requirements inspired an internal NASA study to
determine if economically viable data storage
solutions are likely to be available to support
the Ground Data Transport segment of the Space
Station Mission. This paper summarizes the
results of the study and outlines an internal
NASA effort to prototype a portion of the
required ground data processing system.

I. Introduction

The base and platform elements of the Space
Station will create a number of data storage and
delivery challenges for the information system:

- operation of high rate instruments in
  the 100 to 300 Mbps range;
- near-real-time delivery of data to
  support telescience operations;
- data driven (CCSDS virtual channels and
  packets) instead of time scheduled
  operation;
- reconfiguration of experiments
  on-board; and
- maintainability and growth over a long
  lifetime.

A Digital Data Storage (DIGSTOR) team was
recently formed at NASA/Goddard Space Flight
Center to assess the storage requirements
associated with Space Station data transport and
processing from receipt at the Tracking and Data
Relay Satellite (TDRS) ground station to
distribution to users, and to determine what
technology development, if any, was needed to
meet those requirements. This paper summarizes
the mission requirements, projected storage
solutions, and current prototyping work focused
on meeting these challenges.

Space Station Data Storage Requirements

Space Station data rates were projected from
existing mission models identifying candidate
base and platform experiments by year.
Individual instrument data rates were expressed
in terms of peak rate and duty cycle. Aggregate
rates for all instruments were then mapped into
the data buffering and processing functions
within a high level model of the data transport
system architecture.

Mission Data Profile

The bar chart in Figure 1 shows the peak data
rate from all instruments vs. year for the first
six years of the Space Station Mission, assuming
the current baseline schedule with initial
experiment operations in 1992. Shading within
bars is used to indicate the proportion of
payload data vs audio/video vs fill. (Fill data
results from the difference between the actual
data rate and the TDRS KSA link rate.)

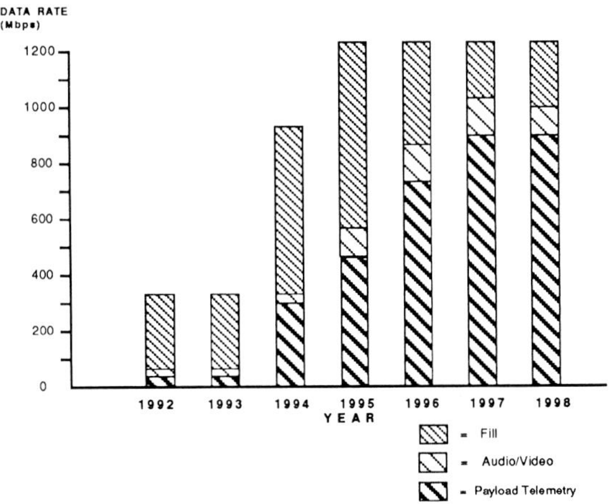

Figure 1. Peak Data Rate Time Profile

Not all instruments will always operate
simultaneously. Therefore, an effective maximum
rate (the "90% availability" rate) was estimated
by computing the joint probability that
combinations of instruments would operate
simultaneously, and from this set of joint
probabilities, a data rate threshold was
determined such that the data rate of
simultaneously operating instruments would fall
within the threshold 90% of the time. The
remaining 10% of the time, instrument operations
would have to be rescheduled, or on-board data
recording would be required, to fit within the
data rate threshold.

As shown in Figure 2, this assumption has a
dramatic effect on the non-fill data rate,
reducing it from 1 Gbps to 500 Mbps in 1997.

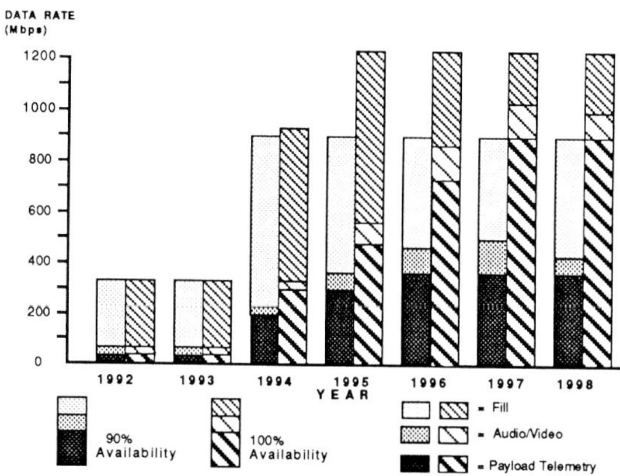

Figure 2. Peak Data Rate Comparison

Estimates of daily data volume were developed for the same six year period in order to predict the impact of volumetric data storage requirements on various elements of the NASA supported data transport system as well as to project the potential impact on end user archival data storage. As shown in Figure 3, the volume of payload data rises rapidly to about 3 terabytes per day during the first four years of the mission and then remains relatively stable.

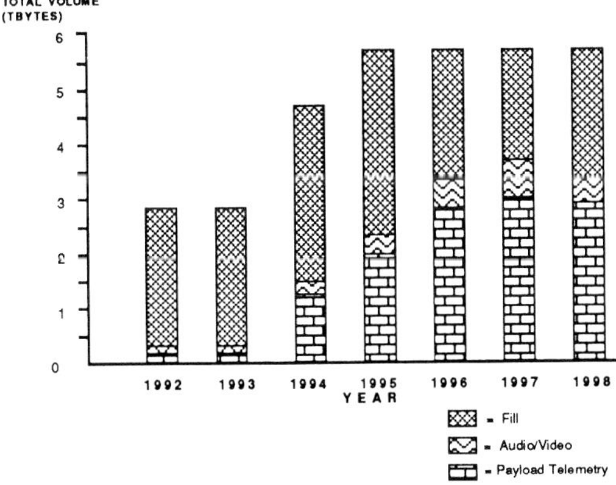

Figure 3. Daily Data Volume Time Profile

Data Storage Subsystem Requirements

In order to derive the requirements for data storage subsystems, several alternative data transport architectures were identified with different degrees of decentralization. Data storage operations at each subsystem were categorized based on access time and retrieval functions, and reduced to the following types of subsystems:

1. ### First In First Out (FIFO) Storage

    Very high data rate capture of large volumes of data is required for protection against downstream system or line outages. Such data storage systems will be required to function only as giant FIFO buffers with no random access requirement. These storage systems must be capable of ingesting continuous 300 Mbs streams for at least 20 minutes (45 gigabytes) and by 1996 must handle total data volumes in the 2+ terabyte range.

2. ### Fast Random Access Storage

    Fast random access storage is required to support level zero processing and to provide data buffering in front of asynchronous and/or reduced data rate communication systems. Level zero processing functions which require fast random access include playback data reversal and time ordering of data received from real-time and playback sources. The prime function of random access data buffering in front of communication systems is to allow use of lower rate dedicated communication channels or asynchronous shared channels in order to minimize communication costs.

3. ### Slow Access With Staging

    Storage systems providing a hierarchy comprising a moderate (10's of gigabytes) volume fast random access component and a large volume (terabytes) slow access component will be required for retention of processed data to support scheduled or demand delivery to Space Station experimenters, thus avoiding the need for real-time data handling at user facilities.

Comparing the requirements for each type of data storage subsystem in each alternative architecture assured that data storage technology selection would not constrain the system architecture design. In fact, none of the architecture alternatives had a significant impact on storage requirements. Separating the storage systems into the above categories allowed comparison of alternatives within each category based on data rate, capacity and technical feasibility.

Finally, the analysis resulted in the derived requirements for peak data rate in Mbps (megabits per second) and volume in GB (gigabytes) for each type of storage subsystem as shown in Table I.

Category	1992 Rate (Mbps)	1992 Volume (GB)	1994 Rate (Mbps)	1994 Volume (GB)	1996 Rate (Mbps)	1996 Volume (GB)
Sequential	300	972	900	1562	900	1853
Fast Access	11	10	300	200	500	600
Slow Access With Staging	11	300	300	2500	900	5400

Table I.
Storage Subsystem Requirements

It is important to note that these are requirements for the buffering and processing in the data transport system between ground receipt of data from the Tracking and Data Relay Satellite (TDRS) and its distribution to users. Trade-offs and interfaces exist between this transport system, the flight data system, and the user data systems. The characteristics of on-board data storage and playback, impact the level zero storage and processing systems on the ground. At the data distribution interface, the storage technology used to meet the "slow access with staging" requirements can also be used for actual data delivery if the media and data formats are readable at user facilities.

Technology Assessment and Projections

In order to deal with the wide variety of magnetic and optical products currently available or under development, potential solutions were grouped in the categories in Table I. Specific devices were considered if their performance was within a reasonable factor of the data rate and volume requirements, such that the requirements could be met through multiple devices. Predictions of future product availability dates were based on vendor projections, however, an additional period of at least four years was added to account for vendor optimism, product stabilization, and NASA's procurement and system installation/integration lead time. Finally, a consultant, with expertise in the data storage industry, was brought in to perform an independent assessment of the predictions.

The results of the analysis are shown graphically in Figures 4 thru 6 for the sequential (FIFO), fast access, and slow access with staging functions, respectively. Requirements for key years are shown as squares on a plot of peak data rate vs. volume. Selected subsystems and projected year of availability are plotted with vectors drawn to the appropriate "requirement" year based on projected development time. Each vector is labeled with the number of drives and/or media units required to meet the requirement.

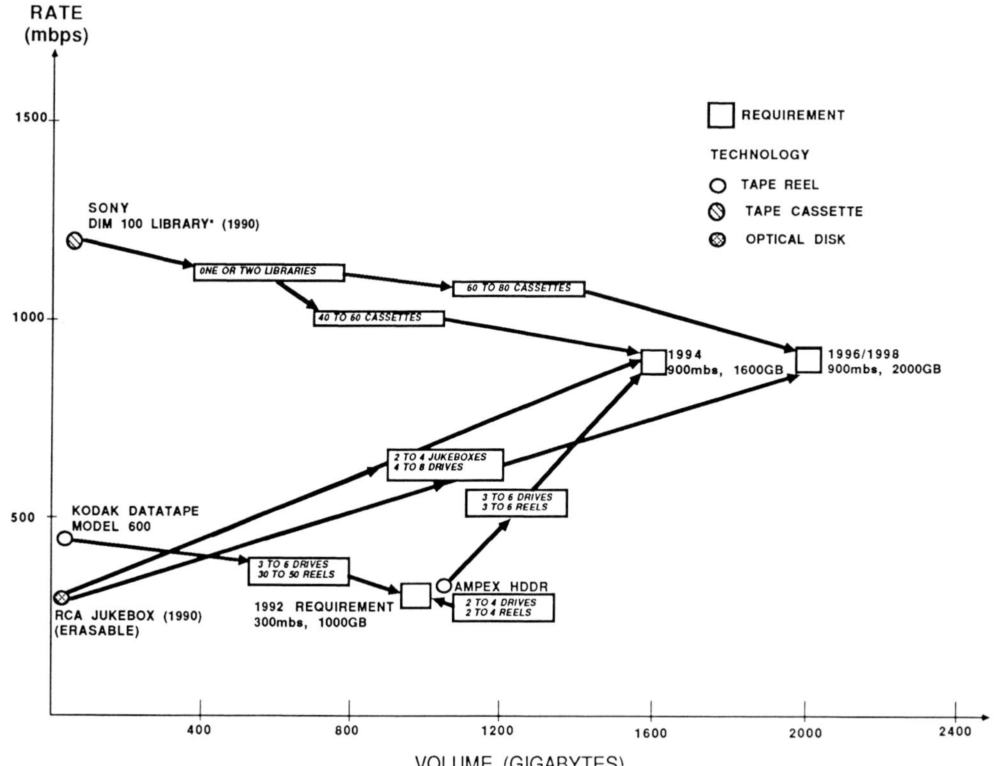

Figure 4 - System Outage / Line Outage Protection - Potential Technology Solutions

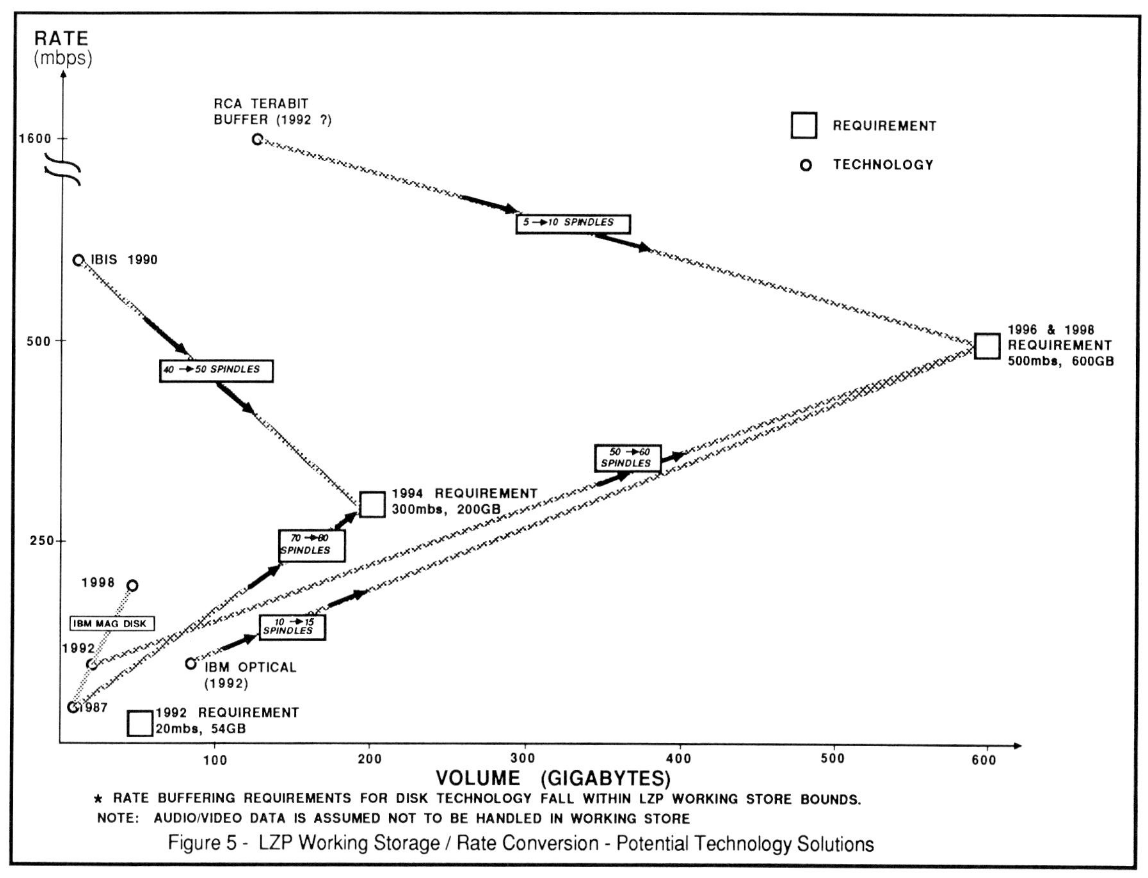

Figure 5 - LZP Working Storage / Rate Conversion - Potential Technology Solutions

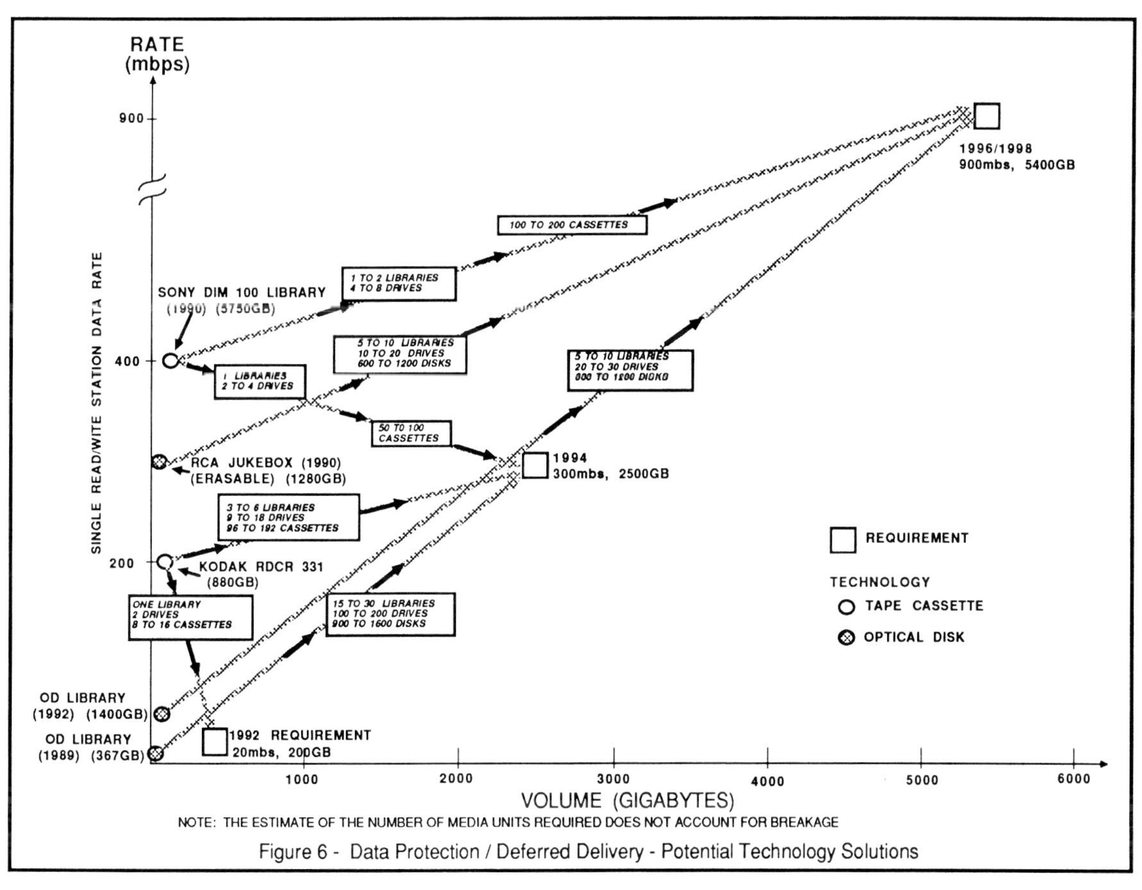

Figure 6 - Data Protection / Deferred Delivery - Potential Technology Solutions

Vendors are identified based on application of technology projections to the characteristics of their current products. The identification of a vendor for a future product does not assure that the vendor will bring the product to market as projected. For example, Figure 5 shows the projected 1996 Level Zero Processing/Rate Buffering Subsystem requirement being met by a generation of magnetic disks projected to be available in 1992 if IBM or some other vendor continues to improve the technology at the current rate. While the volume requirement of 600 GB is at least a factor of 30 greater than the projected disk drive volume, 50 to 60 such drives with appropriate controllers can be configured to meet the requirement, including capacity to account for file management overhead and an operational margin. However, if capacities do not continue to grow, 200 or more drives would be required.

The following is a summary of the conclusions reached based on the analysis:

Sequential Storage (System and Line Outage Protection)
While reel-to-reel high density digital recorders (HDDR) continue to meet data capture requirements, the next generation of digital tape cassettes will definitely be better suited for use in automated tape libraries. Adaptation of 19mm video cassette technology to 300 Mbps digital recording appears to provide the best solution for NASA's future sequential storage requirements.

Fast Access Storage (Level Zero Processing and Rate Conversion)
It is projected that requirements can be met by continued increases in areal density of magnetic disks, however, it is not certain that the commercial market will drive the controller and computer I/O rates as projected. Furthermore, requirements can only be met by unusually large disk farms, with distribution of data over 10 or more loosely synchronized drives to meet data rate requirements. Parallel transfer, erasable optical disk systems such as RCA's terabit buffer may be an alternative depending on product cost and availability for the quantities needed.

Slow Access With Staging (Data Protection/Deferred Delivery)
Potential technology solutions for Data Protection/Deferred Delivery functions include automated digital tape cassette libraries or automated erasable optical disk libraries with fixed media high performance magnetic disk subsystems for staging. Digital tape cassette technology based on 19mm SMPTE and/or ANSI D1 video tape technology appears to be the most likely near term solution for high data rate (100+ Mbps) applications. Although high data rate erasable optical disk jukeboxes appear to meet this requirement, they were not the first choice because of potential cost and schedule impacts, and because they were judged to be unlikely to become commercial products.

Use of removable media for the deferred delivery function, which is essentially back-end file management for electronic distribution of data to users, offers an additional advantage in that individual media units (optical disks or tape cassettes) could serve as a periodic bulk data distribution mechanism if user facilities have compatible reading devices.

Storage Subsystem Development Required

The mission requirements analysis and technology assessment discussed in the previous sections produced the following conclusions:

1. Requirements for all ground data storage functions can be met with commercial disk and tape drives assuming conservative technology improvements.

2. However, in order to meet Space Station data rates with commercial technology, the data will have to be distributed over multiple devices operating in parallel and in a sustained maximum throughput mode.

3. Experience is needed in use of the 19mm D1 video cassette for high rate digital data storage.

4. Fill data accounts for up to 60 percent of the system outage protection requirement. A highly reliable front end for removal of fill data could result in dramatic savings in the sequential storage subsystems.

5. More specific design and cost trades between flight and ground systems, and between subsystems on the ground, need to be made as system design and engineering proceeds. Decision points must be established to track the projected technology improvements and develop alternate plans if necessary.

6. Archival storage requirements are severe, and a broader assessment of erasable and non-erasable technologies needs to be made including the trade-off between distribution of media vs. buffering and electronic transmission of data assumed in this study.

The development of a high data rate storage subsystem based on commercial magnetic disk drives has the highest degree of risk of all of the technology solutions identified in this study. While large disk farms are found in several commercial applications, they are used to provide on-line access to extremely large data bases of periodicly accessed files with no definitive throughput requirement. The systolic operation of such a large farm appears to be unprecedented. A subsystem controller is needed with the following features:

> synchronized distribution and retrieval of both fixed data blocks and variable length packets over multiple drives at maximum drive throughput rates.

- expansion to additional drives to increase capacity.

- independent data paths to multiple devices for simultaneous recording and playback for rate conversion applications.

- dynamic detection of and recovery from drive and controller subsystem failures.

- repairable on-line.

- an adaptable interface to disk controllers to provide vendor independence and technology upgrades.

An initial prototype is being designed to verify this concept by synchronizing four 24 Mbps, 800 Megabyte disk drives to demonstrate maximum (goal is 75 Mbps) attainable sustained data rates (Figure 7). Each drive will be required to sustain approximately 19 Mbps in order to meet the 75 Mbps goal. High performance data capture front end systems are currently under development to perform bit and frame synchronization, Reed-Solomon decoding, and packet reassembly using a pipeline consisting of VLSI gate arrays controlled by MC68020 microprocessors. The initial CMOS implementation of the data capture front end is projected to perform at 30 Mbps. The CMOS version of the frame synchronizer card for this system is now under test at GSFC (Figure 8). Design and testing of other sub-systems, including the synchronizing controllers for the disk subsystem, will be sufficiently complete by the fall of 1988 to complete the magnetic disk sustained data rate tests. Transfer of key VLSI components to ECL and expansion of the controller concept being developed to accommodate 16 disk drives (fewer if higher data rate units become available) will allow data rates to approach the 300 Mbps required for a single TDRS link.

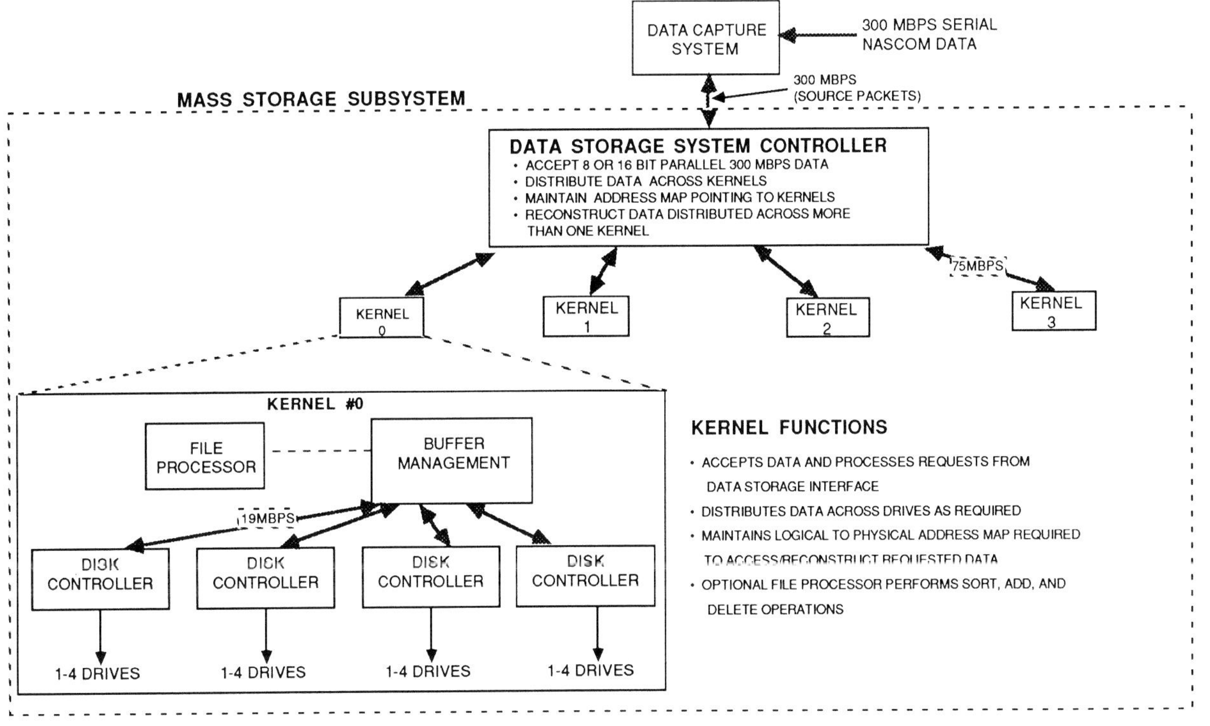

FIGURE 7- FAST ACCESS MASS STORAGE SUBSYSTEM CONCEPT

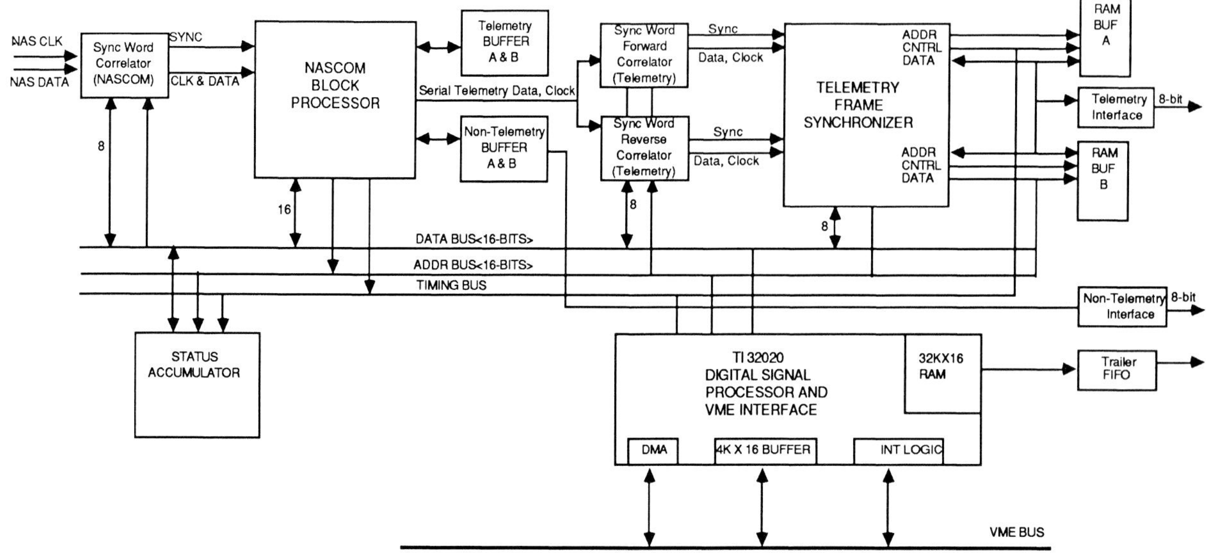

FIGURE 8 - FRAME SYNCHRONIZER FUNCTIONAL BLOCK DIAGRAM

Acknowledgments

The authors would like to especially acknowledge the following members of the DIGSTOR team who were most intensely involved in the analysis: Richard Carper, Systems Management Office; Edward Connell, Data Systems Technology Division; Dr. Stanley Sobieski, Networks Division; Michael Mahoney, Information Processing Division; Gene Smith, Information Processing Division (all of NASA/GSFC); John Martin, Computer Technology Associates; and Pierre Portmann, Mitre Corp. The independent consultant Leonard Laub, of Vision Three, Inc. provided valuable industry information and analysis.

Finally, we would like to thank Hugh Fosque, Richard Costa, and Romo Cortez of the NASA Office of Space Tracking and Data Systems for their program management of Data Storage Technology and VLSI systems development.

DATA CAPTURE AND PROCESSING

John Lyon, Gene Smith, Richard Carper

Mission Operations and Data Systems Directorate
Goddard Space Flight Center
National Aeronautics and Space Administration

Abstract

Increased data volumes and transmission rates, coupled with rising user expectations, create new challenges for data capture, preprocessing, and distribution systems supporting space missions of the 1990s. The systems concept described here has been developed in response to the specific requirements imposed by the Space Station and affiliated instrumentation. This concept has broader applications as an approach for all flight missions with similar requirements. The specific subsystems discussed are those associated with initial data capture, handling, routing, and distribution control for return link data via the Tracking and Data Relay Satellite System. These subsystems will support data handling under a virtual channelization and packet construct, are largely data-driven, and appear to be the most cost-effective solutions to the technical challenges offered. A detailed description of the functions performed is provided. Relevant experience with similar design concepts at the NASA Goddard Space Flight Center is outlined, further supporting the belief that the described system affords a practical and achievable solution. A set of remaining architectural options under investigation is provided.

I.   Introduction

NASA's Goddard Space Flight Center (GSFC) has traditionally provided data capture, processing, flight control, and communications services to Agency flight missions. Most of these services are now provided by the Mission Operations and Data Systems Directorate (MO&DSD), including the NASCOM communications networks, substantial components of the Tracking and Data Relay Satellite System (TDRSS), various spacecraft control centers, and data handling/preprocessing systems.

A key concern within the Agency has been the development of an appropriate institutional strategy that addresses the flight mission needs of the Space Station (SS) era. The SS itself introduces an exceptional set of challenges, not only in the design and assembly of the orbiting elements, but also in broader Agency operating practice. Within MO&DSD , an initiative has been established to solve the technical and operational problems introduced by the SS. In the process of developing such a solution, it has become apparent that an approach that could satisfy not only the SS, but all flight missions, appears to offer an opportunity to retool in several areas of flight mission support. This initiative includes command and control and various data processing and routing functions which typically have been handled on a mission-specific basis.

The conceived approach has been designated the Customer Data and Operations System (CDOS). It is intended to serve as an institutional adjunct to the TDRSS and NASCOM services provided by MO&DSD. The CDOS is thus planned to consist specifically of (1) a Platform Support Center (PSC), which will provide ground-based control of the SS platforms, (2) a Multi-Application Control Center (MACC), which will support control for flight mission investigators of their experiments and payload accommodation equipment, (3) a Data Interface Facility (DIF), which will serve as an interface for all data sent to or returned from spacecraft via the TDRSS, (4) a Data Handling Center (DHC), intended to provide clean and useable data to users, and (5) various other supporting systems. The CDOS will also draw upon other institutional services, upgraded appropriately for SS-era support (e.g., the Flight Dynamics Facility, which supports orbit and attitude computations for all missions).

The SS-era environment is characterized by a substantial (perhaps two orders of magnitude) increase in the proposed data volumes subject to near real-time handling and distribution, when compared with current practice and requirements. In addition, an analysis of existing, often labor-intensive procedures for the disposition of such data conveys the clear message that a simple increase in the scale of such operations would be prohibitively expensive, if not unworkable. User requirements, whether examined individually or in the aggregate, also appear to indicate an urgent need to facilitate, and probably to automate fully, a number of operations conducted in the past with human intervention. With these considerations and others, a substantial revision to the operational approach has been strongly indicated. The concepts described below have evolved in cooperation with other NASA organizations toward the definition of the overall Space Station Information System (SSIS).

The DHC and the DIF within CDOS constitute an approach to the specific technical and operational problems presented by the SS for a set of data handling and processing functions, primarily in the return link from spacecraft via TDRSS. These functions include: data capture, demultiplexing and routing, early preprocessing ("level zero"), and ancillary data handling.

The remaining sections of this paper outline specifically the functions allocated to the DHC and the DIF, the current views of the facilities themselves, a number of key considerations in the system organization, and areas of remaining work. In general, it is believed that the approach described affords an effective response to the requirements of the SS and other missions in the 1990s. Related work within the MO&DSD , specifically in support of the Gamma Ray Observatory (GRO), has already proceeded through the implementation phase on a smaller scale and has borne out substantially the implementation and operational advantages of the proposed approaches.

II. Functions

Data Capture

Upon receipt of a bit-synchronized serial data stream from the ground terminal, the first required function is data capture. In its pure form, this means the simple recording on some storage medium of every received bit. The purpose is to protect against loss of space-generated data due to any failures in the extensive processing, storage, or communications services beyond the point of capture. Data capture has been traditionally implemented in this pure form, utilizing high density tape recorders as the only economical way to meet the combined rate/volume requirements. This is labor intensive, however, and hence a very expensive approach to capturing and recovering data. With the enormous data rates and volumes in the SS era, it is not clear that this traditional approach, with its high cost, is justified. The very high reliability of current and future LSI and VLSI devices encourages the view that functions such as separating and compartmentalizing data (thus enabling fully automated storage, cataloging, and recovery procedures) and eliminating fill ("empty") data (reducing the amount of storage required) prior to capture may be feasible. These options are being actively explored. Whatever the method selected, the data will be protected against loss to a very high level of certainty. The period of retention of such captured data will be limited to that necessary to ascertain that the data has been safely stored within a following data transport or processing system. It is expected that this will fall in the range of three to twenty-four hours.

First-Level Demultiplexing

Virtual Channels - A New Multiplexing Construct. Traditionally, data from a single spacecraft has been received at the ground terminal and either sent to a single first-level destination (e.g., one NASA Center), or the entire bitstream was sent to more than one Center. When the bit rates are low (in the tens of kilobits per second), this is an effective and reasonably economical way of distributing data. In the SS era, however, four major changes will take place.

First, the spacecraft will be very large and will carry complements of instruments for which several different, geographically diverse organizations are responsible.

Second, the space-to-ground bit rates will become so high that it will no longer be economically acceptable to transport the entire downlink bitstream to multiple first-level destinations and then allow each destination to extract only its own data. Instead, it is highly desirable to develop a method that will allow separating (demultiplexing) and forwarding high rate (up to 300 Mbps) data to each first-level destination.

Third, as the sophistication of flight systems increases, there will be a need for different grades of service appropriate to the kind of data being transported. For example, large amounts of data will still be traditional telemetry data from monitor or survey types of instruments. This data can tolerate relatively high bit error rates and occasional small amounts of missing data. The proliferation of microprocessors in orbiting spacecraft, however, creates a concomitant need for suitable data transport services for program loads, memory patches and dumps, data base updates, etc. These strain the "user friendliness" and appropriateness of current practices, which consist of using procedures and techniques developed for simple spacecraft commanding. It is clearly desirable for both user friendliness and operational effectiveness that normal computer-to-computer services be provided between space and ground, allowing the usual quality, reliability, and procedural characteristics of ground-to-ground inter-computer networks to be made available to the user.

Fourth, in addition to different grades (qualities) of service, there will be an increasing variety of data types to be supported, such as traditional telemetry, compressed telemetry (having a low tolerance to errors), audio, video, and computer-to-computer messaging. It is desirable that each of these data types be formatted in a way that is appropriate to the data type and in an international or national standard for that data type. Audio and video in CCITT standard formats, telemetry in CCSDS packets, and computer-to-computer data in ISO protocol data units are typical formatting examples. A transport method is needed that allows these different formats to be kept separate from each other.

The technique selected to enable these functions is called virtual channelization.

Virtual channelization is a multiplexing structure that allows a single physical link to be divided into several apparent or virtual links. Each apparent link (channel) is then assigned three characteristics: grade of service, type of data, and first-level destination. Data from all sources which require the same grade of service, data type, and first-level destination are sent over a single specified Virtual Channel (VC). Multiplexing is performed at the Transfer-Frame level in order to: minimize overhead, by identifying reasonable quantities of data associated with each header; allow very high speed sorting, by steering reasonable quantities of data with each header interpretation process; and allow reasonably easy implementation of grades of service, by attaching forward error correction codes to some frames. In addition, because the current level of technology is not adequate to provide higher-level (e.g., packet) multiplexing at hundreds of Mbps, some VCs may be assigned to specific very high rate instruments. It should be noted, however, that this assignment of VCs will be an option to be decided between the flight data system and ground data system organizations, and will not be an instrument-selectable option.

Figure 1 shows a simple example of the format and generation of VCs. At the top left, a group of instruments that are the responsibility of a single center (e.g. GSFC) and require the same grade of service (e.g., Grade 3) generates CCSDS packets that are multiplexed onto a single VC, which is assigned the VC ID of 1. Below that, a single very high rate instrument (e.g., a SAR) is shown as having been provided with its own VC, identified as VC #2. It might also be Grade 3 but have a first-level destination of JPL. At the bottom is a proprietary or non-standard instrument with a data format totally unknown to the transport system. Only the grade of service (e.g., 3) and the first-level destination (e.g., JSC) would be known. The data itself would be relayed to its destination as a simple bitstream. Only extremely limited accounting, quality, and fault isolation services would be available for this type of data. The right side of the figure shows how the data would look after the three VCs (each with a different rate of generation) have been multiplexed together on to a single physical channel. The exact format specifications for VCs and Transfer

Frames are currently being determined within the international Consultative Committee for Space Data Systems (CCSDS).

<u>Virtual Channel Demultiplexing</u> When a composite data stream is received on the ground, it will be demodulated, bit synchronized, sent to a normal telemetry frame synchronizer, and then to a VC demultiplexer. The VC demultiplexer will separate the VCs frame by frame from the composite stream and route each VC frame to its first-level destination. This will be done through the use of tables within the demultiplexer which will contain the relationship between each VC ID and the first-level destination address for that ID. The tables will be managed and maintained through the Network Control Center (NCC) at the GSFC. Considerable attention has been given to the possibility of dynamically specifying a true network address for the destination through a "call setup" protocol within the VC structure itself. An examination of first-level destinations, however, shows that they are quite static, a few changing perhaps once per resupply flight (90 days apart) for the SS Base and even less frequently for the platforms. Under these conditions, the considerable complexity that would be added to the flight systems seems excessive for the questionable benefits to be derived. Hence, the decision has been made to control the VC ID-to-first-level destination relationships through table management rather than network protocol.

<u>First-Level Routing</u>

First-level destinations are either major data centers or network gateways. Figure 2 shows an imaginary but realistic set of such destinations. In this figure, telemetry, audio, video, and computer-to-computer data has been generated in the flight system, which constitutes the left half of the diagram. This data will be carried through the space/ground transfer system via VCs as described above, each VC containing data of the same type and grade of service and with the same first-level destination. The VC demultiplexer and router will separate the VCs and, through the table-driven process, will send each to its appropriate destination. Fill frames that maintain the physical space/ground frame rate are identified by a unique VC ID number and are discarded at this point.

The first-level destinations shown in the figure are similar to those now envisioned for the SS. They consist of Level Zero Processors (LZPs), spacecraft Control Centers which contain an LZP-like function (although most will elect to have their data come through an LZP, thus obtaining a significant set of standard data services), an audio processing system, a video processing system, and a Grade 1 Service Processor. The latter processor handles the acknowledgements and retransmission requests over the space/ground links for this grade of service, and separates the different protocol data units if more than one computer-to-computer network protocol is supported. A first-level destination could be co-located with the VC demultiplexer and router, or it could be processors, and the Grade 1 Service Processor and its associated Network Gateways, will be co-located. Directly-fed Control Centers will be remotely located. The number and location of the LZPs are under active study; the three LZPs shown are used to demonstrate possible variations in data types and distribution. The top LZP might be a single-instrument LZP (e.g., for the SAR) located at JPL and receiving VC frames of Grade 3 service, appropriate for SAR data. The LZP immediately below it might be one for an astronomy payload with error-sensitive data requiring higher data quality and thus using Grade 2 service (as does the Hubble Space Telescope). The final LZP represents a multi-instrument institutional facility. It would receive multiple VCs (one or more from each of several different spacecraft), some of Grade 2 and some of Grade 3 service. Another first-level destination is the VC demultiplexer and router itself. When an onboard tape recorder containing stored data is played back, the tape recorder output is treated as a bitstream and is routed through a specific VC designated a recorder playback channel. The data on the recorder is in the composite link format--that is, it contains frames that also contain multiplexed VCs. When this recorder playback VC is separated on the ground by the VC demultiplexer and router, its destination is an input to the demultiplexer. Thus the "nested" composite data stream is extracted from the incoming composite stream and is routed back into the demultiplexer (after frame synchronization) where it is treated as any other incoming composite link. This is basically the same concept as that used in the Spacelab data system.

Second-Level Demultiplexing and Routing

Second-level demultiplexing involves separating each end user's data from the VC in which it was carried. The demultiplexing methods and formats of the user data depend on the data type.

The large majority of data will be telemetry data contained in CCSDS source packets. Each source packet, when generated on board the spacecraft, contains an Application Process ID uniquely assigned to that user. These packets are separated and sorted by Application Process ID (user) at an LZP. As with VC IDs, a table within the LZP relates each ID to a destination address for the user. Again, these are managed rather than protocol-driven relationships. Each user's data packets are then either forwarded out through NASCOM-provided circuits in real time, or stored for later merging and processing into user-specified files for entry into the user's database.

The second-level demultiplexing and routing methods have not yet been determined for audio and video. It is intended, however, that techniques be developed that exploit, in general, standard CCITT signalling protocols, and that they be ISDN-compatible. This approach will require the minimum amount of new development and will allow the maximum flexibility, ease, and economy in interfacing to ground audio and video distribution systems. The long term goal is to have audio and video systems that will give access to space from the ground, or to the ground from space, identical to and interoperable with, standard ground-to-ground systems.

The Grade 1 service (computer-to-computer) will use one or more standard international protocols for second-level routing. At the present time, it is hoped that ISO protocols will be of sufficient maturity for use. If not, it may be necessary to use TCP/IP or another widely accepted standard. It seems quite possible that it will be advantageous, at least early in the program, to support more than one protocol, as noted in the discussion above. This would probably be accomplished by encapsulating each of the

Layer 3 protocol data units in a CCSDS source packet, with the source packet's Application Process ID denoting the network gateway appropriate for that protocol data unit. This is the procedure assumed in Figure 2. Since an automatic retransmission protocol will be used at the frame level over the space/ground link, the Grade 1 service is capable of supporting any of the standard network protocols. Again, the objective is to extend standard ground-to-ground network services and procedures through the space/ground links. The result will be to provide the user with the capability to access his instrument in space through a normal computer network, just as if the instrument were only across the campus.

Level Zero Processing

A terminology has been developed over the past ten years to characterize the nature of data produced by a sensor or instrument on board a spacecraft, and the differing stages or levels to which the data sets may be transformed to meet the research or applications requirements. The instrument observation, as passed to the spacecraft data handling system, is said to be "level zero data." No change to the nature of the data is produced through the space-to-ground links for the effects of noise and the addition of data format structures for multiplexing and error detection and correction.

Artifacts of NASA Data Transport Service
Level zero processing provides: on-the-ground recovery from lost, noisy and disorganized data, removal of frame and VC format structures, removal of disordering artifacts of the NASA data transport service, and elimination of replicated packets. Higher-level processing elevates the data to forms more removed from the initial observation, and constitutes the transformations, combinations, and samplings necessary for the analysis of the data or creation of a useful product. Figure 3 depicts the disordering effect of onboard storage on a user's data.

LZP Standard Services and Functions The first objective of level zero processing is to insure the safety of the data received. This is done by verifying successful reception and by protecting against system failure until the processed level zero data sets are passed to the level one processing node. If necessary, retransmissions are requested.

Next, the system is responsible for recreating the original instrument data form; that is, each instrument's data is demultiplexed from the composite data and formed into instrument data sets. Then the data sets are time ordered (intermixing real-time and playback data acquisitions), and duplicated data packets are eliminated. To provide this service:

• Packets are identified and separated by instrument (Application Process ID)

• Packets to be processed as real-time data are passed for immediate transmission to appropriate facilities

• Packets from data recorded on board and specified for special treatment are "priority playback" processed

• All data packets including those transmitted as real-time or priority playback packets are production-processed to reverse playback data, remove redundant packets, and chronologically order data sets

• Data sets are formatted for delivery to users

There are several utility and support services to complete the list of functions of the LZP.

• Error correction (if required) and error checking are done

• Limited real-time data management is performed

• Data acounting and data quality assessments are made

• Data catalog and support directory services are provided

• Short term data storage is provided to stage data for distribution, defer delivery, and hold copies of distributed data for verification of reception and protection against system outage at the user site

Ancillary Data Processing

Ancillary data support computing functions are being identified that would provide standard, user-pertinent core data parameters to a potentially large class of customers. Not all SS instruments are expected to commit to a system to access the flight Data Management Systems to extract relevant core engineering and subsystem data for inclusion in instrument data packets. Instead, ancillary data containing attitude and orbit state vectors, pressures, temperatures, water dumps, gas jet firings, and other environmental information will be grouped into ancillary data packets (possibly several formats) and transmitted to the ground.

Users, such as support centers and operations centers, would receive the data in real time or as priority play-backs and would perform no additional ancillary data processing. If the onboard complement of core parameters is not in units or coordinates convenient to many users, however, ground computing of expanded ancillary data sets, through application of well-defined algorithms, may be necessary to support common requirements. Exact requirements are being assembled for identification of practical trades. This function is not viewed as a real-time support activity, but it would operate on complete data sets assembled in the LZP production processing function and passed over to the ancillary processing function.

Ancillary Data Storage and Interactive Data Base

Related to the Ancillary Data Processing function, but serving a different requirement level, is the Ancillary Data Storage and Interactive Data Base function. This function would allow users who require platform or space base core engineering data on a sporadic or irregular basis to access an ancillary data base from their home sites rather than receiving standard ancillary data sets from the LZP or Ancillary Data Processing functions. In addition, these users could be provided with a temporary work space for shaping unique ancillary data sets using software support tools available in the facility.

The ancillary data would be stored on line for two years to meet the SS Program (SSP) requirements, and potentially stored off line and held for longer periods if a need were identified. These data sets would be received from the LZP production processing and ancillary data processing functions, and logged into the data base. All references would be updated, resulting in a short delay for these activities to be completed and checked. It is expected that data sets would be added to the data base within twenty-four hours of the ancillary parameter's generation on board.

III. Facilities

Data Interface Facility (DIF)

The DIF will provide an interface between the TDRSS and those multiple ground facilities which will communicate with SS-era spacecraft. The DIF will have the potential for providing automated data handling functions not possible in past NASA data systems. The following items, operating in concert, have made this possible:

• Adoption, in principle, of standardized data formats and protocols

• Use of space-based computers for data management and operations control functions

The environment in which the DIF will function will be different from that of the past and will be a result of evolution within NASA, the user community, and scientific and engineering technologies. The following projections represent significant changes considered in developing the DIF configuration:

• Commercial data transport services will soon outpace the NASA-unique system, both in types of service and available bandwidth. Users will expect levels of service from NASA equal to those available commercially.

• Space/ground data messages will become largely computer-to-computer. Future microcomputer-based instruments will appear almost revolutionary compared with instruments used today.

• Inter-operations with international partners at several levels will become a major driver of data formats and data handling procedures.

• The nature of space communications will change dramatically in the next decade with more emphasis being given to "operations in space." The relationships between humans, automatic systems, and data transmissions will continue to evolve, placing greater emphasis on system flexibility and automation. Issues of system transparency, as viewed by the external user (customer), will become increasingly important.

DIF Functions The DIF will function as the centralized capture facility for data returned from SS era spacecraft via the TDRSS and its follow-on. It will be the centralized multiplexing facility for data forwarded to those same spacecraft via the TDRSS.

For return link data, the DIF will receive data streams containing transfer frames via the TDRSS, remove fill frame data, separate the data by VCs, and route that data to several geographically diverse processing facilities, including LZPs, control centers, and network gateways. The DIF will buffer return data to protect against loss of data, to defer data delivery, or to change the rate of data when necessary to match ground communication capability or the ingest rate of certain LZPs. Processing of the return link data stream will provide frame synchronization, forward error correction decoding, sorting and routing of VCs, and issuance of necessary retransmission requests for Grade 1 service. Grade 1 service data will initially be a small percentage of the total data volume.

For forward link data, the DIF will accept data packets from various facilities and control centers, multiplex this data together, and forward it via the TDRSS to the supported spacecraft. To accomplish this, the DIF will provide storage for forward link data to protect against line and link outage, provide rate smoothing, defer data delivery, or retransmit Grade 1 data not received error-free and complete by the supported spacecraft. Processing of the forward link data will also provide access control to ensure that a user has the proper authorization to access the link and authentication of data messages. It will provide frame generation, forward error correction encoding, and routing to the appropriate TDRSS port.

The DIF will accommodate the different grades of data service defined by the SSP. Additionally, the DIF will provide data accounting and quality information; space/ground protocol conversion; routing control; and data administration messages, including fault analysis, status, data control message services, and summary reports for both forward link data and return link data.

Data Handling Center (DHC)

In the evolving concept for SS ground data handling, the DHC is a central facility for shaping platform and space base telemetry data sets -- both instrument and ancillary data -- for most NASA users. The following user classes are expected to have dedicated VCs, which may be separated at the DIF and sent directly to designated sites for processing and analysis:

• Commercial customers with proprietary data

• Experimenters with very high data rate instruments

• International partners

• The Space Station Support Center at JSC

All other VCs will have packets from multiple instruments multiplexed by the onboard Data Management Systems.

DHC Functions The major functions of the DHC are appropriate for a centralized, institutional facility:

• LZP with short term storage

• Ancillary Support Computing (ASC)

• Ancillary Data Storage & Interactive Data Base (ADS&IDB)

• Level 1 Processing (standard procedures — under consideration)

See Figure 4 for a concept and wiring diagram of the DHC functions and subelements.

No "value-added" processing is incurred in the LZP subsystem, and only well–defined, relatively invariant or paramaterized computations are performed in the ASC and level one processing subsystems. The ADS&IDB subsystem serves as a resource for remote users to interactively select ancillary parameters and, with the user's own or supplied tools, alter the data sets. Several functions have been included in the DHC to provide institutional support of standard storage and ancillary services which benefit from the projected data flow through the DHC.

IV. Topologies Under Consideration

A multitude of theoretically possible geographical configurations of the DIF and LZP functions exist. These options were first explored in the early phases of the Space Station Data System Analysis/Architecture Studies. They have been periodically restudied as the data system concepts have become more clearly defined and as the system architecture matured. The field has been narrowed to three significantly different options. These options are: a fully centralized, integrated DIF and LZP (DIF/LZP) located in or near the TDRSS ground terminals; a single centralized DIF in or near the TDRSS ground terminals, and distributed LZPs located at a few prime sites such as the NASA Centers; and a fully distributed system in which all the data from all spacecraft would be multiplexed together at the TDRSS ground terminals and broadcast to distributed DIF/LZPs located at a few prime sites such as the NASA Centers.

A decision among these possibilities must still be made. The decision will be based on an analysis of complex cost and management trades. As a high level example, the first (fully centralized) option is appealing in two major respects. First, it minimizes data link costs by reducing the bandwidth requirements of individual ground-to-ground links through immediately demultiplexing, at a geographically central location, to the individual user level. Second, a fully centralized facility is inherently more easily managed than a distributed one. However, the technical feasibility of performing the required LZP functions at the dramatic peak rates required (600 Mbps to 1Gbps), and the availability of the enormous amount of storage required in such a single facility are being questioned. The second option (centralized DIF and distributed LZPs), appears, at least superficially, to be a good compromise. It reduces the system-wide communication bandwidth requirements (compared with the third option) by demultiplexing to a "trunking" level (that is, to the VC structures) and delivering to each distributed LZP only the data needed by that LZP. Then each LZP services its own community of users. The third option (fully distributed), is appealing because it appears to minimize the amount of equipment and operations in the White Sands area and makes all data available at every distributed DIF/LZP.

It is easy to imagine the complicated questions that arise from a more detailed examination of the above options. In the first option, it could be questioned whether the processing and data storage requirements are reasonable for 1996, considering current cost/performance trends and storage technology projections. In the second option, the data rate requirement for one specific LZP (for example one for earth imaging instruments) might be so high that one could suggest upgrading it slightly to a single centralized LZP. In the third option, one could question the technical and economic feasibility of multiplexing all the space-to-ground data (a peak rate of 600 Mbps to 1 Gbps) into a single broadcast stream, as well as demultiplexing at each DIF/LZP to handle that peak rate.

Obviously, there are many more questions and none has an easy answer. The end-to-end network performance characteristics will, to a significant degree, depend on which option is selected. It also seems probable that the long term life cycle cost to NASA will be greatly affected by the decision. Studies are continuing in this area, and although an early decision would be beneficial for planning, it is clear that reaching the correct decision is more important. In the meantime, the option of a centralized DIF and distributed LZPs is being used as the strawman for planning.

<u>V. Role of Standards</u>

It is intended that a strong emphasis be placed on the use of appropriate standards throughout the SSIS. Because of the geographic separation of the numerous elements in the SSIS, standards for data formats and communications protocols will be mandatory. It is intended that standards be employed wherever available and applicable.

They will be utilized in the following order of priority:

- International standards
- National standards
- U.S. Government standards
- NASA standards
- De facto standards

Representative sources of standards include:

- International Organization for Standards (ISO)

- Consultative Committee for International Telegraph and Telephone (CCITT)

- Consultative Committee for Space Data Systems (CCSDS)

- Government Open Systems Interconnection Procurement specification (GOSIP) (proposed)

- Federal Information Processing Standards (FIPS)

- Federal Telecommunication Standards (FTS)

In the light of broad NASA and international participation, special attention is being directed to the work taking place within the CCSDS to support the application of the current CCSDS space/ground data transport standards to the very high rate, multiple-use environment of the SS era. Frame structures, virtual channelization of physical links, user data structures, telecommand structures, and error correction codes and procedures are being addressed. These emerging standards hold the potential for stabilizing the user interface both in space and on the ground, and for providing the user with a range of data quality services from which to select the services best suited to the needs of the particular payload. The standards are aimed at achieving a balance of functions between space and ground in order to achieve the lowest overall system cost, to promote commonality among systems, and to allow data transport cross-support among the various international partners and other national space agencies.

Plans include giving first consideration to the proposed Government Open Systems Interconnection Procurement (GOSIP) specification for Local Area Network (LAN) and ground-to-ground Wide Area Network (WAN) interfaces. It is possible that the GOSIP specification will not be mature enough to use, and clearly the specification will not be appropriate for certain interfaces (e.g., point-to-point very high rate simplex data circuits). The advantages of utilizing a "state of the art" government-wide procurement specification, however, are fully recognized and will be used if economically and technically feasible.

The application of data format and protocol standards is governed by the type of data rather than the physical facility through which data passes. Figure 2 shows data paths through the SSIS. Note that types of data and grades of service, rather than physical facilities that constitute the system, are emphasized in this figure.

The advantage of considering data paths through the system, and representing these paths by type of data and service, becomes evident in Figure 5 where the system is shown by "domains" of the data format and protocol standards system.

VI. Related Experience

GSFC has the responsibility for the telemetry ground data processing for two current spacecraft using packet telemetry. These spacecraft are the Hubble Space Telescope (HST) and the Gamma Ray Observatory (GRO). GSFC has designed and built two ground systems to meet this responsibility. The first is the HST Data Capture Facility (DCF), which has now completed its integration and test phase and is supporting spacecraft and data flow tests. The second is the Packet Processor System (Pacor). This system is in its final development phase and is scheduled to be operational(i.e., ready for specific tests) in September 1987. The HST DCF is a dedicated system for use with the HST only. The Pacor system is an institutional system to support the GRO and spacecraft that conform to the current CCSDS telemetry recommendation. Neither the HST nor the GRO conforms exactly to the CCSDS recommendation since both reached the design freeze point before the CCSDS recommendations were completed. Both do use packet telemetry, and the GRO format is very similar to that recommended by the CCSDS. The capability to support CCSDS-compatible formats is included in the

Pacor. GSFC also has several development activities related to CCSDS-formatted telemetry data. These include programs in processing extremely high rate telemetry links (tens to hundreds of megabits per second), and transfer frame and VC generators/multiplexers at these rates.

The GSFC has had very positive results using the packet telemetry approach. These results include stable interfaces during development and testing, high reuse of software between systems, less labor-intensive coordination between ground and flight systems developers, increased processing capability for a given computer size, greater user friendliness (where the user is the scientist/instrument developer), and greatly increased manageability. Our GRO experience has provided especially useful insights into the advantages of packet telemetry in real world systems. The following paragraphs summarize lessons learned from that experience.

The level of format standardization inherent in packet telemetry has allowed early, firm, and stable system definition. The self-descriptive character of packets (as to source and length) has made the normally complex and tedious task of tracking telemetry word slot assignments and changes to all but disappear. There have been three or four significant changes to the bandwidth allocations given to the GRO experiments to date, and the complete elimination of one early instrument. In marked contrast to Time Division Multiplexing (TDM) experience, there has been minimal impact on the various flight and ground data handling systems. This is an important point for vehicles with a changing complement of instruments, as in SpaceLab and the Space Station Program Elements (SSPEs). A significant amount of effort goes into making these assignments in TDM systems, their distribution, implementation in software, verification of the implementation, documentation, and then repetition of the process for each change to the assignments. In contrast to this, is the effect on the ground system for packet telemetry, which does not have to be changed each time a user bandwidth allocation is changed. Even when an entire new user is added, the only action required is to enter into a single table the name equivalent of the new source code (for accounting reports), the types of data the user is to receive (real-time, quick- look, production), and the network destination identifier for the user. The system is then ready for end-to-end test and verification. The self-identifying characteristic of the packets frees the ground system from any need to know about the multiplexing/polling system used in the spacecraft system. The spacecraft may use a rigidly cyclical method, demand access, or any instrument polling method, from simple to adaptive, or switch between these, with no need for any changes or updates to the ground system.

We can now safely project that software maintenance for the Level Zero Processing system will be significantly less than for previous systems. The incoming format is stable, and changes in instrument data rate or bandwidth assignments do not require changes in the software. Addition and deletion of instruments, as in instrument changeout, require only minor source-destination table updates. There will therefore be far fewer changes made, and software reliability will be increased. Again, this is an important consideration for systems such as those for SS, and it is even more important if distributed Level Zero processing is being considered.

In GRO, the Principal Investigators (PIs) are being presented, for the first time, with a constant data format throughout the life cycle of their instrument. Previously, a PI has had one format while developing his instrument (raw instrument output), a second format during test and integration (the composite telemetry bitstream or the spacecraft contractor's decommutation of his data on tape or print-out), a third format for real-time data (the composite telemetry stream or a unique Control Center decommutation of his data, perhaps with some housekeeping data), and finally a very different format for his production data (a sophisticated edited, merged, filled, annotated decommutation of his data, ancillary data, and perhaps some housekeeping data). Each of these different formats has required different supporting software and different operation and evaluation procedures. With packet telemetry, the GRO PI sees packets as his instrument output during development, test and integration, and for both real-time and production data. This constancy of format

across data sources and time is a large element in user's definition of "user friendly".

The key word for our experience to date with the GRO packet telemetry system, ground and flight, is manageability. The manageability of the data system, in all its phases and from end-to-end, has been greatly enhanced. This is true whether viewed from the perspective of the spacecraft systems and instrument managers, the ground system manager, the PI, or the operations manager.

It is difficult to overstate the importance of this for SS. Manageability of the SS data system, across all its Program Elements and in all its aspects - instrument development, integration and test, procedures development, certification and recertification, accomodation of an extremely wide spectrum of instrument types and characteristics, instrument changeout - presents an almost overwhelming problem. A careful implementation of packet telemetry as the standard technique for the Space Station Program will greatly improve our ability to achieve our objectives.

References:

1) SSIS Architecture Design Document - 10/86

2) Customer Requirements for Standard Services from the SSIS - 5/86

3) Customer Data & Operations System Concept Definition Document - 3/87

4) Space Station Program Definition and Requirements Document - 10/85

5) The Enhanced Telemetry Throughput Technology Study - 12/85

6) The Space Station Ancillary Data Concept, Requirements, & Operations Analysis Document - 12/86

7) Mature Data Transport and Command Management Services for the Space Station (AIAA-86-1181-CP) - 6/86

8) Mass Storage Systems for the Early Space Station Era - 4/87

9) SSDS Analysis/Architecture Study (MDAC, DR-05) - 12/85

10) SSDS Analysis/Architecture Study (TRW, DR-05) - 12/85

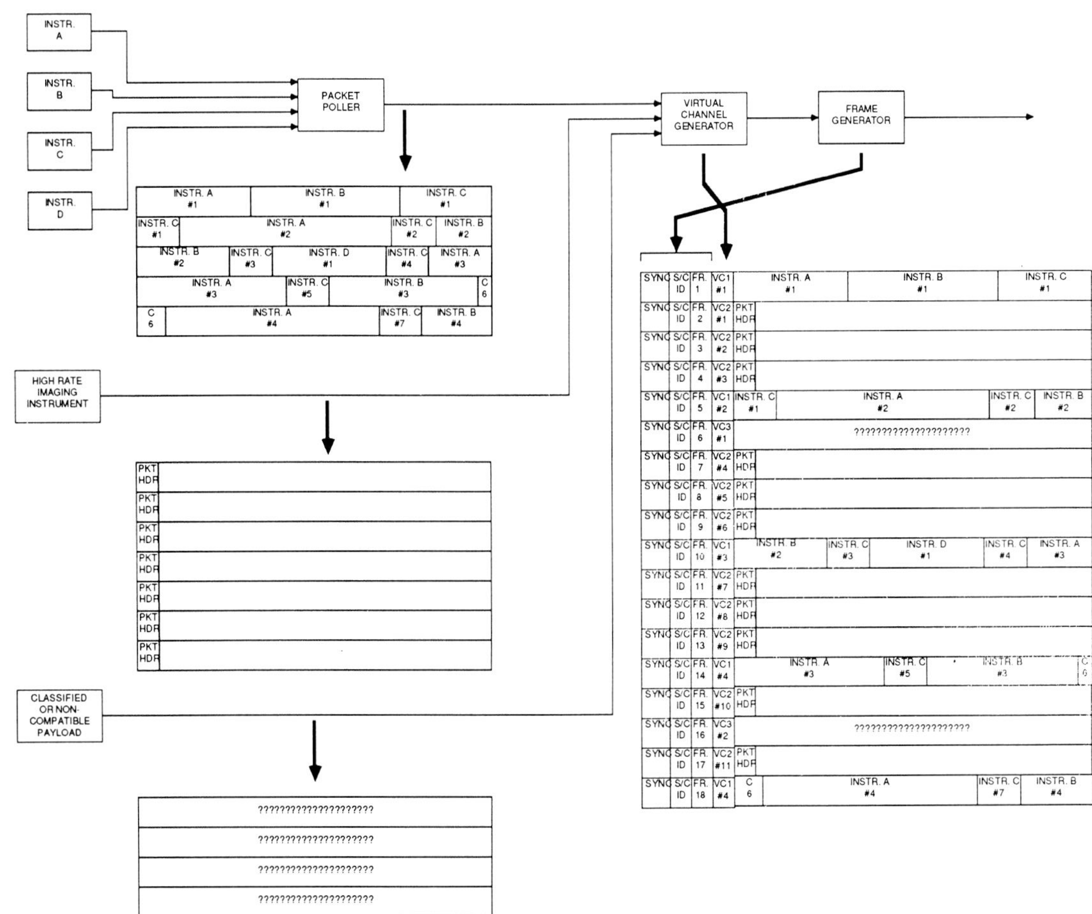

Figure 1:   Generation of Virtual Channels

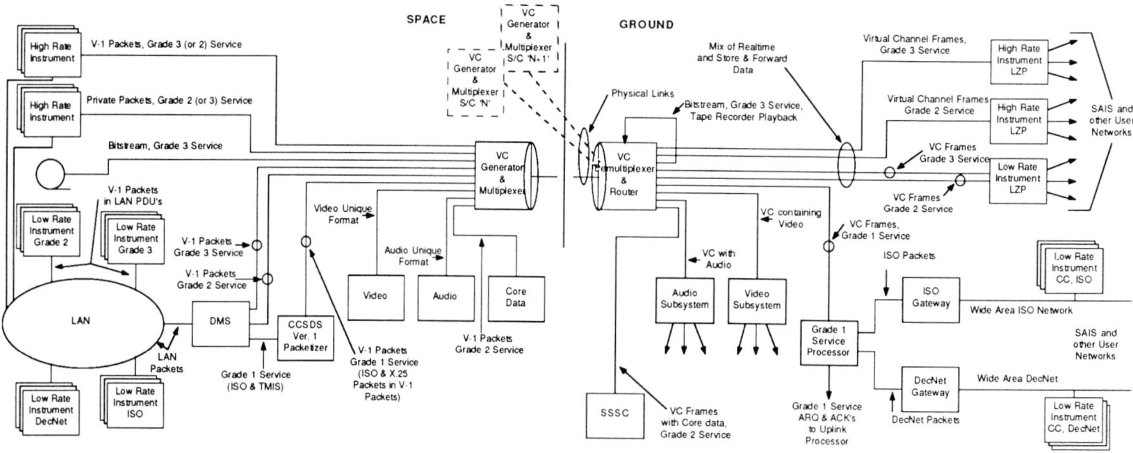

Figure 2:   SSIS Data Paths

USE OF ON-BOARD STORAGE IS REQUIRED BECAUSE
OF THE TDRS ZONE OF EXCLUSION, FOR BETTER
UTILIZATION OF SINGLE ACCESS TDRS CHANNELS,
AND FOR OPERATIONAL EFFICIENCY.

USE OF ON-BOARD STORAGE CAUSES THE STORED
DATA TO BE DELIVERED OUT OF SEQUENCE WITH
REAL-TIME DATA, SOMETIMES OUT OF SEQUENCE
WITH OTHER STORED DATA, AND, USING MOST
CURRENT ON-BOARD STORAGE DEVICES, BACK-
WARDS..

THUS, DATA WHICH CAME OUT OF THE CUSTOMERS
INSTRUMENT LIKE THIS:

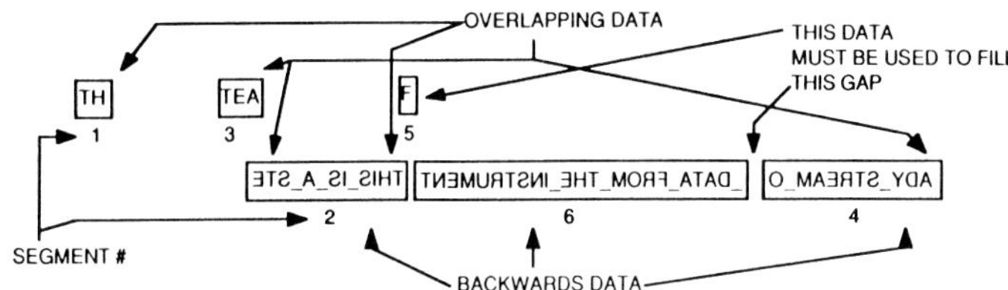

Figure 3:   Artifacts Introduced by the Data Transport System

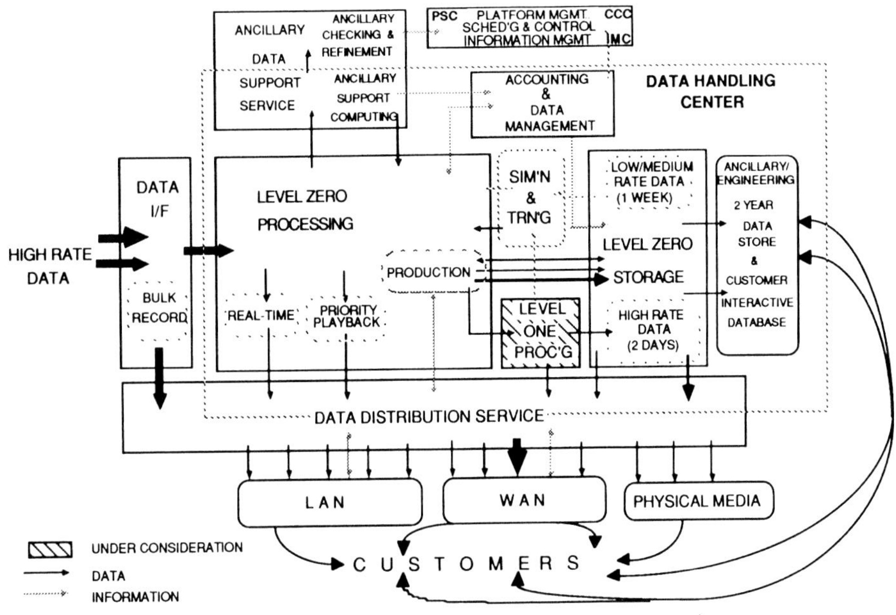

Figure 4:   DHC Concept and Wiring Diagram

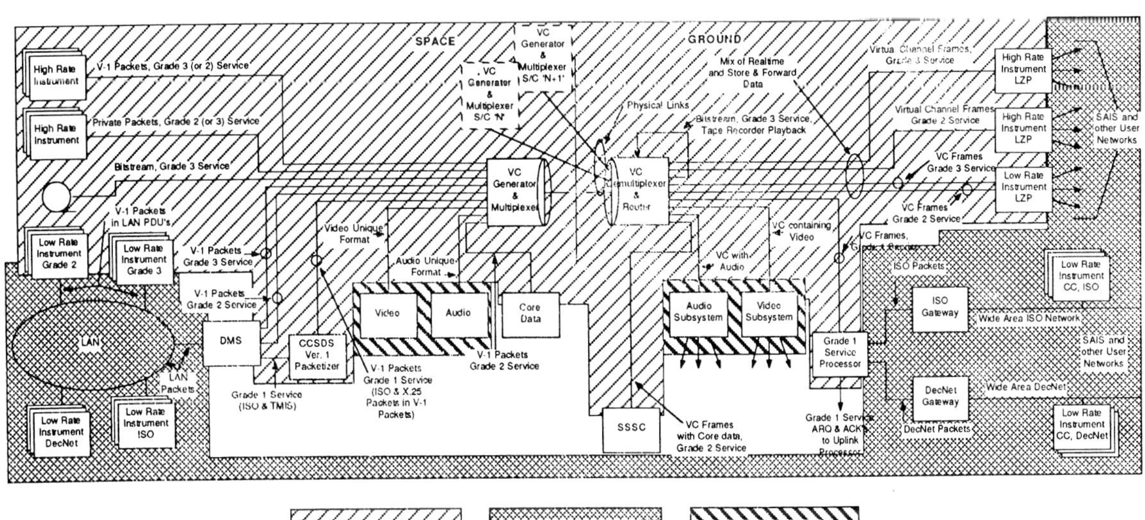

Figure 5:   SSIS Standards Domains

111

Management 1: Standards and Protocols
Co-Chairmen: Dale L. Fahnestock and Hubertus Wanke

THE CONSULTATIVE COMMITTEE FOR
SPACE DATA SYSTEMS STANDARDS PROGRAM

S. Richard Costa
Director, Communications and Data Systems Division
National Aeronautics and Space Administration Headquarters
Code TS
600 Independence Avenue, SW
Washington, D.C. 20546

Abstract

The developmental and operational complexity of the data system required to support space-based science experimentation is increasing dramatically. One approach to reducing the costs associated with this element of space exploration is to take maximum advantage of data system standards. This includes utilizing commercial standards whenever appropriate and developing standards in critical areas where the performance and/or environmental needs dictate special solutions. The latter area has been the subject of intensive activity by the Consultative Committee on Space Data Systems (CCSDS), a volunteer organization comprising most of the world's space agencies. The CCSDS has developed recommendations for space data systems standards. These recommendations provide guidelines for participating agencies to use in the development of their own internal standards and form the foundation for an international space program. The process of developing recommendations involves considerable deliberation and analysis by technical experts to reach local consensus on technical content, followed by agency-wide review and resolution of technical issues. Important recommendations which have been developed include packet telemetry and telecommand concepts, coding designs, ancillary data parameters and formats, and data exchange conventions. Current work of the CCSDS encompasses analysis of applicability of the existing recommendations to the Space Station Program.

Introduction

The use of standards can have significant benefits in the design and operation of data systems. For example, the design effort required to solve a given problem can be reduced substantially by applying proven standards where appropriate. Furthermore, the use of standards provides confidence that the selected solution is correct because similar problems have been solved using identical approaches. The cost to implement the design can be reduced dramatically by using proven implementations of the areas selected for standardization, and once implemented, the solution contains a foundation for future growth. Finally, system operations and maintenance can benefit from reduced need for specialized parts and training.

For large, complex systems the use of standards at key interfaces is particularly desirable because such an approach will allow elements within the standard interfaces to be replaced, modified, or changed as necessary without perturbing the remainder of the system as long as the new elements continue to meet the interface specification. This allows for graceful evolution of system capabilities as well as changeout for repair or to accommodate new technologies.

The value of standards is reflected by the many organizations actively developing them. In the communications and computer systems areas the work is particularly intense, with large numbers of national and international bodies working to identify and agree upon standards that will satisfy the needs of buyers of equipment and services. For data systems that support space missions, many of these standards are applicable and used. For certain aspects of space data systems, however, the requirements are driven by unique performance needs or environmental constraints and the use of industry and/or commercial standards is not appropriate. This area is where substantial work in the development and recommendation of standards has been accomplished by the Consultative Committee for Space Data Systems (CCSDS).

Consultative Committee for Space Data Systems

The National Aeronautics and Space Administration sponsored an international workshop in 1982 to address space data systems issues. A key outcome of that workshop was the recognition that the world's space agencies shared a common interest in developing techniques for enabling mission support interoperability and sharing mission products. The CCSDS was established to address this interest. The official objectives of the organization, as documented in its procedures manual[1], are:

- To provide a forum wherein interested agencies may exchange technical information relative to their internal development or application of space mission data system standards

- To identify those common elements of space data systems which if implemented in a standardized way will result in significant enhancements in the operation of future cooperative space missions or in the sharing of mission products

115

- To develop through consensus the appropriate technical recommendations which guide the development of compatible agency standards so that interoperability is maximized

Membership in the CCSDS is open to any agency willing to participate in CCSDS activities and provide an appropriate level of support. These participants are referred to as CCSDS Member Agencies. Other agencies that are interested in the activities of the CCSDS but cannot dedicate the necessary resources to qualify as Member Agencies are recognized as Observer Agencies. Membership in the CCSDS is, of course, voluntary and the success of the organization depends upon the good will and support of its participants.

Most of the world's space agencies are Members of the CCSDS (see Table 1). Many standards and technical agencies are among the active Observers, listed in Table 2.

Table 1
CCSDS Member Agencies

Center National d'Etudes Spatiales (CNES)/
   France
Deutsche Forschungs-u. Versuchsanstalt
   Fuer Luft und Raumfahrt e.V (DFVLR)/
   West Germany
European Space Agency (ESA)/Europe
Indian Space Research Organization
   (ISRO)/India
Instituto de Pesquisas Espaciais
   (INPE)/Brazil
National Aeronautics and Space Administra-
   tion (NASA)/United States of America
National Space Development Agency Japan
   (NASDA)/Japan

Table 2
CCSDS Observer Agencies

Bureau d'Orientation de la Normalisation
   en Informatique (BNI)/France
Centro Tecnico Aerospacial (CTA)/Brazil
Chinese Academy of Space Technology/
   Peoples Republic of China
Department of Communications/Communica-
   tions Research Center (DOC/CRC)/Canada
Institute for Space Astronautics and
   Science (ISAS)/Japan
National Institute for Telecommunications
   Research (NITR)/South Africa
National Oceanic and Atmospheric
   Administration (NOAA)/United States of
   America
National Research Council of Canada/Canada
   Centre for Space Science (NRCC/CCSS)/
   Canada
Nederlands Institute Voor
   Vliegtuigentwikkeling en Ruimtevaart
   (NIVR)/Netherlands
Norges-Teknisk-Naturvitenskapelige
   Forskningsrad (NTNF)/Norway
Radio Research Laboratory (RRL)/Japan
British National Space Centre/United
   Kingdom
Swedish Space Corporation (SSC)/Sweden

Operating Procedures of the CCSDS

The normal operating procedure for the CCSDS is to agree as a body on the need for guidelines for standards in selected areas, then to appoint panels of experts to perform the necessary technical analysis to arrive at consensus. The results of CCSDS activities are documents called Recommendations for Space Data System Standards; these documents are guidelines for the development of standards rather than standards per se. When a Member Agency agrees to a Recommendation, that agency is expected to use the Recommendation as a guideline in the development of its own internal standards. As stated earlier, the CCSDS operates on the principle of consensus; all Member Agencies must agree upon the contents of a document before it is approved as a Recommendation.

Figure 1[2] illustrates the process of developing a Recommendation within the CCSDS. As shown, the concept is based on an increasing exposure of ideas to broader communities of experts and users as the concept matures. Definable products are generated at several important stages.

The process is initiated by the development of a concept paper by Member and/or Observer Agencies. This paper is reviewed by the full Committee and usually results in the refinement of ideas to incorporate the views of other agencies. If the consensus of the Committee is that the concept has technical merit, is applicable to an area of concern to the CCSDS, and has widespread interest, a panel is formed and assigned the task of developing a White Book on the subject.

When completed, the White Book represents a common technical understanding among the contributing experts and provides a conceptual baseline for the development of specific technical details. The White Book is developed through studies, analyses, and deliberations and may go through many iterations before appearing in its final form. Although the panel may choose to acquire assistance from outside experts in critical areas, the White Book is generally an internal document. When panel members have reached agreement on its contents, the White Book is presented to the CCSDS for consideration. After review by the representatives of the Member and Observer Agencies, the White Book becomes a Red Book and enters the next stage of the development process.

A CCSDS Red Book represents the technical consensus of the members of the panel responsible for its development. By implication, the panel members agree that the material is mature enough to expose to technical experts within the Member and Observer Agencies. Red Books are presented to various technical experts for review and comment and, as with the White Book, there may be several iterations before there is agreement on the contents.

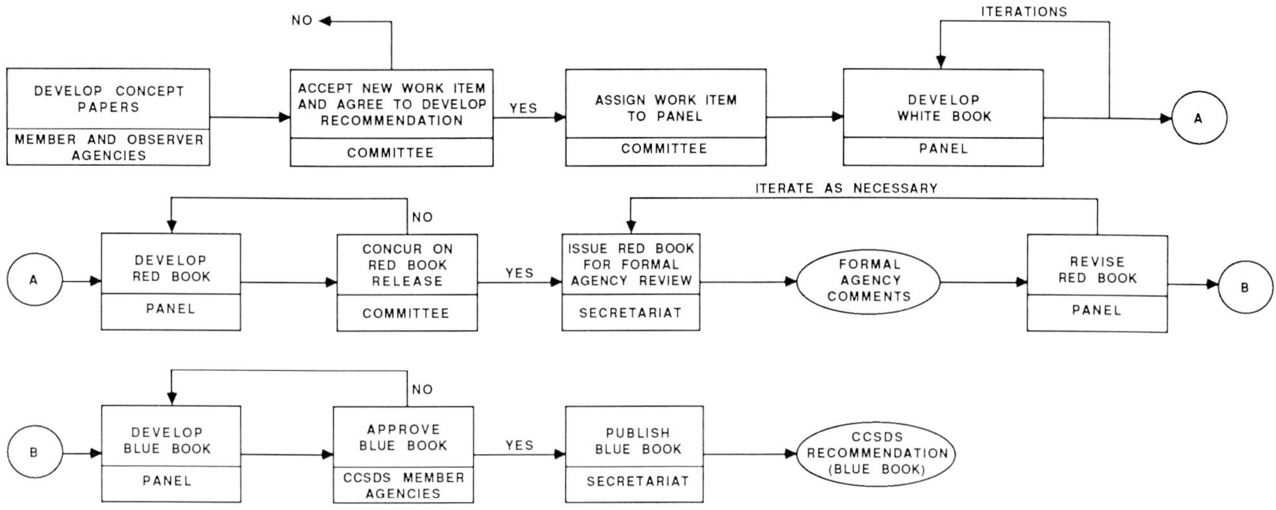

Fig. 1 Development process for CCSDS recommendations

When all issues have been resolved and all comments addressed, the final version of the Red Book is sent to Member Agencies for formal review. Since the formal review process is expensive and time-consuming for many of the agencies, extraordinary effort goes into ensuring that all technical issues have been resolved before this process begins.

When all Member Agencies have completed their formal reviews, the CCSDS meets as a body to vote on the Red Book. This usually involves consultation with the original panel of experts because there are inevitably small changes requested by Member Agencies as a result of their internal analyses. The final consensus among the agencies is documented as a Recommendation in the Blue Book. By its concurrence each agency implies that its future internal standards will be consistent with the Recommendation.

Blue Books are controlled by the CCSDS. Change procedures allow for technology evolution, future growth, or updates based on lessons learned during implementation of the Recommendations. Panels established to develop specific Recommendations usually continue (often as smaller bodies) for maintenance purposes.

A similar but less formal process is followed for the development of CCSDS Reports called Green Books. These books document existing systems, describe operations concepts, or illustrate applications of the Recommendations. These documents are initiated in the same manner as a Recommendation, but final approval for publication lies with the panel charged with the development of the Report. Since Recommendations tend to be terse, the Green Book(s) corresponding to a particular Recommendation are often invaluable in understanding the intended applicability of the guidelines.

A Model for Analysis of Standards Applications

To attempt to develop a model of a system as complex as the NASA data system for mission support would be a difficult task. However, for purposes of describing data flows and for analyzing potential interfaces where standards may have great benefit, the data systems reference model shown in Figures 2a, 2b, and 2c[3] has proven to be more than adequate. This model is shown in three parts with an increasing exposure of the data system's internal elements. At the most detailed level, Figure 2c is drawn in a manner to show "peer" processes at the same horizontal level so that while data flow through the system from one end to the other, corresponding relationships are directly "across" from one another.

The model can be interpreted in the following manner. An experimenter is interested in examining a remote domain of interest. The function of the data system is to allow this examination to occur in an efficient, cost-effective manner that is essentially transparent to the user (Figure 2a). A connection is established by providing a means of sensing and effecting the domain of interest, controlling the effector and analyzing the data from the sensor, and transmitting data between these two functions (Figure 2b). In the particular case of scientific exploration of space the sensing and effecting functions are more commonly known as data acquisition and instrument control. Figure 2c presents the internal processes in more detail and shows where guidelines for standards have been developed or are underway in the CCSDS.

As indicated in Figure 2c, data are typically distributed over a local spacecraft bus. For complex missions there may be a certain amount of local data processing

Fig. 2a   High level reference model

Fig. 2b   Reference model showing major functions

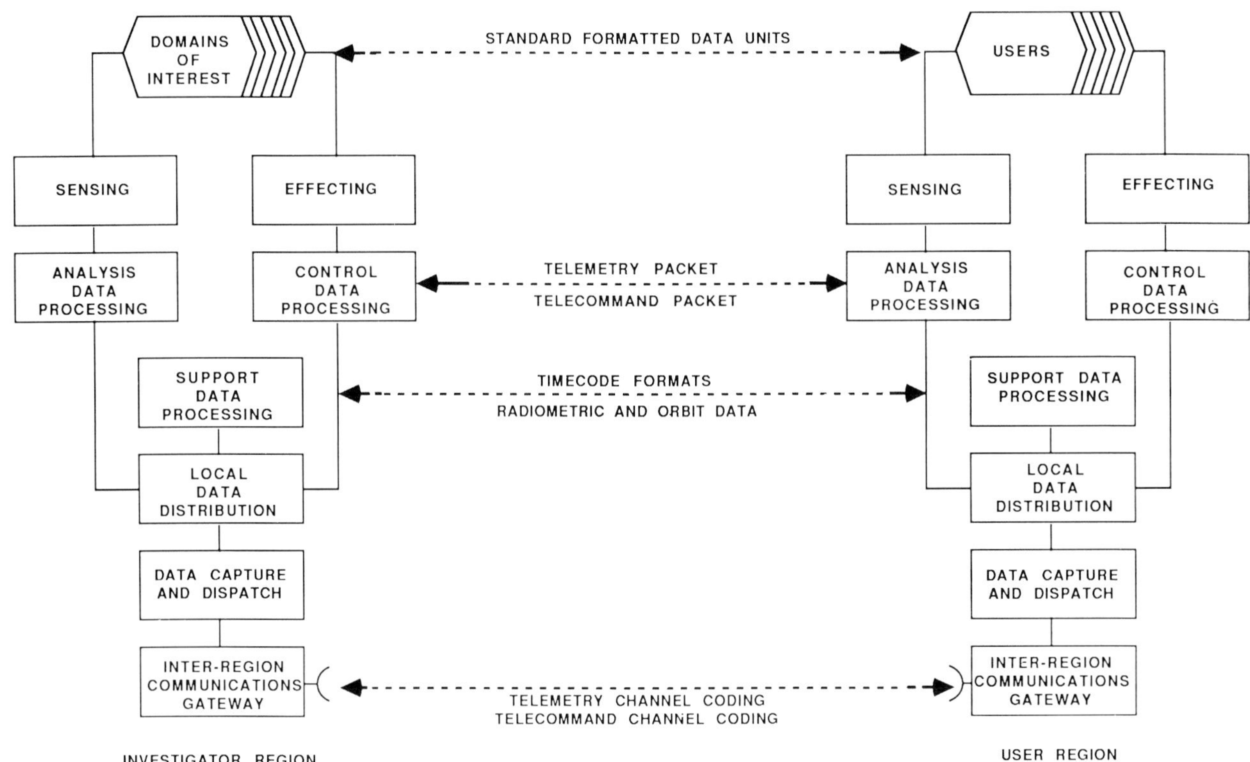

Fig. 2c   Detailed data systems reference model

and, in some cases, local processing of support data (e.g., orbit, attitude). Data are "captured" or recorded for holding until a transmission channel becomes available, at which time it is sent to a communications system which transmits the data to a receiving site. This portion of the system is labelled the "investigatory region" in Figure 2c because it is the portion of the system that historically has been located in space. The "user region" has been on Earth; this portion of the system has functions that mirror the spaceborne part. A communications system receives the transmitted data which is recorded and transmitted to the user over some type of network or distribution system. Many of

the "user region" functions contain the inverse of the "investigatory region" elements to compensate for artifacts of the data acquisition process. For example, onboard tape recorders are not rewound before transmission of data to the ground; therefore, the data are received in reverse time order and must be reordered before the data are processed. There is a two-way data flow through the system, instrument data flow to the experimenter, and command and control data flow from the experimenter to the onboard system. As indicated, the model was developed to reflect traditional space/ground relationships (e.g., instrument in space, experimenter on ground). However, the model is equally appropriate in the Space Station environment where the experimenter might reside in space. In this case the communications channel would not be a space/ground link but would exist nonetheless.

By representing the operation of space-borne instruments in this fashion, important interfaces that might benefit through the use of standards are exposed. As Figure 2c shows, a number of such interfaces have been identified and work is underway (or completed) to develop Recommendations for standards at these interfaces. The specific areas of interest are telemetry and telecommand coding, telemetry and telecommand packetization, time code formats, radiometric and orbit data, and standard formatted data units.

Telemetry[4] and Telecommand[5] Coding

Error performance is an important consideration in any communications system. This is particularly true in spacecraft command functions and when operating a downlink system that is designed to realize maximum efficiency by automatic data processing. The CCSDS has developed recommendations that break telemetry and telecommand packets into fixed-length objects that are protected from channel-induced errors by highly effective coding schemes. These coding schemes are designed to provide virtually error-free performance; this is crucial when automated data handling concepts are implemented because error recovery procedures often require external (manual) intervention. Furthermore, a set of command operation procedures has been defined that guarantees closed loop commanding capability over a broad range of mission complexities and environmental conditions. Together, the coding schemes contained in the referenced Recommendations provide the foundation that allows the implementation of advanced data handling concepts (such as packet telemetry) in the hostile environment of space.

Packet Telemetry[6] and Packet Telecommand[7,8]

The principle of the telemetry/telecommand packet concept is to provide data system users—instrument designers and science experimenters, for example—with a flexible data structure for packaging downlink instrument telemetry and uplink commands that would remain fixed over the instrument life cycle. This would mean that a principal investigator would have the same data interface when building an instrument in his laboratory as when integrating it for launch or operating it in space. The CCSDS Recommendations provide for efficient, application-independent structures that are conducive to automated operations while meeting a broad range of user and system requirements. Both telemetry and telecommand packets (which can be variable in length) utilize an underlying fixed-length frame structure that accommodates high performance coding schemes and uses the space/ground link efficiently. Thus, the overall approach provides an end-to-end data structure that satisfies the experimenter's need for a consistent data interface while maintaining careful utilization of scarce communications resources.

Time Code Formats[9]

When processing data from a space-based instrument, information other than the telemetry data is often required. For a number of reasons the most critical ancillary data is time. Time is often the only link between an instrument measurement and some other important data, such as the platform pointing vector [e.g., at time T measurement M was taken, and at time T the platform pointing vector was (x, y, z); therefore measurement M was taken when the platform was pointed in direction (x, y, z)]. Furthermore, the execution of complex instrument activities is accomplished frequently by generating or executing commands at precise, predetermined times. Analysis of instrument data is often performed based on a sampled sensor time series. Finally, spacecraft operations are generally scheduled and executed based on time.

There are many time code formats in use in the world today. The objective of the CCSDS activities in this area is to select or develop a small set of time code formats useful in space mission operations. To that end, two broad categories of formats have been developed: (1) unsegmented time codes, which are pure binary counts of time units and fractional time units from a starting time called an "epoch" and (2) segmented time codes, wherein the count of time units and fractional time units is accumulated in two or more cascaded counters which count moduli of various bases and start from the epoch. The formats each include an identifier with additional important information such as the epoch. The CCSDS Recommendation identifies three unique codes with several allowed variations. This set meets the needs of the space community, is small enough to be manageable, and can be auto matically interpreted in real time.

Radiometric and Orbit Data[10]

Agencies which operate space missions must devote considerable resources to orbit computation (e.g., extrapolation, reconstitution, etc.). The goal of the CCSDS work in radiometric and orbit data is to come up with standard parameters for representing this data when exchanging it with other Agencies. The Recommendation in this area includes radiometric data, spacecraft orbital elements, and solar system ephemerides along with tracking station locations, astrometric data, reference systems, astrodynamic data, and spacecraft dynamics parameters. The document includes coordinate definitions, parameters and their units of measurement, and reference system identifications.

Representing these types of data in a standard manner can provide significant benefits in preflight planning for mission support, for interagency comparison of estimated trajectories, and for tracking predictions to support radiometric data interchanges.

Standard Formatted Data Unit[11]

The purpose of the standard formatted data unit is to implement a standard data structure for interchanging data in a uniform, automated fashion. The CCSDS work to date has been focused on developing the structure and construction rules. A fundamental structural element based on Type-Length-Value encoding has emerged. The operations concept for systems supporting, standard formatted data units follows. At the time an experimenter (for example) defines his data format, he registers that format in a machine-interpretable form with a control authority. The data description is given a unique identifier. All data in standard formatted data unit form contain this identifier so that a receiver of the data can determine where to look for information on how to interpret the data. Furthermore, this information will be stored in machine-interpretable form so that the experimenter can obtain the required data definition information, load it into his computer, and initiate automated parsing and linking with applications programs. Although substantial work needs to be done before such a scenario will be generally practical, the benefits of this concept are expected to provide significant payoff by enabling efficient, automated data processing.

Applications of CCSDS Recommendations

The CCSDS Recommendations were developed with relatively small, low data rate missions in mind. Recently the Space Station has emerged as the focus of a major international program and two subpanels have been formed within the CCSDS to address the application of the Recommendations to missions such as this. A key portion of the work was completed by the ad hoc Space Station Advisory Group. This team analyzed the Recommendations compared to the Space Station requirements and found that the general concepts and approaches of the CCSDS Recommendations provided excellent solutions to the difficult data system challenges that the Space Station faces.[12] The new subpanel efforts are targeted at examining the Recommendations to ensure that specific requirements, such as addressability, efficiency, etc., are properly addressed and that the concepts are adequate to handle unique Space Station needs for multiple-instrument, high data rate configurations.

Status

The CCSDS has recently published a number of Recommendations, as noted in the References section of this paper. This group of Recommendations along with an earlier set provides a foundation for future data system development. The status of various CCSDS documents is shown in Table 3. The efforts of the two new subpanels will focus the Recommendations on the specific needs of the Space Station. The panel addressing standard formatted data units is attacking the difficult problem of automated data exchange. The latter panel has released a Red Book on structure and construction rules for Agency review.

Conclusion

A substantial effort has gone into articulating a concept for efficient space mission operations and Recommendations to support that concept have been developed. The resulting body of work represents many person-hours of labor and international participation. This effort can provide the cornerstone of a cooperative effort to continue the exploration of space in the most cost-effective manner possible.

References

1. "Procedures Manual for the Consultative Committee on Space Data Systems", Issue 1, August 1985.

2. Ibid.

3. desJardins, R., "Program Chairman's Address", Conference Proceedings of the Consultative Committee on Space Data Systems, October 2-4, 1982.

4. CCSDS Recommendation for Space Data System Standards: "Telemetry Channel Coding", Blue Book, January 1987.

5. CCSDS Recommendation for Space Data System Standards: "Telecommand Part- 1: Channel Service, Architectural Specification", Blue Book, January 1987.

6. CCSDS Recommendation for Space Data System Standards: "Packet Telemetry", Blue Book, January 1987.

7. CCSDS Recommendation for Space Data System Standards: "Telecommand Part-2: Data Routing Service, Architectural Specification", Blue Book, January 1987.

8. CCSDS Recommendation for Space Data System Standards: "Telecommand Part-3: Data Management Service, Architectural Specification", Blue Book, January 1987.

9. CCSDS Recommendation for Space Data System Standards: "Time Code Formats", Blue Book, January 1987.

Table 3. Status of CCSDS documentation

DOCUMENT TITLE	DOCUMENT STATUS	TECHNICAL APPLICABILITY	PROGRAMMATIC EXAMPLES
• PACKET TELEMETRY	COMPLETED	DEEP-SPACE & NEAR-EARTH TELEMETRY LINKS	GRO, MGCO, SPACE STATION
• TELEMETRY CHANNEL CODING	COMPLETED	POWER-LIMITED TLM LINKS	GALILEO, EURECA, MGCO
• RF & MODULATION: PART 1, EARTH STATIONS & SPACECRAFT	COMPLETED	TRACKING, TELEMETRY, & COMMAND	ALL MISSIONS SUPP'TD BY GND STATIONS
• RADIOMETRIC & ORBIT DATA	COMPLETED	ORBITS & TRAJECTORIES	ALL MISSIONS
• TELECOMMAND: PART 1, CHANNEL SERVICE	COMPLETED	LOW RATE COMMANDING	GOES, ST, GRO, MGCO
• TELECOMMAND: PART 2, DATA ROUTING SERVICE	COMPLETED	LOW RATE COMMANDING	GOES, ST, GRO, MGCO
• TELECOMMAND: PART 3, DATA MANAGEMENT SERVICE	COMPLETED	UPLINKING	ALL MISSIONS
• TIME CODE FORMATS	COMPLETED	SPACECRAFT, NETWORKS, MISSION OPS	ALL MISSIONS
• STANDARD FORMATTED DATA UNITS--STRUCTURE & CONSTRUCTION RULES	AGENCY REVIEW	DATA CONSTRUCTION RULES	ISTP, MGCO, SPACE STATION
• CONTROL AUTHORITY PROCEDURES	INITIAL DISCUSSIONS	GLOBAL FORMAT REGISTRATION	ALL MISSIONS
• DATA DEFINITION LANGUAGES	INITIAL DISCUSSIONS	AUTOMATED DATA INTERPRETATION	SPACE STATION
• DATA DICTIONARIES	INITIAL DISCUSSIONS	DATA FORMAT BROWSING	ALL MISSIONS
• DATA CATALOGS	INITIAL DISCUSSIONS	DATA BASE BROWSING	SPACE STATION
• SPACE STATION: APPLICATION OF CCSDS RECOMMENDATIONS	COMPLETED	MULTI-INSTRUMENT SPACE PLATFORMS	SPACE STATION
• SPACE STATION: MODIFICATIONS TO CCSDS RECOMMENDATIONS	INITIAL DISCUSSIONS	MULTI-INSTRUMENT SPACE PLATFORMS	SPACE STATION

10. CCSDS Recommendation for Space Data System Standards: "Radiometric and Orbit Data", Blue Book, January 1987.

11. CCSDS Recommendation for Space Data System Standards: "Standard Formatted Data Units--Structure and Construction Rules", Red Book Issue-2, February 1987.

12. CCSDS Report Concerning Space Data System Standards: "Space Station: Application of CCSDS Recommendations for Space Data System Standards to the Space Station Information System (SSIS) Architecture", Green Book Issue-1, October 1985.

DATA MANAGEMENT STANDARDS FOR SPACE INFORMATION SYSTEMS

Richard desJardins
Computer Technology Associates, Inc.
7927 Jones Branch Drive, Suite 600W
McLean, Virginia 22102 USA

and

Carlo Mazza
European Space Operations Center
Robert-Bosch-Strasse, 5
D-6100 Darmstadt, West Germany

Abstract

Data management -- that is, storing, describing and retrieving data -- is a special problem for the high performance bit-efficient information systems required for space missions. This paper presents a summary description of data management for space information systems, and describes four specific problem areas that can benefit from data management standards in the Space Station era: data description, data capture, data interchange, and data interpretation. In each area, a recommended modern data management standard or related technique is described as an example recommendation for future space information systems. The paper concludes with a recommendation that space agencies develop testbed validations of these "new" approaches to data management.

1. Introduction:
The Data Management Problem

"Data management" means **managing data:**

o **storing data** as it arrives in a system, including identifying the data, capturing it reliably, accounting for its handling, and cataloging it properly so it can be later retrieved;

o **describing the data,** including translating the data to alternative formats that are also described;

o **retrieving the data,** either on a periodic or demand basis, including controlling access to the data, interpreting the data according to its description, and interchanging the data (transferring it to or from another system).

For space data, data management presents several special problems including three that are generally system design drivers:

o **high volumes and high rates of data,** exacerbating the storage and retrieval problems;

o **emphasis on bit efficiency,** that is, on encoding the data in the fewest number of bits, leading to low interpretability of the raw data;

o **need for data interchange** across projects and agencies, due to the universal nature of much space data.

These data management problems are now being faced in the planning for the Space Station era. The Space Station Base being built by NASA, the Columbus laboratory module being built by ESA, and the Japanese Experiment Module, all contribute to and must help solve the Space Station data management problems.

In this paper we present an overview of data management for space information systems in the Space Station era, and we examine how data management standards may be applied in some specific areas of data management.

2. Overview of Data Management

2.1 Program Data Management Policy

In the Space Station era, data management for space information systems will (we would hope) begin and end at the agency and program policy level. As a part of each new program request preparation, the expected data return from the program would be considered. The data sets to be produced by the program would be identified and analyzed for scientific interest,

---

Richard desJardins is a senior systems engineer currently with Computer Technology Associates (CTA), formerly with NASA Goddard Space Flight Center, and is Chairman of the ISO Subcommittee on Open Systems Interconnection (OSI).
Carlo Mazza is Head of the Data Processing Division at European Space Operations Center (ESOC), and is Chairman of the ESA Board for Software Standardization and Control (BSSC).

anticipated coverage, and compatibility of formats and contents as compared with related data.

From the program preliminary data analysis, a **program data management policy** would be developed that describes these data sets, identifies the general program approach to their production and the expected production delay before they are made available to the scientific community, and proposes the nominal lengths of time that the data sets will be held online in working sets or offline in archives before being retired.

Such clearly expressed policies will replace the sometimes informal procedures used today. When authorizing a new program, agencies will essentially be offering to contract with project scientists to provide mission data sets of assured scientific interest and known quality, accompanied by adequate descriptive and accounting documentation.

2.2 Data Description

Following the data management policy formulation, project scientists and engineers must prepare the detailed **data definitions** of the data sets to be produced. Typically either a project data czar or a project data team will be assigned this responsibility.

Data definitions are prepared prior to the operational phase to serve as the basis of all data management operations and to be associated with the data throughout the entire data management life cycle.

Data definitions consist primarily of detailed descriptions of data items and data sets, sometimes accompanied by descriptions of data processing operations that will be performed on the data, identification of ancillary or correlative data required for processing, and descriptions of formats to be used for presenting or interchanging the data.

Data definitions for a project are frequently organized into a **project data dictionary** that serves as the data management configuration control mechanism throughout the project data management life cycle.

2.3 Data Operations

The operational phase of data management begins with **data generation** at the data source. This includes primary data generation in the scientific instrument, data formatting as either a data stream or as source packets, and onboard data processing and storage as required prior to transmission to the ground.

Data staging and transmission includes the communications between onboard and ground, including any data accounting performed to enhance operational reliability. At the ground, the data are prepared and staged as "level zero" data sets and subsequently transferred to users on either a scheduled or a demand basis.

Data staging and transmission will be much more complex in the Space Station era than heretofore due to the large number of participants and the wide variety of data systems and networks that will be involved.

Data preparation and processing is the last part of data operations. This includes data ingest, which covers both data accounting and data capture on a reliable medium; preprocessing, that is, artifact removal and calibration, the output of which is a "level one" data set; and processing, that is, interpretation and manipulation of the calibrated data, the output of which is a "level two" data set. A level two data sets contains the physical measurements of interest to scientists, as well as derived data products such as summaries, graphs and maps.

Major data management problems in the operational phases of past missions have been the lack of rigorous data accounting and the proliferation of poorly controlled data formats, that is, difficulty in assuring that all the data have been captured and difficulty in interpreting the data following capture.

2.4 Data Archiving

Data archiving includes both storage and retrieval aspects. **Data storage** involves creating the data sets in appropriate storage formats on permanent storage media. **Data retrieval** involves finding the data on demand, including the data and format definitions, and supporting the interchange of the retrieved data to users via physical or electronic media.

The data archiving phase lasts until the end of the data management life cycle. When the life cycle ends, the status of the data becomes "undefined" or at least "unmanaged". At this point, the program declines further responsibility for being able to retrieve or interpret the data for users.

2.5 Using the Data

Data use is data management as seen from the user perspective. It includes **finding** the required data, including identifying, locating and retrieving the primary data sets, ancillary data sets, correlative data sets, data item definitions, data set definitions, and the transfer syntaxes (formats) for data exchange.

It may be necessary for a user or user agent to **subset** or **reformat** the data to obtain a useful working data set.

The working data set is **interpreted** into scientifically meaningful forms and elements, **integrated** with ancillary and correlative data, and **analyzed** by batch, interactive or manual scientific data analysis programs and procedures.

The results of analysis are **presented** to a user on a display device or hardcopy media, or sent to a remote user via magnetic media or electronic data interchange protocols.

3. Data Description

A **data description** is a statement in some language (which could be some "natural language" such as English or could be some "formal language" such as Ada) that provides declarative information about some data.

Because a data description is "data about data", it is frequently referred to as **metadata.** In this paper, we prefer to use the term "data description" because it is more restrictive (that is, "metadata" includes any kind of information about the data, not just descriptive information).

For space data especially, because of the "bit-efficiency" driver, data received does not usually contain any information about its format or its representation. Also, the standards for information interchange that do exist (such as ASCII) tend to be bit-inefficient and therefore tend not to be adopted by projects. Instead, project software engineers write **interface control documents (ICDs)** defining all the data formats that may cross each of the interfaces of the system.

The ICD approach worked well when the interfaces were sufficiently simple and few, in particular, when there was a single interface: "project to users". However even in cases where there were a small number of interfaces, there frequently developed a proliferation of "slightly different" data formats and structures because of the bit-efficiency driver.

Because the bit-efficiency driver is not going to go away in the Space Station era (no matter what you may hear from enthusiastic data communications people, who see only one aspect of the bit-efficiency driver), the tendency will continue for raw space data to be "dense" and to lack full information about its format and representation.

3.1 Standard Formatted Data Units

To alleviate these bit-efficiency problems, space agencies have gotten together under the aegis of the Consultative Committee for Space Data Systems (CCSDS) Panel 2 on Standard Data Interchange Structures. This Panel has developed the concept of packaging space data in a **Standard Formatted Data Unit**

(SFDU) that provides a mechanism for associating data with its description.

An SFDU is a unit of data, represented as an octet-string of a definite length and a well-defined encoded format, that is transferred from an originator to a recipient. Each SFDU so transferred consists of a series of **labels** and a **data content.** In this respect an SFDU is very similar to an **interpersonal message (IPM)** as defined in Recommendation X.420 of the International Telegraph and Telephone Consultative Committee (CCITT); an IPM consists of a **heading** and a **body,** and is transferred from an originator to a recipient represented as an octet-string of a definite length and a well-defined encoded format.

The differences between an SFDU and an IPM are precisely that the SFDU labels and data content are not the same and are not encoded the same as the heading and body of an IPM. However, since both SFDUs and IPMs are represented as octet-strings, they could be interchanged between "user agents" (that is, software acting on behalf of ultimate users) by the same underlying data transfer protocols. In particular, it would be possible to transfer SFDUs in standard CCITT IPM "envelopes"; the SFDU encoding format would be considered as a "Pc" protocol in the CCITT X.400 Message Handling System Reference Model. This aspect is considered further in Section 5.

The essential characteristic of an SFDU is that its labels provide for the identification of the data description of its data content. The data description is contained in a **data descriptive record (DDR)**, which is a series of statements written in a formal language called a **data descriptive language (DDL).** The DDR provides enough information about the data to enable the recipient user agent software to "parse", that is, interpret, the SFDU data content.

The DDR is intended to be available from a **data dictionary** maintained by a **control authority** responsible for the configuration control of the data description information.

Conceptually, when an SFDU is received, the recipient user agent looks at the labels (the first 12 octets) and reads the Authority and DDR Identifier (ADI). The information contained in the ADI is then sufficient to enable the recipient to request that specific DDR from that specific control authority. When the DDR is received from the control authority, a data interpretation program can use the data description contained in the DDR to interpret the SFDU data content.

Conceptually, the entire process described above could be performed automatically at run time. In practice, especially in the early stages of implementation of the SFDU concept within agencies, the DDRs will be accessed and compiled in advance so that any particular data interpretation program will be able to interpret only those SFDU formats which have been prearranged.

CCSDS Panel 2 is now examining the question of what DDLs may be used in early implementations of the SFDU concept. Both traditional programming languages, such as FORTRAN, Pascal, and COBOL, as well as newer languages oriented toward abstract data types, such as Ada, General Data Interchange Language (GDIL), and Abstract Syntax Notation One (ASN.1), are being considered. Panel 2 is also considering whether a new Data Interchange Language (DIL) is required to be developed.

4. Data Capture

Data capture in space information systems is defined as the assurance that data intended for transmission from one place to another has arrived at the intended destination and has been reliably stored and catalogued there. Data capture is a building block for any phase involving handover of data from one system element to another, for example, from the onboard data management system to the ground data staging and level zero processing facilities.

In the past, data capture for many missions has involved manual steps such as data labeling and time tagging at an operator terminal or tape labeling station. In the Space Station era, given the high volume and complexity of the data, the entire data capture process will have to be automated.

The data capture process involves essentially two steps: storing the data and cataloguing the data. Storing data at high speed is commonly done on high-density multi-track magnetic instrumentation tape or on newer optical media. The focus of automated data capture is not so much on the physical residence of the data but instead on the accounting, cataloguing and configuration control of the data and storage volumes. ·

The "nerve center" of automated data capture would be the data capture management system. This would be organized around a modern database management system (DBMS) that could quickly create the data management records of accounts, physical locations and volumes, references to related data, status of processing and handling, distribution lists, and so on, for all incoming data. These data management records would then serve as the management information base to direct the data capture operation, from data ingest to ultimate disposal of the data to various destinations.

The data capture data management system could be built on a flexible DBMS such as one based on the international standard relational data language SQL. If it is necessary to provide sophisticated access to the data, for example, spatial, entity-relationship, or inheritance models as used in artificial intelligence, more powerful data modeling constructs and facilities would have to be added to the basic system.

Elements of a modern data capture data management system are implemented in the example shown in Figure 1. The ESOC Spacecraft Control and Operations System (SCOS) is a configurable system for capturing and manipulating real-time data for spacecraft control. SCOS is driven by a relational database that contains the description of the telemetry parameters, commands, display proformas, network set-ups, and system tables according to which the SCOS run-time system is configured.

The SCOS database is implemented on a VAX under the ORACLE relational DBMS. SCOS system programs on the VAX generate internal tables from the database. These tables are used directly by the SCOS control and operations application programs to capture and manage the incoming data in real time.

The information in the database is also exported to the Spacecraft Performance Evaluation System (SPES), which is an IBM-compatible system running its own version of ORACLE. SPES also uses run-time internal tables generated from the database for fast response time in interactive data extraction and interpretation.

The method used in SCOS and SPES -- that is, maintaining the data management information in a modern DBMS while the high-rate data is actually managed using internal tables generated from the database -- is the key to using the power and flexibility of DBMSs to help manage large volumes of high-rate data in real time.

5. Data Interchange

Data interchange is the transfer of data and information about data from one user (the "originator") to another user (the "recipient"). The data and information are transferred using some communication networks and protocols that assure reliable transfer of the bit stream (actually an octet string in modern 32-bit systems) that represents the data. Therefore just as for data capture the physical residence of the data is not the main issue, but instead the main concern is with the information about the data and its physical residence; just so in data interchange the data communication networks and protocols are not the main issue, but instead the concern is with the information about the data and the communication networks and protocols that are being used.

Two methods of data interchange have prominence in discussions for the Space

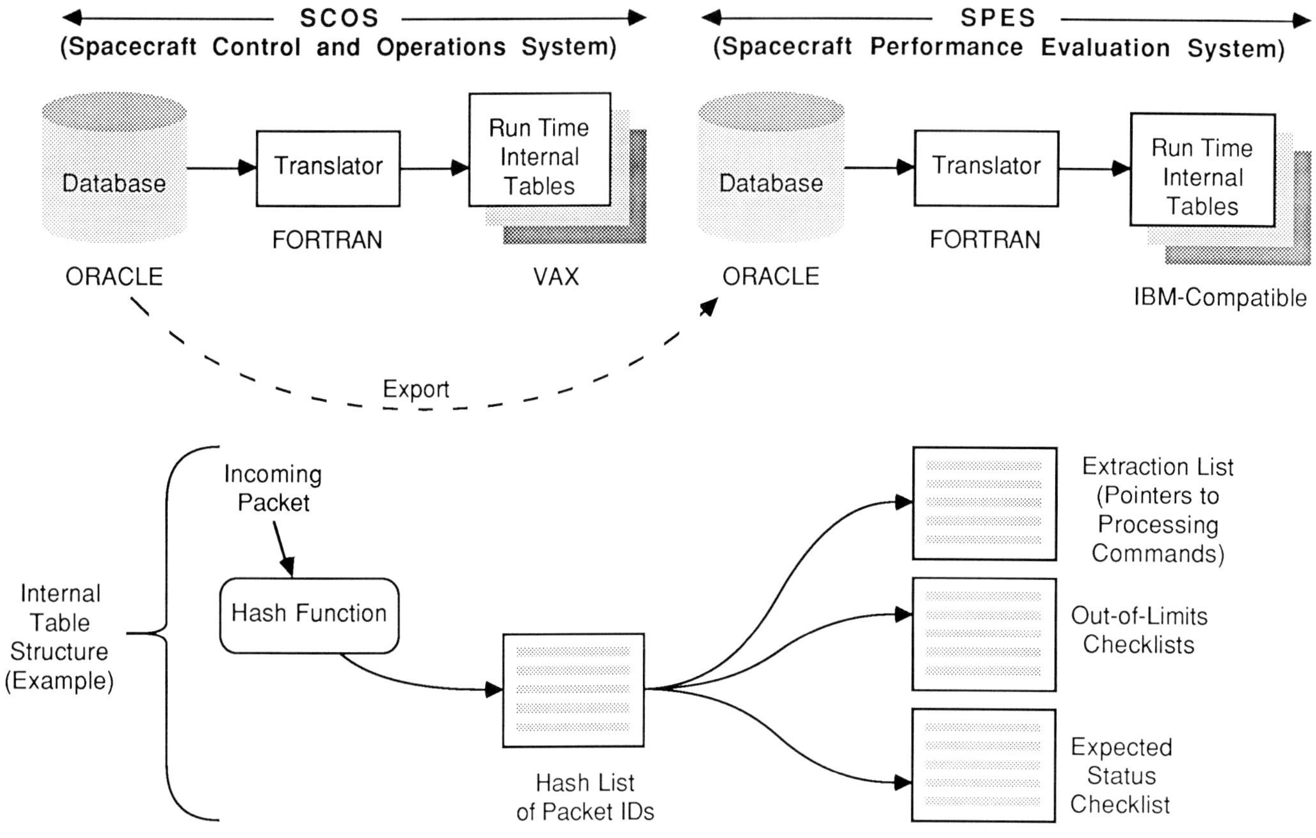

Figure 1. SCOS/SPES System Architecture

Station era: these may be called "SFDU" and "OSI".

SFDUs (see Section 3) are standardized data packages that contain labels allowing for an identification of the data content. The identification is registered with a "control authority" responsible for the data description. The data description may be obtained from the control authority on request and used to interpret the data.

OSI -- Open Systems Interconnection, a protocol standardization program jointly administered by the International Organization for Standardization (ISO) and the CCITT -- refers to the use of the ASN.1 (Abstract Syntax Notation One) language to define the structure and optionally the encoding of a data content. The ASN.1 language is being used for the specification and optional encoding of all OSI application protocol data units.

SFDUs and OSI are not incompatible. SFDUs by themselves are a means of identifying bit-efficient data representations ("objects"). SFDUs can either stand alone or can be embedded in an ISO data structure. For example, SFDUs can be carried in an X.400 message envelope, or can make up data units in an ISO virtual file, or can be used as a building block (or as the entirety) of an application protocol data unit, for example, for communication between an instrument and its remote ground controller. OSI provides all

necessary means -- through data typing, presentation context, or the object identifier primitive -- to distinguish an SFDU from any other object represented as an octet string.

6. Data Interpretation

Currently, **data interpretation** is left to each individual researcher. Typically, each researcher either writes his or her own data interpretation program from scratch, using the data description as a specification, or obtains a data interpretation program from another researcher or from the owner of the data.

In the Space Station era, it is anticipated that this method will be too inefficient to be widely useful. Instead, we imagine two methods of providing data interpretation tools to assist the researcher.

One method will be by use of the data descriptive record (DDR). If the DDR is expressed in a standard data descriptive language (DDL), it is conceptually feasible to write a general purpose interpreter that could use the DDR as a template to display the data in a human oriented or database oriented format.

Another approach would be by requiring or at least encouraging data description agencies to produce a data interpretation program written in a standard programming language such as Ada. Such a data interpretation program could be used as part of a general purpose toolkit including analysis and display programs to display the data or to populate an interpretive database.

The DRUID (Data Recognition Using Interpretable Definitions) experimental system developed by ESOC is an example of a general purpose data interpretation tool. As depicted in Figure 2, on the host machine, data specified in a DDR is accessed from a database and used to generate an SFDU corresponding to the DDR. The DDR and the data instances requested are then passed in the form of SFDUs to a user machine where the inverse process is performed: the DDR is used to extract the data from the data instances SFDU for interpretation.

Note that in DRUID the SFDU is just used as a data identification and packaging technique. Therefore the importance of the method demonstrated in the DRUID system is independent of the SFDU, but instead, exemplifies the general concept of use of a DDR to interpret a received octet string.

In DRUID, the DDR is expressed in the same programming language in which the user machine software is written. This approach simplifies the design of the demonstration system, and can be generalized to a heterogeneous environment using some of the newer techniques that are available, such as the use of Ada to manipulate octet string representations of objects, or the use of the OSI Presentation Layer to identify and transform object representations between heterogeneous machines.

7. Conclusion:
A New Approach to Data Management: Need for Validation Programs

In this paper we have described some fairly straightforward modern techniques for achieving much more automated data management in the Space Station era than has been achieved routinely heretofore. These methods depend strongly on the widespread application of sophisticated new standards such as Ada, SFDUs, ASN.1, and the OSI Presentation Service and Protocol.

Since there is today no "standard" experience base in "standard" applications of these sophisticated standards, it is imperative that before being widely used in the expensive and far-sighted Space Station era programs, the technical and application aspects of these standards should be widely understood. This calls for experimentation, demonstration, and validation of these techniques in open, scientifically replicatable, quantitative settings.

This would imply that the major agencies should be developing validation programs based on the testbedding environments that are now being developed for the Space Station program in the various agency centers. These validation programs should focus on the data description, data capture, data interchange, and data interpretation building blocks as described in this paper, leaving to each agency to apply the validated building blocks to its own space information systems data management architectures.

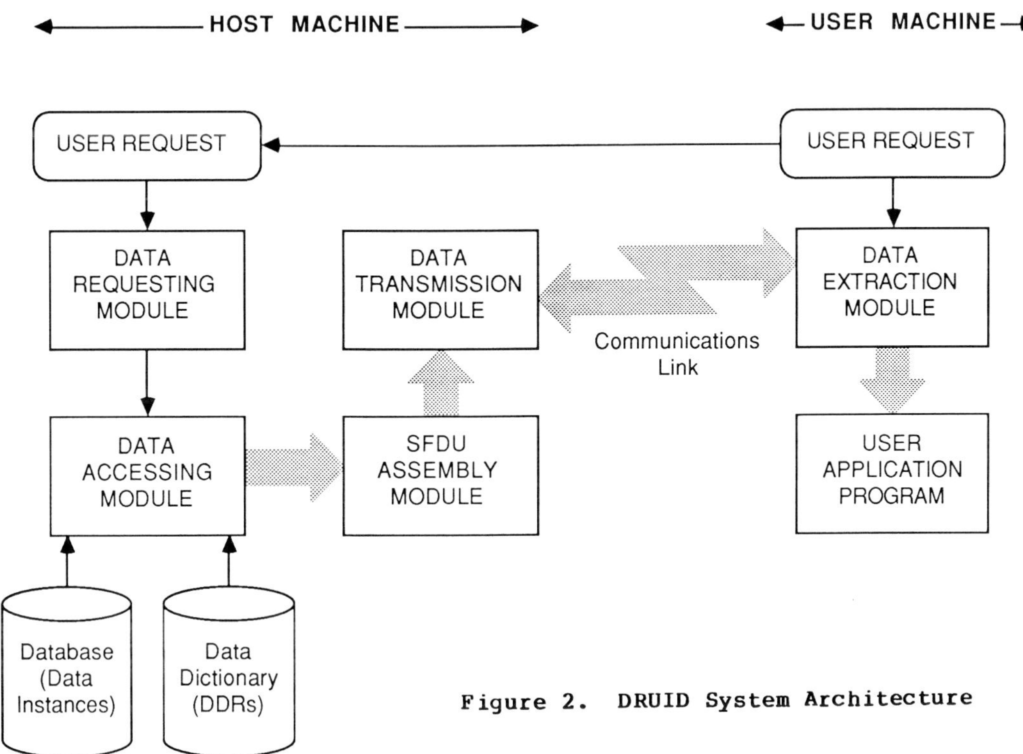

Figure 2. DRUID System Architecture

GROUND DATA NETWORK STANDARDS AND PROTOCOLS

A. Frederick Fath
Manager of Architecture and Development
Boeing Computer Services Company

Abstract

This paper addresses the impacts of new communication technology on both standards and protocols used on ground networks for distribution and use of data and information collected or generated as part of the space program. The basic theme is that protocols and network architectures should be different depending on the type and use of the data being transmitted. No universal standard will exist which will satisfy all data networking requirements. Standards should be established within communities of interest at the higher layers and network architectures chosen that support the intended use of the data. Finally, system architectures will need to be established that will take advantage of the advances in communication technology and support the various standards needed within a community.

Classification of Data Transmission Types

Before considering network architectures, it is important to look at the characteristics of the data intended for transmission. One of the most important characteristics is how the data changes in time with respect to content and value. For example, data including product specifications, catalogues, reports, test results, application programs, etc. can exist for long periods of time without significant change. Surely additions can be made, but the basic content is the same. Other data, however, can change rapidly with time. As examples consider data representing the status of any scheduled or contended resource such as airplane seats or telephone trunks. The point of the above discussion is that data content can change with time over a wide range and that systems designed to access the data can and should take advantage of this characteristic. The second characteristic of data variability is that of value. Naturally the time value of the data is a function of the intended use. A weather map needed for forecasting realizes an hourly decline in value as forecasts generally span hours to days. The same map used for instructional or analysis purposes can retain its value for indefinite periods. Likewise, data used to make a decision loses value after the decision is made but the same data used to justify a decision retains value as long as justification is needed. Finally, data that is needed for a step in some process that cannot proceed until the data is obtained has a time value related to the cost or value of the process.

Realizing the time variability of data in both content and value, it is important to consider two different network objectives for the transmission of data. The first network objective is that of minimizing network delay in the delivery of a unit of data from a source to a destination. The type of data transmission that has this objective is termed "Response Time Critical" or RTC data transmission. The reason delay is important is that for this type of data transmission, it is assumed that people or processes become non-productive while the data is in transit. Thus by minimizing the network delay, the end user productivity is increased. The second network objective is that of minimizing cost or the time needed to complete a transmission. If network transmission costs are usage sensitive with respect to time, then minimizing the time to complete a transmission is the same as minimizing the cost. Thus maximizing the utilization of a resource through high efficiency is the network objective for a class of data traffic termed "Throughput Critical". For this class of traffic, costs can also be impacted by using resources during low rate periods or alternate means of transmission such as mailing a laser disk with data that is not rapidly changing or decreasing in value with time.

Present Networking Approach

Most data networks in existence today share an architectural trait that they rely on a fixed network topology. This means that the networks are composed of network processors (that switch and route data packets) and fixed trunks that carry the transmissions between the processors. Local area networks are an exception as they can share the transmission media among many processors. The major reason for this architecture is that transmission costs were the predominate cost of the network and because of analog technology, the only method to obtain reasonable bandwidth was with dedicated circuits. Once the step was taken to obtain the dedicated bandwidth, then the issue of how to get reasonable utilization had to be addressed if any type of cost control was desired. The packet network technology allowed the traffic of different types from a wide variety of users to be mixed into the network and this in turn led to the utilization needed. Thus the architecture of most data networks today use packet technology on a fixed topology to carry both response time critical and throughput critical traffic.

This present data network architecture has some characteristics which cause some inefficiency. First, the issue of mixing the two different types of data on the same network causes some problems. For response time critical data, the trunks should never be designed for greater than 50% to 60% utilization. This insures that the average wait for a packet to have access to a trunk circuit is very low. Further, the packet size for response time critical traffic should be small so that forwarding at network nodes and queuing delays due to other traffic is minimized. Throughput critical traffic on the other hand should maintain as close to 100% trunk utilization as possible. For efficiency, the packet size should be large to minimize the overhead and the product of window size and packet size should be high enough to allow high efficiency even with acknowledgement delays due to buffering or satellite circuits. There are possible protocols that will lessen but not eliminate the problems of

gateway to other networks if the desired destination of the traffic is not on the same network. Sometimes, it may be necessary to tandem through multiple networks in order to reach the desired destination. For this to happen, it is critical that standard protocols and standard addressing be selected otherwise the gateway function will tend to be very complex. While the gateway approach to reach destinations not on the originating network has been successful, the network design and optimization task is quite difficult. Traffic engineering for networks with tandem type traffic and alternate routes outside the network usually results in low utilization of transmission resources if performance levels are to be maintained.

Figure 1 is a representation of the architecture of the present network approach. It shows an end user system connected to a local network with a gateway to a global network. Assuming that the destination system was yet another network away, a gateway between global networks is shown. At each gateway and at each node within the global networks, the packets must be buffered and queued for the next transmission. Further, even with alternate routing, the availability of gateways and global network nodes determines the overall reliability. The more gateways and nodes that a packet needs

Present Network Approach

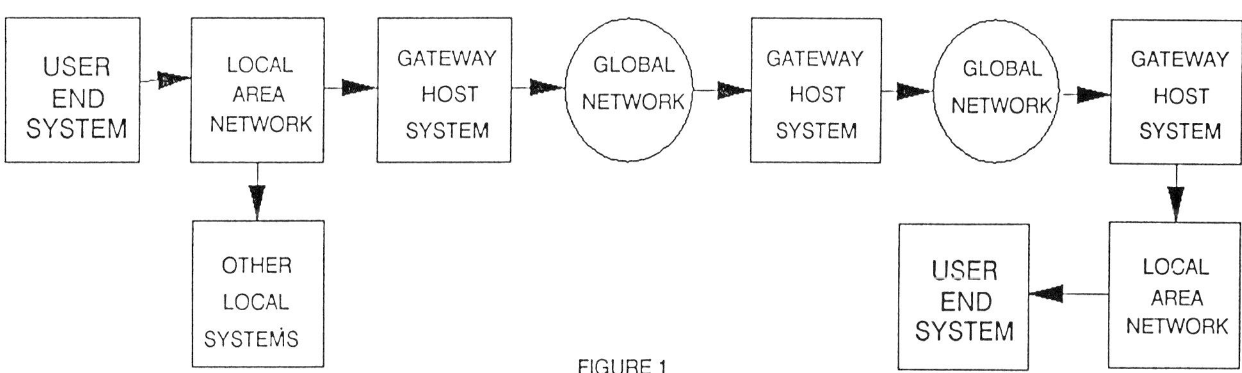

FIGURE 1

mixing the two types of traffic. Queuing priorities, transmission aborts, dynamic flow control, and other routing mechanisms could be implemented as a compromise to mix the traffic. Few existing network implementations are very successful at efficiently carrying the mix.

The last issue on the present network approach is that of gateways. The fixed topology architecture generally means that users or systems connected to one network will have to use some

to traverse, the more delay and chance for failure.

Emerging Networking Approach

The networking approach that seems to be emerging with the advances in both computing systems and communication technology is one that separates the response time critical traffic and the throughput critical traffic. The characteristics of each type of traffic are taken into account in the selection of the network approach for each.

Response time critical traffic with its generally short statistically distributed messages seems a natural for current packet switched networks. The X.25 interface standard is the most universal standard for the lower three layers of the ISO reference model and one for which many manufacturers offer various products. Other packet network protocols, however, such as TCP/IP and vendor specific implementations such as IBM's SNA and others will continue to exist meaning that no single packet standard will subsume all others. The changes that will be seen in the future center on the ability of the network processors to use dynamic bandwidth for trunking. In periods of low traffic, low bandwidth would be used. When traffic levels increase, more bandwidth could be added. Processors may very well evolve that would be capable of not only adding bandwidth but of creating new direct trunks to nodes as the traffic patterns changed.

The changes in communication technology that would allow the dynamic addition of bandwidth or the creation of new packet network trunks is essentially the move towards total digital switching and transmission. As the dependence on voice grade analog circuits is decreased, the ability to provide higher bandwidth digital paths increases. If network designers have the ability to use circuit switched digital paths in an efficient manner, the opportunities for optimized networks are enhanced. With digital circuit switching, throughput critical traffic can be carried directly from source to destination with circuit utilization at close to 100% with proper protocols. Since direct connection from source to destination is possible, the protocols need not be burdened with the network layer (ISO level three) overhead associated with packet routing. The link layer with appropriate error control can provide the necessary transmission protocol. The old IBM 2780 and 3780 Remote Job Entry protocols are examples of such a link layer protocol without a packet routing overhead. These specific protocols suffer because they use a window size (the number of packets that can be sent before an acknowledgement is received) of one which make them very sensitive to transmission delays and thus inefficient but the concept is most appropriate for circuit switched throughput critical transmissions. The same protocol approach with a product of window size times the block size greater than the transmission speed times the round trip transmission delay could yield the desired 100% utilization. If costs are related to usage and there are multiple destinations, then circuit switching will generally minimize the network costs for throughput critical data. If there is a high concentration of traffic to one location, dedicated facilities at a fixed cost may be required. Note that within the realm of throughput critical data, concepts of off hour and alternate means of transmission should be used for cost mini-

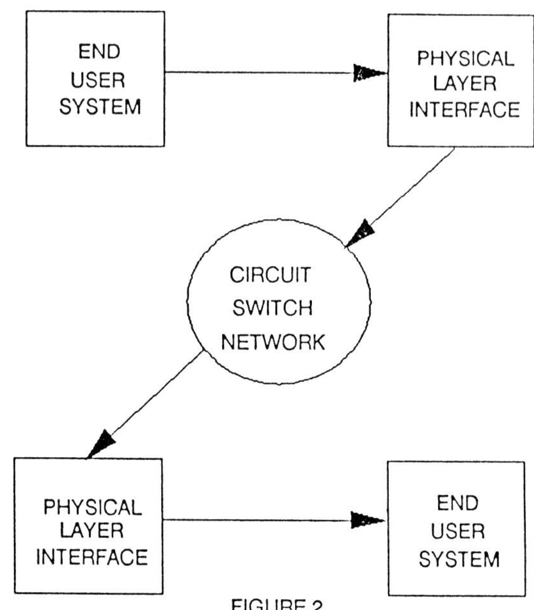

FIGURE 2

mization where the time value of the data permits such transmission.

Figure 2 illustrates the approach using circuit switching. Once the circuit is established, the impact is the same as the physical interface being extended between the two end user systems. Depending on the communication media used within the circuit switch network, the delay could be significant. If satellite technology is used because of its cost advantage, the bit delay from input to output can be around 250 milliseconds. If multiple hops are encountered as sometimes is needed for international service, the delay will increase. Thus for throughput critical traffic, it is necessary to use protocols that have the product of window size and block size great enough to keep the link busy.

There are advantages of the circuit switched approach for throughput critical data besides cost. Since circuit switching is essentially a physical layer interface and bit transparent, as long as the end systems agree on a protocol suite the connection can be made. There is no requirement for intermediate protocols or gateway functions. While protocol standards are encouraged, the network does not require them. Thus even vendor specific protocols could be used when appropriate between similar end systems. Further, protocols appropriate for direct connect through such media as satellites could be chosen without impacting other network functions requiring packet level routing. Since circuit switching does no intermediate buffering and the full bandwidth is available, the time to transmit a file once the connection is made is minimized. The network does not require any intermediate host or gateway availability which makes operations more simple. Finally, the characteristics of throughput critical data follow that of other circuit switched services such as voice, facsimile, teleconferencing meaning that integrated services can be combined on network resources to gain the economy desired.

If one follows the separation of traffic types onto different classes of network, it will mean that the end user systems will have to have multiple network access. Since every user will likely have need for various functions in both the response time critical and throughput critical arenas, it will be necessary for the system to support both types of network connections. Further, there will be local functions supported on various types of servers which will, in all likelihood, need a local area network connection. The result of this is that systems designers will need to address the different types of network connections to obtain desired performance and cost objectives. Figure 3 is a schematic of this structure.

Trends in Carrier Services

One of the most talked about directions in the public carrier arena is that of ISDN or Integrated Services Digital Networks. This concept of digital circuit switched and packet switched service that allows multiple communication requirements to be satisfied over a standard interface is supported by virtually every carrier both domestic and international. International standards committees have begun setting standards for the user interface at essentially two levels. The basic rate interface would provide two 64Kbps circuit switched channels (called

B channels) used for voice and data and one 16Kbps packet switched channel (called the D channel) for signaling, control, and other packet data services. These logical channels would share a physical channel connecting the interface to the carrier network. A higher speed primary rate interface follows the same structure but with 23 B channels and a 64Kbps D channel. Carriers are planning to offer 64Kbps, 386Kbps, and other circuit switched service as well as the packet switched service. If one needed higher speed packet service, one could use a circuit switched service to access a higher speed packet switch. Carriers may elect to provide some of the circuit switched services on a packet basis but the interface seen by the user is the synchronous bit transparent circuit specified under ISDN.

Given that market demand and tariff pricing will make the ISDN an economic approach, then again the systems will need to be designed to take advantage of this network type of network service. This is exactly what was suggested in the previous section.

Network Standards

Ideally it would be nice if standards existed and were followed for all systems that we use. In reality, it would be a great step forward if even a limited set of standards could be agreed upon for such systems. Most successful standards, however, have evolved from a defacto or dominate supplier system. What results is that standards are chosen for communities of interest as needed and no one standard satisfies all communities. While pursuit of standards is to be encouraged, dependence on standards may be a disaster. It would be nice if everyone in the world spoke English (certainly a

OPTIMAL END USER SYSTEM STRUCTURE

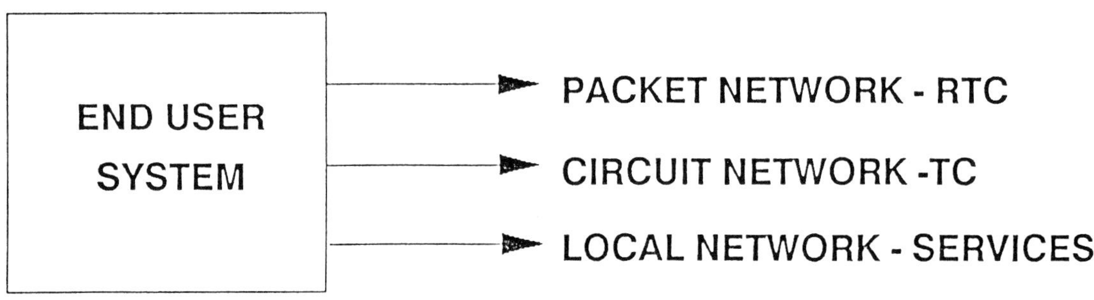

FIGURE 3

recognized standard) but a system depending on it would certainly fail.

The design of networks, especially those carrying integrated services, do depend on standards even though many options exist. Part of the success of the telephone industry has been the acceptance of standards at the low levels and the acceptance of differences at the higher levels. The emergence of circuit switching for data transmission follows this in that physical layer connectivity will be provided with the functions of the higher layers provided by the end user systems. This means that community standards for operability can be set without impact on other communities with different requirements and standards. This type of approach extends to other low level standards such as IEEE 802.3 where multiple high layer protocols can share a transmission resource.

While circuit switching and other low layer protocols allow network sharing, they do little for the higher layers especially when connectivity is needed between different communities of interest. Several approaches exist for solving this problem. First, get everyone to accept a single standard for an entire protocol suite. While possible within a community of interest or a single company or agency, it may be quite impractical for larger communities. Second, create a network function that will gateway or convert the protocols to allow access across different standards. Again this may be impractical as the network costs will increase to add this function which many communities may not desire and be willing to pay for. A practical option may be to add this type of service as separate offering paid for by those who needed it. The third approach is to create, at the end user system level, an interface with multiple standards. This means a user system would need to emulate or interface with systems of different standards for the set of communities that connectivity is desired. This means that costs for connectivity is directly related to the number of different community standards that one system must access. If communication is within a single standard, then no additional cost for interface is needed. If a community chooses a standard not supported by many systems, then connectivity may be quite expensive. In this approach, the network would still supply the lower layer communication services but would rely on the end user systems to correct the higher layer differences. In the world of personal computers, this is equivalent to having different terminal emulation packages that can be loaded into the PC depending on the system being accessed.

Conclusions

The general conclusions that can be drawn as to the trends in data communications are as follows. Packet networks will continue to be the prime method of switching and transmitting response time critical traffic. Circuit switching will emerge as the most cost effective method for carrying throughput critical traffic. In the near term, common low layer networks will support multiple high level protocols to similar communities of interest. Since the characteristics of throughput critical traffic are similar to voice and facsimile traffic, circuit switched data will be integrated with voice, facsimile, and even video traffic in an ISDN environment. End user systems will evolve to support multiple network interfaces for the various types of data transmission. Many end user systems will emerge that will integrate other user communication services beyond data such as voice, facsimile and mail. Standards will be chosen within communities and intercommunity connectivity will be accomplished via emulation or protocol matching at the end user system level. Standards within communities for the higher layers such as data, file, document, and job formats will need to be agreed upon so that activities at the application layer can take place among the various members of the community. It is not imperative, however, that every community agree to the same standard. Standards that satisfy the scientific community interested in space applications with emphasis on numeric data with fine precision could be inappropriate for other fields such as administration or natural sciences. Finally, the real challenge in the future will be the system designs that make use of the advances and standards that have evolved in the communications field and those advances in the computing field to provide the end user a simple cost conscious approach to accomplishing his job.

ESA SOFTWARE ENGINEERING STANDARDS FOR FUTURE PROGRAMMES

Dr C. Mazza
European Space Agency (ESA)

Abstract

ESA has established since several years a board
for software standardization which has developed
and promoted the ESA Software Engineering
Standards. These Standards have been in use now
for 3 years on several software projects and
subsequently reviewed. The last issue will
constitute the baseline for software standards for
all ESA future missions. The essential principles
on which the standards are based are explained.
ESA has also formulated a policy for the choice of
programming languages for future projects and for
the selection of a European Space Software
Developmen Environment (ESSDE) which will support
the above standards and the selected languages.

1. Introduction

The amount of software which will have to be
developed from now to year 2000 to support the
European Space Agency programme is hard to predict
and no reliable estimates can be produced at the
moment. It is certain, however, that we are
talking of several thousands of man-years of
effort.

Ambitious projects such as those intended in the
ESA infrastructure programme (Ariane V, Hermes,
Columbus and the Data Relay Satellite (DRS)), in
the Scientifc Programme, in the Earth Observation
Programme, in the Microgravity Programme and in
the Telecommunications Programme will require an
unprecedented amount of software.

Thinking to the number and the variety of areas
where software will be needed to support such a
programme and to the millions of lines of code
which will have to be developed reasonalbe doubt
raises, will we be able to master the
implementation of all this software?

In order to dispel this doubt and be prepared to
face the problem on good engineering foundations,
the Agency has promulgated a software engineering
policy and a plan to implement it.

This paper illustrates both of them.

2. The ESA Software Engineering Standards

Several years ago the European Space Agency
established a Board for Software Standardization
and Control (BSSC). This Board defined a set of
software engineering standards tailored to ESA
needs. This set of standards, after having been
tested in several projects for about three years,
was issued at the beginning of 1984[1]. Since then
the standards have been adopted for any important
software project developed by ESA or by Industry
for ESA.

As a feed-back from the various projects which
used the standards several improvement proposals
were collected. In 1986 the BSSC undertook an
important revision of the standards which led to
the issue in early 1987, of a new document now
called PSS-05-0 which replaces entirely the
BSSC(84)1[2].

The document is presented in two parts.

Part 1 is mainly concerned with the product
standards, while Part 2 is mainly concerned with
the managerial procedures related to the process
of producing software.

The standards are classified, according to the
IEEE practices for software engineering standards,
into three different categories, ie:

- **mandatory standards**

  Mandatory standards cannot be waived

- **recommended practices**

  Recommended practices require, in case they are
  not followed, the granting of a special waiver
  by the appropriate level in the Agency's
  hierarchy.

- **guidelines**

  Considered to be only useful practices.

The ESA Software Engineering Standards is one of
the very few software standards which covers all
the phases of the software life cycle and
addresses all the aspects of a software project.

It is based on the definition of the software life
cycle detailed in the figure below which is
extracted directly as well as the following
description from the document PSS-05-0.

....."Figure 1 specifies 6 phases:

UR phase	Definition of the user's requirements	Preliminary Phase
SR phase	Definition of the software requirements	
AD phase	Definition of the general architecture of the system and detailed development plan	Software life cycle
DD phase	Detailed design and production of the software software	
TR phase	Transfer of the system to operations	
OM phase	Operations and Maintenance	

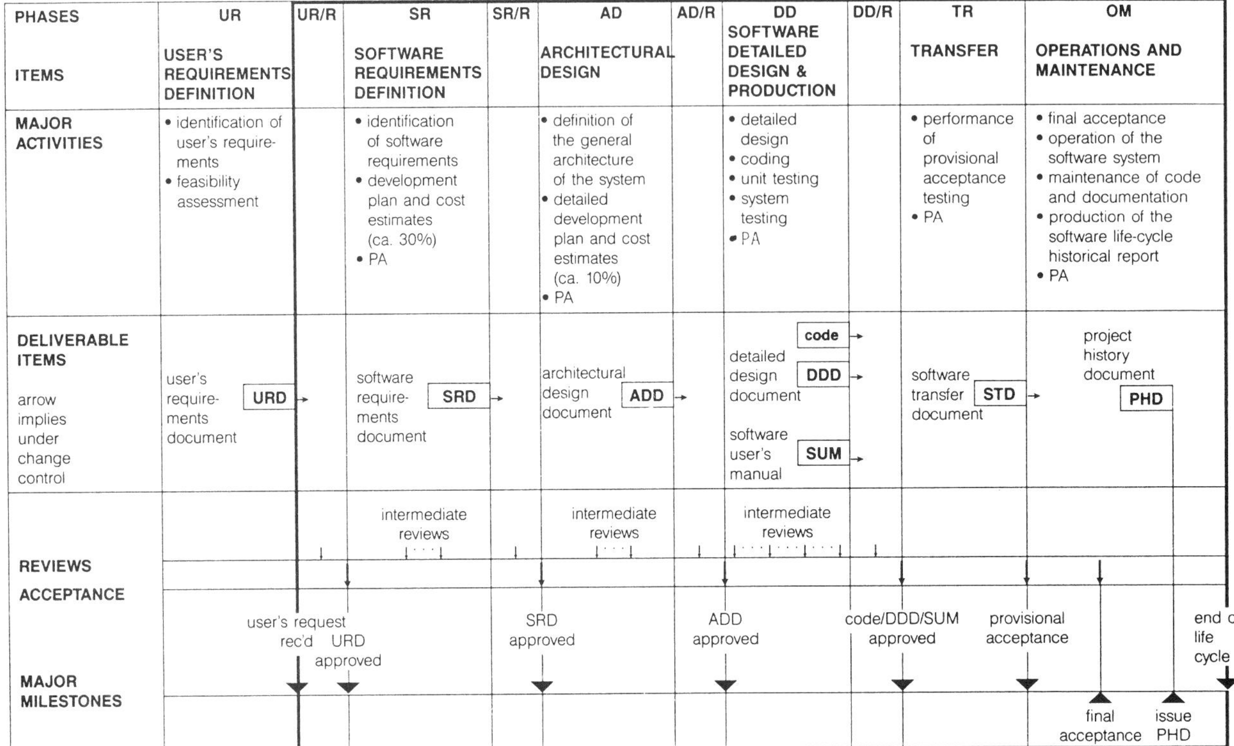

Figure 1.1 Software life cycle management scheme

The phase of definition of the user's requirements is not considered part of the software life cycle, but constitutes a necessary step before a software project can begin. The software life cycle begins with the notification of a user's requirements. The first activity consists of the review and approval of the user's requirements.

At the end of each of the following three phases of the software life cycle, the plan foresees a separate activity consisting of the review and the formal acceptance of the phase concerned ....
.......................Figure 1 indicates the major milestones of a software life cycle between the inception of the development and the point when the software is dismissed. The major milestones, which should always be present, even in a small project, to allow the progress of development to be monitored, are:

- the software inception, coinciding with the notification of a user's request;

- the approval of the User Requirement Document (URD);

- the approval of the Software Requirements Document (SRD);

- the approval of the Architectural Design Document (ADD);

- the statement of readiness for provisional acceptance testing, ie the acceptance of the Detailed Design Document (DDD), of the Software

User's Manual (SUM) and of the code for provisional acceptance testing;

- the statement of provisional acceptance;

- the statement of final acceptance;

- the issue of the Project History Document (PHD)."

It is to be noted that the figure presents the phases of the life cycle as sequential activities for sake of simplicity. The text however indicates clearly that particularly in the initial phases, there could be several iterations when alternative requirements or designs have to be considered and evaluated.   A specific phase is considered concluded only when the related output has been reviewed, approved and put under change control.

3. The Software Engineering Policy

All software projects will have to follow the ESA Software Engineering Standards. Document PSS-05-0 is the top document of the software engineering standards and is applicable to any development of either on-board or ground software supporting the ESA missions.

This section (section 3) lists all the items which are mandatory standards in each of the phases of the life cycle and indicates which further detailed standards are expected to be added.

User Requirements Definition Phase

- a written User Requirements Document **(URD)**;
- requirements are classified as essential or desirable;
- a test must be specified to verify each of the essential requirements;
- numbering of requirements;
- the URD must contain a section describing the man/machine user interface;
- SMP1, ie the initial version of the Software Management Plan must be prepared during the UR phase.

Software Requirements Definition Phase

- Software requirements must be numbered to facilitate tracing through implementation and test;
- Software requirements are classified as essential or desirable;
- A list of acceptance tests is compiled during the SR phase;
- a Software Requirements Document **(SRD)** is produced with the following table of contents.

Service Information

a - Abstract
b - Table of Contents
c - Change-control procedures
d - Document-change log
e - Reference Documents

1.0 Introduction
  Is used to give an overview of the entire SRD. It should state briefly the purpose of the software task, the participants and the time frame.

2.0 Reference to applicable documents

3.0 General Description
  3.1 Project in perspective with other systems/projects
  3.2 Predecessor and successor projects
  3.3 Function and purpose
  3.4 Environmental considerations
  3.5 Relation to other systems
  3.6 General constraints

4.0 Specific requirements
  4.1 Functional requirements
  4.2 Performance requirements
  4.3 Interface requirements
  4.4 Operational requirements
  4.5 Resource requirements
  4.6 Verification requirements
  4.7 Acceptance testing requirements
  4.8 Document requirements
  4.9 Quality requirements
  4.10 Safety requirements
  4.11 Reliability requirements
  4.12 Maintainability requirements

5.0 Glossary and references

- SMP2, ie a revision of the Software Management Plan is produced.

Architectural Design Phase

- the Architectural Design Document (ADD) must be sufficiently detailed to allow the Project Leader to set up a detailed implementation plan.
- the following activities are performed during the AD phase:

1. Decomposition of the software system into its hierarchical components
2. Functional definition of the components
3. Definition of the data structures, data flow and control flow
4. Trade-off between alternative designs
5. Computer resource utilisation definition
6. Choice of programming language(s)
7. Management activities
8. Test definition

- the decomposition of the system is done according to the two following design principles:

  - top-down approach
  - maximisation of module independence (modularity)

- a specific methodology for high level design is selected (PSS-05-0 does not impose one specific methodology but requires one to be selected in each project).

- for each component of the system specified in the ADD the following detailed information is supplied:

  - the data input
  - the data output
  - the process to be performed

- the flow of control is specified

- the computer resources for both development and operations are estimated

- SMP3, ie a revision of the Software Management Plan (SMP) is produced

- an Architectural Design Document **(ADD)** is produced with the following table of contents:

Service Information

a - Abstract
b - Table of Contents
c - Change-control procedures
d - Document-change log
e - Reference Documents

1.0 Introduction
  A description of the functional requirements of the system to be designed making reference to the preceding URD and SRD.

2.0 System Design
  2.1 High-level design methodology
  2.2 Decomposition of the system into its components
  2.3 Brief functional definition of each component
  2.4 Data structures and interfaces between components
  2.5 Control flow (state transitions)
  2.6 Data flow

3. **Component design** (repeated for each
   component)
   3.n.1 Detailed functional description of
          component
   3.n.2 Control flow within component
   3.n.3 Data flow within component

4. **Verification- and acceptance-tests concept**
   Defines the approach to be taken in such a
   way that the verification and acceptance
   plans can be derived during the detailed
   design phase.

Detailed Design and Production Phase

- programming standards are defined for each
  project (detailed programming standards for the
  language selected must be adopted but are not
  imposed by PSS-05-0)

- each module has a standard header

- outputs from the phase are:

  o code
  o the Detailed Design Document (DDD)
  o the Software User Manual (SUM)
  o the Software Test Plan (STP)

- the table of contents of the **DDD** is:

  Service Information

  a – Abstract
  b – Table of Contents
  c – Change-control procedures
  d – Document-change log
  e – Reference Documents

  **Part 1 – Project standards, conventions and
           procedures**

  1.1 Low-level design standards
  1.2 Documentation standards and conventions
  1.3 File-naming conventions
  1.4 Programming standards
  1.5 Exception-handling conventions
  1.6 Development procedures
  1.7 Operational procedures
  1.8 Configuration item list
  1.9 Build procedures

  **Part 2 – Detailed design specifications
           organised in levels according to
           the system breakdown structure**

  For each component the following information
  should be produced, organised in sections
  labelled as follows:

  A Design assumptions
  F Functional description
  S Structure description
  I Interface description
  D Data specification
  R Resource utilisation
  T Testing procedures and test results

- the table of contents of the **SUM** is:

  Service Information

  a – Abstract

b – Table of Contents
c – Change-control procedures
d – Document-change log
e – Reference Documents

1. **Purpose** of the software system (ref to the
   URD)

2. **External view** of the system (user's
   viewpoint)
   2.1 Input data
   2.2 Output data

3. Reference to **applicable documents**

4. **Operations environment**
   4.1 Hardware configuration
   4.2 Software configuration
   4.3 Operation constraints

5. **Operations control procedures** and
   instructions (Operator Manual)
   5.1 Set-up and installation
   5.2 Starting
   5.3 Mode selection and control
   5.4 Normal operations
   5.5 Normal termination
   5.6 Error conditions
   5.7 Recovery runs
   5.8 Housekeeping

   Annex 1 – Error messages. List of all possible
           errors, diagnosis and procedures to be
           applied for each of them

   Annex 2 – List of operator commands, with
           possible parameters, etc

- the Table of Contents of the STP is:

1. **Purpose of the Test Plan** (description of the
   approach and of the principles on which the
   acceptance tests are based)

2. **STP, Part 1.** List of acceptance tests

3. **STP, Part 2.** Design of tests and of neccesary
   data sets
   3.1 Definition of the hardware and of the
        environment on which tests are to be run
   3.2 Software Test Files

4. **STP, Part 3.** List of tests to be run and
   related procedures

5. **Reference to relevant documentation**

Transfer Phase

- for each problem found from now onwards a
  Software Problem Report (SPR) is written

- a Software Transfer Document **(STD)** is delivered.
  Typical table of contents of the STD is:

  Service Information

  a – Abstract
  b – Table of Contents
  c – Change-control procedures
  d – Document-change log
  e – Reference Documents

1. **STP.** Pre-acceptance version

2. Checked **hardware configuration** and list of tests

3. **Log of acceptance testing** with a list of appended test data, results and anomaly reports.

4. List of **special test software** and additional materials used.

5. List of items delivered with **statement of provisional acceptance.**

6. The **final version** of the **STP** including amendments made during acceptance testing.

Operations and Maintenance Phase

- written procedure for problem reporting and correction

- problems are corrected in new releases of software

- new releases are always accompanied by a Release Document, listing which problems have been rectified

- a software which has not been dismissed has a designated software maintenance engineer

Quality Assurance Activity

- the following QA functions are established:

  . all mandatory reviews are organised and related checklists are used

  . forward traceability from requirements to design and from design to code is established during implementation

  . backward traceability from code to design and from design to requirements is established during verification and validation

  . the correct application of the standards is subject of an audit during the AD and DD phase

- a Software Quality Assurance Plan (SQAP) is established. Contents of the SQAP is according to IEEE Standard for Quality Assurance Plan[3].

- within the SQAP a Software Configuration Management Plan (SCMP) is established

- according to the IEEE Standard for Software Configuration Management Plans[4], each configuration item is uniquely identified by a configuration item number.

- the software being developed is organised in libraries as follows:

  - Development library
  - Integration library
  - Master library, composed of Tested Modules library and of Verified Version library

- all releases of software are done by the Librarian from the Master library

- the Librarian is responsible for the security aspects of all libraries

- an SPR contains as a minimum the following information:

  - SPR number
  - SPR title
  - Name of originator
  - Date
  - Configuration Item (s) affected
  - Problem Description
  - Description of Environment (hardware, operating system, etc)
  - Recommended solution
  - Software Review Board decision and date
  - List of attachments for evidence and analysis

- a log of the status of each SPR is maintained by the Project Leader

In the framework of these mandatory standards detailed ad-hoc standards will have to be added according to the specific requirements of each project.

For example, in the context of the Columbus project, the BSSC has established a special committee for the selection of suitable detailed standards, eg programming standards for the languages to be used on the Columbus supporting software.

4. The Software Development Supporting Policy

All the software to be produced by ESA or for ESA for support of future programmes will be developed according to the standards specified in Section 3.

This section (Section 4) explains the policy adopted by the Agency in terms of programming languages and of tools for software support.

The policy is based on the following considerations:

- the need for minimizing the cost of the training programme of software engineers on new languages

- the necessity of increasing productivity of software teams

- the necessity of enhancing reliability of the software produced

- the necessity of achieving portability of software

- the necessity of harmonizing the software development activities in particular for COLUMBUS and HERMES

- the necessity of promoting inter-Agency commonality.

4.1 Programming Languages

A programmatic decision has been made in the Agency that ADA will be the ESA programming language for COLUMBUS, HERMES and related projects.

137

As the same decision has been made for the Space Station by NASA, ESA believes that this commonality will bring in the long run distinct advantages in the international space community.

Nevertheless, ESA has still some reservations as concerns the performance of ADA programmes for real-time tasks with critical response time requirements and the policy recommends that an effort be undertaken in the Agency for the selection of a suitable language for these tasks.

Possible candidates are C, Real-time Pascal (also called LTR-3), Modula and of course ADA itself if it can be demonstrated that the reservations are not justified. It is not excluded that other candidates may be evaluated before the final selection is made.

The above policy for the choice of programming languages is stated in a BSSC recommendation[5].

4.2 The Software Development Environment

As stated above two of the goals to achieve with the ESA policy for software development are an increase in productivity and an enhancement of reliability.

The traditional obstacles in the direction of achieving these goals are:

- lack of proper automated tools for the various phases of the software life cycle

- different man-machine interfaces to the variety of host machines and target machines used

- different man-machine interfaces for the various tools even within the same hardware environment

- lack of flexibility for incorporation of new tools

- lack of standardisation of the programming languages used

In the context of its Esprit programme, the European Community (EEC) has developed an interesting concept called PCTE (Portable Common Tool Environment) the aim of which is to give an answer to the problem of common interfaces for tools to be used for software development support. ESA has considered this approach sufficiently interesting to include it in its specification for a European Space Software Development Environment (ESSDE).

These specifications consititute one part of the BSSC recommendation quoted in section 4.1 (ref [5]) and states:

"the BSSC recommends

A. that a common **European Space Software Development Environment (ESSDE)** be selected by the Agency and promoted within the Industry.

   The ESSDE shall be adopted for the development of all the software realted to COLUMBUS and HERMES and other future ESA programmes.

The ESSDE shall be based on the Portable Common Tool Environment (PCTE) concept developed under the EEC Esprit programme and have the following characteristics:

1. Support all the phases of the software life cycle.

2. Have a standardized, hardware independent, man-machine interface.

3. Support as a minimum Ada and a suitable real-time language for performance-critical software.

4. Be flexible enough to allow incorporation of new tools as they become available.

5. Exhibit, where possible, commnality with the Space Station Software Support Environment.

At the moment ESA has placed contracts with several European companies which should lead to detailed specification for the ESSDE which will be used to build software development workstations qualified as ESSDE.

ESSDE's workstations will be adopted for the development of all the software needed for supporting ESA future programmes.

REFERENCES

(1) Doc. ESA BSSC(84)1 - Issue No 1
     ESA Software Engineering Standards

(2) Doc. ESA PSS-05-01 - Issue No 1, January 1987
     ESA Software Engineering Standards
     (supersedes ref [1])

(3) IEEE Std. 730-1984
     IEEE Standard for Quality Assurance Plan

(4) IEEE Std. 828-1983
     IEEE Standard for Software Configuration Management Plans

(5) Doc. ESA BSSC(87)1 - Issue No 0
     BSSC Recommendation for the Selection of a European Space Software Development Environment (ESSDE) and Choice of Programming Languages for COLUMBUS, HERMES and Other Future ESA Programmes

THE SPACE STATION SOFTWARE SUPPORT ENVIRONMENT
NOT JUST WHAT, BUT WHY

by John R. Garman

Director, Information Systems Services

NASA Space Station Program Office

ABSTRACT

The limits to automation or functionality of data systems are established primarily by the limits of our ability to create higher and higher levels of software to perform the processing necessary to provide those functions. In turn, higher levels of software functionality or automation create more and more complex data systems - both in hardware and software. It is with this line of reasoning, then, that one might conclude that complexity becomes a limit in functionality, particularly if it is the type which increases risk of error, poor performance, or cost.

Methods to "combat" that form of complexity in data systems must therefore be pursued as one of the most fundamental approaches to obtaining higher functional data systems with reasonable return on investment.

It is this author's belief that "layering" and "distribution" of functions and services are today's best understood methods to combat complexity. Along with commonality in tools and methods in the lower layers of any software architecture, they form the cornerstones of this battle.

The tools and methods associated with development of procedural software-based applications control the lower layers of complexity in software architectures. It is the common Software Support Environment (SSE) effort within the Space Station Program which embodies National Aeronautics and Space Administration's (NASA) attack on enabling higher levels of functionality and complexity while maintaining high standards of reliability and commensurate cost.

The SSE isn't so much a "high tech" new start in the software "tools and rules" arena; it is rather a codifying, for a given domain, of all the "lessons" and experiences in the software development that the industry has picked up over the last 15 years. It is the thesis of this paper that only by "cementing" that layer, by "closing if off" and putting it behind us, can we move on to the higher levels of software architecture requisite for "safe complexity" and high functionality or automation.

THE NASA ENVIRONMENT

Each industry and environment has its own unique problems and issues with respect to the development and use of data systems. NASA is perhaps one arena that can claim several unique considerations, and it is from within the manned spaceflight projects that the thoughts forming this paper are based. Although I feel that the observations and conclusions herein are applicable in many domains, it is important to understand the context of NASA's current mission data systems evolution for any reader to agree.

Mission Data Systems in NASA

For this paper, "mission data systems" in the National Aeronautics and Space Administration are computer-based process systems which actively participate in conducting the business of flying space missions. Mission data systems in NASA are going through several significant changes, and the most significant are the growth and longevity of these systems.

In the past, when the Agency was purely Research and Development, one could track the use of data systems in projects almost like a set of sawteeth (see Figure 1). As each program came to an end, all the software and data systems were effectively discarded and the new programs would start virtually "from scratch." With the introduction of the Space Shuttle Program, this luxury in systems development has ended. In fact, the newest program, Space Station, is totally dependent on the continuation of the Space Shuttle - and the Shuttle, following recovery from the Challenger tragedy, will continue to be the backbone of the Agency's Space Transportation System for many years to come. As a result, the Shuttle-related data systems, both flight and ground, have and will continue to evolve -and the Space Station systems will be built in addition to those Shuttle systems.

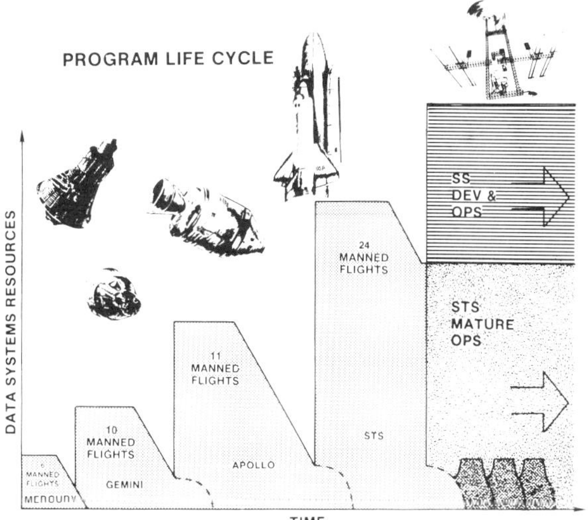

Figure 1

This "building waves" saga continues with the Space Station itself. The initial capabilities in the Space Station are meant to be incremental, with development and continued addition of function and process to the Space Station contemplated (though not approved) to continue for years beyond the "first permanent manned presence" by this country. The Space Station is also unique in that the basic systems can not be "brought home" for update and refurbishment. Figure 1 not only characterizes this effect, but the additional effect of growth in the amount of data systems resources that has been involved in each succeeding space program.

The Strategic View

Consequently, many of us in the Agency have concluded that we must take a more global and strategic view of computing and data systems within our manned spaceflight program - particularly with an eye to the long lifetimes and evolutionary nature

"

that the current and new systems have. The strategies associated with the Space Station data systems were created under this aegis and it is the purpose of the paper to attempt to describe and communicate some of them.

CHARACTERIZATION OF SPACE STATION DATA SYSTEMS

It is worthwhile to characterize these flight and ground mission data systems which will support the Space Station. First, with a view of the Space Station itself, some of us have taken the view that the vehicle is more of a "facility" than a spacecraft in several ways. There are flight safety and critical systems and structures associated with every aspect of space flight. However, it is important, with respect to the flight data systems (known within the Agency as the "Data Management System" or "DMS"), to note that a much larger share of the onboard computing resource will be devoted to information management and networking than has ever been the case in the past.

In prior spacecraft, the onboard software and computers were embedded elements within the avionics system whose primary functions had to do with guidance, navigation, and flight control. While these are still prime functions, their role is diminished because the Space Station has no ascent and entry (takeoff and landing) phases. Also, because the flight crew of the Space Station is there principally to do the processing and user servicing functions that such a platform will afford, there is a significant increase in the requirements for automation both in the management of onboard systems and flight operations. Likewise, the predictions of the amount of information that could potentially flow from the Space Station to ground-supporting facilities is so large as to lead us to project a great increase in onboard information management and processing of information. Such "payload" information processing will help compress the information that must be networked to the ground in a "live" sense.

Finally, and most important in the goals and design of the Space Station - its principal function is to provide service to its users. The flight and ground mission data systems, acting together as a single end-to-end Space Station Information System ("SSIS"), provide no small fraction of the real usability of the overall system. Figure 2 depicts the the scope of the SSIS. Altogether these requirements and characteristics will fundamentally change the proportions and character of onboard processing.

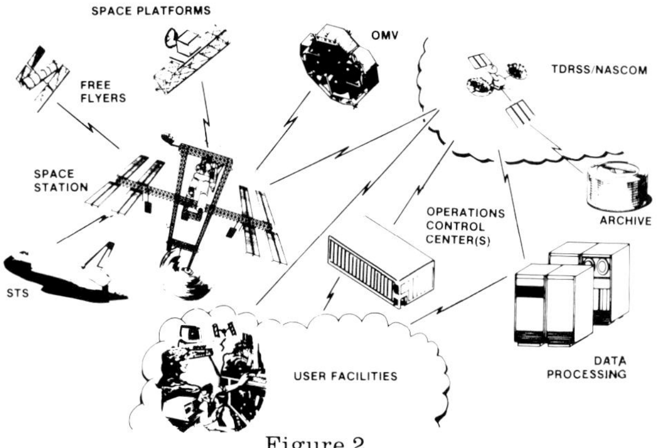

Figure 2

The principal characteristic changes in onboard and ground processing for Space Station, I feel, can be listed in the following categories:

1) Highly distributed (federated) data systems with many dependencies on networks and connectivity.

2) Increased functionality and therefore complexity in architecture.

3) Terminal and session oriented user interfaces with much less of the "dedicated" instrumentation that characterizes embedded flight and ground command and control systems. Characteristic of such systems there will be an incredibly larger number of active end-users of the SSIS, compared to equivalent data systems in the Shuttle and prior programs.

4) A significantly more distributed (with respect to the numbers of heterogeneous organizations) software development effort coupled with a much more evolutionary life cycle than past programs.

Distributed Data Systems

Past space projects have involved distributed data systems; the Shuttle is a prime example. However, the Space Station is the first time that a space program has been initiated in the technological framework of distributed data systems. In this framework, people, both engineers (providers) and scientists (and other users), tend to view all the data systems associated with the operation of the Space Station and the services it provides as one larger homogeneous information system (see Figure 3).

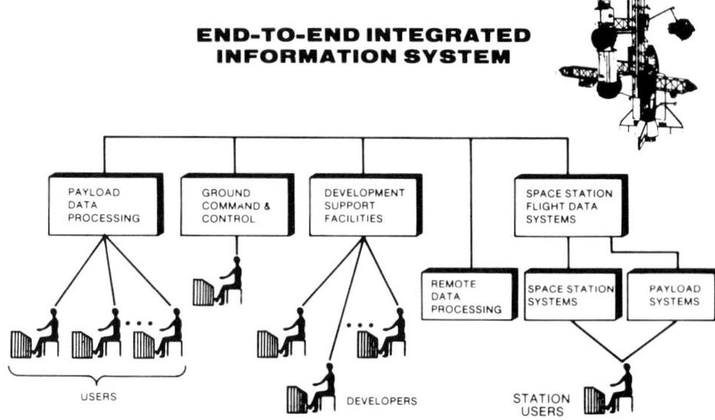

Figure 3

In this model, processing can be physically separate from data and sensors, as can users, review, control, and monitoring. In the Shuttle framework, data systems were autonomously developed with minimum interfaces and dependencies on other data systems associated with mission operations. While this led to fairly "clean" interfaces, it also led to a great deal of redundant information and processing. Redundancy itself is not bad or wasteful, but when great effort is required to keep separate "islands of information" with otherwise equivalent information consistent with each other both in content and processing, then the inefficiencies of such an architecture across a life cycle of operations becomes apparent. These inefficiencies waterfall into human as well as data processing resources waste as the number of managers, engineers, operators, support personnel, training programs, software maintenance and test facilities, and so on, remains large and largely overlapping even in the more mature phases of the systems' life cycle.

Some of this is quite necessary when the embedded and flight safety nature of the software and data systems is so ubiquitous as to virtually dictate replication and redundancy as forms of verification and certification. But while the Shuttle Program wrestles seriously with this dilemma, it has occurred to several of us that the Space Station Program provides an intrinsically different environment, in both development and operations, in which to attempt great leaps in commonality and the commensurate potential in consolidation across the program's life cycle. It is the interdependency of physically separate data systems, which is derived from policies of minimum replication in information data basing that is one of the principal characteristics of distributed systems.

Functionality and Complexity

One of the real enigmas of today's endeavors with computers is that the more we can get them to do, the more complex they become (see Figure 4). It is intuitively obvious, but in a pragmatic sense the level of complexity is growing faster than the functionality it affords.

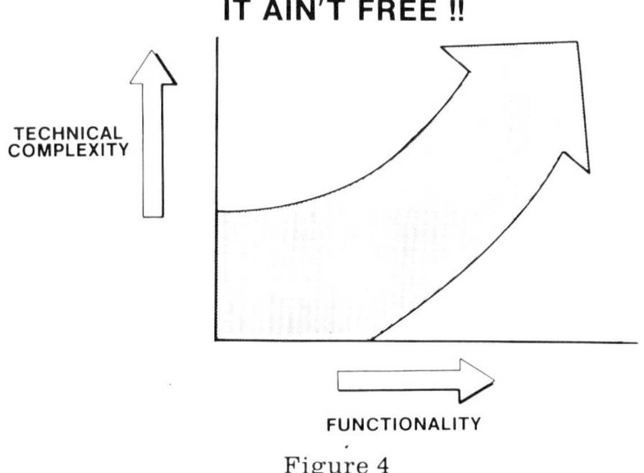

Figure 4

In particular, the expectations of the using community, and the "promises" from the tools-industry continue to create a gap between user expectation (and requirements) and the ability of tools and services to provide "easy" applications development.

Figure 5 models several types of complexity in data systems, any one of which can contribute to the "real" complexity of a system: that associated with risk or error.

The figure shows a triangle representing, at its base, the service level functions of machine and network. Near its apex, the boundary between what the computer application performs and what the user does is shown as a break. In this context, the "application" is defined to be the aggregate functions performed by the data system for the user. In virtually all data systems today, those functions are represented by the software within the computer or computers which comprise the processing components of the overall data system.

The entire triangle, therefore, represents the real function to be performed and the highest level "gap" or complexity is the one which is fundamental to all functions: the gap between what the entire function is.

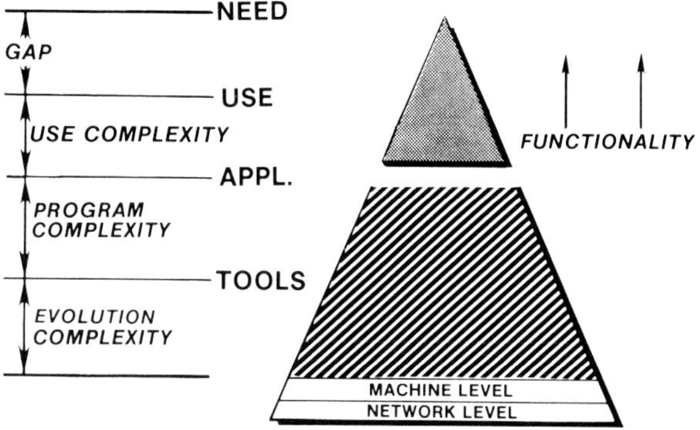

Figure 5

and the level at which the human user, aided by the machines underneath, can actually perform. If that function is "flying an airplane," then we as passengers on that airplane would like to think that the gap is near-zero. But we would still like technology to increase the functionality of the data systems supporting the pilot so that he or she can work at higher and higher levels within the overall "triangle" and thus diminish this very fundamental "gap." This level (at which humans operate) grows not only with increased functionality, or automation, of the underlying data systems, but also with training, experience, and intelligence. Unfortunately, the "gap" always seems to be there - no matter how well we do, with computers or people. It seems that in our reach for high productivity and "purpose" or quality of life, we always add more function to those who are performing well (make the triangle larger) assuming that such performance is a signal that the user has "graduated" to higher levels.

The "use complexity" is, on the other hand, the gap between the level at which the user actually performs the function, and the level of the underlying application and system. A fully "automated" system would, in theory, have a very slim area or gap here in fact. The concept of "user friendliness" also contributes to this gap of course.

The next "gap" is now within the machine. And indeed, it is the one which we in the software industry focus virtually all of our energies. It is the "program complexity" - the gap between the functionality of the application developed and the tools which are used to develop it.

If an application is a simple "spreadsheet" and the tool is one of the popular personal computer spreadsheet programs, then this gap is very slim as represented by the ease with which the application is developed. On the other hand, if the same tool is used to build a database, then the gap is quite large. While the tool can almost perform the job, there are many authors making quite a profit in the books they've written explaining how to do just about anything with such "macro"-languages.

The final and lowest "gap" has to do with the level of the tool itself with respect to the machine environment in which it operates. For example, software tools which don't support networking but are being used on a distributed system will display an ever-growing complexity as the application evolves into more and more distributed objects or "islands of information" (which would be better integrated).

Session Oriented User Interface

The concept of "online" computing, of "logging on" to a large data system and interacting with it in "real time" is much newer to computing than many people realize. With the advent of the popular microcomputer, immediate and interactive access to computational power is understood and expected in today's using community. However, to provide that kind of access where information being processed is distributed and heterogeneous is not a straightforward concept.

The SSIS promises to be one of the first integrated collection of mission data systems which establish this form of real time online interaction with its users - and to establish that interface independent of the user and application location. The reasons for such a user interface are many and include factors such as:

a) avoiding the need for redundant collections of unique dedicated displays and controls.

b) precluding physical location requirements for users to perform functions, whether the users be astronauts or ground crews operating the space base, or scientists and engineers operating experiments or manufacturing processes on that base or a free-flying platform.

c) precluding the physical location requirements for information in the same way; which avoids expensive replication of data and processing.

It is difficult to imagine a Space Station command center in the image of the cockpit of the Shuttle or a modern airliner. The constraints to operations that such vast arrays of dedicated displays and controls would engender is immense - not to mention the reliability issues associated with such an "all your eggs in one basket" approach to interfaces which can't be brought home for refurbishment or repair. It is far easier to imagine a limited amount of such dedicated interfaces and instead, a series of "workstations" consisting of common input and output media for users, flight or ground, to interact with the Space Station systems via networks, computers, and applications.

Distributed Software Development

Unlike prior manned spaceflight projects, the Space Station Program is planning a decentralized development of the software within the mission data systems. This is particularly new to the onboard or flight software arena where prior programs have tended to centralize such development under single organizations and contractor teams for aggregate accountability and reliability. Unfortunately, centralization of actual software development also causes the link between systems requirements for software processing and change, and the accountability for schedule and cost to become "disconnected." Furthermore, accountability for the end-to-end performance of a subsystem which uses the onboard data system in its functionality, can not be placed solely on the subsystem developer.

The very practical solution to this dilemma is to have the organization responsible for a subsystem develop not only the subsystem, but the applications which run it. This is what we mean by distributed software development.

Impacts of Distributed Development

However, distributed development has the potential effect of creating hundreds and hundreds of "islands of information" in terms of data bases, software development tools, software engineering practices, and so on.

It also creates the need for a software integration process of potentially heterogeneous software elements. When many applications come together in a given Space Station mission data system, onboard or ground, they will share data system resources. The degree of sharing could be as little as data buses, networks, and storage, all the way to co-residency in a virtual machine, virtual memory computing environment. In the former, the attention to physical interfaces and reconfiguration and recovery from data system component failures is very high in the requirements and definition process. In the latter, much of that failure recovery process can be imbedded as a service with data systems components, including computers, allocated to applications dynamically as a function of need. In either case, operational integration and test of all the applications which reside in the data system is required independent of the integration and test of the subsystems which they support. This software integration and certification step can also be significantly exacerbated if there are multiple standards in software development tools and methods. Many good components do necessarily make a good system.

WHAT IS THE SSE?

The concept of a Support Software Environment, or SSE, within the Space Station Program addresses this issue of distributed software development (and many others as well). In this context, the development of the rules and tools supporting the development of software in mission data systems is centralized for purposes of cost (avoiding replication) and commonality (enhanced life cycle maintenance and evolution).

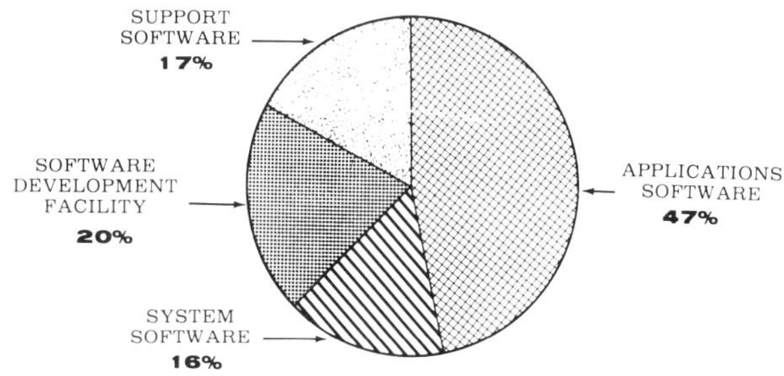

Figure 6

The Leverage of Commonality

We have some statistics from the primary avionics software development project for the Shuttle Program (see Figure 6) which shows that less than half of the costs attributable to the development of the flight software actually went to the development of the applications. The rest went toward the functions associated with the Shuttle equivalent of the SSE. As such, the leverage to the Space Station Program in creating a single SSE in an environment that can have the equivalent of many, many flight software development projects, is enormous. It provides a single development of tools and rules, rather than many. It also contributes substantially to the Agency's ability to consolidate contractors and skills in sustaining engineering of Space Station "software" following the main development phases.

SSE Components

Specifically, the SSE is a set of tools and rules for the development of mission software, and other software as appropriate, for the program. Its goals are to enable easier, feasible integration and affordable maintenance of mission software. The specific "components" of the SSE include:

a) Software Development Tools - including programming languages and compilers, software development management and control tools, and simulation and analysis models and systems with which to verify software as developed. The language selected as the base for Space Station software development is Ada (a trademark of the Ada Joint Program Office).

b) Hardware Tools Support Software Test - Particularly for flight software, where the actual mission environment can not be duplicated for system and software testing purposes, the SSE supports the development of simulation interface buffers which allow the interaction of the actual target flight computers with simulation models of the environment in their normal operating environment.

c) Operating Systems Interfaces - one of the key attributes of the environment of any applications software is the "operating system" that it runs under. Operating systems, much different than their name implies, are simply the collection of software services which interface applications to the resources available in data system environment in which they operate. The resources normally include time, memory, and storage, although the services afforded by operating systems can often be quite sophisticated and lengthy. There are several popular versions of operating systems in the microcomputer world which generally go under variations of a name including the phrase "Disk Operating System" or DOS, for example. The SSE will provide the specifications or interfaces for operating system services for applications that are developed with the SSE.

d) Management Plan and Standards ("rules") - No effort at commonality, ease of integration, or reduced life cycle costs could call itself serious without a frontal attack on a common definition for software life cycle, programming standards, documentation standards, and so on. Such standardization is intrinsic to the SSE approach, and perhaps the most difficult to implement.

THE CONTEXT OF THE SSE STRATEGY

I mentioned above that I believe that "layering" and "distribution" of functions and services are today's best understood methods to combat complexity, and that along with commonality in the lower layers of any software architecture they form the cornerstones in the battle to control complexity while reaching for higher and higher levels of functionality. To properly communicate the rationale for a common SSE within the Space Station Program it is necessary to describe two models of the context within which we believe it will operate.

a) A software technology model and its layers of abstraction.

b) A strategic model for integrated data systems and information resource management.

Software Technology

Figure 7 has a vertical axis depicting one view of layers of technology in software architecture. In this figure, the layers of software architecture are:

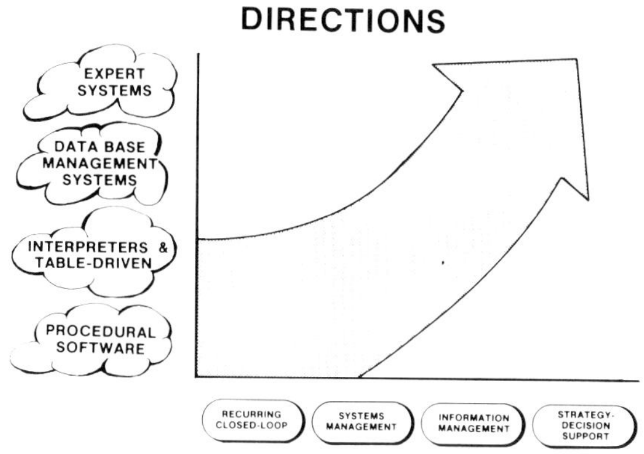

Figure 7

a) Procedural Software - The approach toward programming and tools that most of us were "brought up on." Ada and the software development technology efforts associated with it represent perhaps the most ambitious and far-reaching development in tools and software engineering methodology at this level.

b) Interpreters and Table-driven - This is an approach to software architecture which begins to be non-procedural (have you ever tried to write a flow diagram of the processing in your spread sheet application?). It involves the use of extremely rich language and user interface constructs which are processed "online." It includes all forms of procedural software architecture, but can reach far beyond into the type of "programs" associated with today's data base management systems.

c) Data Base Management Systems (DBMS)- Not really "systems," this term as a software architecture refers to the concept of designing software around the information it processes rather than around the "flow" of the software doing the processing. In designing this way, the access to information and the processing of relations between elements of information are embedded as services - layers of abstraction from an architectural vantage point. All commercial DBMS products are built primarily using procedural software technology and bring with them interpretive language constructs to implement both procedural and non-procedural software architectures in applications built with them.

d) **Expert Systems** - This software architecture, a branch of the artificial intelligence arena in computer technology, involves rule-based information processing and decision support. It is dependent on the technologies below it in the diagram, and represents the most promising direction in achieving higher and higher levels of functionality and automation via software and data systems. But it is, way on top of its technology neighbors discussed thus far, very different. For the first time the concept of "maybe" is embodied in the heretofore "yes/no" world of computer calculations... and the information or "rules" on which answers are dependent are not embedded inside packaged software systems - nor must the control of the content of those rules be as "tight."

Each of these layers represents a higher level of complexity and capability in engineering and developing applications. Furthermore, each layer is dependent on the one below it, and for many applications, can provide all the services of the application and mask or "hide" the use of capabilities in lower layers. There are certainly many ways to draw those layers on this axis, and this particular model is meant to represent increasingly complex and capable software architectures which are dependent on their predecessors.

The very same figure, on its horizontal axis, depicts increasingly complex functionality from a user-perspective. While the terms used on the axis can vary a great deal as a function of the industry or application, they represent the higher and higher levels being reached today in the boundary of functionality between people and computer systems that was described earlier in the discussion of complexity "gaps." The shaded area of the figure represents a "directions" concept which is key to the arguments presented in this paper. Applications developed outside that area represent unacceptable gaps in complexity - a disparity between the functionality of the application and the architecture used to implement it. "High" function applications can not be reasonably carried out using "low tech" architectures (e.g., a data base application using nothing but a procedural language). Similarly, the use of "high tech" architectures for "low function" applications is potentially inefficient and over-complicates the resultant system (e.g., an embedded sensor decoding process using a rule based expert system approach).

Staying within the shaded area as we move up to higher and higher functionality is critical to controlling complexity. That is, layers of services and architecture need to be developed and understood; moreover they need to be embedded far below the level of functionality at which the application operates and interfaces with users. This is absolutely key to the success of data system intensive efforts such as the Space Station. The "understanding" aspect of that hypothesis is more than technical; it includes our abilities to manage the development of such systems. Figure 8 shows the same domain but represents the level of "management experience" (measured perhaps in number of systems built, or simply

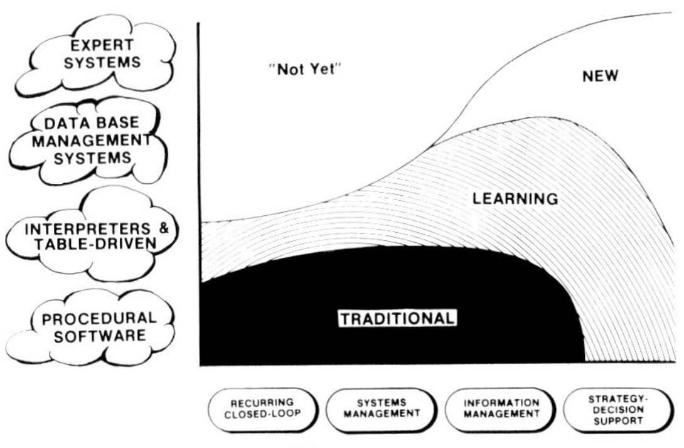

Figure 8

intuitively in the number of software managers articulate in the fields represented). The area labeled "traditional" is not "out of date" - it is absolutely fundamental to reaching the other areas and success in implementing, as a matter or course, the high function applications of the future. The Space Station's SSE today is aimed first squarely at that "traditional" area - not because we don't know how to do it... but because we know so many different ways.

Strategic Model for Integrated Data Processing

In reference 4, a model was defined for bringing the technology of software and data systems to bear on real problems and functions. While this model was developed for information management in a given domain (that associated with a single NASA field center), it also forms the basis for strategy within the SSIS. It contains two system architecture and two management components - a total of four "layers" as depicted in Figure 9 which are briefly discussed below:

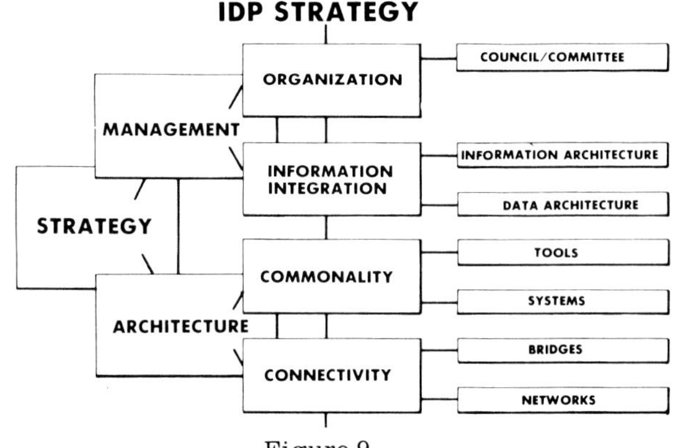

Figure 9

a) At the bottom, "connectivity" or networking. Embedding the services required to access remote information and processing is absolutely the most fundamental requisite to high functionality in today's world of data processing. The separation of users and applications from the information and decisions being processed will always exist - the issue is whether that separation is handled, architecturally, as a visible and driving design force, or appears simply as virtual interfaces to hidden layers of services and processing (much the way today's electronic mail systems present mail independent of the location of the source).

b) The second layer is "commonality", which in Space Station will take many forms. The most important, perhaps, is the concept of "common tools for common functions" and is the one the SSE concept is based on. However, commonality reaches much further into our thinking. Viewing mission data systems as institutional capabilities, themselves "tools" in the larger sense, can lead to a much more fundamental interpretation of "common tools for common functions" - within programs, across programs and certainly across the Agency.

c) The third layer, the first "management" layer, is information integration. It is recognition of another "dictate" in software architecture - "tools do not products build." No more than a personal computer spreadsheet program builds an accurate financial model for a corporation does a DBMS or expert system avoid the hard work of determining the information to be processed and the decisions to be made - the REQUIREMENTS of the applications. The focus in this strategy, however, is not on the processing (the "traditional" approach) but rather on the data and information to be processed. Identified and defined in a standard way across a program, the information base will live and evolve far, far longer than any hardware component, than any software program or application.

d) The last and top layer is organization. The term here refers to the process of efficiently distributing responsibility and accountability in such a way the standards can be kept and evolved across their necessary sphere of influence. Management councils, working groups, and other forms of committee, properly chartered, continue to be the mechanisms which work the best in NASA's environment of physically distributed and largely autonomous field centers. One of the most important aspects of forming division of responsibility developed in this strategy, is the concept of "user" versus "provider." While every organization plays both roles in some domain, the key is to recognize that providers shouldn't tell their users what they need, and users shouldn't tell their providers how to provide it. To do otherwise defeats the gains of dividing the work, and invites incredible inefficiency in management and decision-making.

The "Punch Line"

The SSE is one of three major information systems services efforts within the Space Station Program. The SSE itself focuses on the "commonality" layer in the strategy, particularly with respect to the highest leverage and highest risk components in attaining the kind of Space Station envisioned and demanded by its sponsors and potential users: the software in Space Station mission data systems. The second is a multi-program effort known as "SSIS Integration." It is focused on the connectivity and information integration layers of the strategy across the many different mission and support data systems which will, in the end, make up the environment of use and operations of the Space Station. Third is the Technical and Management Information System (TMIS) which focuses on the commonality and information integration layers in the support systems and processing required by the NASA engineers, managers, and interfacing contractor teams to develop the Space Station itself. This, then, is the context of the SSE effort, both within a strategy, and with respect to other major information systems efforts today in the program.

CONCLUSION

The SSE, in the end, will do more than support "traditional" software architectures and management. The thesis of this paper was that selecting a set of standards and tools at that level is a prerequisite to truly moving onto higher levels of architecture and system functionality. All this isn't to say that there aren't great strides yet ahead of us in the technology of software engineering of basic applications and services. Instead, I feel that we are beginning to focus too much of our energies on the wrong target. The industry has captured the best in methodologies and techniques for creating the kind of software that most of us were "brought up" on. As we continue to invest in pushing that frontier, the return on our investments is diminishing. On the other hand, the very promising and newer non-procedural and higher level software architectures are at hand.

While these approaches to software construction are dependent architecturally on traditional software, they involve remarkably new challenges in engineering and verification which are very foreign to our past methodologies. These approaches range from distributed data base applications with the verification challenge in assuring homogeneous information from heterogeneous and distributed data to expert systems with evolving rules and inference logic which belies traditional software checkout and verification approaches. Some of us are beginning to get the feeling that all the tools and methodologies we've been working on are simply simplified subsets of something much more complex and powerful in the creation of applications and data systems.

Not too long ago, I saw a chart (figure 10) which modeled productivity as a function of time in industry. The productivity moved upward as a series of overlapping "S"-curves. This graphically demonstrated the common incongruity wherein the end of one "era" in technology yields higher productivity

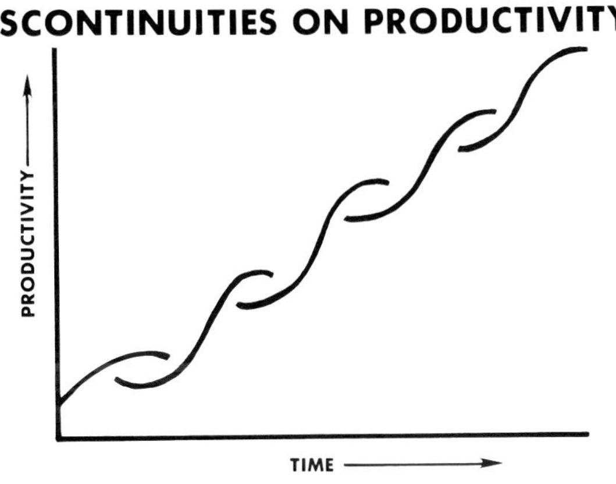

Figure 10

than the beginning of a new one. This creates the situation which the "keepers" of the old can feel comfort and "prove" the superiority of the "traditional" technology, while the defenders of the new can only speak of promises of incredibly higher functionality and productivity - someday.

Industry's movement from the steam engine to the internal combustion engine, the electronic industry's movement from vacuum tubes to transistors and from transistors to integrated circuits, the software industry's movement from assembly languages to methodology-based high order programming languages, and many more, exemplify this effect. However, in most of these cases, the new technology was totally dependent on the prior. As fundamental as the changes in approach might be, the infrastructure represented by the science, engineering, and operations of the older technology were absolute prerequisites to comprehending how the new technology could be efficiently applied to the problems they eventually solved.

It is within this context of technological evolution that I speak of "closing off" the era of procedural software development. The industry is far from done - and there is no little disagreement about the correct approaches, tools, and methods. However, it is my opinion that success in this endeavor is the most enabling and failure the most debilitating, of all efforts we could undertake to increase the functionality and automation of our systems and processes with computers and software. Put simply, today's highest technologies such as AI and expert systems will be useful only if the layers of tools and software architectures underneath are stable and well understood. Only then can we really build such systems, and only then can we push the layers of technology even higher!

REFERENCES

1.   Garman, John R., *"Data systems for the Space Station and Beyond"*, AIAA Conference - Computers in Aerospace V, San Diego, California, October 21- 23, 1985.

2.   Garman, John R., *"Forecasting Trends in NASA Flight Software Development Tools,"* AIAA Conference - Computers in Aerospace IV, Hartford Connecticut, October 22-26, 1983.

3.   Simanton, Don F., and Garman, John R., *"Improving Management Decision Processes Through Centralized Communication Linkages."*, PROCEEDINGS OF THE CENTER FOR ADVANCED PROGRAMS, University of Houston, Clear Lake, August 1985.

4.   Dunseith, Lynwood C., and Garman, John R., et al, NASA Report #JSC 22178, *"Report of the JSC ADP Strategic Planning Committee"*, Johnson Space Center, April 1986.

5.   Garman, John R., *"The Bug Heard 'Round the World"*, ACM Software Engineering Notes, Vol. 6, Number 5, October 1981.

This paper was submitted to the AIAA/NASA International Symposium on Space Information Systems in the Space Station Era.

STANDARDS FOR THE USER INTERFACE: DEVELOPING A USER CONSENSUS

Karen L. Moe[*]
Dorothy C. Perkins
Martha R. Szczur

NASA/Goddard Space Flight Center
Data Systems Technology Division, Code 520
Greenbelt, Maryland 20771

Abstract

In preparation for the Space Station era, NASA has been exploring the benefits and challenges of providing a common user interface for payload and station operation activities. The User Support Environment, or USE, is proposed to provide the software system which controls the interactions between the Space Station Information System (SSIS) and its human users (crew members, payload operators and ground support operations). The USE concept includes a User Interface Language (UIL), a run time environment and user interface management system, support tools and standards for human interaction methods. This paper discusses the goals and challenges of the USE, and a methodology based on prototype demonstrations for involving users in the process of validating the USE concepts.

Introduction

With the announcement of the Space Station as the next major NASA program, scientists and engineers began to look at new concepts for performing experiments in space in the 21st century. In particular, Space Station planners are looking for a way to improve the effectiveness of the human role in the daily operations of stations, payloads, and support facilities on the ground. As more powerful workstations continue to appear on the market with ever increasing capabilities, the application of new techniques for building user interfaces based on sound human factors principles becomes more achievable. The Space Station program has proposed the USE as both a standard for user interfaces, and a mechanism for aiding designers in building human factors concepts into their user interface systems. This paper presents the major questions posed and proposed approaches for developing a user consensus on user interface standards.

What is the User Support Environment (USE)?

The USE is a set of software tools for a flexible but standard interactive user interface to the Space Station systems, platforms and payloads. It supports ground and on-orbit integration, test and operations, for use by engineers, crew and scientists. The User Interface Language (UIL) is a subset of the USE and provides an English-like language for familiarity and readability, and a short form for real-time interactive control procedure capability. The UIL differs from traditional programming languages (such as Fortran or Pascal) which control general purpose computers, in that the UIL is a higher order procedural language for controlling space systems.   In addition to the UIL, the USE includes software utility support for graphics, display and dialogue interface control.

Why a Common User Interface?

One of the key concerns of science users of Spacelab was that the data system for Shuttle attached payloads was really a collection of different data systems separated by mission phase, NASA center and Shuttle system configuration. There were multiple interfaces that the users had to adapt to, causing duplication of effort, especially in writing test and operations procedures at the different integration sites. Another user concern was the desire for more remote control of their payloads. They recommended that NASA support user control of payload systems from remote locations (i.e., telescience) and that uniform data handling interfaces be implemented to facilitate this objective.

Standard high level languages to make the different mission phases and site locations transparent to users and payload command procedures were requested. There is significant similarity in the operations requirements at the development site, integration site, launch site and installation and operations on-orbit to drive commonality. The need for specialized programmers to support test and operations engineers would be reduced by introducing one, common, higher order control language in which engineers could write control procedures directly. A common UIL supports standardization of the user interface with multiple subsystems, creating a structure for improving comprehension of interface requirements. This is particularly important for crew members who will operate independently developed payloads. Finally, a common UIL will provide a consistent self-documenting capability for space operating procedures.

---

[*] Member AIAA

The user community of the USE includes Space Station customers, operations personnel and flight crew. Customers come from science research, operations (as for weather satellites), and commercial interests. Operations personnel support space operations during all phases of a mission life-cycle (i.e., design, development, test, operations, servicing) for payloads, platforms and Space Station subsystems. This includes test engineers, flight controllers, and a wide range of ground support functions. Finally, the flight crew will be composed of both investigators and operations personnel. In addition to the end user community, who uses the USE to control an external system (such as a payload or subsystem), the USE will have users who indirectly support customers, operators and crew. This group includes application programmers, human-computer interface designers, and USE system developers. Therefore, the USE also is intended to provide the user interface to the Software Support Environment (SSE) and the Technical and Management Information System (TMIS).

Where does the USE fit?

In early 1987, the NASA Johnson Space Center completed a draft User Support Environment Functional Requirements document which will be baselined within the Space Station Program later in the year. The USE is planned to be designed within the SSE effort, and one of the first implementations will be for the Space Station Data Management System (DMS) crew interface (including interfaces to the lab modules as well as the base subsystems). In addition to the DMS, its use is anticipated for the TMIS, the SSE, the Customer Data and Operations System (CDOS) where selected payload and platform control will reside, and the Ground Data Management System (GDMS) at the test and launch site. As a user interface standard, the USE will be controlled by an Information Systems Control Board. A USE User Working Group is planned to advise the control board by reviewing documents, validating requirements, and making recommendations. As presented later in this paper, the use of prototypes in the review process has been suggested as a key mechanism for communicating concepts identified in the USE functional requirements, both from an end user's viewpoint as well as the designer's viewpoint.

When did USE get Started?

During the past four years, a group made up of several NASA centers and user representatives has been involved in defining the scope and requirements for the USE. One step in the process was a thorough review of existing relevant languages. This study included comparing environments, sample scenarios, their strengths and weaknesses, and their applicability to the Space Station development.

Three languages currently in use within the NASA community were reviewed.

- GOAL, the Ground Operations Aerospace Language, developed at the Kennedy Space Center for Shuttle integrated test, checkout and launch operations, payload integration with the Shuttle, and the European Space Agency Spacelab operations;

- STOL, the System Test and Operations Language, developed at the Goddard Space Flight Center for instrument and subsystem development and test, payload and spacecraft integration, test and operations, and Payload Operations Control Center ground system control; and

- TAE, the Transportable Applications Executive, developed at Goddard for instrument and payload data analysis. TAE provides services for binding a system of application programs together and supporting user operation of applications through a standard interactive user interface.

In addition, several other languages/standards were identified and reviewed. These included the European Space Agency's Experiment Test and Operations Language (ETOL, which is very similar to STOL); the IEEE's draft control language standard, the System Control and Operations Language (SCOL); and a prototype crew workstation for a Space Station mock-up developed at the Johnson Space Center were examined. Several unique languages that were developed for use at single locations were also reviewed, such as the Crew Activity Planning System at Johnson, and the Mission Information Planning System at Marshall Space Flight Center.

The conclusion of this study was that no one language met all of the requirements of the UIL. Each existing operational language was designed for its specific environment of integration and test, operations or analysis. The prototype crew workstation did provide evidence that graphic displays are needed to convey complex system configurations, status and control. The study recommended the following:

- Start a focused USE development effort to provide a consistent user interface from systems integration and test through operations;

- Provide a common user interface to on-board flight and payload systems, and between flight and ground payload operations, to minimize redundant development by Space Station contractors;

- Include graphics control as an integral function of the user interface system; and

- Include features found to be essential in the mission functions to be supported (e.g., the integration and test features from GOAL, operations from STOL, menus and object oriented displays from Shuttle experience).

As a result of these recommendations, requirements were generated at four NASA centers (Kennedy, Goddard, Johnson, and Marshall), and then reviewed and integrated into the recently released USE functional requirements document. Also, prototype efforts, particularly in integration and test operations, and payload operations scenarios were developed and reviewed by end users, and factored into the functional requirements. The current status is that the USE requirements will be baselined by the Space Station Program and provided to the SSE contractor for generating specifications and designs.

<u>What are the Challenges?</u>

Four major challenges face the establishment of a common, broadly applied USE:

1. New Space Station Challenges

Foremost among the challenges that the Space Station brings to technology and science is its unprecedented life expectancy. A major implication of this longevity is that it is not possible to fully specify ahead of implementation the lifetime requirements and expectations of the system. For example, the full complement of instruments and payloads to be installed across the life of Space Station is unknown. Therefore, the demands to be placed on processing systems and resources, and user and system information needs are also unknown.

2. Remote User Access

Among potential Space Station science users, there is a growing momentum to institute a method of user access to payloads called telescience, as mentioned previously. The major thrust of this operational concept is that users be allowed to command their instruments and receive their instrument data both remotely (e.g., from their home institutions) and in real or near-real time. This is not a new concept to NASA. Precedents exist in earlier satellites such as the International Ultraviolet Explorer, which allows real time instrument manipulation, and in instruments on geostationary satellites which return quicklook data for engineering purposes. In general, however, user interaction with payloads, particularly where those payloads share resources with other payloads,

has been limited and performed from a central location. Operations are, therefore, well controlled and predictable.

To provide this sort of access to all users of Space Station is unprecedented in its scope. The method and breadth of such interaction has major implications on the institutional systems that must support it. Perhaps most important, they must have the flexibility to react to dynamically changing circumstances and to support a diversity of user needs, built through intelligent automation and powerful tools to support human operators of Space Station.

3. User Involvement

Extraordinary effort must also be made to involve users in the ongoing design and development of the Space Station systems. While it is user-proposed ways of doing business that will drive the NASA design, user acceptance of the system, as with any system, is critical for its success and for retaining user support. Past experience reinforces this goal. A classic scenario of system design process for users is that they are asked first for requirements, requirements are frozen, designers translate requirements, design and development occur, and the user then lives with the results -- which may bear no resemblance to preconceived notions of what he/she wants. Often this occurs because requirements evolve over the time when development is taking place, and there is no mechanism in place to insert new or modified requirements into the development process. In the Space Station era, however, when requirements are <u>expected</u> to change, this cycle is obviously unacceptable.

In any effort at standardization, there are also perceptions which must be overcome -- perceptions that imposed systems and standards increase user costs, limit flexibility, limit independence. The benefits of having such standards must outweigh costs in resources and loss of control.

To its credit, the Space Station program office has given high priority to user involvement in planning. Committees, task forces, workshops, research dollars all exist to encourage user action in Space Station design. To date however, much of this involvement is through paper -- review of proposed designs and requirements. Because of the uncertainty of operational impacts and the inability to visualize benefits of various approaches, the major challenge facing developers is establishing an ongoing and effective process for user involvement.

4. Impacts on Major Contractor Efforts

Considerable effort has already been expended in NASA to develop new Space Station concepts. For USE, several centers have been actively involved in defining user requirements for the types of users

149

that predominate at their sites. Goddard, for
example, has a large complement of science users
and has worked to document their needs as part of
USE requirements (reference CDOL document).
The work in defining requirements and concepts
has gone beyond mere paper statements, however.
Work to develop detailed scenarios of Space Station
use and to instantiate portions of those scenarios in
prototype form has also been accomplished. These
prototypes have served to refine both requirements
and concepts, as users view and react to their
operation. A challenge for Space Station managers
is to assure that the results of this work are not lost
to major new contractors.

What are the Implications?

Prominent among the implications of meeting
the Space Station challenges is that Space Station
systems must be designed with change -- addition of
new features and functions, replacement of
component parts -- in mind. Evolution of technology
and requirements is inevitable and must be planned
into the system. Evolution of the user population
must also be assumed -- both different people using
systems across their lifetime, and different user
information needs as the environment and understanding of the systems change. Unknown future
operational impacts on the system must be planned
for, and software design must be innovative and
allow for selective enhancement when new system
requirements become known. At Goddard, we
believe that the construction and active use of user
prototypes are powerful methods for assuring that
flexibility is built into the NASA systems.

What is Meant by Prototyping?

Prototypes can be constructed with various levels
of sophistication and fidelity. At their simplest they
are mockups, frequently on paper though more
often now on graphics terminals, of the item to be
built. A prototype can be a dynamic simulation of
an activity or a sequence of events, with simulated
control between steps. It can also be a working
model of a system, one which can evolve into an
operational system or be used for research
purposes.

In our view, prototypes serve two major
purposes. First, by modeling some portion of the
system, they can be used to refine and validate
requirements and interfaces, either by offering
several alternative approaches to solving a problem,
or by causing a user early in the system life cycle to
work with actual system interactions. Realizations
caused by such interactions often cause new
insights into concepts. Often such concepts become
more fully understood and refined; sometimes the
prototype is discarded with the realization that the
concepts are flawed.

The second purpose a prototype can serve is to
explore alternative engineering design solutions to
a problem. These solutions can apply to user
interfaces as well as internal system processing.
Often a prototype can point the way to a good
solution without incurring the expense of a full
implementation.

At Goddard, we have developed a "methodology"
-- a series of steps -- that we follow in developing
prototypes. Our focus is on user interface
prototyping, and the steps are as follows.

1. Hold a "talk session" with the intended users
   of the system, establishing concepts to be
   demonstrated and audience of the
   demonstration.

2. Build a paper prototype of a series of static
   displays that the user would view. Talk these
   through with the intended prototype
   audience.

3. Build a rapid prototype on a graphics
   workstation to acquire early user feedback.
   This prototype typically implements a
   scenario developed with the users, and is
   typically not robust. That is, it is possible to
   sequence from one screen to the follow-on
   screen, but it is not possible to move
   arbitrarily among displays as you would in a
   complete system.

4. Build a robust prototype, incorporating
   underlying applications (themselves either
   prototypes or complete packages),
   communications among processes, and
   flexible maneuvering. This prototype, while
   lacking perhaps the full functionality of the
   system, would give an accurate view of what
   the system would look like and how it would
   react to user commands. Users would be able
   to exercise this prototype independently.

5. Replace stubs in the prototype with real
   subsystems and components.

In fact, these steps are not always sequential.
Feedback and reconsideration of ideas may cause
steps to be repeated. Figure 1 shows the iterative
nature of prototyping.

Similarly, the route chosen after the fourth step,
robust prototyping, may vary depending on whether
the prototype is being used to validate requirements
or to validate designs. If the former, it is not likely
to be carried forth to an operational system. Its
purpose has been served. If, on the other hand, it is
being used to validate designs, the chosen approach
is then likely to be carried through to completion.

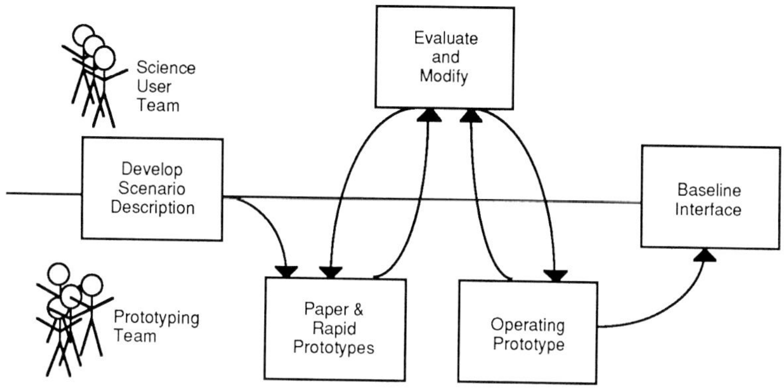

Figure 1. The iterative process of the Prototyping Methodology

Conclusions

At Goddard, this prototyping approach has been exercised on several projects related to Space Station operations. The TAE has been enhanced to facilitate development of graphical rapid prototypes and is the major tool in the demonstration systems. Experience has shown that rapid prototypes significantly facilitate user comprehension of system functions, merely by demonstrating how a mock-up (from step 3) of the user interface works. A USE User Working Group has been established by the Space Station program to assist in the process of involving user review. We have recommended that prototyping become a key element in the contractor development of the USE. Prototypes are needed to communicate specifications from NASA to the contractor, as well as to validate contractor interpretation and design by NASA's user community.   Prototyping key concepts and salient features of proposed user interface standards, applied in typical operations scenarios, greatly enhances the user's ability to respond and have their concerns understood. Thus, a prototyping methodology holds great promise for user acceptance and adoption of standards.

References

1. Space Station Information System User Support Environment Functional Requirements, Final Draft, NASA/Johnson Space Center JSC 30497, April 1987.

2. Space Station Operations Language, NASA/Kennedy Space Center, L. Dickison/DL-DED-22, March 1987.

3. Space Station User Support Environment/User Interface Language: Customer Data and Operations Requirements, NASA/Goddard Space Flight Center, Data Systems Technology Division, December 1986.

4. Space Station Information System User Interface Scope, NASA/Johnson Space Center JSC 32004, May 1986.

5. Telescience Operations and Systems Concept for Command and Control of Space Station Payloads, NASA/Goddard Space Flight Center, K. Moe, M. Rackley, Data Systems Technology Division, and M. Neely, Computer Technology Assoc. Inc., August 1985.

6. A NASA Transportable Applications Executive for Interactive Applications, D. Perkins and D. Howell, Proceedings of the Workshop on Digital Image Processing in Remote Sensing, Imperial College of London, March 1984.

Systems 2: Integrated System Architectures
Co-Chairmen: Richard G.Schell and Bernard Curbelie

SPACE STATION INFORMATION SYSTEM
INTEGRATED COMMUNICATIONS CONCEPT

John Muratore *
James Bigham **
Virginia Whitelaw +
Walter Marker ++
National Aeronautics and Space Administration
Lyndon B. Johnson Space Center
Houston, Texas

Abstract

This paper presents a model for integrated communications within the Space Station Information System (SSIS). The SSIS is generally defined as the integrated set of space and ground information systems and networks which will provide required data services to the Space Station flight crew, ground operations personnel, and customer communities. This model is based on the International Standards Organization (ISO) layered model for Open Systems Interconnection (OSI). The requirements used to develop the model are presented and the various elements of the model described.

Introduction

The Space Station will be one of the major NASA missions in the 1990's. A major concern during the early phases of development of the Space Station Information System (SSIS) has been the specification of SSIS communications services. Early in the Space Station planning process it was recognized that a communications services model was required to allow different organizations to discuss and evaluate service requirements. The integrated communications model developed by the Johnson Space Center Space Station Data Flow Working Group in response to that need is presented in this paper.

It is valuable to discuss the SSIS model for two reasons. First, because the Space Station represents a major long-term NASA committment, many future space systems will require information system interoperability with the SSIS. As such, we must evaluate the the SSIS to insure it incorporates our best ideas on information systems interoperability because it will affect the design of space information systems for the next thirty years.

Second, the SSIS is in many ways the leading edge of information system technology development for NASA. It represents the largest integrated information system that NASA has ever attempted to assemble and a number of significant lessons have already been learned in this area and incorporated into the model .

The model presented in this paper is intended to serve as an end-to-end integrated communications model for the NASA Space Station Information System (SSIS) . This model has been developed by the SSIS project to serve as a common framework for discussions and evaluations of key SSIS issues during development phases. It also provides a common terminology that SSIS developers can utilize to describe and test requirements.

This model is not a design for the SSIS. It serves as a reference model for evaluating concepts and identifying key interfaces.

The model is based on the seven layer International Standards Organization (ISO) model for Open Systems Interconnection (OSI). The adoption of "OSI-like" models for layered communications is a key concept for Space Station. Use of this model does not require use of the ISO protocols which have been proposed to implement the ISO model, although the ISO protocols are natural candidates for consideration.

User Scenarios As Requirements

In order to define end to end services requirements which the SSIS must support, a set of scenarios describing typical uses for end to end services was developed. These scenarios were originally developed from a systems operator's viewpoint (flight controller) and were then broadened to include system user's (scientific/commercial) requirements. These scenarios are shown in Table 1.

---

* Flight Controller, Mission Operations Directorate
** SSIS Project Integration Manager, Space Station Project Office, Member AIAA
+ Research Engineer,Avionics System Division, Member AIAA
++ Research Engineer, Tracking and Communications System Division, Member AIAA

Table 1 - End to End Service Scenarios

Scenario	Endpoints			
	Ground to Space	Space To Ground	Space To Space	Ground To Ground
Electronic mail	X	X		X
Remote system access	X	X		X
Transferring computer memory loads/dumps/ programs/textual information	X	X	X	X
Retrieving historical data from databases	X	X		X
Commanding remote systems (non time critical)	X	X	X	X
Commanding remote systems (time critical)	X	X	X	X
Teleconferencing	X	X		X
System performance telemetry		X	X	X
Scientific instrument serial data		X	X	X
Scheduling messages	X	X		X
Initiating/controlling remote automated processes	X			X

Definition Of End To End Services

Based on analysis of the user scenarios, and examination of services provided by modern computer networks, the following generalized set of end to end services were identified for Space Station:

File to file transfer - The ability to move an information file from one processor in the SSIS to another remote processor in SSIS.

Virtual terminal - The ability to interact with a SSIS application from a remote terminal as if the terminal was directly connected to the application.

Real time telecommand/telemetry - The ability to exchange a stream of commands and measurement values between two remote SSIS systems under specified time constraints.

Non-real time telemetry delivery - The ability to deliver a complete set of remote measurements collected at one SSIS endpoint to another SSIS endpoint under a specifiable delay.

Database access - The ability to access and change remote databases from endpoints in SSIS.

Remote job entry - The ability to control remote automated processes from endpoints in SSIS.

Application to application messaging - The ability to exchange messages between applications located at different SSIS endpoints.

Interpersonal messaging - The ability to exchange electronic mail between SSIS users.

Audio/Video - The ability to transfer audio and video signals between SSIS endpoints.

Multi-media teleconferencing - The ability to connect multiple SSIS endpoints in multi-media (voice, video and data) conferences.

Multicast message distribution - The ability to broadcast information of common utility from one SSIS endpoint to multiple SSIS endpoints.

These analyses led to the following definition of end to end services for the Space Station program :

End to end services are those capabilities provided by SSIS that provide for coordination of activities at multiple separate SSIS endpoints or data transfer between multiple separate SSIS endpoints.

SSIS Endpoint Definition

A discussion of the end-to-end model requires definition of a SSIS endpoint. For the purposes of this model, the SSIS endpoint is the initial source or ultimate destination of data being transported by the SSIS. It may be roughly described as "where the user is located". A SSIS endpoint may exist anywhere in SSIS.

The initial source or ultimate destination can be a person or an application process running in a computer. In some cases (e.g a Synthetic Aperture Radar (SAR)) it may be a hardware device.

The SSIS Primary And Secondary Domains

The Space Station Program (SSP) currently defines the SSIS as all of the information systems and data links that process Space Station related information. The definition is very useful from a programmatic standpoint because it shows the SSP committment to an open system that includes users at their own facilities. In many ways it is the ultimate statement of interoperability. It is however, a very difficult definition to use in specifying end to end services.

For the purposes of the integrated communications model, the SSIS can be divided into two parts. These parts are the Primary and Secondary SSIS Domain. The Primary Domain consists of those subnetworks of the SSIS which agree to exchange information according to SSIS standard protocols for the purposes of guaranteeing successful completion of end to end services. The Secondary Domain consists of those subnetworks which do not agree to use SSIS standard protocols.

Information systems in the SSIS Secondary Domain are not "second class" information systems. Rather the distinction between the Primary and Secondary Domain information systems reflects whether the information system has invested in the implementation of SSIS standards. It is expected that all critical "backbone" information systems supporting life and facility safety will be part of the Primary Domain. Users may choose to operate in the Secondary Domain according to their own standards for their own reasons (efficiency, performance or cost). A critical challenge to SSIS interoperability is to insure that Secondary Domain users have access to quality end to end services.

SSIS Service Access Points

In order to provide access to end to end services to Secondary Domain users , the SSIS provides SSIS Service Access Points (SSIS SAPs). SSIS SAPs are entry points into the SSIS.

If a user's endpoint is in the Secondary Domain, the user may perform remote access to the SSIS by attaching a Secondary Domain endpoint to a SAP. The user is responsible for providing any communications services between a SAP and the Secondary Domain endpoint.

When a user's endpoint is in the Secondary Domain and it is attached to a SAP then the SSIS will guarantee end to end services to the SAP. The user will be responsible for continuing the service beyond the SAP. The SSIS is not responsible beyond the SAP. Figure 1 illustrates this arrangement.

For example, if a user at a university wishes to perform a file to file transfer to an onboard station instrument

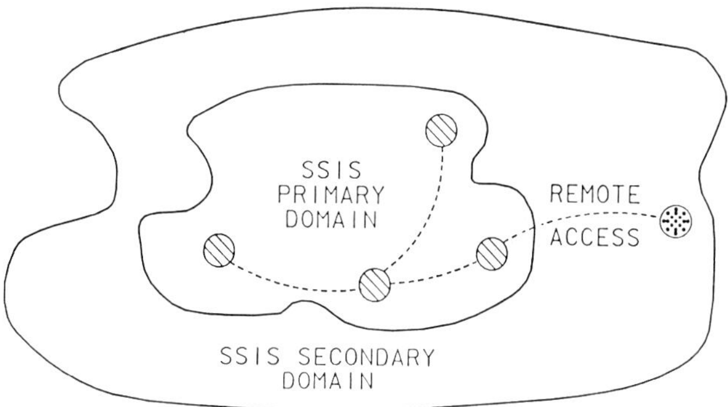

Fig. 1 SSIS Primary and Secondary Domains

computer, the SSIS would be responsible for insuring error free service from the instrument's onboard interface (an onboard SAP) to a ground SAP (such as a NASA Payload Operations Control Center). The user would be responsible for error control on the links between the university and the ground SAP (for example error control on a noisy university phone system). In this way the user is free to develop whatever arrangements which best suit his needs, without burdening the SSIS design with responsibility for guaranteeing service over networks it does not control (e.g. a university phone system).

It is important to note that a SAP may be implemented in hardware (i.e. a hardware connection for a high rate payload or in software (Data Management System (DMS) supplied interface to a payload or core system application).

The concept of Primary and Secondary Domain and service access points can be applied across multiple NASA information systems and could be one of the key elements to information systems interoperability. Figure 2 shows how the concept could be applied in the SSIS context to provide interoperability between the SSIS Primary Domain and a number of subnetworks in the SSIS Secondary Domain that contain valuable management, engineering and scientific data.

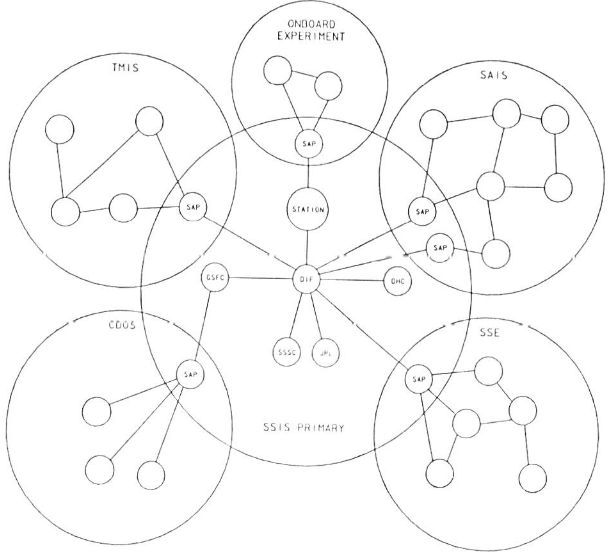

TMIS - Technical Management and Information System
CDOS - Customer Data Operations System
SAIS - Science and Applications Information System
SSE - Software Support Environment
DIF - Data Interface Facility
DHC - Data Handling Center
SSSC - Space Station Support Center
GSFC - Goddard Space Flight Center
JPL - Jet Propulsion Laboratory

Note: This is an example Topology and is not intended to show a specific approved design.

Fig. 2 - Example SSIS Topology showing Primary and Secondary Domains

During the development of the integrated communications model, a large number of arguments centered on the anticipated importance of different service types that SSIS must support. In our opinion, these discussions generally divided into two major advocacy groups.

The "Telemetry advocates" stated that most of the bulk of SSIS data will be traditional telemetry, audio and video. The "Telemetry advocates" showed estimates that this type of data may be as much as 98 % of the bulk of space station data. They further argued that because these services comprise most of the bulk of the data, and must be capable of operating at high rates (greater than 100 Mbps), that the system must be optimized for these type of services.

The "Telemetry advocates " further pointed out that telemetry type services require minimum network routing capability. The possible maximum number of destinations is on the order of 500. Typically (although not always) these are "first level destinations" where significant offline processing is performed on data before it is passed onto the users.

The "Telemetry advocates" did not express any interest in the higher OSI model services such as session or presentation services .

Because of all of these arguments the "Telemetry advocates" argued for access in the SSIS at the lowest possible layers of the ISO seven layer model. Typically the argument was that ISO type network routing services are unnecessarily general (capable of handling thousands of destinations) and that telemetry type services required access to "stripped down" routing services at the network layer of the ISO model. Some groups maintained that telemetry users should provide their own routing identification and SSIS should provide access at the data link layer (layer two).

The "Computer Applications Services advocates" argued that remote operations of science instruments (telescience) and the station subsystems requires access to advanced computer applications services such as virtual terminal access, file to file transfer and remote job management. These services require all of the layers of the ISO model.

The "Computer Applications Services advocates" also argued that the possible number of destinations in SSIS is very large because there will be thousands of users of SSIS who need access via remote terminals to accomplish program management, sustaining engineering, operations and scientific research in the geographically distributed Space Station program. Computer applications services require routing capabilities to these final destinations in real time. They also pointed out that their transactions are typically small in length and not continuous so that they are a small amount of the total bulk.

Finding some resolution to these points of view is critical for the modeling of SSIS. Naming and address identification, routing, error control, and location and size of buffers are typical of the types of issues that depend heavily on the resolution of these issues.

The current integrated communications model recognizes that both types of services will be important and efficient mechanisms must be provided for both kinds of services. The communications model defines two major types of services : Low service requirements (typically telemetry) and and High service requirement (typically computer to computer) traffic.

Low Service Definition

Low Service requirement information in space station can be loosely characterized as large volume one way data streams. They comprise the majority of the bulk of data transported by SSIS. They typically require routing only to a limited number of first level destinations . This number is anticipated to be approximately 500 which is similar to the number of destinations supported in the current NASA Communications (NASCOM) network.

Another characteristic of Low Service users is that they require minimal processing services during a communication session (they may require large offline services such as Level Zero Processing to reverse data ordering, but these services are not required during the communication).

Low service users are generally only interested in routing to first level destinations where significant offline processing takes place prior to forwarding the data to users.

High Service Definition

High Service Traffic typically consists of Computer to Computer sessions as normally seen in a remote terminal environment where high layer applications services (such as file to file transfer, electronic mail, or virtual terminal services) are required. These are relatively low volume two way information exchanges between applications processes (computer software) which utilize "open" protocols for message transfer. In this definition "open" refers to the fact that the protocols are published and any user that wants to join in transactions only has to implement the protocol.

High Service (computer to computer) transactions are generally interested in moving information between final destinations and not just first level destinations. Typically this would require the capability to reach up to 10,000 destinations. A key characteristic of High Service Users is that they require extensive high level services during a communication session (e.g a file to file transfer protocol).

The High Services users require most of the services provided by all of the layers of the OSI model. For example a file to file transfer user requires session and transport layer services. Telemetry users typically want stripped down services and only require the lowest layers of the OSI model (physical, data link and network).

In order to provide maximum efficiency for these two uses, two major types of SSIS SAPs are anticipated,a high service SAP and a Low Service SAP.

High Service SAP

Typically a layer seven (Applications) SAP will provide access to the High service user. The following services will be available at the Applications Layer SAP:

File to file transfer
Virtual terminal
Database access
Remote job entry
Application to application messaging
Interpersonal messaging
Multi-cast message distribution

A "Full Service SAP" supports all of these services. It is expected that most Applications SAPs may only implement some of these services.

SSIS Logical Channels and Low Service SAPs

Low Service users will typically be serviced by a Network layer (Layer three) SAP. A layer three SAP will provide access to the "SSIS Logical Channel " (SLC) mechanism . The SLC mechanism is a managed association between airborne and ground elements that provides an efficient mechanism for routing information to first level destinations. The SLC provides for separation of information intended for different destinations even though the information is travelling on the same physical link.

Each SLC has certain characteristics such as Grade of Service, Destination, Error Control, Priority , Latency and Timing etc...

At minimum, the following services will be provided to the user via layer three SAPs,

Real Time Telecommand/Telemetry
Non-Real Time Telemetry Delivery
Audio/video
Multi-Media Teleconferencing

Service Diagram

Figure 3 illustrates how the two approaches can be used to allow full access to services for computer to computer users, while allowing stripped down services for telemetry users.

It is expected that the SSIS will provide at least two kinds of SLC service . A bit stream SLC service will be supplied so that users can provide any bit stream to a SLC and remove it at the other end. This provides a "logical channel switched bitstream" service to users. A packet SLC service will also be provded so that a user will be able to place packets into a SLC for routing to a first level destination.

It is important to note that traffic from High Service users located at layer seven can be combined into SLCs at layer three if there is a common destination. For example traffic from the various subsystem computers onboard the station that is intended for the Space Station Support Center (SSSC) could easily be combined into a single SLC destined for the SSSC.

Another critical observation from the model is that High Service naming and address identification techniques can be completely independent of Low Service techniques. This allows Low Service to adopt very limited efficient identification techniques while High Service can use more complex more flexible identification techniques.

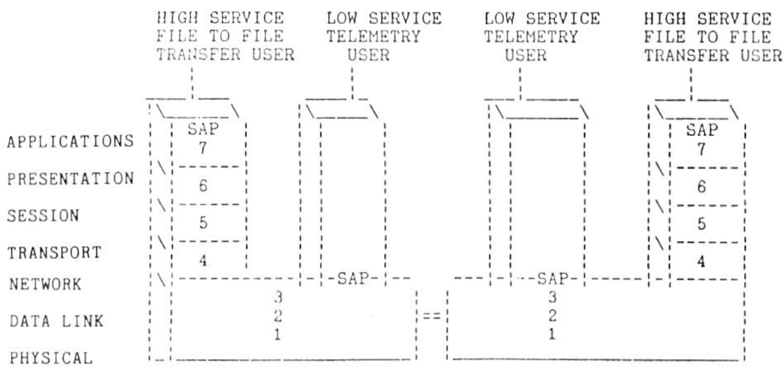

Fig 3 - Alternative Approaches to Service for Telemetry and Computer Applications Services.

<u>Implications For Interoperability</u>

Several techniques used in the construction of the SSIS Integrated Communications Model are applicable to other NASA information systems. For example, the use of the Primary and Secondary Domains concept is a way to logically separate responsibilties for completing end to end services between interoperable information systems.

The use of the SSIS Service Access Point is general and could be used in a number of information systems. The observation that space systems require two major types of access, network layer and applications layer, may be valid for other space systems and will be interesting to observe over time.

A large part of the constructs used in the communications services model were borrowed from the Open Systems Interconnection work being performed under the auspices of the International Standards Organization. Use of the ISO work as an initial baseline has greatly assisted the Space Station program in developing a integrated communications model that "promises interoperability". The SSIS experience suggests that this is a good starting point for NASA-wide interoperability studies.

Finally, the baseline set of end to end services identified for SSIS maps very well to current commercial standards development. This list could be viewed as a services menu which future programs could examine for implementation.

<u>Status of the SSIS Integrated
Communications Model</u>

The SSIS Integrated Communications model has been included in the Space Station Information System Architecture Definition Document[1]. In this capacity it is not binding on the various areas of the program, but does serve as a reference . This reference has been used to construct the various Requests For Proposals for the Space Station project. In addition a number of change requests to the formal Space Station Program Requirements Document[2] based on the Communications Model have been written. These change requests allocate functions from the communications services model to flight and ground systems. These are currently under review by the larger Space Station design, operations and user community.

<u>Acknowledgements</u>

This work has been the result of the following individuals outstanding support to the Data Flow Working Group : Glen LeBlanc, Carl Stroud, Tom Ohnesorge, Donald LaCombe, NancyLi Landi, James Dashiell, Kelly Hunter, Edward Dong, Henry Wyle, Richard Carper, Adrian Hooke, Edward Greenberg, Lee Neitzel and Richard desJardins . We would also like to thank Robert Castle , David Whittle and Jack Knight from JSC Mission Operations for their support and comments on the integrated communications modelling activities.

<u>References</u>

1. <u>The Space Station Information System Architecture Definition Document</u> , Johnson Space Center Document 30225, Revision A, October 9, 1986

2. <u>The Space Station Program Requirements Document</u>, Johnson Space Center Document 30000, September 1986

SPACE STATION INFORMATION SYSTEM REQUIREMENTS
FOR INTEGRATED COMMUNICATIONS

Dr. Walter Marker [*]
Dr. Virginia Whitelaw [**]
John Muratore [+]
James Bigham, Jr. [++]

National Aeronautics and Space Administration
Lyndon B. Johnson Space Center
Houston, Texas

Abstract

Space Station Information System (SSIS) requirements for integrated end-to-end communications are presented. The SSIS is defined as the integrated set of space and ground data and information systems and networks which will provide required data services to the Space Station flight crew, ground operations personnel, and customer communities. This model is based on the International Standards Organization (ISO) layered model for Open System Interconnection (OSI). These SSIS requirements include: grades of service; priority classifications; systems management; flow control; bandwidth allocation; and standard SSIS data services.

Introduction

Roughly a year ago, the Space Station Information System (SSIS) Data Flow Working Group (DFWG) was formed by the Johnson Space Center to address the problem of identifying the end-to-end communication services needed by the SSIS and defining how these services could be provided. Their work resulted in a SSIS Data Flow Model (Appendix J of Reference 1) as well as a set of requirements entered into official Space Station Program documentation. These requirements were further broken down and made binding on NASA's flight portions of SSIS by becoming part of the JSC Space Station Phase C/D Requests for Proposals (RFP's). The set of SSIS requirements and the derived requirements in each portion of SSIS (i.e., referred to as SSIS subnets) have also been submitted as a Change Request (CR) to the Space Station Program Definition and Requirements Document.[2] This paper summarizes the requirements that were derived from the Data Flow Model, how they were molded by various user's inputs, and how they impact the various SSIS subnets.

SSIS Requirement Process

The DFWG performed a top down analysis of the SSIS data communications requirements. The fundamental driving force which initiated the effort was a set of user scenarios. These scenarios originated from the perspective of a JSC Mission Control Center Flight Controller, but were gradually broadened through wide review to include a variety of scientific users' scenarios. The fundamental question that was asked in all scenarios was what end-to-end SSIS data communications services are required to support the scenario. All such identified services were documented and submitted to the Space Station Program as high level SSIS requirements. However, within the Space Station Program, SSIS is not a program deliverable. Rather the program deliverables are SSIS subnets such as the Data Management System (DMS) or the Communications and Tracking System (C&T). High level SSIS requirements are actually implemented in the Program by being specifically introduced into the various SSIS subnets. Hence, it was necessary to decompose the high level SSIS requirements into specific detailed requirements that could be allocated to four major SSIS subnets and elements: DMS, C&T, Data Interface Facility (DIF), and the Ground Distribution System. The overall approach followed was to identify SSIS-wide service and each subnets role in providing them; identify SSIS-wide standards that could be adopted and implemented within each subnet; and identify SSIS-wide network management responsibility. Not surprisingly, the details of the allocation revealed that each of these systems have highly complimentary service requirements. For SSIS to provide end-to-end services, it is necessary for each system in the chain to support a given service according to tightly interlocking requirements.

The overall effect of the subnet requirements was not to impose radically different requirements on any of the SSIS subnets. Instead the effect was to specify sufficiently detailed requirements to insure that all subnets work smoothly together as part of a larger SSIS picture. The SSIS CR's have been formally incorporated into the Space Station Program elements that are under direct JSC control; DMS and C&T. These are now part of the Request for Proposals (RFP) of Work Package 02. The DIF and Ground Distribution portion of the CR's have been submitted to NASA Headquarters, Office of Space Science and Application (Code E) and Office of Space Operations (Code T), for inclusion in their appropriate documentation.

SSIS Data Flow Model

One of the principle services of the Data Flow Model was to provide a common framework and terminology to allow personnel from a variety of disciplines to discuss the issues. The Data Flow Model is based on the International Standards Organization (ISO) Open System Interconnection (OSI) model and can be used to discuss any data

---

*   Research Engineer, Tracking and Communications Division, Member AIAA
** Research Engineer, Avionics Systems Division, Member AIAA
+   Flight Controller, Mission Operations Directorate
++ SSIS Project Integration Manager, Space Station Projects Office, Member AIAA

communication system. While numerous concepts and
terms are discussed in detail in the report of the
Data Flow Model, the following definitions are
critical to the subsequent section of this report.

o SSIS Primary Domain - consists of SSIS
subnets that agree to exchange information
according to SSIS standard protocols and management
functions.

o SSIS Secondary Domain - consists of all
other subnets in SSIS.

o A SSIS end point is "where the user is."
It can be either in the Primary or Secondary
Domain.

o SSIS services are only guaranteed by SSIS
between end points in the Primary SSIS Domain.

o SSIS services are accessible from the
Secondary Domain, but the quality of the services
cannot be guaranteed by SSIS since they are
provided, in part by networks over which SSIS has
no direct control.

By these definitions, the requirements that
follow apply to the SSIS Primary Domain.

<u>Areas of SSIS Derived Requirements</u>

The high level SSIS derived requirements were
broken down into 8 specific areas that were applied
to the SSIS subnets.

1. Grade of Service
2. Priority
3. Systems Management
4. Flow Control
5. Bandwidth Allocation
6. Services and Service Access
    Points
7. Naming and Addressing

Each of these areas will be discussed in turn.

<u>Grades of Service</u>

The SSIS shall provide the following grades of
service to the user on each SSIS subnet (DMS, C&T,
DIF, and ground distribution systems):

1. Grade I Service: Data output from a
subnet are guaranteed to be complete, error-free,
with input sequence maintained, and without
duplication. All necessary checks will be
performed to guarantee these characteristics.
Error-free is defined to be a bit error rate (BER)
of no more than 10e-12 (undetected, uncorrected
errors). Expedited data shall not request Grade I
Service.

2. Grade II Service: Data output from a
subnet are possibly incomplete, possibly with
errors, possibly out of sequence, and possibly with
duplication. Sufficient forward error correction
shall be provided to achieve a BER of no greater
than 10e-8. Network data units or frames in which
errors cannot be corrected are not discarded, but
are flagged as being in error. Network data units
or frames that are out of sequence are still
transmitted. No check on completeness or
duplication is required. Users who desire that
their non-real time Grade II data be guaranteed in

sequence, or without duplication, or that
uncorrectable errors not be sent, or any
combination of these, may obtain these options
through use of standard ground-based SSIS Level
Zero Processing (LZP) facility. The LZP is a new
NASA facility which will provide basic data
processing capabilities for the user community.

3. Grade III Service: Data output from a
subnet are possibly incomplete, possibly with
errors, possibly out of sequence, and possibly with
duplication, with a BER of no greater than 10e-5.
No checks are required on completeness, sequence or
duplication. Users who desire that their non-real
time Grade III data be guaranteed in sequence, or
without duplication, or that data known to contain
errors not be sent, or any combination of these,
may obtain these options through use of a standard
ground-based SSIS LZP facility.

These Grades of Service are modifications of
the Grades of Service originally proposed by The
Consultative Committee for Space Data Systems
(CCSDS). The original CCSDS recommendations
contained 4 grades of services. SSIS modified
these recommenations because of strong inputs from
the user community. One of the old CCSDS Grades of
Service required discarding any data in which the
communication system detected a transmission error.
Many scientists objected to the communication
system discarding their data. They reasoned that
some errors are critical to data integrity and some
errors are tolerable. However, only the scientist
has the knowledge to make this decision.
Therefore, SSIS agreed to flag any data in which an
error had been detected, which could not be
corrected. The data would then be delivered to the
end user, and they could make the decision on
whether or not to discard the data. Hence, one
CCSDS Grade of Service was eliminated.

To minimize the misdelivery of SSIS network
data units, headers used in routing shall be
protected to a BER of 10^{-12} for all Grades of
Service. Clearly a packet with an uncorrectable
error in a header may have a corrupted distribution
address. Hence it is not clear where to route that
packet. Thus, packets with header errors will be
discarded.

Table 1 summarizes the characteristics of the
Grades of Service.

The idea is to satisfy the Grades of Service
on each subnet. In other words, DMS guarantees the
Grade of Service on the DMS Local Area Network
(LAN) and C&T, and the DIF working together
guarantee the Grade of Service on the RF link. The
Ground Distribution System guarantees the Grade of
Service on the ground. Each subnet informs the
adjacent subnets of the required Grade of Service.
When a communication service requires
retransmission because of errors (i.e., for Grade I
Service), the retransmission will be limited to the
subnet that encounters the error. There are
several reasons that this is desirable. It is
anticipated that the RF links will be the
bottleneck of SSIS. There is a definite incentive
to keep retransmission requests from other subnets
from propagating to the RF link. Furthermore, the
amount of time a message must be held until
acknowledged is directly proportional to the
distance over which the acknowledgement takes
place. Limiting retransmission requests to subnets

Table 1 Grade of Service Characteristics

GRADE	I	II	III
Completeness Guaranteed	Yes	No	No
Bit Error Rate	10e-12	10e-8	10e-5
Error Detection	Yes	Yes	No
In Sequence Guaranteed	Yes	No	No
Duplicates Possible	No	Yes	Yes
Delivery of non-correctable errors to user	No	Yes	Yes
Retransmission strategies if required to achieve BER	Yes	Yes	No
Forward Error Correction (if necessary to achieve BER)	Yes	Yes	No
Header BER	10e-12	10e-12	10e-12

and their interfaces, greatly reduces buffer requirements throughout SSIS.

There are some qualifications of these Grades of Service guidelines. First of all, a subnet is allowed to substitute Grade I Service for any requested Grade of Service, provided performance and latency requirements can be met. Second, each SSIS subnet must be designed to handle 1 Mbps of user requested Grade I Service. Third, if a user requires a better BER than Grade I Service (i.e., better than BER of 10^{-12}), they are permitted to utilize their own end-to-end retransmission protocols between their own equipment (e.g., their computer onboard to their computer on the ground). However, SSIS will not offer end-to-end acknowledgement as a standard SSIS service.

Priority

SSIS will provide at least 4 levels of end-to-end priority within the SSIS Primary Domain. The characteristics of the priority levels are:

Emergency	Limited to messages which directly involve the health and safety of the Space Station
Expedited	Limited to traffic in which maintenance of timing characteristics is essential, examples include real time audio, real time video, and real time data involved in tight control loops
Normal	Normal real time or near-real time transmissions
Background	Bulk file transfers

Each communications path in each SSIS subnet need not support all levels of priority. A given communication path need only support the levels of SSIS priority that users connected to it require. For example, if a dedicated leased line is employed to link a scientist to SSIS only background and normal priority might have to be supported on that line.

Each priority level will correspond to a particular latency range (mean and variance) within which that class of messages must be delivered. The specific values are not specified at the current time. However, it is clear that a high priority message must gain access to SSIS resources faster than a low priority message. Similarily, under conditions of resource shortage, traffic will be handled according to priority. It is of course possible to over engineer any subnet so that all messages can be handled within the performance criteria established for emergency traffic. In this case, the SSIS priority system need only be activated during off-normal conditions.

Systems Management

The SSIS Systems Management functionality is an area that is now undergoing rapid development. SSIS plans to provide a high level, distributed system management function called the Network Integration Manager (NIM). The idea is that the NIM will receive sufficient event reporting from all SSIS subnet managers to provide high level coordination of SSIS network management. Specifically, the NIM is responsible for intersubnet coordination of:

Configuration management
Fault management
Performance Management
Security Management
Accounting Management

Each subnet in the SSIS Prime Domain (DMS, C&T, DIF, and Ground Systems) will implement its own system manager in the above areas so that it can both maintain operational integrity of its own subnet, independent of the NIM, and report to the NIM in these areas.

Flow Control

The SSIS must perform flow control and congestion control on network transmissions. The only way this can be achieved in a distributed system is by interaction between the NIM and the SSIS subnet managers. Each SSIS subnet must detect and, as much as possible, correct its own congestion problems. Any load throttling must take priority of the traffic into account. In addition, each SSIS subnet must be able to exchange, interpret and respond to flow control information from the NIM and from adjacent subnets. The NIM will maintain a SSIS-wide view of network data flow and bottlenecks.

On-Demand Bandwidth Access

The DFWG has received strong input from the user community on the necessity of SSIS being able to respond to targets of opportunity (e.g., providing more resources to a solar observing payload during a solar flare). The DFWG is also aware of the problems of integrating synchronous

traffic (such as audio and video) with asynchronous
traffic (computer or experiment data). Clearly,
prearranged, fixed bandwidth allocation is simply
not acceptable. All SSIS subnets are required to
provide on demand bandwidth access and bandwidth
reconfiguration consistent with operational and
resource management practices. The SSIS NIM shall
globally authorize bandwidth reallocation and
convey reconfiguration information to each of the
affected subnet system managers. Each subnet must
be rapidly reconfigurable, although specific timing
requirements have not yet been established. The
DFWG welcomes inputs from the user community that
can quantify exactly how quickly this
reconfiguration must take place.

SSIS Services and Service Access Points

There are clear tradeoffs in data
communications between high services and high
performance. High services result in either
reduced data throughput or in increased hardware
speed and usually power to maintain the same
throughput rate. Numerous meetings with a variety
of users have convinced the DFWG that SSIS users
will tend to fall into two distinct populations;
those users who require high services from the
communication system and those users who require
simple services but high throughput (referred to as
low services). To attempt to satisfy both user
groups, SSIS will provide standard user interfaces
that access both Layer 7, the Application Layer,
for high services users and layer 3, the Network
Layer, for low service users. In addition, it is
also planned to allow unnecessary protocol layers
to be bypassed as much as possible. Thus, while
all users on the DMS LAN will see a standard Layer
7 user interface, a direct drop through to ISO
Layer 2 or 3 will be provided.

The services included under High Service are:

    file-to-file transfer
    virtual terminal
    remote job entry
    telemetry/telecommand - real time
    control/monioring
    nonreal-time telemetry delivery
    application-to-application messaging
    database access
    interpersonal messaging (electronic
    mail)
    multicast message distribution
    multi-media conferencing (Growth service
    only, no requirements at this time)

The services included under Low Service are:

    real time telemetry
    nonreal-time telemetry delivery
    audio
    video

These high and low SSIS services shall be
provided using SSIS-wide communication standards
within the SSIS Primary Domain. The generality of
these services shall be limited to supporting
specified SSIS defined standards such as standard
file structures and standard virtual terminals.
The SSIS standards and protocols are not specified
at the current time. However, international
standards and protocols are strongly preferred.
Conversely company proprietary standards and
protocols are not presently considered desirable.

SSIS Naming and Addressing

SSIS wishes to make the user as independent as
possible from the details of the SSIS networks.
Hence, SSIS will support global naming. Only a
SSIS global name will be required to route a
message from any node in SSIS to any globally named
endpoint. It is acceptable for a user to host
several applications behind a single global name,
but the user must then extend the naming and
addressing. Similarily, SSIS will allow local
names (i.e., "Nicknames") on individual SSIS
subnets. SSIS will establish a global naming
authority to allocate and reallocate global names.
When an application or user is no longer associated
with the Space Station Program, its SSIS global
name may be reassigned. However, the SSIS naming
authority will maintain an archive which associates
global names with users as a function of time.

Conclusion

SSIS has identified five major areas
pertaining to communications that it intends to
standarize across the SSIS Primary Domain.

1. The types of Service Access Points and the
mechanism by which users will invoke standard SSIS
services.

2. RF Link Formats

3. Services supplied at ISO Layers 3 to 7 and
the mechanisms for implementing these services.
All protocol implementations shall generally comply
with public standards; in order of precedence:
international, national, industry, SSIS specific.

4. Level Zero Processing services and the
user interface to them.

5. A SSIS network data unit used to exchange
data between SSIS subnets.

The full extent of the SSIS Primary Domain is
not yet known. As stated earlier, the requirements
discussed in this paper are only binding on DMS and
C&T at present. They are under review for the DIF
and Ground Distribution Systems. The number of
Ground Distribution Systems that will adopt them
and their acceptance among the international
partners is still to be determined. A companion
paper in this proceedings, Interoperability in the
Space station Information System, explores the
implications of the span of the SSIS Primary Domain
on international interoperability.

References

1. Space Station Information System Definition
Document, Johnson Space Center Document 30225,
Revision A, October 9, 1986.

2. Space Station Program Definition and
Requirements, Johnson Space Center Document
30000, September 1986.

ANALYSIS AND IMPLEMENTATION OF AUTOMATION ASPECTS

IN THE COLUMBUS AND HERMES END TO END SYSTEMS

M. Sainz[*]

Matra Espace

Toulouse, France

Abstract

The challenge of a fully integrated european space infrastracture in the next decade is promoting an increased contribution of automation. Lessons are learned from two significant operational experiences : Spacelab and Spot. Most promising automation candidates are highlighted : operations management, contingencies recovery and crew interface. Incremental automation is proposed through structural provision in the architecture of the end to end systems.

I - European space operations profile

I.1 - Background

The past decade has been driving significant steps forwards in european space technology and operational experience like :

- launching capabilities with Ariane,
- manned flight environment and experiments operations with Spacelab,
- spacecraft operations with Spot satellite.

Ahead in this decade is a development effort of even larger endeavours. The apparent break in space operations turns out as a major benefit :

- lessons learned are formalized
- true needs are identified
- integrated understanding of programs is generated.

I.2 - New space programs

Already today, the overall scheme for the next decade is set up for Europe with :
- the Ariane V launcher
- the Columbus elements with the Attached Pressurized Module, the Man-Tended Free Flyer and the Polar Platform
- the Hermes spaceplane
- the european Data Relay Satellite

Other programs like the Spacelab follow-on microgravity platform Eureca may attach to these programs.

I.3 - Space infrastructure

The related operations are definitely a matter of spacecrafts, but also of entire systems to support them. Europe's industry is taking the challenge of building a full integrated space infrastructure as sketched in figure 1.

[*] Senior engineer, Space Transportation and Manned Flights Programs Direction.

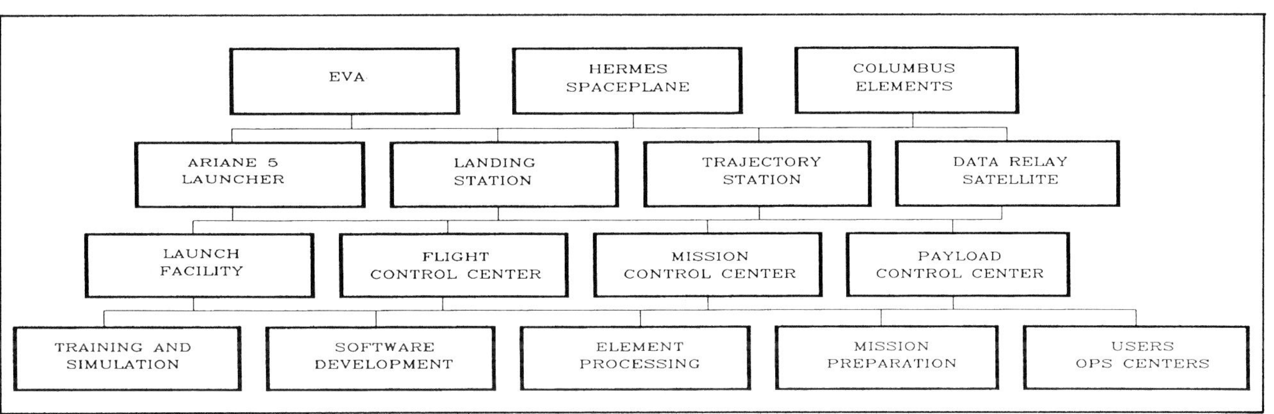

Figure 1 : Simplified Infrastructure

I.4 - Constraints

The foreseen operations scenarii induce three types of constraints :

a) Operational constraints

- mission complexity promotes an integrated effort beyond scattered performance,
- parallel processing implies strong operational planning,
- potential hazards demand safety rules and contingency management,
- variety of elements induces resources allocation effort,
- elements performance is mandatory for success.

b) User constraints

- Missions goals are not only technological demonstration, but also production.

- Experiments are needing to be assessed and reoriented in real time on a interactive basis

c) Program constraints

Companion budget is beyond all european ventures ever and puts pressure on operational costs reduction.

I.5 - Automation

Automation is going to be the answer for some of these constraints. But no selection can be made until specific lessons be learned from past experience.

II - Lessons learned

II.1 - Lessons learned from Spot 1 satellite

a) Introduction to Spot

The Spot satellite is a french earth observation satellite characterized by a high spatial resolution in the visible. A steerable mirror enables imaging to be performed nadir or off-nadir viewing. This capability offers several advantages : access to preselected zones, repetitivity of the observation for given zones, stereoscopic observation.

Spot has been launched in February 1986 and is in operations.

b) Operational constraints

The number and diversity of users requests is such that the generation of the daily operations plan in compliance with the spacecraft capabilities, operational constraints and user priorities is often a challenge :

- recorders dump periods distributed independently of the users requests
- evolving purchase trends toward isolated elementary pictures, that would best be grouped in constant strips
- parallel processing of the two instruments on two different targets
- conflict between picturing and satellite platform engineering activities
- transition times between payloads operations modes
- conflicts over recording time of the recorders.

c) Lessons learned

- Ground management of resources goes beyond classical algorithms, especially in view of unexpected requests outside mission standards
- Operations conflict resolution goes best through interactive automation, where operators guide the assessment of possible solutions
- Operations management is as important for overall success as technological performance.

II.2 - Lessons learned from Spacelab

a) Introduction to Spacelab

Spacelab is the manned laboratory built by ESA and operated in the NASA shuttle cargo bay. It has flown successfully four times, providing a wide variety of scientific results. Many more missions are ahead.

b) Lesson learned

The operations of the four flights have brought a full crop of lessons :

- Experimenters take the greatest benefit in interactive operations. The telescience concept is now well introduced for Columbus,
- Crew workload and resources management requires near real-time permanent replanning,

- On-board configuration management is a major operational burden, with no direct productive output, but great expense of engineering, integration and operations time.

- Equipment failure should be confined at the lowest possible level. Failure propagation and single point failures should be avoided. Although no such failure occurred during Spacelab flights, it has been clearly identified that centralized data processing can foster too much of a sensitivity to overall success.

- Great simplicity can be drawn from local preprocessing at experiment level before connection to the data management system.

- Crew is highly demanding on operational interfaces. Engineering interfaces shall be restricted and true handshake dialog shall be promoted for initiated actions.

III – Automation candidates on-board

Automation on-board Columbus and Hermes is going to settle on three cornerstones :

- lessons learned
- operational constraints
- adaptation to evolutive needs.

Four area of automation have been clearly identified :

- contingency management
- configuration management
- operations management
- crew interface

III.1 – Contingency management

Contingency management goes through fault detection, fault confinement and recovery.

a) Fault detection

It is provided by :

- Built-In Test Equipments
- Cross-surveillance at different layers of the Data Management System network.

Both are foreseen for the Columbus elements.

For Hermes spaceplane, coherence checks on data will be used extensively in addition.

b) Confinement

Physical means are used with distributed architecture. This is true for Columbus with Local Area Networks and for Hermes by segregation between Mission Management System and Guidance, Navigation, Control System.

Logical means are introduced with software driven layers that segregate engineering data from transaction.

c) Recovery

In Hermes spaceplane, recovery is performed by majority voting so that faults be confined to the faulty level and recovered before propagation.

In Columbus, recovery goes through the use of local redundancies internal to the affected layer and, if needed, upgrading of the resolution to the next upper logical layer.

d) Main issue

Automation in contingency management takes place through the architectural solution of segregation. Once segregation is set-up, automation appears as an accessible issue limited to simpler decision trees and appropriate crew interface.

Automation in contingency management goes through highly structured design

III.2 – Configuration management

Configuration management automation requires a steady effort in mode representation and high level commanding.

While this is now baseline for telemetry and multiplexer formats switching, configuration management still goes through heavy crew intervention on engineering level.

The SGASAA[**] study for ESTEC has resulted in a proposal of representation using layers, modes, goals and milestones. This proposal is now being first evaluated on a Hermes case with an object-oriented langage.

[**] SGASAA : Standard Generic Approach for Spacecraft Autonomy and Automation.

III.3 - Operations management

Interactive operations planning is to stress with the use of predefined free-floats for short term periods. However it is not planned to have full on-board replanning automation until incremental experience be gained on ground.

III.4 - Crew interface

Crew interface shall favour skill-based behaviour over knowledge-based behaviour.

Astronaut thinking should be stimulated by information and action structures promoting situational representation.

Such representation shall match the mental models of the spacecraft as it may result from the astronaut training.

This adequacy between natural or induced mental model and information structure is identified as key issue for automation with crew interface.

III.5 - Adaptation to evolution

Automation on-board will be incremental in nature, based on operational experience. The evolution in the use of Columbus fully justifies the multi-purpose architecure of the Columbus Data Management System oriented toward services offerings.

IV - Automation candidates on-ground

Performance and budget are the initial drivers for automation.

Columbus elements will be operated continuously for years and Hermes intervention will spread accordingly. Longlasting pressure on experts as well as alertness for routine operations shall be addressed.

Contingency management, telecommands generation and operations planning are of major concern.

IV.1 - Contingency management

Contingency management is a matter for experts intervention. The point is that :

- expertise can quickly fade unless the expert pratises constantly to maintain proficiency,

- the turn-over of personnel challenges the permanence of expertise.

Automation for contingency management is shaped around expert systems.

Initial experience has been gained with Spotex, a prototype expert system for Spot satellite survival mode recovery. It has been recognized that knowledge base is built best progressively : expertise lies ultimately at a highly compiled level difficult to extract.

Expert systems for contingency management are not meant to replace experts but to guide on finding the failure origin and recovery on an interactive basis.

IV.2 - Telecommand generation

More classical automation is foreseen in this area :

a) Commands block generation

The automated generation of goal-oriented command blocks is now well introduced, for instance in the Control Center of Inmarsat 2 satellites family, where up to five satellites can be controlled simultaneously.

b) Command/Control loops

A command/control loop between telecommands and resulting effects through telemetry is especially interesting for actions requiring a large processing beyond on-board allocated capacity.

An example for the Columbus Payload Operations Control Center is the uploads of the thermal control parameters to obtain a high level of thermal stability.

Such control loop shall be provided through command sequences interleaved with telemetry checks to form subroutines similar to those used for Spacelab ground operations with Goal language.

c) Resources compatibility check

The check of telecommands is of concern for Columbus User Operations Centers to avoid resources overstep or hazardous actions.

Checks will have to be as straightforward as possible because of the large amount of commands to be checked and of the little acceptable delay in telescience. The payloads shall be designed in such a way that the relation between command and resource consumption is always easy to anticipate.

IV.3 - Payload operations planning

Automation shall go beyond a classical algorithm tool of plan generation using for instance simplex method.

Expert capabilities shall be added in two areas :

a) for short term operations planning

- postponing, advancing or compressing an activity
- locating a reconfiguraiotn or repair activity
- assessing the impact of an abnormal event on the operations schedule

b) for resources monitoring

- assessing the projected impact of resources actual consumption
- assessing the robustness of plans options and issuing recommendations to the operator.

IV.4 - Interaction

The three above-mentioned candidates for on-ground automation have been selected because of true experience gained on Spacelab and Spot. Interactive effort shall be stressed, to take into accound the highly creative capacity of human operators as well as their learning capacity to adjust strategies to new situations.

V - Initital implementation

Both Hermes and Columbus programs are in the definition phase. Investigation has been initiated in three area :

- Columbus Fault Detection and Initial Recovery using a Columbus Data Management System test bed,

- Mission mode representation on Hermes using Smalltalk, as object-oriented langage

- Expert planning system, using KEE as expert system.

VI - Conclusion

The issues addressed in this paper have not been robotics or control, but automation candidates that have proven to be truely applicable operational burdens of conflicts resolution.

The challenges of the next space ventures will not permit to carry out these tasks without automation.

Provision shall be made in the projected designs to accomodate new automation needs as they will come up from experience analysis or new operations.

THE HARDWARE/SOFTWARE ARCHITECTURE OF
THE COLUMBUS PRESSURIZED MODULE ELEMENT

G.C. Cassisa, L. Sarlo
AERITALIA
Space Systems Group
Torino, ITALY

ABSTRACT

The Columbus Pressurized Module Element will be launched in the 1990s with an Hardware/Software configuration supporting an advanced data processing and mission management capability based on the extensive use of distributed intelligence. The Pressurized Module Avionics shall not only serve and meet requirements usual to the state-of-the-art satellites as the ones imposed by data handling and transmission, attitude control, power conditioning and distribution and so on, but shall actively support functions required by the man presence on board. In particular the Avionics is designed to assume and exercise the management of the environmental control functions, the failure detection, isolation and recovery, the contingency operations, emergency planning, the man-machine interface, etc. The objective of the paper is to provide a consistent basic design description of the Pressurized Module Avionics architecture starting from a system level approach down to the general subsystem consideration leading to the identification of the software functions and hardware configuration.

INTRODUCTION

The Pressurized Module (PM) is designed to provide environment and facilities for space laboratory work and to provide operational support to the crew.

Two operational scenarios are foreseen:

1. Four-Segment Pressurized Module (PM4) permanently attached to the International Space Station (ISS).

2. Two-Segment Pressurized Module (PM2) connected to a Resource Module (RM), to make-up the Man Tended Free Flyer (MTFF).

The PM4 is considered as an extension of the manned core of the ISS and it is operated permanently berthed to a free port of a berthing node of the ISS as one of the laboratory modules. The PM4 receives resources from the ISS still maintaining its own capability to operate and control the payloads either European or International.

The MTFF-PM2 offers a free flying "Quiet Laboratory" where very precise microgravity experiments can be conducted with a minimum of disturbance. Functionally the PM2 receives resources from the RM including power and consumable supply. The on-board intelligence is spread over the PM2 and RM subsystems. The overall mission management and control is performed by the PM2 dedicated processing resources. MTFF servicing operations are performed by several vehicles including NSTS and HERMES.

This paper will address only the PM4 Avionics considering that, for commonality reasons and design approach, the architecture of the PM2-PM4 Hardware and Software can be considered similar.

THE PRESSURIZED MODULE WITHIN THE OVERALL SSIS

The Space Station Information System (SSIS) is designed to support the overall End-to-End information management and processing. The on-orbit Columbus elements and the European Ground Facilities configure a Columbus Information Management System (CIMS), in which the Pressurized Module is one of the nodes during any operation and for any Columbus phase. The CIMS subset related to the attached Pressurized Module configuration is physically and functionally part of the overall SSIS.

The Pressurized Module Avionics, in particular, provides services to SSIS End-to-End users in supporting:

- The on-board information processing and distribution;

- the functional communication paths among different elements and ground control/user facilities;

- the Electrical Ground Support Equipment (EGSE) and Simulator facilities;

- all Software Engineering, including the Columbus Software Development Environment (SDE);

- interfaces to other components of the SSIS and CIMS.

A functional CIMS overview is provided in Fig. 1., in which arrows represent potential information flows, and boxes the virtual nodes of the basic network.

Attached PM functions and services relevant to CIMS include:

- System Level Management
- Mission Management and Control
- Subsystem Level management
- Man Machine Interface
- External Interfaces Management
- On-board Simulation, Planning, and Trend Analysis
- Data Distribution
- On-Board Check-Out
- Servicing Support Facilities

SYSTEM LEVEL MANAGEMENT

This is part of the on-board System and Mission
Management (SMM) and it is designed to:

- Exercise the management and coordination of any
  system oriented activity, including redundancy
  management and monitoring.

- Perform the management and control of non-intel-
  ligent subsystems.

- Provide facilities to back-up any Independent
  Caution and Warning function.

MISSION MANAGEMENT AND CONTROL

This is part of the on-board SMM and it is designed to exercise the including control of the sys-
tem/flight configurations fault management, to ensure that mission objectives are achieved.

SUBSYSTEM LEVEL MANAGEMENT

This involves all products designed to exercize
the management of each Attached PM intelligent subsystem. In particular it includes the Subsystem
Operation Plan support and the Subsystem monitoring, control and check-out.

MAN MACHINE INTERFACE (MMI)

This involves a quite large area of software facilities which are concerned with crew access to sys-
tem/performance control and it includes the following:

- Command languages.
- Command validation and classification.
- The user authentication.
- Encryption/Decryption systems (for specific pay-
  load activity or private communications).
- Any intelligent Front End to the underlying S/W
  interactive applications.
- Alphanumeric and graphic data display and manage-
  ment.
- The on-board Software Development Environment
  (On-Board SDE).
- Input/output modules and devices.

EXTERNAL INTERFACES MANAGEMENT

This includes any facility designed to control the
communication paths between the Attached PM and external elements. In particular it includes the
following software functions:

- Telemetry/Telecommand Ground Interface through
  ISS and the TDRS/DRS.

- External Command Filtering and Authentication.

- The Gateway management between the PM Data Management Subsystem (DMS) and the International
  Space Station (ISS).

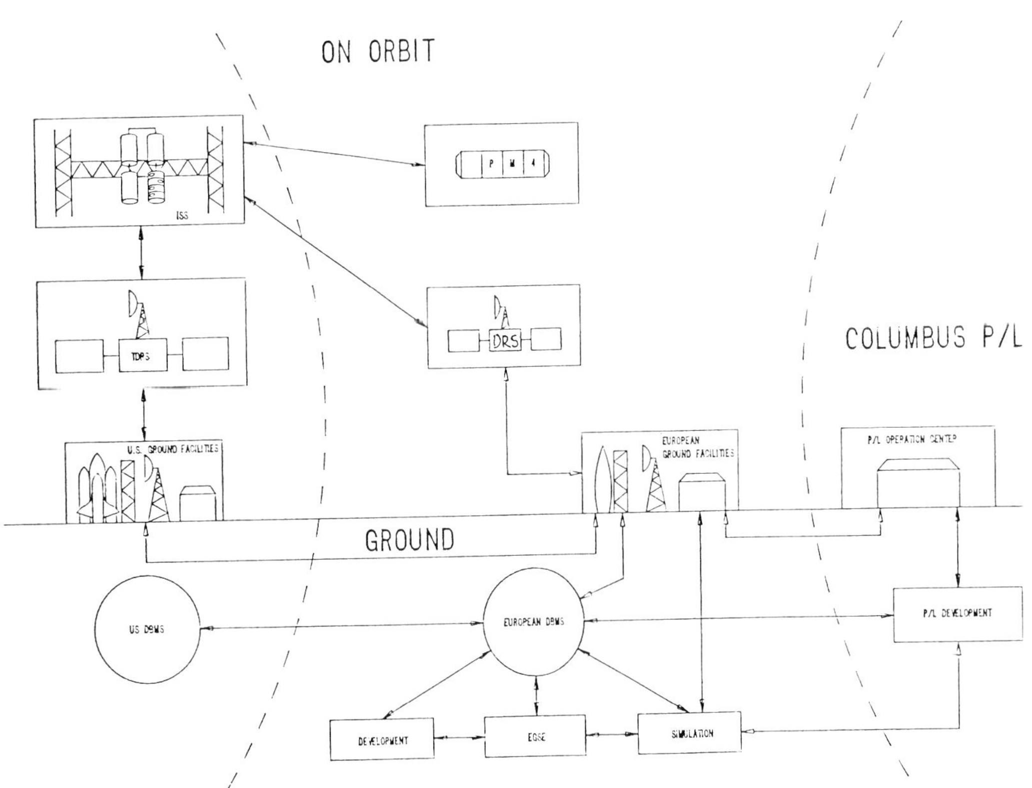

FIG. 1 ISS/CIMS OVERVIEW

SIMULATION, PLANNING, AND TREND ANALYSIS

This concerns any application software, at SMM level or Subsystem level, designed to assist in processing and/or simulating a mission plan on the basis of a predefined Mission plan and of specific trend analysis results.

DATA DISTRIBUTION

The on-board data distribution involves the activation of the Attached PM related services and makes use of a Local Area Network, designed to link any Attached PM processor and equipment.

CHECK-OUT

This includes any function designed to perform checks on subsystem and payloads to support on--board check-out.

SERVICING SUPPORT

This includes services provided to assist crew, subsystems, and payloads in both nominal and contingency operations (automatic procedures revision, implementation, submission, maintenance assistance, etc...).

CIMS HIERARCHICAL ORGANIZATION

In relation to the functions to be performed and/or controlled, the following hierarchical CIMS levels can been envisaged:

- LEVEL 0 : High level decisions and safety critical functions (VITAL LAYER).

- LEVEL 1 : Intelligent Data Processing at System level .

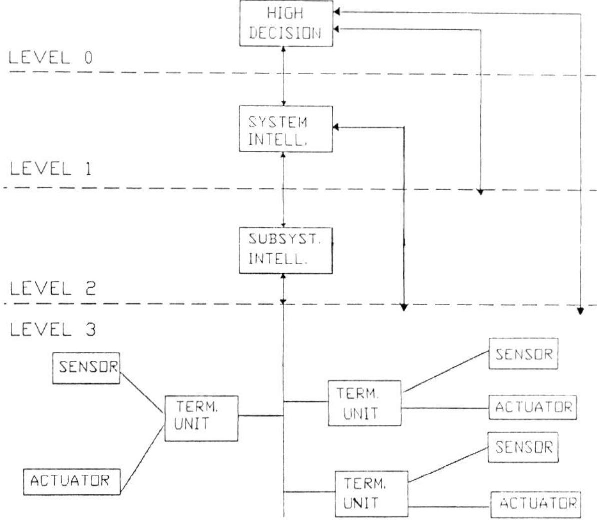

FIG. 2   INTELLIGENCE HIERARCHY

- LEVEL 2 : Intelligent Data processing at Subsystem level

- LEVEL 3 : Non intelligent Sensor/Actuator Subsystem level

A schematic diagram of the hierarchical CIMS structure is shown in Fig. 2.
This kind of organization supports an almost complete independence from ground control and minimize the on-board crew involvement, reducing the ground support effort and the response time of on-board operations, allowing at the same time the manual override capability and intervention at all level functions.
The communication between level 0 and the lower levels will be, as far as possible, via dedicated physical and/or functional links besides the normal channels.
The two intelligent levels (1 and 2) communicate by means of structured data packets running on the system network.
Communications between the level 2 (Subsystem Processors) and the level 3 will be via the same system Local Area Network (LAN) through terminal units interfacing the subsystem non intelligent devices (sensors/actuators).

LEVEL 0 FUNCTIONS

The Vital Layer (Level 0) is in charge of the overall control and management of those PM functions that have safety critical implications; its decisions will take the highest priority within the PM4 system.
Included at this level are:

- The caution and Warning Functions

- The Initialization Functions

- Any Ground/On-Board Man Intervention (Overriding) for safety critical operations.

LEVEL 1 FUNCTIONS

The level 1 (Intelligent System Level) consists of the on-board System and Mission Management (SMM) and shall have the responsibility on the entire mission during nominal and contingency operations. During nominal operations, this level shall exercise the management of the execution of the mission in accordance with the planning and shall control the performance of the subsystems higher level. It shall also generate the system status monitoring to the higher decision level allowing at any time the manual override capability.
During contingency operations, the level 1 shall analyze detected problems and produce a recovery plan, taking into account the mission impacts, to perform the management of the required system reconfiguration.

LEVEL 2 FUNCTIONS

The level 2 (Subsystem intelligent level) is in

charge of the control and management of dedicated
subsystems, including the Failure Detection, Isolation, and Recovery (FDIR), up to the point where
the available information is sufficiently detailed.
At this step the higher level system processor will
take over.
The level 2 provides the subsystem status monitoring to higher levels allowing at any time the man-
nual override capability.

LEVEL 3 FUNCTIONS

The level 3 essentially consists of subsystem
distributed non intelligent devices (sensors/actua-
tors) controlled by relevant level 2 subsystem intelligence.
They provide means for health status monitoring and
for any actuator performance (e.g. Motor Rate).

FAILURE DETECTION, ISOLATION, AND RECOVERY

The failure detection and recovery is kept as
local as possible.
This means that each subsystem processor level has
the capability to detect the failure and to perform
the redundancy management of its own subsystem.
If no local actions are suitable, the responsibility of detecting and recovering the failure is
taken by the next higher level (System and Mission
Management).
If the failure affects more than one subsystem, the
redundancy management is allocated at the system
level (SMM), and it will be performed either on the
basis of a fault tolerant processor approach, or by
the crew intervention .

CAUTION AND WARNING CONCEPT

The Caution and Warning (C&W) system is capable
of detecting, announcing, and safing the emergency,
warning, and caution situations that can arise during any Attached PM active mission phase.
In particular, Caution and Warning intervention involves the following tasks:

- Control of Pressure and Atmosphere Composition
  out of limits

- Fire Detection and Suppression

- Control of the Contamination and Radiation levels
  out of limits

- Docking and Berthing latches and inhibit monitoring.

The major design requirements and the relevant implementation approach of the Caution and Warning
are shown in Table 1 and Fig. 3.

REQUIREMENTS	IMPLEMENTATION
INDEPENDENT C&W FUNCTIONS : i.e. H/W and S/W supporting these functions are not sup- porting any other function.	The attached PM C&W is implemented as a sepa- rate and independent functional assembly
SEPARATE and INDEPENDENT H/W C& LINK	Use of a dedicated link (separate lines or de- dicated bus)
C&W FUNCTIONS ARE SAFETY CRITICAL AND FAIL-OP-FAIL SAFE (FO/FS) REQUIREMENTS APPLY. However, a redundant C&W assembly with a suitable back-up by SMM is adequate. Therefore, monitoring and control function of the SMM shall include C&W parameters.	The FO/FS requirement is met by a single fai- lure tolerant C&W sub- system consisting of C&W assembly, plus de- dicated links, sensors, etc., with SMM as a back-up. The health status of C&W subsystem is performed at high level by SMM; in the event of a com- plete C&W failure, the SMM will take care of any C&W function.

TABLE 1

CAUTION & WARNING REQUIREMENTS AND IMPLEMENTATION

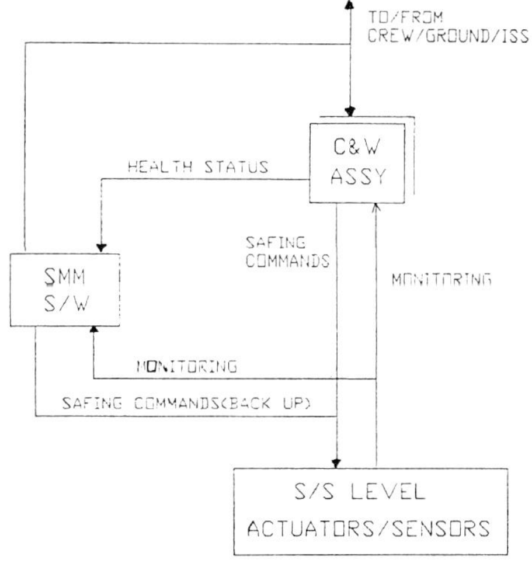

FIG. 3   CAUTION & WARNING
IMPLEMENTATION

AVIONICS ARCHITECTURE

The functions accomplished by the PRESSURIZED
MODULE electronics are identified as follows:

SYSTEM PROVISION

- Environmental control management
- Power management and distribution
- Signal conditioning, processing and distribution
- Internal, to ISS, and GROUND communication sup-
  port
- Docking and Berthing functions support
- Data storage and processing
- Voice processing, recording, and distribution
- TV images processing, digitalization, storage,
  and distribution
- External interfaces support
- Overall system performance control

SAFETY, REDUNDANCY, and VERIFICATION

- System/subsystem monitoring and control
- Redundancy management and verification
- Automatic control protection
- Automatic function override support
- Caution and Warning provision
- System/subsystem fault detection, isolation
  and recovery
- System/subsystem checkout support
- Safing operation support
- Emergency planning

MISSION

- PM initialization functions
- Mission support
- Payloads support
- Mission planning and scheduling
- P/L data management and processing (on request)
- Housekeeping and time reference distribution
- Mission S/W storage

MAN INTERFACE

- Interactive controls
- Crew training
- General purpose program language support
- User friendly MMI support
- Alarms and alerting signal provision

The responsbility for the functions performance
is spread throughout the subsystems :

- Data Management Subsystem :
                    . Data distribution
                    . Central Data Base Management
                    . System management

- Electrical Power Distribution

- Caution and Warning

- Communication Subsystem :
                    . Ground data multiplexing-
                      demultiplexing
                    . Video and Audio signals
                      distribution

- Environmental Control Subsystem

- Thermal Control Subsystem

- Mechanism Subsystem

- Crew Work Station Subsystem

Each of the above subsystems has it own central intelligent level within a dedicated computer plus
distributed controllers performing locally specific
close loop management functions.

PHYSICAL ORGANIZATION

The PM avionics system is based on a distributed
intelligence concept functionally organized in order to fit within the hierarchical approach previously defined.

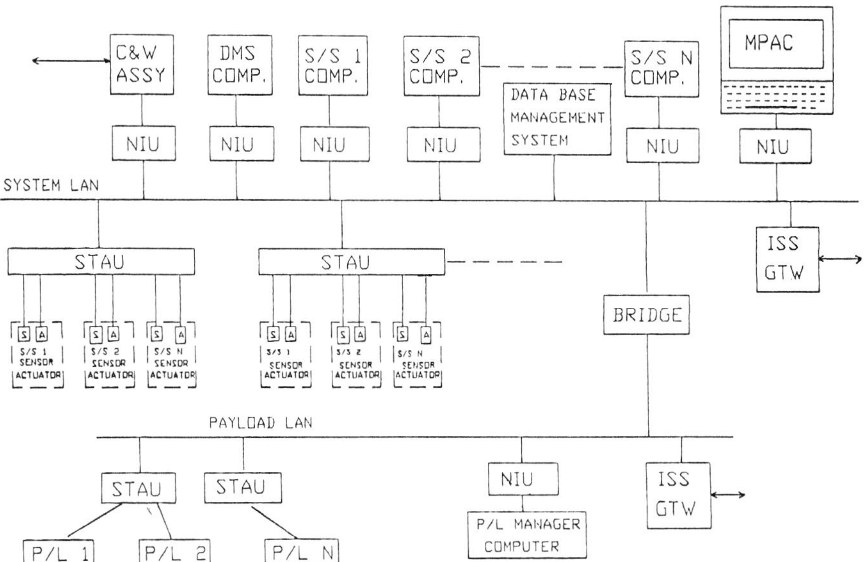

FIG. 4 PHYSICAL IMPLEMENTATION APPROACH

The distribution of data, necessary for the control and management of the overall system (including Payloads), is supported by two Local Area Networks (LAN°: a system LAN mainly dedicated to the distribution of data for system and subsystem performance and a payload (P/L) LAN dedicated essentially to the Payloads needs.

The two LAN's are connected by a dedicated device to provide the system intelligent level with the monitoring and control capability of the overall system.

In order to minimize the harness routing and the relevant EMC problems, standard remote I/O devices (STAU) have been considered.

Each of these I/O devices supports the communication between different Subsystem (S/S) controllers (S/S Control Unit) and their relevant remote devices (mainly sensors and actuators) interfacing the LAN.

A schematic diagram is shown in Fig. 4.

THE INTERCONNECTION SYSTEM

Geographically distributed subsystem equipment and Payloads are physically and logically interconnected by means of a Local Area Network (LAN).

The interconnection systems consist of the following components:

- The physical medium
- the interconnection stations (I/S)
- the Network Interface units
- the Bridge (BR) for interconnecting the System and the Payload LAN's
- the Gateways for interconnecting PM external LAN's
- the Standard Acquisition Units (STAU)
- the necessary software for communication management and host computer interfacing.

PHYSICAL MEDIUM :   The current proposed type is based on coaxial-cable technology but usage of a optical-fiber medium is envisaged and under study. In order to fulfill the "single failure tolerant requirement", each LAN system (the System and the Payload ones) uses a physical redundant medium; only one cable is used at a time while the other is the back-up.

The foreseen protocol is based on a Token-bus concept while the estimated needed bit rate is 10 MBPS.

INTERCONNECTION STATION :   It is a small package installed at each location of a possible node attachment performing the following functions: - Diverts signals from/to the cable to/from the node user - Allows an easy connection/disconnection of node user - Provides protection/isolation between the user node and the medium.

NETWORK INTERFACE UNIT :   The Network Interface Unit (NIU) is essentially a communication interface board located within the node user device provi-

ding the capability to the user for interfacing the medium.

The communication functions are implemented both in hardware and in software based on a microprocessor system and relevant memory.

BRIDGE :   It is used to interconnect the System and the Payload LAN.

It consists essentially of 2 head-to-tail simplified NIU's and of a bridging function for extending addressing capability over both the networks. Bridge is transparent in the overall architecture with respect to the users end-to-end links.

GATEWAY :   It is used to interconnect the Pressurized Module Networks to the any other communication medium (e.g. other ISS element). It consists of two head-to-tail NIU's and of a gateway function that performs the protocol translation.

STANDARD ACQUISITION UNIT (STAU) : These devices are used to interface the subsystems distributed sensors and actuators to the LAN completing the functional link with the relevant controller.

The assignement of a STAU channel to a remote device is based on geographical and redundancy considerations rather than on subsystem basis. As a consequence one STAU handles commands and monitors relevant to different subsystems and conversely the signals of a single subsystem are managed by several STAU's.

The provided signal interfaces include:

-   5 V Command Output
-   28 V Command Output
- 120 V Command Output
- Serial Input/Output
- Two-Level Status Monitor Input
- Analog Monitor Input
- Variable Resistor Signal Input

COMPUTERS

The PRESSURIZED MODULE computers shall all belong to the same family and shall include the following main peculiarity:

- Modularity   : memory size, I/O controller, etc
- Compatibility : task migration between computer shall be possible
- Fail safe concept : fail stop philosophy will be implemented.

The main computer features are:

- CPU : 80286 based
        500 kips
        Floating point processing capability
        16 Mbit addressing capability
        Direct Memory Access provision

- RAM : 512 kbytes expandible

- I/O : Network Interface Unit is implemented as separate board.

NOMINAL OPERATION

The system and subsystem set-up (parameter
thresholds, resource distribution, permitted operations, etc.) is generated in accordance with the
planned mission operation activities. Each subsystem computer provides the controlling of the
subsystem function including the failure detection
isolation and recovery exchanging data with the
remote devices through the LAN. Continuous monitoring of the subsystem status and health is provided to the CREW and to the system mission management software that controls   of the overall
system and mission overview.

DATA BASE MANAGEMENT SYSTEM

A centralized Data Base Management System has
been foreseen providing servicing for data and
software storage and retrieval for all the on-
-board users. The foreseen mass memory technology
is winchester disks (3 X 32 Mbit).

MAN-MACHINE INTERFACE

The system man-machine interface is through a
Multi Purpose Application Consol. It is based on a
computer system managing the different I/O devices
and providing dedicated processing capability.
The main output device is a color CRT with multis-
creen capability fed by video memory board. Text
and graphic display capabilities are included. A
keyboard is the main device for data entry, however the use of other I/O devices as jysticks,
touchscreens, etc., is under investigation.

SOFTWARE ARCHITECTURE

The attached PM Software System is designed to
support Development, Integration, Test, Check-Out
Operations, and Maintenance of the Attached PM E-
lement during all phases of its life.
IN particular its support is required to perform
the following tasks at Space/Ground Segment level:

- Mission Planning and Execution
- Communication Services
- Payload Operations
- Subsystem Control and Operations
- Maintenance and Servicing

At the top level, the following sets of Software
have been identified to cover the above objectives:

- Software for Development, Maintenance, and Check
  -Out
- Flight Software
- Mission Support and Preparation Software

SOFTWARE FOR DEVELOPMENT, MAINTENANCE, AND CHECK-OUT

This Software is designed to support Management
Development, Continuous Development, Integration,
Verification, Qualification, Preventive and Reme-
dial Maintenance of the whole Attached PM software
and to support any Configuration Management Level

activity for Hardware and Software components.
The Columbus Software Development Environment (SDE),
mainly based on commercially available tools, and the
Electrical Ground Support Equipment (EGSE) Software
are parts of this Software set.
The SDE Software in particular includes the following products :

- Operating Systems and associated packages
- Compilers and associated packages
- Architectural Design tools
- Detailed Design tools
- Software Integration and Test Development tools
- User Interfaces Development tools
- Tools for Software Metrics
- Software Engineering Management tools
- Software Documentation tools
- Software Maintenance tools

The EGSE Software is designed to support the following activities :

- Subsystem Test for mutual non-interference
- Element External Interface Verification
- Testing of high-level management funtion of the
  on-board Data Management System (DMS)
- Integration/Verification Test.

FLIGHT SOFTWARE

The flight Software consists of a set of on-board
Operating Systems, of the Data Base Management Facilities, and of the Application Software and related data/tables.
The application Software is designed to assume and
exercise the management of a complete flight configuration, monitoring and controlling the intelligent subsystems and performing data processing for
non intelligent subsystems and payloads.

ON-BOARD OPERATING SYSTEMS

The On-Board Operating Systems, that are part of
the Data Management Subsystems (DMS), are designed
to support data processing and distribution at the
Element level.
Two sets of On-Board Operating System Software are
identified:

- The Network Operating System (NOS)
- The Local Operating Systems (LOS)

The NOS is a collection of Software and associated
protocols designed to allow communications among
the nodes of the on-board Information System.
In particular the NOS provides services to :

- Allow users to access remote resources
- Control access requests
- Build-up a Virtual System in which local and re-
  mote resources appear to be identical
- Packetize and format data
- Perform priority management

The LOS is in charge of the following tasks:

- Overall System Security

- Process Management (e.g. scheduling, syncroniza-
  tion, data exchange)

- Storage Management (e.g. virtual/real memory management, paging, swapping).

ON-BOARD APPLICATION SOFTWARE

   The attached PM Flight Application Software can
be divided into the following areas:

- System Oriented Activity Support Software
- Subsystem dedicated Software
- Man Oriented Activity Support Software
- Payload Support Software

The System Oriented Activity Support Software included the System and Mission Management (SMM)
Software.
The Subsystem Dedicated Software includes all software packages designed to perform the management
of each subsystem.
The Man Oriented Activity Support Software is designed to assist Crew access to System/Performance
control.
Essential features are:

- Artificial Intelligence
- Man Machine Interface (MMI)
- The On-Board Software Development Environment
- User Command Languages
- Encryption/Decryption

The Payload Support Software includes all software
designed to support any non intelligent payload
for data processing and DBMS services requests.

MISSION SUPPORT AND PREPARATION SOFTWARE

   This Software is designed to support the Attached PM mission preparation and planning.

It is composed of:

- The Mission Preparation Software (MPS), designed
  to assist in Mission Plan and Data generation

- The Flight/Ground specific Software generation
  Software

- The Software Simulators

- Payload software support facilities

This software is designed to be executed on ground
based processors.

ACKNOWLEDGMENT

   The Columbus Pressurized Modules Hardware and
Software development and follow-on enhancements
will require the dedicated effort of a large group
of persons over a period of many years.
This paper represents a synthesis of the
Aeritalia Columbus Avionics and Software Teams effort during the early phase B2 of the Program.
Many contributions were received from these teams,
which we gratefully acknowledge.

Michael T. Ward*
Ford Aerospace & Communications Corporation
College Park, Maryland

Abstract

This paper examines the problem of integrated scheduling during the Space Station era. Scheduling for Space Station entails coordinating the support of many distributed users who are sharing common resources and pursuing individual and sometimes conflicting objectives. This paper compares the scheduling integration problems of current missions with those anticipated for the Space Station era. It examines the facilities and the proposed operations environment for Space Station. It concludes that the pattern of interdependencies among the users and facilities, which are the source of the integration problem, is well structured, allowing a dividing of the larger problem into smaller problems. It proposes an architecture to support integrated scheduling by scheduling efficiently at local facilities as a function of dependencies with other facilities of the program. A prototype is described that is being developed to demonstrate this integration concept.

Introduction

One of the many demanding challenges facing Space Station planners is the development of planning and scheduling concepts that support the integrated utilization of a highly complex and distributed system by many users with varying, and sometimes conflicting needs. This is certainly one of the major challenges facing information system designers. On the one hand information systems are a part of the problem, as they are themselves major components of the highly complex and distributed resources that must satisfy demanding and diverse user needs. On the other hand, information systems are the principal elements of the solution to this problem.

As with current space projects, the planning and scheduling process is driven by the operation of the spacecraft. In current missions, this typically entails the operation of a free-flyer scientific research spacecraft or the Space Shuttle. In the Space Station era, it will entail the operation of the Space Station base, with its individual modules, operated by NASA and the international partner agencies, and the operation of highly complex and diverse space platforms, in some cases supporting the requirements of a number of projects.

In the broadest sense planning and scheduling is the process that translates customer objectives into the coordinated operation of the Space Station base, the platforms, their modules and payloads, along with the various ground elements that support this operation. The process includes these scheduling functions.

- Scheduling of instruments on platforms and the Space Station base

- Scheduling of platform and Space Station base core systems, including onboard resources, such as

power and data storage, that are necessary to carry out the instrument and system operations

- Scheduling of Space Network (SN) or other tracking and data acquisition (T&DA) support required to carry out the flight operations

- Scheduling of communications services required to carry out the above flight operations

- Scheduling of various ground facilities, such as control centers and data capture facilities.

Integration of the planning and scheduling process is required only where there are activities scheduled that do not operate independently of other activities. We will find that this dependency occurs among the facilities or elements of the Space Station program and the institutional facilities of NASA and the international partner agencies. Dependency may occur when there are resources shared among two or more elements, or an activity at one element imposes a constraint on activities at one or more other elements. The constraints may be negative in nature, as when the operation of one instrument interferes with the operation of a nearby instrument; or positive, as when science objectives require data from two or more instruments operating in concert.

This paper takes as its point of departure the Space Station Information System Architecture Definition Document (SSIS ADD, October 1986). Our approach is to develop a scheduling integration concept that can be applied to the information system elements and functions described in the SSIS ADD. Figure 1 is a high-level representation of the SSIS architecture. (At the end of this paper is a brief glossary of SSIS ADD terminology.)

We begin by comparing the mission scheduling environment in the Space Station era with that of current space missions. We then examine the scheduling integration problem of the Space Station era and analyze the dependencies that cause it. Based upon that analysis, we derive the essential features of an integrated scheduling system and propose an information system concept with these features. We conclude with a brief scenario illustrating the concept.

A Comparison of Current Era With Space Station Era Planning and Scheduling

Flight Operations

Today flight operations are characterized by free-flyer spacecraft controlled by operations centers with a single mission focus. In the space Station era, there will be a diffusion of operations focus in the space elements, the ground elements that control them, and the organizations that sponsor and support them.

In current missions, the spacecraft payload is composed of a package of instruments with a specific project or mission focus. Though a number of principal investigators may have science responsibility for the individual instruments, the spacecraft and its payload are in operation in the service of the project's typically well-focused mission objectives.

---

*Senior Systems Engineer

GPS CONSTELLATION — 2.12

TDRS SATELLITE 2.13 · OMV 2.4 · SPACE STATION 2.1 · EMU/MMU 2.9 · OTV 2.5 · US COP 2.2 · ORBITER 2.11 · FREE FLIER 2.6 · MSC 2.15 · US POP 2.3 · EUROPEAN RELAY SAT 2.14 · JAPANESE RELAY SAT (TBS)

TDRSS GROUND STATIONS 3.46 · INTERNATIONAL PARTNERS FACILITIES 3.49 (JAPAN), 3.50 (ESA), 3.51 (CANADA) · INTERNATIONAL COMMUNICATIONS NETWORK 3.48 · DATA INTERFACE FACILITY (DIF) 3.45 · DIRECT SPACE - GROUND ELEMENTS 3.61 · INTERNATIONAL GROUND ELEMENTS RELAY AND DIRECT BROADCAST 3.47 · PUBLIC NETWORK 3.58 · CUSTOMER FACILITIES 3.52

NASCOM 3.56

JSC: NSTS MCC 3.3 · SMS 3.4 · POCC 3.5 · CDSF 3.7 · EAF 3.6 · SPACE STATION OPS MGT FUNCTION 3.66 · SSSC 3.1 · SSTF 3.2 · CUSTOMER COORD CAPABILITY 3.39 · NASDA BRANCH 3.49.2

GSFC: MULTIAPP CONTROL CENTER 3.12 · DHC 3.19 · CDSF 3.17 · EAF 3.20 · FLIGHT DYNAMICS FACILITY 3.64 · US PLATF'M SPT CTR 3.11 · NASCOM CONTL CTR 3.16 · NETWORK CONTL CTR 3.18 · CIVF 3.70 · CCN 3.80

MSFC: POCC 3.21 · CDSF 3.27 · INTEGR LOG SPT CTR 3.65 · OMV CONTL CTR 3.69 · PSCN CONTL CTR 3.23 · EAF 3.27 · CIVF 3.71

LERC: POCC 3.23 · CDSF 3.33 · EAF 3.3

KSC / VAFB / NSTL: NSTS MCC 3.3 · NSTS MCC 3.3

JPL: POCC * 3.41 · CDSF 3.42

ARC: POCC * 3.75 · CDSF 3.76

LARC: POCC * 3.85 · CDSF 3.86

TBS

PSCN 3.57

SSE SYSTEM 3.8 · TMIS 3.10 · C/D CONTRACTOR FACILITY 3.62 · NON-SPACE STATION INFORMATION SYST 3.40 · GPS GROUND ELEMENT 3.91

* CAN INCLUDE CARRIER CONTROL CENTER, INSTRUMENT CONTROL CENTER, OTHER

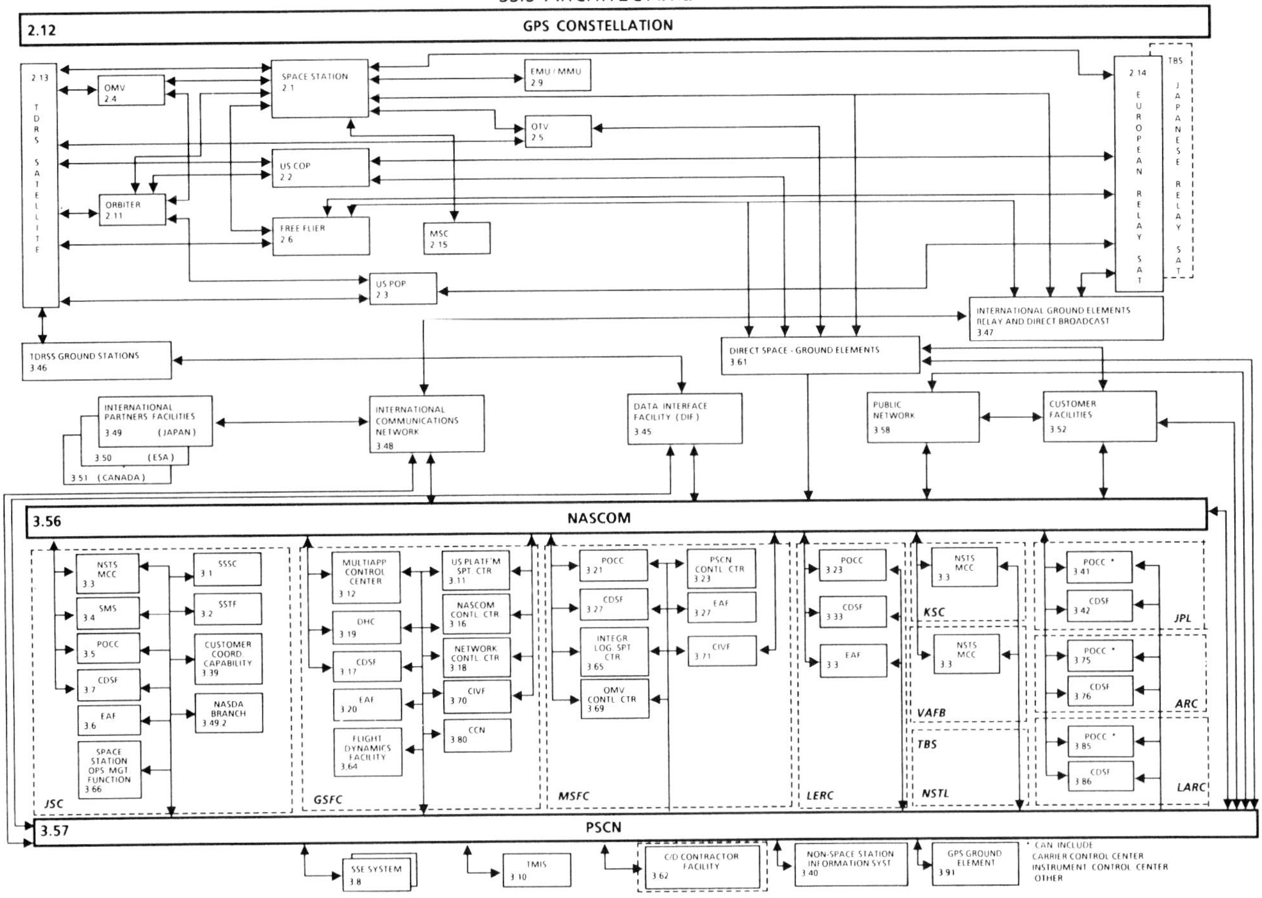

Figure 1. SSIS Architecture

The major space elements of the Space Station are the unmanned platforms and the manned base. Each will carry a number of modules and instrument payloads, representing various customer objectives, with multidiscipline as well as international perspectives. The common resources of the platforms and the base will be shared by a number of customers whose objectives are not necessarily in concert, indeed may have no relation to each other.

Unlike the existing scenario, where operations control for a spacecraft and its instruments is typically focused in a single project control center, in the Space Station era operations control of the platforms, the base, the modules and instrument payloads will be distributed. In a sense, there is a mirroring on the ground of the diffusion of mission related systems in space.

The SSIS ADD describes a number of operations control centers for Space Station. The Platform Support Center (PSC) and the Space Station Support Center (SSSC) will exercise operations control of the platforms and Space Station base, respectively. But mission-focused operations control, where payload operations are derived from science objectives, are in a number of Payload Operations Control Centers (POCC's), which have subordinate hierarchical relationships to the PSC and the SSSC.

A comparison with flight operations typical in current space missions suggests analogies with both free-flyer spacecraft and attached Shuttle payloads. In the former comparison, the individual modules and instrument payloads of the Space Station era are analogous to the free-flyer spacecraft of today, in that the modules and payloads are the focus of the "mission" objectives. In the latter view, they are analogous to the attached payloads carried onboard the Space Shuttle, in that their operation is subordinate to a host "platform," the Space Shuttle.

Figures 2 and 3 represent the concepts of resource sharing in space and the relationship to mission objectives in current and Space Station-era missions.

Communications

In today's missions, tracking and data acquisition (T&DA) and the directly coupled data transport are typically the major shared resources. NASA provides T&DA support through its Tracking and Data Relay Satellite (TDRS) Space Network (SN) and the remaining stations of the Ground Network (GN). Because many spacecraft are competing for these services, the scheduling the T&DA support among these users is a significant component of the integration problem in current mission operations.

An expanded SN will provide T&DA support to the Space Station base and platforms. Data relay satellite systems provided by the European Space Agency (ESA) and the National Space Development Agency (NASDA) of Japan may also contribute to the T&DA service of Space Station flight elements.

The current thinking is that SN services would be allocated to the Space Station program in major chunks. Thus, the SN will support the Space Station base with one equivalent K-band Single Access (KSA) 300 mbps link and

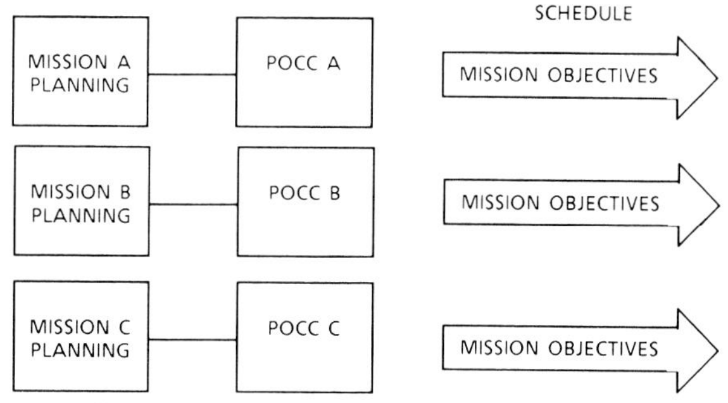

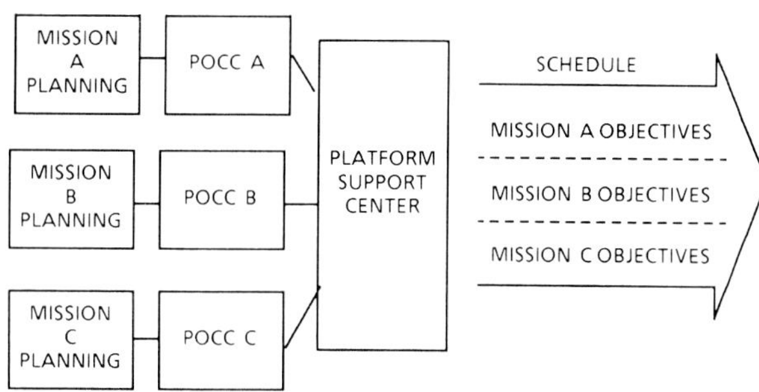

Figure 2. Mission Objectives and Scheduling

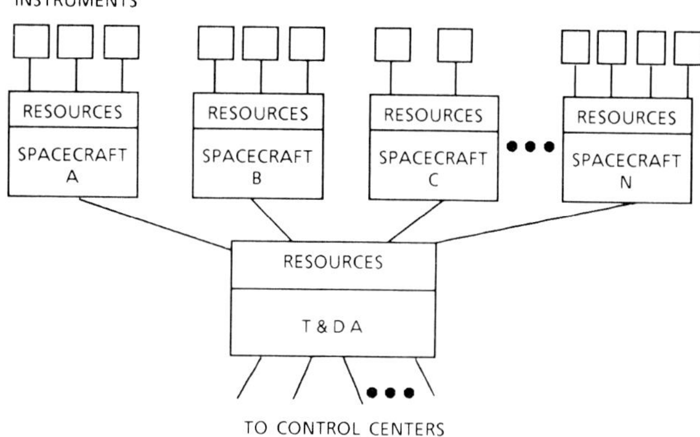

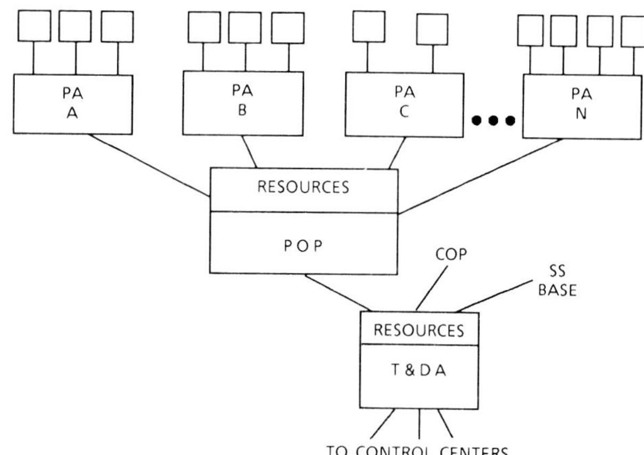

Figure 3. Resource Sharing in Current
and Space Station Eras

the total of all U.S. platforms with one equivalent KSA link. Where today the individual free-flyer spacecraft requires a dedicated T&DA link during support periods, Space Station communications support will employ a technique in which a number of payloads will each be allocated individual "virtual channels", effectively dividing the bandwidth of the 300 mbps KSA SN link.

The interactive role of international data relay satellite systems with NASA's SN is yet to be established. The specifics of this interaction could have significant effects upon the integration of T&DA support for Space Station.

Figure 4 illustrates T&DA resource sharing in both the current and the Space Station era.

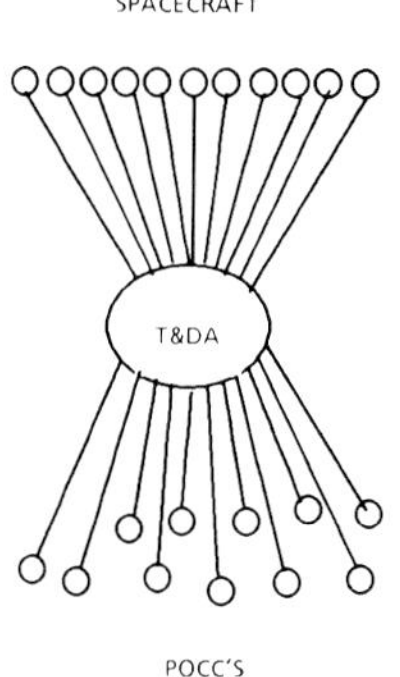

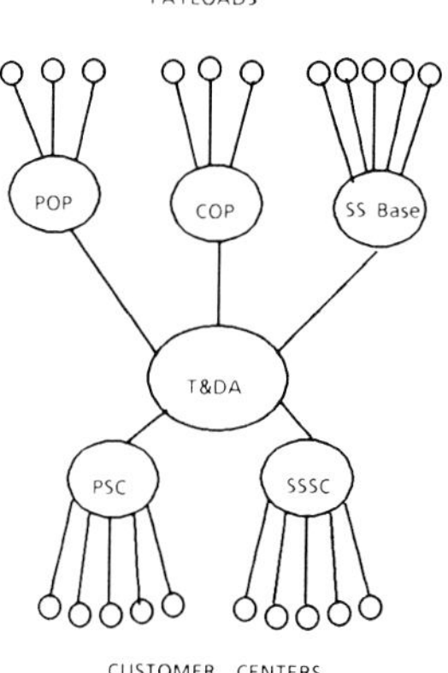

Figure 4. T & DA Resource Sharing

Data Capture and Processing

Higher data rates and many more users characterize the Space Station data capture and processing services. Current planning is that these services be supported by data driven facilities in the Space Station era, not subject to dynamic scheduling integration.

Customer data processing during the Space Station era remains typically removed from the dynamics of the operations scheduling process, though, as with current missions, there will likely be occasional exceptions.

The Space Station base and platforms will be in service for many years, resulting in the need for periodic missions to service or replace payloads and systems, and to replenish consumables. Each servicing visit will disrupt operations and will require integration into the platform and base schedules. Because the service depends upon transportation provided by the National Space Transportation System (NSTS) Space Shuttle, its schedule must also be integrated into the servicing and logistics schedules.

Planning and Scheduling Integration Features

The principal characteristic contributing to the efficient integration of planning and scheduling in the existing process is that there are not numerous pools of shared resources. The T&DA resources are the major shared resources. In the Space Station era, on the other hand, there are numerous pools of shared resources. Significant among these are the common flight resources that must be shared by many customers, customers who have, in many cases, different mission objectives.

The diffusion of flight operations control will also increase the challenge of integrating planning and scheduling. The major effect here is associated with the attempt to include science objective considerations as an integral part of the planning and scheduling process. Prevalent in current mission operations is a focus of science objectives and mission scheduling with a specific tie to a free flyer spacecraft. This feature is not a part of the proposed Space Station era operations scenario.

In summary, with existing space missions scheduling integration is characterized by these features:

- The customer relationship is generally defined through an individual flight project

- Flight operations, the driver of the process, are focused on free-flyer spacecraft generally dedicated to the scientific goals of a specific mission

- The T&DA service represents the only shared major resource

- Integration in the existing scenario is essentially the sorting of the T&DA schedule among competing users.

Planning and scheduling in the Space Station era is anticipated to have these significant features:

- Customer relationships to the program are not exclusively defined

- A number of payloads on the platforms and base share a common pool of space resources

- Control of operations is diffused

- Ground systems are highly distributed among elements and subject to interelement dependencies.

The Space Station Scheduling Integration Problem

In this section we consider in more detail the scheduling integration problem of the Space Station era. We examine the SSIS elements that play roles in the scheduling integration process and analyze the interelement dependencies that drive the need for integration.

Elements with Integration Roles

Of the many SSIS elements, the following have significant roles in the scheduling integration process.

Customer Facilities. Customer Facilities (CF's) are sources of requests for services that must be integrated. We anticipate various types of CF's. Some CF's will operate Instrument Control Centers (ICC'S) and have the additional responsibility of scheduling instruments on a platform or the Space Station Base. CF's representing discipline facilities could have the responsibility for coordinating multi-instrument science campaign schedules.

Customer Coordination Capability. The Customer Coordination Capability (CCC) facility coordinates customer scheduling issues that cannot be resolved by an operation control center.

Instrument Control Center. An ICC schedules and controls the operation of an individual instrument.

Payload Operations Control Center. We use this term generically to designate a center with a "mission" or "project" focused operations responsibility for a payload, group of payloads, or module on the base or platforms. There will be a number of POCC's at various NASA and international partner facilities.

Platform Support Center. The PSC is responsible for scheduling the Space Station platforms and for coordinating the schedules of the payloads on the platforms. International PSC's will carry out these scheduling roles for international platforms participating in the Space Station Program.

Space Station Support Center . The SSSC is responsible for scheduling the Space Station base and for coordinating the schedules of the modules and payloads on the base.

Integrated Logistics Support System. The Integrated Logistics Support System (ILSS) develops resupply schedules.

NSTS Mission Control Center. The NSTS Mission Control Center (MCC) is responsible for scheduling the NSTS, which provides transportation for Space Station servicing and logistics.

Network Control Center. The Network Control Center (NCC) schedules SN services for the Space Station base and platforms as well as for spacecraft not associated with the Space Station Program. International NCC's will perform these scheduling responsibilities for international data relay satellites.

Interelement Integration Roles

Of the above elements, the CF's are a part of the scheduling integration process because they are the sources of the requests for resources and services that must be integrated. As such, their role in the integration process is mostly reactive, although CF's with discipline coordination authority could have an active integration role within the customer community.

The CCC's role in the integration process is that of an active coordinator among customer needs, where those needs cannot be coordinated among the customers, or through an operations control center.

The remaining elements are participants in the integration process because they have operations and scheduling authority for Space Station systems whose operations must be coordinated with customer requirements and with the operation of other elements.

All the above elements have fairly well defined local scheduling responsibilities. Likewise the interelement dependencies and associated scheduling integration interactions are also well defined. A POCC, for example, integrates "locally" the schedules for the payloads it controls on the Polar Orbiting Platform (POP), and coordinates with the PSC to integrate those payload schedules in terms of platform resources. The PSC manages its own local schedule, including, at this higher hierarchical level the schedules of the POCC payloads, and interfaces with all other elements as required to obtain and coordinate services external to the platform, e.g., SN support, servicing, and logistics resupply. The NCC has "local" responsibility for developing the SN schedule, including schedules for all of the facilities, such as the New Mexico ground station, that are a part of the SN, and interfaces with other elements to coordinate SN support for them.

The above description indicates some characteristics of a hierarchical structure, in the relationship of the POCC's to the PSC and the SSSC, in terms of the scheduling responsibility for payloads on the platforms and the base. In other aspects, such as the relationship of most of the elements to the NCC, for example, the structure is not hierarchical.

Managing customer imposed scheduling dependencies seems likely to present a new sort of challenge during the Space Station era. NASA's flight projects have traditionally supported the scheduling and integration associated with deriving space and support system operations to achieve the science objectives. The Space Station era may present more extensive requirements associated with the integration of discipline and campaign objectives, where in previous projects the focus has been more exclusive, concentrating on the free-flyer spacecraft and the requirements of the Principal Investigator and the instrument.

It is not clear yet what specific organizational or facility mechanisms will perform this larger focus discipline and interdiscipline customer integration. One example is the Earth Observing System (EOS) study project, which has been organized with a multiplatform and interdiscipline perspective that could accomplish this. On the other hand, it is not clear how pervasive this larger focus science integration will be during the Space Station era. For example, planning scenarios also indicate that the co-orbiting platforms (COP's) will support a limited number of science disciplines or programs, or in the case of the Advanced X-ray Astronomy Facility (AXAF), support only one program.

In the analysis that follows, we will assume discipline scheduling coordination functions reside in CF's with specific discipline roles.

<u>Features of a Viable Information System Approach</u>

The above discussion suggests that a viable information system approach to integrated scheduling for the Space Station Program will have these features:

<u>Supports Local and Global Scheduling</u>. Because Space Station facilities and users are diverse and distributed, scheduling must support both local and global applica-
tions. Scheduling of an individual instrument is typically a local process. Scheduling the end-to-end system support for an instrument data acquisition event is a global process.

<u>Supports Autonomous and Interactive Scheduling</u>. Many local operations have only local interest. Some instrument data acquisition events are of interest to a single investigator. The system should permit the investigator to schedule the operation with a measure of autonomy consistent with that of the observation itself. Other operations require interaction among a number of facilities and users, as when a discipline-oriented investigation requires coordination of data acquisition by a number of instruments. The system should support this also.

<u>Supports Micro and Macro Scheduling</u>. The requirement to support both fine and coarse scale scheduling is a function of both time – where one is in the planning cycle – and place – where one is in the system. Early phase planning must support coarse scale scheduling; near and real-time phases require detail scheduling. On the other hand, two different facilities may have a need to see the "same" schedule at different levels. An ICC will develop a detailed instrument activity schedule. At the PSC this instrument schedule would be supported in terms of a higher level resource envelope schedule (refer to Glossary).

<u>Provides Access to Information</u>. Effective scheduling requires access to schedules and schedule-related information. Scheduling "in the dark" is not a viable option in the Space Station environment of distributed and interdependent facilities.

<u>Supports Dynamic Rescheduling</u>. Customers and Space Station operators will require a measure of flexibility in operation. The ability to reschedule local operations with minimum impact to the system at large allows this flexibility without upsetting the integrated schedule of the total distributed system.

<u>Is Evolvable</u>. Space Station is a long-term program, whose users, facilities, and their scheduling needs and dependencies will change over time. The scheduling system, and its integration features, must be able to evolve in kind, always satisfying the needs of the users and the program.

<u>Supports Coherence of the End-to-End System</u>. Though implied in earlier features, this feature is of sufficient importance to merit separate mention. The distributed nature of Space Station facilities and users, with their often autonomous roles, implies a decentralized scheduling approach. However, such an approach carries with it the risk that the Space Station scheduling information systems, will lack coherent control as an end-to-end system. System-wide coherence is essential, because, while the autonomy of local scheduling realms is important, there are also aspects of Space Station operations that have system-wide breadth. These include the coordination of customer activities in general, which should always recognize that the horizon is the end-to-end system rather than just an individual node; the coordination of communication and servicing support, which are system-wide concerns; and finally, the fact that Space Station is an international program, with interests that are concerned with the total system. All of these characteristics argue for a feature of the scheduling integration concept that ensures system-wide coherence.

182

A Recommended Architecture
for Integrated Scheduling

Space Station is composed of a number of distributed facilities, each having a set of general autonomous local scheduling responsibilities. But, the whole is loosely coupled due to a set of well-defined interfacility dependencies. This arrangement in itself establishes the general spatial definition of the architecture. What remains to be determined are: the allocation of scheduling functions among the Space Station facilities and the characteristics of the scheduling functions.

Allocation of Scheduling Functions

Scheduling function allocation applies to functions that perform the actual scheduling as well as functions that ensure the integration of the schedules. The functional aspects of Space Station operations, facilities, and user roles provide part of the answer to the question concerning the allocation of scheduling functions. Each facility becomes a node with an "application" scheduler responsible for the scheduling of all systems and operations falling under the responsibility of the facility. For example, an ICC performs the scheduling of an instrument on a Space Station platform; it also schedules all of the local resources, such as data processors, terminals, printers, etc., that support the operation of the control center itself.

The second part of the functional allocation question concerns the allocation of functions that support integration of local schedules among the various nodes. In our architecture we propose at each node a scheduling interface module that supports transactions with other nodes to integrate local schedules as a function of the internode dependencies.

One final allocation is required to complete the architecture concept. We propose an allocation of information as well as of functions. Our architecture concept has a "central" schedule information management system, a facility that supports access to schedules and schedule-related information at all of the nodes. The individual nodes query the schedule information management system to obtain information they need to perform local scheduling in terms of internode dependencies.

It should be stressed that the "central" aspect of the schedule information management system is a function that provides query, location, and access to information. The information itself does not necessarily have to reside at the central facility; it could well reside at the scheduling nodes themselves.

Figure 5 illustrates an example architecture composed of SSIS element nodes and a schedule information management system.

The key to the success of the concept is that the nodes perform local scheduling efficiently, with scheduling tools "well tuned" to the local application; and that they do this with knowledge of and as a function of internode dependencies. We now describe the scheduling functions that provide this capability.

Characteristics of the Scheduling Functions

Node Application Scheduler. The application scheduler performs the scheduling of local resources and operations. We expand upon the ICC example suggested

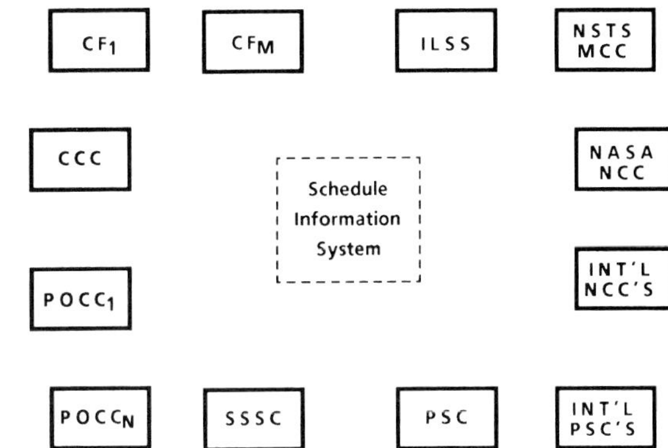

Figure 5. Example Architecture Concept

above. The ICC node has an application scheduler specially designed to schedule an instrument on a Space Station platform. Its functions include:

- Develops instrument activity schedules from science requirements

- Derives resource envelope schedules from instrument activity schedules

- Sends and receives scheduling messages from the node interface module

- Identifies internode dependent information in schedule messages received from the interface module

- Manipulates instrument schedules as a function of local and internode dependencies. This feature includes a number of "what if" capabilities to aid in the scheduling process.

These capabilities are used by the ICC node both to develop schedules from given science requirements and resource allocations; and to reschedule as a result of targets of opportunity or other contingency. The last of the five functions is essential to the rescheduling process It provides the ICC the capability to reschedule with knowledge of and as a function of the extra-node effects. In any rescheduling scenario, there may be a number of schedule options that achieve the user's objectives; this tool allows an evaluation and selection of the option that results in the least disturbance to the schedules of other users.

The node application schedulers are candidates for expert system schedulers. There has been promising research recently on the development of expert system approaches for scheduling applications, including applications to support space mission scheduling. For example, Ford Aerospace has developed a prototype knowledge base system to demonstrate planning and scheduling of instrument activities on the Upper Atmospheric Research Satellite. (See The Mission Operations Planning Assistant, James G. Schuetzle, 1987 Goddard Conference on Space Applications of Artificial Intelligence and Robotocs, May 1987.) For a survey of expert system schedulers see Expert Systems for Scheduling: Survey and Preliminary Design Concepts, J. Liebowitz and P. Lightfoot, GSFC, 1986.

Node Interface Module. The purpose of the interface module is to provide a communication interface between the application scheduler at the local node and other nodes and the central schedule information management system. It sends to other nodes and the schedule information management system schedule messages it receives from the local application scheduler. It queries the schedule information management system to obtain information relevant to local node scheduling, receives that information and sends it to the local application scheduler. Finally, it sends revised scheduling information from the local node to the schedule information management system. Figure 6 illustrates the above concept.

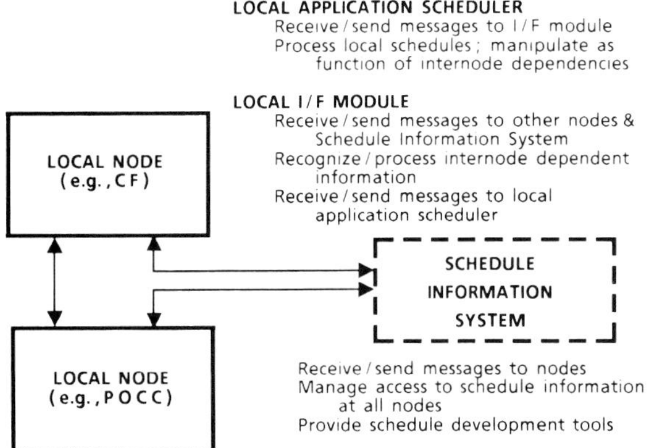

Figure 6. Typical Node Functions

It is proposed that the node interface modules at all of the nodes be standardized units, employing standard formats and protocols. Unique scheduling features at each node should be confined to the application processor. This characteristic promotes adaptive flexibility needed when the scheduling architecture itself must grow and evolve as scheduling requirements and responsibilities change.

Schedule Information Management System. The schedule information management system is the key to maintaining end-to-end system coherence. It also provides users immediate access to information they may require in order to carry out their scheduling needs. It maintains schedule information from all nodes in sufficient detail to indicate scheduling dependencies among the nodes. It is freely accessible by all nodes and users. Examples of information maintained at the schedule information management system include:

- Schedules for Space Station Base and platforms, including
  - Orbit/attitude
  - Core systems
  - T&DA support
  - Servicing/logistics

- Space Network and International Data Relay Satellite Schedules

- National Space Transportation System Schedules
  - Launch dates
  - Payloads
  - Destination

- Customer Schedules
  - Campaign/event
  - Discipline coordination
  - Supporting Space Station Elements
  - Supporting Institutional Facilities.

Evaluating the Architecture

Local and global scheduling are supported by the marriage of specialized local node schedulers, capable of scheduling as a function both of local and internode dependencies, and global availability of scheduling information. These characteristics also provide the needed capabilities to support local autonomous scheduling as well as scheduling interactively with other nodes.

The flexibility of the nodal approach, scheduling as a function of local and internode dependencies, provides the capability to support micro as well as macro scheduling. Dependencies can generally be defined at any level of detail desirable, and the system can be designed accordingly. This feature is essential to the resource envelope concept as well. A customer develops detailed instrument activity schedules locally and sends a resource envelope schedule, a macro schedule, to an external node for integration. The customer continues to manipulate the micro schedule locally, doing so with ongoing knowledge and control of its dependent relationship to the macro envelope schedule. This capability also supports the ability to reschedule dynamically.

Access to information is provided through the ability to query the schedule information management system as well as other nodes for scheduling information.

Access to information concerning internode dependences teamed with the ability to manipulate local schedules as a function of the internode dependencies support dynamic rescheduling. This allows rescheduling to be performed with full knowledge of the impact to other uses and facilities, and with the ability to minimize these impacts.

Evolution of the system is supported through the modular nature of the architecture. Because of its modularity, its feasibility is not dependent upon a specific assignment of Space Station elements and functions. So long as the overall architecture permits a division into generally autonomously scheduled realms, with well-defined interdependencies, the concept is not tied to specific information system nodes and functions. New elements can be added, those that are no longer needed can be eliminated. The adding of a new element requires that the local application scheduler be designed after a thorough requirements analysis to determine what internode dependences apply, and assuring that it be developed with the ability to schedule locally as a function of these. An interface module, which we propose be a standard module employed across the program, would of course be part of the local system design also.

Finally, the architecture concept also provides the capability to support end-to-end system coherence. Key to this is the free availability of schedule-related information, coupled with the ability to schedule and employ this knowledge functionally in terms of dependencies applying at the local node. Information systems capabilities are not in themselves sufficient to ensure system coherence, as management and procedural policies must be employed with end-to-end system coherence as a specific objective. This is particularly important in the highly distributed nature of the SSIS.

Our discussion of the architecture concept has generally assumed the existence of an already integrated schedule. The concept has powerful mechanisms for *maintaining* a coherent integrated schedule as the local schedules are refined and modified throughout the planning process. Since these mechanisms provide capabilities to manage the schedule as a function of the system-wide integration-driving dependencies, they may be employed to support *establishing* the coherent integrated schedule as well. Direction is required during the early planning phase that ensures the establishment of an initial schedule that is coherently integrated, at least at a very high level, in the beginning. We assume that NASA and the international partner agencies give this direction during the Strategic Planning phase, when plans and resource allocations are established in top down fashion. Using these plans and allocations, the scheduling architecture supports the development of the initial integrated schedule. From that point, it supports the refinement and maintenance of the integrated schedule. This mixing of management, procedural, and systems functions ensures the coherence of the end-to-end system.

Integration Strategies Employing This Concept

The operational strategy behind this system concept can be summarized as follows:

Establishing an Integrated Schedule. Planning begins with the strategic planning phase, during which NASA and the international partner agencies determine high-level operations objectives and resource allocations. Plans, not schedules (schedules feature timelines), are the principal product. At this point, the principal integration goal is not detailed coordination among elements and systems based on timelines, but assurance that even at this very high level, the plans are in fact supportable by the Space Station elements and system.

The following tactical phase supports the planning necessary to implement the strategic plan in terms of Space Station flight elements, support systems, and the responsible customer and operator organizations. The major objective of the tactical planning phase is to derive operations plans and schedules from the customer and operator objectives. Customers begin the development of the long-term schedules of instrument observations or other payload activities required to carry out the objectives. The resource envelope schedules are developed from these payload schedules. The Space Station and institutional system operators begin the development of the schedules of the major core system.

At this point, the full complexity of the coordination problem resulting from many functional organizations and many systems and facilities makes itself present. This is the first occasion where a specific allocation is attempted in terms of scientific and operations objectives, resource allocations, and constraint effect coordination.

Working from these allocations ,each node develops local application schedules (e.g., an instrument activity schedule) and derives a resource envelope schedule from this. If part of a hierarchical scheduling structure, the envelope schedule is submitted to the next level node, where a number of envelope schedules are integrated, and manipulated until conflicts are removed. When it reaches the highest level, that node (e.g., the PSC), extracts all internode dependencies (e.g., communications and servicing schedules) and coordinates with the appropriate nodes. Further manipulation is performed, until a high-level envelope schedule is integrated across the system.

During the above process, the local node schedulers employ their expertise to develop the most efficient local schedules as a function of the internode dependencies. This capability can be used both to schedule within the bounds of known internode restraints, and to derive internode dependency requirements from developed local schedules.

Maintaining the Integrated Schedule. After the initial development of a system-wide integrated schedule as described above, the local nodes refine the local application schedules as required up until the time the schedule is executed. During this period, the objective is to maintain the integrated schedule. Local application schedulers support this, by providing the capability to refine the schedule as a function of the envelope schedule, which now defines the internode dependencies. The local scheduler should support rescheduling in terms of these dependencies, and should provide "what if" capabilities to evaluate proposed changes to the local application schedule in terms of the dependencies. In this fashion local activities can be rescheduled with minimum impact outside the node. Such a feature will be indispensable for targets of opportunity and other short term rescheduling support.

An example:

Figure 7 illustrates an example scenario of the concept. We assume an existing SSIS-wide integrated schedule at the beginning of the scenario. Because of unexpected results of analysis of previous observations, a customer at a CF desires to modify an upcoming observation schedule for his instrument on a Polar Orbiting Platform (POP). In order to obtain needed information to support planning, the local application scheduler and node interface module are employed to query the central scheduling information system for latest schedules for the POP and SN.

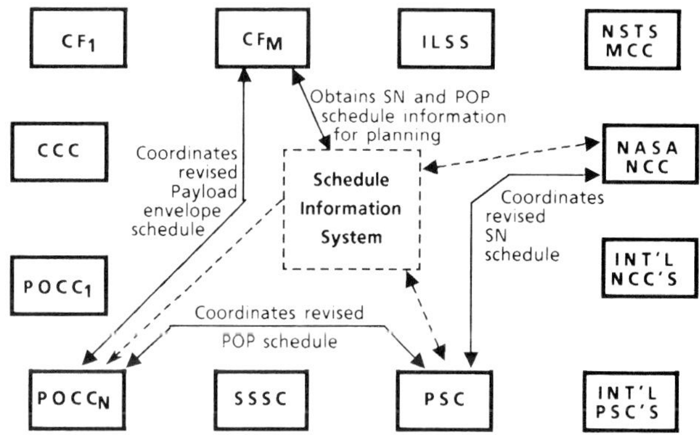

Figure 7. Schedule Transaction Scenario

Knowing the instrument data requirements to support the desired new observation schedule, the customer develops candidate instrument activity schedules that satisfy his needs. Depending upon the science requirements and instrument characteristics, there may be a number of schedules that meet his needs, in terms of instrument modes, time of day of observation, orbit and attitude pointing specifications, etc.

The customer then employs the system to evaluate the options in terms of the internode dependencies. His objective is to develop a rescheduling option with the least impact outside the node. He uses the system "what if" features to determine if there are options to reschedule the instrument within the confines of the resource envelope schedule already in use. This option would impact no other users. The system finds no rescheduling options that achieve his objectives within the current envelope schedule.

The customer then attempts a rescheduling with resource impacts confined to the "payload accommodation system" that supports the instrument. If successful, this rescheduling would affect very few additional users. The customer determines that the desired rescheduling will affect resources on other payloads as well.

As a result of additional analyses of scheduling options, the customer determines that modification of the SN schedule is also required.

The customer has now evaluated a number of rescheduling options. He has determined an option that satisfies his objectives with the least impact to the schedules of other users. He submits the request through the POCC that controls the payload accommodation for the instrument. The POCC, using its scheduling system evaluates it against other instrument users (likely it repeats some of the evaluations already performed at the CF) and submits a request to the PSC.

The PSC evaluates the reschedule request, coordinates with other POCC's as required to reallocate resources, and coordinates with the NCC for a modified SN schedule.

Assuming the customer had the required priority for the rescheduling, it is carried out in the above fashion. Note that the modular rescheduling allows efficient management of integration during the rescheduling process. Note also that if at any point in the above process the desired schedule change was denied, the local scheduling tools would support rescheduling of the instrument to maximize partial success in terms of new objectives within the bounds of the given resources.

Demonstration Prototype

Ford Aerospace is currently developing a prototype to demonstrate the scheduling integration concepts described in this paper. This prototype will model the scheduling functions and interactions of a number of nodes, including a POCC and a Customer Facility Instrument Control Center, demonstrating the capability of these nodes to schedule locally as a function of their internode dependencies. The first phase of this prototype is scheduled for completion in November 1987.

Conclusions

Because of its richness and diversity, Space Station presents a challenge to the integrated scheduling of operations unlike those we have faced in previous space flight programs. Understanding the nature of the interdependencies inherent in that diversity is essential to success in the design of information systems to support integrated scheduling. An examination of the Space Station planning and scheduling environment, as currently envisioned, indicates promising solutions with a marriage of distributed control and advanced applications schedulers at local facilities.

Acknowledgments

The analysis supporting this paper has been supported by NASA Contract Number 5-2750, managed by Goddard Space Flight Center, Greenbelt, Maryland. The demonstration prototype is being developed under this contract. The conclusions discussed in this paper are the responsibility of the author.

Glossary

Co-orbiting Platform (COP) – Unmanned Space Station Program spacecraft in a circular orbit consistent with formation flying with the Space Station base.

Customer Coordination Capability (CCC) – A facility to coordinate Space Station customer support.

Customer Facility (CF) – Home facilities of Space Station customers for local payload operations and data analysis.

Data Interface Facility (DIF) – Gateway system at the Tracking Data Relay Satellite (TDRS) ground station that links NASA ground networks and the space-ground network.

Execution Plan – A detailed plan covering SSP operations for a term of hours to a few days, usually presented with a granularity of minutes.

Ground Network (GN) – The collection of NASA tracking and data acquisition (T&DA) ground stations.

Instrument Control Center (ICC) – A facility that performs operations control of a platform or Space Station base payload instrument.

Integrated Logistics Support System (ILSS) – Facility that manages inventory and requirements for logistics support of the Space Station elements.

Network Control Center (NCC) – Facility that schedules and controls the NASA Space Network (SN).

National Space Transportation System (NSTS) – The NASA Space Shuttle system.

NSTS Mission Control Center (NSTS MCC) – The control center for NSTS operations.

Payload Accommodations (PA) – Platform and base facilities providing interface and services to attached payloads.

Payload Operations Control Center (POCC) – Facilities that provide operations control of payloads or modules on the platforms or Space Station base.

Polar Orbiting Platform (POP) – Unmanned Space Station Program spacecraft in polar orbit.

Platform Support Center (PSC) – Facility that controls the operation of the Space Station platforms.

Resource Envelope – A predefined set of resources and constraint allocations that permits an "envelope" of possible payload operations. It is proposed that users schedule resource envelopes rather than schedule the actual payload events. Once a resource envelope is scheduled, the user is free to develop a detailed schedule for his payload so long as it is within the resource and constraint bounds defined by the envelope.

Space Network (SN) – The NASA Tracking Data Relay Satellites and associated systems.

Space Station Support Center (SSSC) – Facility that controls the operation of the Space Station base.

Strategic Plan – A long-term, top-level, nondetailed plan covering SSP operations, growth, and service pricing, usually presented with a minimized emphasis of chronological sequencing. A strategic plan results from policies set by the international partners.

Tactical Plan – A plan having an intermediate level of detail, covering SSP operations, SSP element utilization, and logistics over a term of weeks to years, usually presented with a granularity of days.

References

Earth Observing System Data and Information System Report of the EOS Data Panel - Volume IIa, National Aeronautics and Space Administration, 1986.

Easton, R. "SSIS Planning and Scheduling White Paper Overview," Presentation to NASA Johnson Space Center, 1986.

Ferns, D.C., Lytton, D.W., and Dillon, M., "Payload Operations on the Columbus Polar Platform," *Journal of the British Interplanetary Society, Volume 40*, 1987.

Garman, John R., "Data Systems for the Space Station and Beyond," *A Collection of Technical Papers: AIAA/ACM/NASA/IEEE Computers in Aerospace V Conference*, 1985.

Liebowitz, J., and Lightfoot, P., "Expert Systems for Scheduling: Survey and Preliminary Design Concepts," NASA Goddard Space Flight Center, 1986.

Space Station Information System Architecture Definition Document (SSIS ADD), NASA Johnson Space Center, October 1986.

Schuetzle, J., "The Mission Operations Planning Assistant," *Proceeding of the NASA Conference on Artificial Intelligence Applications*, 1987.

Zoch, D., "Maintaining Consistency Between Planning Hierarchies: Techniques and Applications," *Proceedings of the NASA Conference on Artificial Intelligence Applications*, 1987.

INTEGRATION OF MULTI-DISCIPLINE DATA PROCESSING FOR EARTH OBSERVING SYSTEMS

Ralph Kahn
Visiting Senior Scientist
Earth Science and Applications Division
NASA Headquarters
Washington, DC 20546

Robert Chase
Technical Director
Advanced Engineering Laboratory
Woods Hole Oceanographic Institution
Woods Hole, MA 02543

ABSTRACT

Philosophically, there are two approaches to establishing a data and information system; it can be designed and built *ab initio* from a complete systems engineering analysis, or it can be assembled from existing hardware and software within the purview of a given organization. In the era of Earth System Science, characterized by research on increasingly multidisciplinary problems, a disparate collection of discipline-specific data systems must be made interoperable to the extent that a researcher hosted on one facility can easily access data, information, and a suite of services provided by and/or resident at other facilities. To meet this goal, it is unlikely that either of the two traditional approaches to developing data systems will be adequate by themselves, particularly given the complexity of the problem, the available time, the limited potential for large new increments in funding, and the substantial investment of resources and experience represented by existing facilities. Consequently, the NASA Earth Science and Applications community has been working toward greater interoperability of its data handling facilities from both perspectives simultaneously. In this paper we describe the first steps taken to provide for the controlled evolution of existing facilities toward greater interoperability and sharing of resources among the NASA-supported Earth Science and Applications data systems.

INTRODUCTION

Historically, data handling for NASA Earth science has been supported through data systems associated with individual flight projects, through centralized facilities like the National Space Science Data Center (NSSDC), and relatively recently by discipline-oriented data systems (Pilot Land, Pilot Climate, NASA Oceans data systems). The resultant heterogeneous data handling environment arose naturally at a time when the frontier of scientific analysis of spacecraft data lay within the boundaries of individual disciplines, often extending no further than the observations of a single experiment. To a degree, limitations in telecommunications and computer technology contributed to the insular nature of the developing data systems. A parallel management structure formed to support numerous individual projects requiring NASA Earth Science and Applications data.

Today, both the scientific analysis requirements and technological capabilities extend beyond those early borders. The National Academy of Sciences Space Science Board (1982) and the Eos Data Panel (Chase, 1986) expressed the growing demands for 'greater interoperability' and greater sharing of resources among spacecraft data handling operations. These demands are motivated by increased scientific awareness of the interrelationship of physical systems. Scientists must cross traditional disciplinary lines, to intercompare data from multiple sources, in addressing key questions in Earth System Science. Research opportunities afforded by more sophisticated data manipulation capabilities, and the necessity of studying ever increasing volumes of data from new missions and retrospective sources, fuel the increased interest.

Existing data systems will have to perform an ever increasing level of routine operations that are common to all elements in the global network of facilities. As the scientific community adopts a new research methodology for Earth System Science, many data handling procedures will be streamlined; some will become relatively routine. It is likely that economies can be realized by developing and maintaining generic hardware and software tools, and standard data exchange and telecommunications protocols, in support of these new activities. Properly designed and developed, these tools can be incrementally adapted to support particular applications at lower cost than developing entirely new tools, while preserving a high level of interoperability. Within NASA's Earth Science and Applications data systems (ESADS), the process of developing such a data handling environment has begun, building upon the considerable talent, interest, and other resources committed to these existing facilities.

TOWARD CONVERGENCE IN DEVELOPMENTAL APPROACHES

Accessing information and data resident in diverse data bases is a central part of the goal of interoperability. The requisite integrated data handling environment will preserve geographically-distributed, dissimilar elements with capabilities and responsibilities specialized to the needs of specific groups of scientific users. A prime objective will be to provide reasonable remote and interactive electronic access to information and services throughout the community. There are two types of approaches to establishing this data handling environment. It can be designed and built *ab initio* from a complete systems engineering analysis, or it can be formed from existing, heterogeneous data handling facilities by incrementally adjusting their operations, tools, and services.

The Eos Data Panel and the Science and Applications Information System User's Working Group have specifically addressed the conceptual design of integrated, multidisciplinary data and information systems (viz. Chase, 1986, 1987). Since the functional design of data transport facilities and services is a fundamental factor, their architecture is based upon space data communications standards. To ensure that the resulting integrated system meets scientific needs, it is designed following two basic principles; layering and standardization (cf. "standards"). Together, these principles are used to create an information system that is adaptively flexible, transparent, and robust, providing the foundation for diverse data processing and archival needs.

Nevertheless, an interoperable data handling environment which would allow Earth science data to be more fully utilized requires years to develop; it must evolve with increasing knowledge and experience in Earth science and in data management (a "bottom-up" approach). To varying degrees, the evolutionary process can be guided toward a common conceptual architecture. Since the Eos Data Panel and the Science and Applications Information System User's Working Group have addressed conceptual architectures (using a "top-down" approach), it seemed reasonable to pursue a bottom-up approach to interoperability, particularly given the complexity of the task, the limited time available before deployment of new spaceborne sensors, and the limited likelihood for large new funding initiatives in the near term. In doing so, we strive for convergence in systems developmental approaches.

ACHIEVING INTEROPERABILITY AND
SHARING OF RESOURCES: Bottom-up Approach

Using a bottom-up approach, our intention was to bring existing talent, interest, and other resources to bear upon the problems of developing commonality among the existing Earth Science and Applications data systems. The task was circumscribed to focus on data systems directly relating to NASA Earth sciences. Retrospective data issues,

long term planning matters (greater than 3 years), and initiatives requiring major new funding were excluded from immediate consideration.

A group at NASA Headquarters, led by Kahn, and including James Dodge, Eni Njoku, and Anthony Villasenor, agreed to develop a Headquarters consensus for this initiative. In mid-November 1986, we began planning for an ESADS Workshop which would bring together data systems personnel, scientists, and management representatives from the Earth Science and Applications community to help define the next steps in data system coordination and evolution. The principal goal of the Workshop was to develop a set of specific, near term action items which would describe the next steps in moving the existing distributed, heterogeneous data handling environment toward greater interoperability and a greater sharing of resources. A second, less tangible objective of the Workshop was to encourage communication within the ESADS community.

Community involvement in the Workshop preparations was critical, both to lay the groundwork for negotiations over specific tasks, and to provide feedback about the scope and content of the exercise. Eight task groups were selected (Table 1) to address pertinent areas. An electronic mail telecommunications system (GTE's TELEMAIL) was used to distribute information to all participants. Electronic bulletin boards and mailing lists were created to support the Workshop. A library of reports and documents relevant to ESADS activities was established and its holdings were posted on the bulletin boards (with copies of documents available to members of the ESADS community upon request). Initial sets of questions, covering each of the eight task group areas, were distributed over this network for comment. The questions asked what the first steps might be in developing greater interoperability and a greater sharing of resources in specific areas.

Kahn and Villasenor, together with the task group chairmen, re-cast these questions into action items, forming a strawman implementation plan that would become the basis of a Workshop Report. Soliciting an additional set of 'user requirements' (the traditional approach for a systems engineering design) was avoided, ensuring that the effort focused on the realities of existing facilities and achievable near-term goals. The National Academy and Eos Data Panel reports provided the necessary overall guidance.

Feedback from the greater ESADS community was solicited over TELEMAIL at each stage of preparation. Benchmark systems questionnaires and scientific user questionnaires were distributed; the collected answers are being used to assess the current status of the ESADS and to measure our progress toward the goals previously defined. Many new issues were raised and "discussed" over TELEMAIL. One such issue was whether there would be adequate programmatic follow-up to assure the Workshop Report would be acted upon.

An ESADS Committee (Table 2) was formed to provide a user perspective to the Report and follow-on activities. This Committee is reviewing the Workshop Report, assuring the recommended action items are consistent with user-specified goals of greater interoperability and sharing of resources. They

Table 1. ESADS Workshops Task Groups

Task Group Focus and Issues	Group Chairman and Affiliation
R. Kahn, Workshop Chairman	
Directories and Catalogs - data system lexicon - catalog interoperability - interactive browse function	John Estes University of California - Santa Barbara
Data Archives - data description language - automated data ingest - data access - data tracking	Mark Abbott Scripps Institution of Oceanography
Data Manipulation Software - interface languages - applications software standards - software catalogs	Robert Chase Woods Hole Oceanographic Institution
Computational Facilities - status information - central facilities access	Robert Price NASA Goddard Space Flight Center
Data Storage Media - media standards and testing - software interfaces - data compression	Mike Martin Jet Propulsion Laboratory
Data Base Management Systems - ISO-model paradigm - functional layer specifications, implementation, and testing	Randy Davis University of Colorado - Boulder
Networks - protocol and interface standards - connectivity - standard network services	Peter Shames Hubble Space Telescope Institute
Scientific and Operations Management - tracking and tuning development - management coordination issues	George Ludwig University of Colorado - Boulder

Table 2. EARTH SCIENCE AND APPLICATIONS DATA SYSTEMS COMMITTEE

Wayman Baker	NASA Goddard Space Flight Center
Bruce Barkstrom	NASA Langley Research Center
Otis Brown	University of Miami
Jack Estes	University of California - Santa Barbara
Roy Jenne	National Center for Atmospheric Research
Ralph Kahn (Chairman)	NASA Headquarters (on leave from Washington University)
Dave McAdoo	National Oceanographic and Atmospheric Administration
Eni Njoku	NASA Headquarters
Ernie Paylor	Jet Propulsion Laboratory
William Rossow	NASA Goddard Institute for Space Studies
Murray Salby	University of Colorado - Boulder
Stewart Smith	IRIS
Roy Spencer	NASA Marshall Space Flight Center
Victor Zlotnicki	Jet Propulsion Laboratory

will also play a role in reviewing the implementation of the action items, suggesting alterations
of the specific objectives, and defining issues for
subsequent ESADS workshops.

A NASA Code EE Data Systems Committee, composed of
Headquarters managers, was created to provide programmatic review of the Workshop Report. This
group, chaired by Eni Njoku, will follow the progress of the ESADS effort, providing top-level
coordination for the evolution of the data systems
they manage, and assuring that the directions taken
are consistent with the goals of their programs.

THE ESADS WORKSHOP

The ESADS Workshop took place from February 24
through 26, 1987, in Easton Maryland. About 85
individuals, representing the full spectrum of NASA
Earth science activities related to data handling,
attended. An initial half day of general presenta-
tions set the tone for the Workshop. Subsequently,
most of the activity centered around the eight task
groups. These groups were asked to produce no more
than about 5 pages of text, so the majority of
effort would go into negotiation and cross-
fertilization of ideas rather than into writing.
Plenary sessions were called daily, with each task
group chairman given the opportunity to present
issues to the assembly. Emphasis at the plenary
sessions was given to those topics which fell
within the purview of several task groups. Joint
task group meetings were set up to negotiate sections of the Workshop Report that crossed these
lines. Key to the success of this activity was the
preparation of the strawman implementation plan
prior to the Easton meeting, and the involvement of
most of the Workshop participants in that preliminary effort.

At the end of the Workshop, we had produced a Report describing where near-term efforts are most
urgently needed to move toward the stated goals of
the ESADS initiative. The Report includes specific
action items, with many names and dates filled in.
Among the issues addressed was the need for information collection in many areas. More coordination
across the ESADS was suggested for catalogs, data
ingest software, user interfaces, networking plans,
and standards in general. The need for more
sharing of expertise in data systems management,
including system security, was stated. We identified the first steps toward achieving these objectives.

The immediate follow-up plan calls for a 30-day
review (over TELEMAIL) of the Workshop Report by
the entire ESADS community, with a subsequent review (lasting two weeks) by the ESADS Committee and
the Code EE Data Systems Committee. Benchmark questionnaires will be reissued in fall, 1987. The
results will be compared with the previous bench-
marks. The ESADS Committee will evaluate the results, and will decide in what areas progress over
the year warrants widening the scope of topics for
an ESADS Workshop in early 1988. It was agreed
that the ESADS Committee and Headquarters Managers
participate as members of the community during the
one month of general revision, to minimize the
controversies generated by the final reviews of
these groups.

To ensure that the Workshop had the intended impact
on the ongoing activities of the ESADS, the action
items from the Report have been entered into a NASA
OSSA Information Systems Office data base, to be
used directly in generating the FY 1988 call for
Research and Technology Objectives and Plans
(RTOPs). In a sense, the ESADS initiative can be
viewed as a management tool, filtering the community's input to provide management with the information it needs to make informed programmatic decisions without becoming overwhelmed with a mass of
unsorted details.

RESULTS AND PROGNOSIS

Although the Workshop Report is still under revision by the community at this time, there were
several key issues and related actions recommended
by each of the task groups. We outline below
several of these items.

Directories and Catalogs Panel

Data System Lexicon. A requisite first step in
coordinating the disparate data handling and computational facilities supporting NASA Earth science
research is the establishment of a common set of
terms. This will be accomplished by creating a
lexicon through iterative negotiation and discussion, largely over TELEMAIL. These terms will form
the basis for communication regarding development,
evolution, and negotiation of shared resources.

Catalog Interoperability. A necessary step towards
facilitating optimal scientific use of available
data is the ability to find descriptions of data
held at various sites, and where possible to provide electronic connections to remote data centers
holding needed data.

Browse System. An appropriate set of browse functions enables a scientist to interactively define a
subset of data granules for subsequent analysis.
Browse capabilities can be an integral part of an
interoperable catalog function. Since several
groups are developing discipline-specific browse
capabilities, custom-tailored to their data, it may
be advisable to proceed in a more generalized
fashion, minimizing redundant developments and
maximizing synergy between these efforts.

Data Archives Panel

Data Description Language. A major problem for the
operation of most archives is the expenditure of
resources needed to take an incoming data set and
extract sufficient information to turn the data
into a useful archival product. If the primary
burden is shifted away from providing adequate data
structures toward providing adequate descriptions
of the data, a minimal packaging structure will be
required, and the computational burden will be
significantly reduced. This suggests the adoption
of a common data description language to be employed by all elements of the ESADS.

Automatic Data Ingest. The costs of including new
data sets into existing archives could be substantially reduced if the collective experience in data
ingest were applied to building general tools for
this purpose. Only incremental adjustments would

be required to tailor these tools to specific applications. Properly developed ingest routines extract sufficient attributes to enable the description of a data set for catalog and directory entries. The standard formatted data unit lends itself to this task. It is a likely candidate for incorporation into archival functions.

Data Access. Data access consists of both the user's request for data and information as well as the distribution and delivery of data. No current NASA data system provides all the access functions required for multidisciplinary research. This suggests that pertinent, common data access functions should be defined and tested. Existing pilot data system projects form ready candidates for testbedding these functions.

Data Manipulation Software Panel

Interface Languages. The prime need in "data manipulation" software is for interface languages (user and interchange) and file utilities. This software would enable users to access information and/or data from remote sources in a uniform fashion; the particulars of the communications links and data subsystem architectures will be transparent to the user. One means of solving this problem lies in the standardization of user interface and data interchange languages.

Applications Software Standards. Existing applications code such as inversion techniques, averaging methods, and interpolation algorithms should not be standardized unless they are to be maintained and supported for the greater community (e.g., generic tools developed for NASA data handling facilities and deliverable Flight Project software). It is appropriate to encourage modern software development methods such as modular construction, top-down design, and structured, well documented code; these practices will not limit creativity and productivity. Organizations responsible for writing and distributing general-purpose code must be particularly sensitive to these issues and must provide significant levels of documentation (including those appropriate to a software catalog) to accompany their code. Guidelines for generating adequate documentation should be developed.

Software Catalog. Since much existent software is non-standard and unverified, its inclusion into a software catalog seems of marginal value. However, there is an increasing level of activity directed toward generation and distribution of common software tools for the greater community. Since this software is generic, it should be listed and detailed in a common software catalog accessible to all NASA-sponsored researchers.

Computational Facilities Panel

Status Information. In the heterogeneous ESADS environment, information about configurations and patterns of utilization for existing facilities would be of use in planning further developments. Such information would also be of interest to groups investigating options for upgrading local facilities, and possibly to outside parties seeking to provide services to the ESADS community.

Accessibility of Facilities. As part of the objective of sharing resources across the ESADS, additional effort is recommended to improve access to major computational facilities supported for community use.

Data Storage Media Panel

Media Testing. Media testing must be performed on a regular basis to monitor possible deterioration of stored data sets. Techniques should be identified, then incorporated into the regular operating schedules of each of the affiliate data systems. In doing so, the scientific value of a particular data set will increase, since it will then be available to create extended, uninterrupted time series from which long-term changes in the Earth system can be assessed.

Uniform Software Access. End users would benefit significantly from the development of a single software interface for access to data on various media (e.g., WORM, CD-ROM, magnetic tape). This unified access software should be extendable to future storage technology improvements, and be updated as mass storage technologies evolve.

Data Compression. Optimum coding and/or spatial data compression has a significant potential utility in reducing the volume of data stored in archives while also reducing transmission times for exchange of these data. These factors translate directly into cost savings for media, storage, and communications. Consequently, potentially appropriate techniques should be identified and tested to determine their validity and assess the potential cost impact.

Database Management Panel

ISO-model Paradigm. Using the simple layered model developed by the International Standards Organization (for general purpose protocols within the framework of "Open System Interconnection") as an example, the ESADS community could significantly benefit by conceptually extending the model to include all activities embraced within the integrated system. In doing so, standards would be set at the interfaces, allowing researchers the maximum level of independence for creativity within each function.

Functional Layer Specification, Implementation, and Testing. A key to creating a useful model will be to map existing data system capabilities (e.g., SFDU, CDF) onto the system model and to identify conflicts and compatibilities in the standards and approaches currently in use. This mapping can then serve as a guide in establishing priorities for the development of extended, requisite capabilities. Selected functional layers and interfaces defined within the model should be subjected to evaluation prior to distribution across the ESADS community. Although it may take five or more years to achieve a high level of interoperability, this procedure will have minimal impact on current levels of productivity, while maximizing the utility of the model implementation.

Network Panel

Protocol and Interface Standards. Procedures and methods for access to the disparate archives within the ESADS should be simple and as uniform as possible, to encourage a broad spectrum of uses. By adopting a standard protocol (such as TCP/IP) for the network, ESADS will ensure connectivity

throughout the system. The network standards
should support electronic mail, remote terminal
log-in, file transfer, and task-to-task interaction
in a manner that is transparent to the user, and
with response times consistent with available resources and user requirements.

Connectivity. Both logical and physical connections among data bases, archives, and computational
resources are required of the integrated network.
The NASA community of researchers must be supported
by this federated system, and reasonable access to
outside data sources must be provided.

Standard Network Services. Standard network services should include electronic mail, directories,
network information center, and trouble reporting
and user assistance. Operationally, traffic flow
must be monitored, and access to applications and
other resources must be provided. Information
derived from the monitoring should provide the
foundation for planning for network augmentation.

Scientific and Operations Management

Tracking and Tuning Overall Development. In order
to facilitate more effective use of existing resources, management must track the development of
generic tools and data systems standards. It
should provide a single-point clearing house for
current information, and a means to respond to
changing technology and new ideas. Mechanisms
should be established to direct suggestions to
appropriate recipients, and to assure that both the
facility for communication and the interest in
cooperation across the ESADS are maintained.

Management Coordination Issues. There are a suite
of activities which require management attention to
coordinate the development of interoperable data
handling and computational facilities. These include addressing system access, data integrity and
security issues, establishing data exchange policies, ensuring the general adoption of appropriate
standards, providing for a documentation library,
establishing appropriate charging and accounting
procedures, and providing guidelines for Flight
Project data management. These issues should be
addressed with the cognizance and council of the
active scientific research community.

For each of the issues outlined above, there are a
set of specific, near term action items in the
ESADS Workshop Report for consideration by NASA
management. Adequate follow-up procedures have
been defined to ensure that these tasks will be
acted upon, consistent with available resources.
Many of the recommended actions parallel existing
developmental efforts. This suggests that there is
sufficient interest and expertise in the community
to begin moving the existing heterogeneous data

handling and computational environment closer to
functional connectivity and interoperability.

CONCLUSION

We found a great willingness on the part of data
systems professionals, scientific users, and management personnel to work together on a structured
project with reasonably well defined goals. Forming a coherent group, they have carefully and
thoughtfully defined a set of specific near-term
action items which are designed to provide a logical next step in federating the existing heterogeneous elements of NASA's data handling and computational facilities. These actions are fully consistent with the conceptual architecture defined by
both the Eos Data Panel and the Science and Applications Information System User's Working Group.
The convergence of these developmental approaches
promises to provide the research community with the
access to data and information which it requires,
and to do so in an orderly and cost effective
manner.

Acknowledgments. We wish to thank the other organizers, participants, and contributors to the ESADS
Workshop activities for their insights and assistance in moving NASA's Earth Science and Applications data systems one step closer to interoperability and a greater sharing of resources. We
particular thank Dixon Butler and Anthony
Villasenor for their support and the NASA OSSA
Information Systems Office for sponsoring the ESADS
Workshop. R.K. is funded by the Visiting Senior
Scientist Program of the Jet Propulsion Laboratory,
and is assigned to NASA Headquarters in support of
the Eos Project. Funding is provided to R.C. by
the National Aeronautics and Space Administration
under Grant NAGW-946. This is Woods Hole Oceanographic Institution contribution number 6460.

REFERENCES

Chase, Robert R.P. (Chm), 1986. Earth Observing
    System Data and Information System: Report of
    the Eos Data Panel. NASA Technical Memorandum
    TM-87777, 62 pp.

Chase, Robert R.P., 1987. A Systems-approach to
    the Design of the Eos Data and Information
    System. IEEE IGARSS 87 Digest. 6 pp. (in
    press).

National Academy of Sciences Space Science Board,
    1982. Data Management and Computation, Volume
    1: Issues and Recommendations. National
    Academy Press, Washington, DC.

COORDINATION OF DATA RELAY SATELLITE SUPPORT

by

S. Sobieski
GSFC/NASA
Greenbelt, Maryland   USA

A. Dickinson
ESTEC/ESA
Noordwijk, Netherlands

M. Kajii
NASDA
Tokyo, Japan

K. Lenhart
ESOC/ESA
Darmstadt, Germany

Abstract

The European Space Agency and the National Space Development Agency of Japan are currently conducting conceptual definition studies for data relay satellite systems, much like the NASA Tracking & Data Relay Satellite System (TDRSS). It is therefore reasonable at this time to consider modes of interoperability among these three systems so as to achieve mutually synergistic benefits once they are operational in the mid-1990's. Inter-Agency discussions have been underway since October 1985 to explore this topic. A number of interoperability benefits have been identified including: back-up in emergency situations, extension of communication coverage, load levelling or sharing, and enhanced data exchange. The purpose of this paper is to present and discuss nominal technical prerequisites for implementing the different levels and modes of interoperability.

TDRSS, as a system near to operational status, provides a quantifiable reference and point of departure. Preliminary consensus is that some subset of the TDRSS S-band (2000-2300 Mhz) services could provide a core capability for international interoperability. Other forms of enhanced compatibility, such as operation at Ka and Ku bands, direct spot-casting for data exchange, and intersatellite cross-linking are being considered but will not be discussed here. There is no consideration of achieving compatibility of the respective relay satellite telemetry, tracking and control functions.

One adverse aspect of compatibility, particularly at the low frequencies of S-band, is the potential for mutual interference. Independent studies show that the magnitude of this interference for nominal space mission scenarios mandates mitigation procedures for relay satellites with orbital spacings of less than approximately 12 degrees. Under certain conditions these effects can appear at even wider spacings. The implication is that as more data relay satellites become operational, coordination will be necessary even if there is no formal

program of cross-supported interoperability. The conditions and parameters for such coordination are discussed in this paper. Above and beyond the usual frequency coordination, other RF signal parameters become important and there are technical and adminstrative conditions which must be considered in the pre-mission planning and the mission support phases. For cooperative programs, there are additionally operational prerequisites related to quality, assurance, accounting and fault management.

As the ESA and NASDA data relay satellite systems become better defined, the concept of a coordinated S-band service will be further refined. This refinement would cover details of the RF signal structure, planning and scheduling procedures, and format and protocol standardization for both control messages and user data. This defintion of a coordinated service will serve as a model for future, more enhanced forms of space communications interoperability.

1.0 Introduction

At this time both the National Space Development Agency of Japan (NASDA) and the European Space Agency (ESA) are considering deploying data relay satellite systems at geosynchronous altitudes as a means to enhance telecommunication support for their respective future space missions. Operations could commence as early as the mid-1990's. The National Aeronautics and Space Agency (NASA) is currently in the process of making its tracking and data relay satellite system (TDRSS) operational. It is, however, also beginning preparations for the second generation version of TDRSS which is planned to become available approximately in this same time frame. Since these systems represent sizable investments of resources, it is natural at this planning juncture to consider ways of maximizing their operational and performance benefits for system users. Interoperability among these systems appears as one such way and inter-Agency meetings have been held since October 1985 to explore this possibility.

NASA's Tracking and Data Relay Satellite
System has been described extensively in the
literature and only a high level overview is
presented here. The baseline systems is
comprised of two operational satellites stationed
at 41 and 171 degrees West longitude,
respectively, both controlled from the White
Sands ground complex in New Mexico. TDRS
telemetry and control data and all user data are
transmitted on the space-to-ground link between
White Sands and the respective TDRS; terrestial
communications are used to complete data
transport to the user's ground facilities. Space
communications are provided at S-band, 2000-2300
Mhz, and the Ku-band, 13.4-15.2 Ghz. Range and
range rate information is generated using the S-
band. The current plans are to implement a
second ground station at White Sands by 1993 and
to augment the space configuration by stationing
an additional satellite in the vicinity of each
of the baseline TDRS' in the mid-90's as demand
for service increases.

NASDA has a plan to implement its own data
relay and tracking satellite system (DRTSS) in
the mid-90's (tentative target; 1995) to support
spacecraft operations for national and
international programs such as its remote sensing
satellites, Space Station, etc. ETS-VI, which is
now planned for launch in 1992, will carry
intersatellite communications equipment as a
testbed for DRTS hardware and system
technologies.

As part of the European in-orbit
infrastructure forseen for the 1990's and beyond
- including the European contribution to the
International Space Station, the Columbus program
and the European spaceplane, Hermes - Europe
plans to implement a global Data Relay System
(DRS). This will be necessary to handle the
greatly increased data flow requirements
associated with these programmes. It is intended
to ensure an operational DRS on a timetable
compatible with the operational timetables of
these users and thus a first satellite will be
launched in mid-1994, a second in early 1995 with
system commissioning being completed by mid-1995.

Both NASDA and ESA plan to implement S-band
services which will be compatible with that
provided by TDRSS. This becomes the key common
denominator for the concept of coordinated
service and is the major topic presented in this
paper.

2.0 Definition of Interoperability

Interoperable means that elements from
different systems can be employed interchangeably
without altering functionality. The three major
functional categories of any data relay satellite
system are: the space communication links for
user support, the space to ground terminal data
transport, and the data relay satellite control
and management. Since each Agency must
independently assure reliable operation of its
own relay satellite system, relay satellite
control has not been a subject for discussion on
interoperability. Although data transport, the
second major functional area, is not discussed in
this paper, it does represent a fruitful area for

international interoperability. In fact, the
Consultative Committee for Space Data Standards,
CCSDS, is presently involved in defining formats
and protocols to facilitate user data exchange.
This paper will deal only with the first
category, i.e the radio frequency (RF)
compatibility between user spacecraft and
different relay satellite systems, specifically
at S-band, and the technical and adminstrative
prerequisites for achieving operational
compatibility.

2.1 Modes of Interoperability

In the broadest of operational terms,
interoperability can range from providing
occasional or episodic support, through routinely
scheduled cross-support, to an operation
requiring the full integration of the separate
networks' functions. The first does not involve
much routine coordination or standardization; it
would be invoked on an ad hoc basis and could be
handled using special procedures. Scheduled
cross-support, on the other hand, implies that
even as early as the mission planning phase,
arrangements are made to provide service from
another Agency, this service being on a routine,
predictable basis. Formal procedures and close
coordination are needed; this is the mode
emphasized in this paper. A mode intermediate to
these two involves back-up support of one Agency
by another for some extended period of time, say
prior and during the activation of a replacement
for a failed relay satellite. Finally, the fully
integrated mode is one comprised of the combined
resources of all participating Agencies
cooperating in all operational aspects from
scheduling to data exchange. This system needs
to be engineered and administered jointly.
Clearly as one progresses through each of these
increasingly more tightly operationally coupled
modes, the greater the degree of technical and
management coordination required among the
various Agencies.

2.2 Benefits of Interoperability

Inter-Agency discussions have tentatively
highlighted a number of specific operational
benefits for these different modes of
interoperability. These include: back-up
communications in emergencies involving either
the user spacecraft or the relay satellites, load
sharing or levelling for synergistic benefit,
and extension of coverage, i.e. minimizing the
impact of the zone of exclusion communication
outage. Calibration and verification support
have not been considered as interoperability
drivers and the question of range/orbit
determination is still open. In no case were the
vital interests and basic system objectives for
any Agency to depend upon resources provided by
the other Agencies.

Emergency support provides useful back-up
particularly for the low data rate housekeeping
and telecommand services. It would apply in such
cases as when the nominal supporting relay
satellite were not available (not in line-of-
sight), for contingencies, and, possibly during
certain incompatible operations such as maneuvers
or launch phases. Another form of support is
when adequate channel time is not readily
available for a user and would be provided as an
unscheduled service.

For routine operation communication support within the zones of exclusion will not normally be required since outages are brief and spacecraft normally provide on-board data storage for this contingency. There are exceptions, however, when support would be highly beneficial. An example is the support of manned flight missions during certain critical operations, such as docking or preparation for re-entry while in the zone of exclusion.

Load levelling may well be one of the most beneficial aspects of routine cross-support. Although it entails a significant amount of coordination involving both users and relay satellite operators, it has the potential for easing user scheduling constraints and relay satellite replenishment requirements. More knowledge is needed, however, of the respective Agencies' mission plans and subsequent system loading before the full benefit of this mode of interoperability can be established.

2.3 Interference and Interoperability

The orbital locations of the data relay satellites is not arbitrary. They must maintain high availability against weather on the space-to-ground link for command and control and user data transmission which imposes minimum elevation angle constraints depending on the ground station siting. For ESA, in addition, satellite location must meet link availability requirements for distributed client user locations since direct data transmission is being planned. Link availability argues for minimizing the differential longitudes of the satellites with respect to the central longitude of the respective ground complex. On the other hand, the desire to maximize the coverage afforded by a two satellite configuration, i.e. minimizing the extent of the "zone of exclusion", argues for maximizing the differential longitude. The tentative baseline locations resulting from this trade yield satellite separations of 100, 110, and 130 degrees for the NASDA, ESA and NASA systems, respectively. These separations along with the central longitude requirements leads to close proximity for two pairings of satellites, NASDA/NASA and NASA/ESA and this, in turn, raises the question about possible operational interference between adjacent satellites. Since beamwidth is inversely proportional to frequency and antenna size, operation at S-band with the small diameter antennae dictated by launch/technology considerations further exacerbates this potential problem. Independent studies have established that the mangnitude of the interference becomes significant for orbital separations of less than approximately 12 degrees, assuming representative scenarios in which low earth orbiting users supported by different Agencies are conducting proximity operations. Under certain conditions, primarily for low antenna gain and/or high powered user systems, interference may be noticeable for much wider spacings. The apparent angular separation between data relay satellites as seen from low earth orbit varies by only several degrees due to orbital motion and is not a strong function of the latitude of the user spacecraft. Thus polar orbits are as affected as low inclination ones. Both forward and return links may be affected and, again for representative scenarios, signal

to interference spectral level ratio may actually be degraded to less than one-to-one. Thus if orbital locations are relatively immutable, possible mutual interference must be addressed.

This situation, of course, is not unusual. Frequency management of the RF spectrum has been and is a continuing, necessary activity and operates at various organizational levels. In the case in question, however, other mitigation approaches must be used since the band is by definition co-populated and frequency coordination will be impractical as the number of user spacecraft increases. Mitigation can be achieved by coordinating other RF signal parameters such as pseudo-random noise (PN) coding and polarization, and/or by mission management, primarily, scheduling. These will be addressed in later sections.

In summary, coordination is mandatory if not under the auspices of interoperability then in order to mitigate mutual interference.

3.0 S-Band Signal Compatibility

This section introduces parameters of the S-band communication system which can become the foundation for a compatible, interoperable user service. TDRSS is used as the reference since all Agencies are currently planning to implement some level of compatible RF systems.

The TDRSS Users Guide, STDN 101.2, Rev 5 provides a detailed description of the full range of capabilities offered by the TDRSS. The options include a wide variety of signal structures and rates; the breadth of these do not impact the TDR spacecraft since it operates primarily as a simple repeater but it does impact the extent of the ground communication processing requirement and complexity of the user transponder. It is of interest, therefore, to determine if some limited subset of the options could be used to establish a viable international "standardized" service which would minimize the cost and complexity of new ground system implementation and normalize the design of the user transponder. A full specification of such a standardized service is beyond the scope of this paper; we attempt only to establish a foundation for future in-depth studies. Since the ability to support a majority of existing or planned TDRSS users within any proposed standardized service is an important boundary condition, it is reasonable to use the characteristics of these supported services as a guide in defining a representative subset for adoption. Tables 1 and 2 show the pertinent signal characteristics for the forward and return links, respectively, at S-band for current or planned TDRSS users.

The specified RF signal structures listed in the Tables are not necessarily optimal. They represent, however, a robust response to the nominal communications environment; optimization would only be possible if the environment could be fully specified, which it cannot. What general comments can be made from examining these Tables and what issues are involved?

1. Most current TDRS users have elected to operate at the fixed Multiple Access (MA)

Table 1: TDRS S-Band Parameters for the Return Link

	STS	HST	HST	GRO	UARS	COBE	ERBS	SMM	LNDSAT	SME	EUVE[3]	GRM[3]
TYPE	SSA	SSA	MA	SSA/MA	SSA/MA	SSA	MA/SSA	MA/SSA	SSA/MA	MA	MA/SSA	SSA
FREQ(MHz)	2287.5/ 2217.5	2255.3	2287.5	2287.5	2287.5	2287.5	2287.5	2287.5	2287.5	2287.5	2287.5	2287.5
MODULATION	BPSK	BPSK	QPSK	QPSK	QPSK	QPSK	QPSK	QPSK	QPSK	QPSK	N/A	
POLARIZATION[1]	LHC	LHC	LHC	RHC/LHC	RHC/LHC	LHC	LHC	LHC	RHC/LHC	LHC	N/A	
CONV CODE	1/3	1/3	1/3+RS[4]	1/2	1/2	1/2	1/2	1/2	1/2	1/2	N/A	
INTERLVNG[2]	N	Y	N	Y	Y	N	N	Y	N	N	N/A	
E.I.R.P(dBW)	16.7	20.5	20.5	29.5	27.3	3.6	25	18.2	22.8	19.8	N/A	

(1) FWD & RTRN LINKS MUST USE SAME POLARIZATION WITHIN SERVICE INTERVAL. MA IS ALWAYS "LHC"

(2) NO INTERLEAVING FOR MA OR SA FORWARD. INTERLEAVING FOR DATA RATE > 300Kbps

(3) UNAPPROVED (Feb. 1987)

(4) RS = REED SOLOMON OUTER CODE

Table 2: TDRS S-Band Parameters for the Forward Link

	STS	HST	GRO	UARS	COBE	ERBS	SMM	LNDSAT	SME	EUVE	GRM*
TYPE	SSA	SSA	SSA	SSA	SSA	MA/SSA	MA/SSA	MA/SSA	MA	MA/SSA	SSA/MA	
FREQ	2106.4/ 2041.9	2106.4	2106.4	2106.4	2106.4	2106.4	2106.4	2106.4	2106.4	2106.4	2106.4	
MODULATION	BPSK	QPSK	QPSK	QPSK	QPSK	QPSK	QPSK	QPSK	QPSK	N/A		
G/T (dB/^{o}K)	-26.8	-33.2*	-20.2	-22.3	-32.7*	-20.2	-19.7	-27.3	-32*	N/A		
CONV CODE	1/3	NO	NO	NO	NO	NO	NO	NO	NO	N/A		

* OMNI-ANTENNA

frequencies (2106.4 and 2287.5 Mhz — forward and return) even for single access (SA) service to achieve cross-service support without requiring dual frequency transponders. TDRS, in order to mitigate interference between the multiple and single access services therefore requires circular polarization diversity. Even so there is concern about overuse of the MA frequencies. The present strategy of supporting MA users with the SA service depends on the high availability of the SA; this situation will most likely not persist.

2. Rate 1/2 convolutional encoding is the most used error correction approach. For higher data rates interleaving is also required to achieve the nominal TDRS Bit Error Rate performance level of 10-5. This encoding is relatively inexpensive to implement and provides the capability of achieving good bit transition densities using a 2-complemented symbol scheme. Its robustness is enhanced by interleaving since this provides protection against events affecting adjacent symbols. A possible issue would be how to achieve a standardization here without inhibiting the future development and use of more bandwidth efficient coding schemes.

3. Quadrature Phase Shift Key (QPSK) modulation is almost universally employed. It provides good bandwidth efficiency and is moderately robust against system, particularly, phase noise. It affords the user substantial flexibility in that it provides a doubling of the data rate for a fixed bandwidth but as importantly facilitates data segregation without the need for multiplexing. Range and range rate can be accomplished simultaneously with data transport. If higher performance modulation schemes, currently being studied, are implemented in the future, the standarization must be procedurually flexible to accommodate such improvements.

4. Although only one user (Space Telescope) has implemented the Reed-Solomon concatenated coding, there is ample analysis to indicate its potential benefit. Serious consideration should be given to adopting the CCSDS proposed coding in the definition of future standardized services. Implementation requires minimal hardware and, being a transparent code, affords compatibility with older non-coded systems.

5. TDRS users, except for STS, operate with a 3 Mchip pseudorandom noise (PN) coding in order to

satisfy power flux density constraints. At
relatively high data rates (>300 Kbps) it is
unncessary because of the broader corresponding
spectral densities and antenna pattern effects.
Chip rate selection depended on matching this
spread spectrum to channel bandwidth, with the
latter being concerned with optimizing the TDRS
figure of merit, G/T, for a wide range of users.
In addition to its use for spread spectrum
purposes, the PN code is used in the MA system
for user identification and supports signal
acquisition. These codes must therefore be
controlled by a management function and
standardization would imply coordination in this
area.

Many of the parameter options for the RF
signal are driven by the radio frequency
environment; some, however, depend on link
considerations related to a) user space terminal
limitations, and b) the performance capabilities
of the respective data relay satellite.
Standardization in these areas impacts system
flexibility. Thus it will be impractical to
standardize transmit power levels (E.I.R.P.) or
the terminal's figure of merit. It will be more
beneficial to adopt minimum performance levels in
these areas as a prerequisite for compatible
support.

In addition to signal structure
compatibility, there is need to have effectively
compatible signal processing particularly to
accomplish signal acquisition. First, updated
ephemerides are required, typically on 9 second
centers for TDRSS, in order to compute pointing
vectors and Doppler shifts. Second, some
provision must be made to capture the user
carrier; establishing a "Doppler window" or
frequency scanning the relay satellite receiver
may be used with subsequent Doppler compensation
of the received carrier. The question of range
and range rate service provision is left as an
issue for further discussion.

If "anti-compatibility" is required in order
to mitigate interference the following are the
principal ways of achieving RF isolation:

Frequency selection - Assigning different
frequency channels within the band to
different user spacecraft is the most direct
way to avoid interference. Fixed bandwidth
channels can be allocated to the separate
agencies or coordinated on an individual
basis. The former simplifies the
coordination while the latter affords the
most flexibility for re-use or service
expansion. Interoperability in any event is
not impacted since all relay satellites must
be capable of operating within the entire
band.

Polarization - Different "handedness" of the
circular polarization can provide as much as
20-30 dB of isolation and is therefore a
useful mitigation approach. The use of
linear polarization is not viable due to the
changing orientation of the user in his
orbit as seen from the relay satellite.
Since only binary states are available, this
precludes isolation of more than two users
in close proximity. As with frequency
mitigation, the need to also be occasionally

compatible could require commandable
configuration of the antenna feed system.

Coding - The use of code division multiple
access for S-band operation can provide
marginally adequate isolation (10-20 dB),
particularly if the E.I.R.P. levels are
constrained to minimum values. PN spectrum
spreading is already used by TDRS to meet
power flux density constraints at low data
rates and code division is used
operationally to provide multi-user support
on the MA system. The presently used Gold
PN code formulation would need to be
extended to meet higher data rate
performance but this would be at the expense
of signal acquisition speed and requires
additional study. For operation at, for
example, 1 Mbps, chip rates more like that
used by the Shuttle system (11 Mchips/sec)
would be required to minimize
intermodulation effects. Other system
(primarily ground) and user spacecraft
impacts are minimal; present non-coded users
can be isolated from coded users by virtue
of the despreading process and hence
spreading vs. non-spreading affords another
degree of freedom for mitigation.

4.0 Prerequisites for Coordination

We introduce in this section the
administrative and operational prerequisites for
interoperability; some also apply for mitigation
control. Although they are based on experience
with the TDRSS, they are sufficiently generic to
be applicable to any system providing analogous
functionality. Again our purpose is not to make
specific recommendations on how to implement
these requirements, it is merely to identify and
qualify them.

4.1 Pre-mission Phase Coordination

Compatibility or cross-support involves the
coordination of a number of preparatory
activities in the pre-mission phase. These
include:

Procedures Specification - The communication
support requirements for each mission are
documented in a "contractual" agreement which
specifies the type/mode of support, performance
parameters, timeline information, data transport
requirements, special procedures (such as tape
dump operations), etc. Interface Control
Documents (ICD's) are used to define details of
the physical interface. It may not be necessary
to define such a complete mission specification
for a coordinated service but some subset will be
needed thus implying a need for standardized
management control documentation.

Compatibility Assurance - It is standard practice
to assure compatibility of the user's RF system
with the TDRS and compliance with the STDN 101.2
Users Guide by analytical modelling of the entire
system. The Communications Link Analysis &
Simulation System, operated from GSFC, is
employed during the design phases to achieve
this. A similar analysis will be needed in the
coordinated service to verify compatability of a
user space terminal with each of the

participating relay systems. This implies the
need to expand existing models integrating
features from the other relay systems.
Indirectly it also suggests that standardizing
nomenclature and analytical procedures could be
beneficial.

Administrative Procedures - An important aspect
of coordination will be access control or the
identification of authorized users.
Identification codes, both at the PN bit level
and the baseband frame level address, must be
managed on an international level either by
exclusive assignment or by coordination on a
mission by mission basis. Establishing priority,
although not too critical for emergency or backup support, may apply for more routine cross-
support and the mechanism for defining and
maintaing these priority lists remains to be
defined. In this same category are costing and
accounting arragements. These require data base
management and configuration control across the
participating systems.

4.2 Mission Phase Coordination

During the actual support period, handovers
between data relay satellites must be
coordinated. This requires the transfer of
mission data base information and schedule
coordination among the user spacecraft control
center and cooperating data relay satellite
control centers. Certain operations currently
require interaction of the user control center
with the communications system. For example,
tape recorder dumps from the spacecraft depend on
control center verification of data lock and
quality before initiating the actual dump. This
mode of operation implies a duplex communications
capability which, in turn, implies the need for a
standardized protocol. Another example, not
directly related to the user payload operation,
is fault detection and diagnosis. Although it is
expected that much of this operational activity
will be highly automated under the control of
expert systems in the future, there will still
nevertheless be a requirement for voice
coordination among the cooperating control
centers - a primary method employed today to
accomplish fault management.

If interference avoidance, rather than
interoperability, is desired then an effective
approach is to schedule mutually exclusive
contacts. This still requires a substantial
interaction between the respective control
centers. Two situations apply. One is where the
proximity of the affected users is not
accidental, i.e. their proximity is a result of a
joint campaign, in which case, coordination has
probably been established already. The other
case is one in which the proximity is an
accidental juxtaposition. Interference avoidance
is very difficult in this situation since it is
unlikely that the respective network control
centers would have sufficient information to
anticipate such an occurence. It is preferable
to achieve avoidance by judicious selection of RF
parameters which would then require no further
"management".

4.3 Orbital Slot Planning

As discussed earlier, the orbital locations
of the data relay satellites comprising each
system result from a compromise between coverage
and space to ground link (feeder link)
availability. NASDA is considering feeder links
to Tsukuba Space Center (36 deg. N lat, 140 deg.
E long) from its data relay satellites and, to
allow for rain attenuation, would like to
maintain a minimum feeder link elevation angle of
approximately 20 degrees. Based on this
constraint, the satellite locations would be 170
deg. W and 90 deg. E giving a separation of 100
degrees. If a Ku band feeder link is used
instead of the currently planned 30/20 Ghz, this
might be reflected in lower permitted elevation
angles and hence wider satellite separations.
The ESA system has a more serious feeder link
availability constraint in that its plans call
for direct downlinks to user ground stations over
a wide area of Europe. A further factor being
taken into account is the possibility of
providing direct links to the United States and
Japan from the western and eastern ESA
satellites, respectively. With these objectives
in mind the baseline locations planned for the
ESA satellites are 44 deg W and 61 deg E. The
operational TDR satellites will be located at 41
deg W and 171 deg W. However, NASA's plans are
to increase the satellite constellation by adding
two addition TDR satellites in the mid-1990's
time frame and thus orbital slot planning for
registration is of concern to NASA also.

Both interference mitigation and
interoperability become drivers to consider in
orbital slot planning. In general terms, the
former argues for maximizing the separation
between adjacent relay satellites whereas the
latter tends to favor reducing the separation.
Although use of location diversity is impractical
for mitigating S-band interference (due to the
broad antennae patterns involved), it can be
quite effective if the Agencies decide to use the
same Ku or Ka band frequencies. The planning
would obviously be affected if interoperability
means direct data tranmission via feeder links to
cooperating partners. But it could also be
impacted if by relying on the coverage extension
or cross-support capability afforded by
interoperability, an Agency favored feeder link
availability in the trade of availability versus
extent of the zone of exclusion. A unique need
for slot planning arises in the most extreme case
of coordination considered in which adjacent
relay satellites stationkeep to maintain both
within the 3 dB contours of a user's high gain
dual frequency Ku/Ka band antenna. Additional
studies are needed to address all these facets of
mitigation and inter-operability and how they may
affect orbital slot planning.

5.0 Summary and Future Work

The NASA, NASDA, and ESA inter-Agency
meetings which have been conducted since October
1985, along with the resulting study activities,
are leading to a mutual understanding of both the
benefits and issues related to the operation of
multiple data relay satellite systems. A key
result to date is the appreciation of the need
for coordination in order to minimize intersystem
interference. Frequency coordination is required
at a minimum. Coordination for registration of
satellite orbital slots is also indicated. But

the important new perspective is that the
prerequisites to achieve adequate interference
mitigation also serve as the basis for
cooperation, i.e. interoperability, at least in
the use of the S-band frequencies. A preliminary
set of both technical and adminstrative
conditions have been identified and await further
study to define an international coordinated S-
band service.

     Areas for future technical discussions
include:

   1. More detailed definition of the RF signal
      structures for achieving a coordinated S-
      band service,

   2. In preparation for joint Space Station
      operational support, work toward defining
      reference data base needs and the protocols
      for their exchange,

   3. Examining in greater detail schemes, such as
      baseband coding, which minimize
      interference while permitting joint use of
      the S-band and also promoting
      interoperability,

   4. As NASDA and ESA refine their approach for
      communication service management,
      investigating suitable protocols for
      internetwork service scheduling and
      coordination, and

   5. Establishing common nomenclature and tools
      for describing and analyzing relay satellite
      links.

Acknowledgements

We are pleased to acknowledge the many
suggestions and inputs provided by Claudio
Soprano, ESTEC/ESA, and Robert Godfrey,
GSFC/NASA.

Technology 2: Software and Automation
Co-Chairmen: Howard L. Yudkin and Rudolf J. Lauber

Larry Druffel
Software Engineering Institute
Carnegie Mellon University
Pittsburgh, PA 15213

ABSTRACT

This paper provides a structure for assessing
trends in software development environments. It
first addresses three issues in software engineering: quality, managing the process, and productivity. The paper then describes four trends in
software environments: the tool kit framework, the
language-centered approach, structure-oriented approach, and the method- or process-based approach.
After describing work of the Software Engineering
Institute on defining levels of maturity of the
software process, the paper considers the concerns
of software users, such as conceptual integrity
and the ability to add new tools.

During this presentation, I will discuss a structure for assessing trends in software development
environments. In one sense, software development
environments have been around for many years. They
evolved from operating systems to which we added
tools, often in an ad hoc manner. For example, a
tool was added because a project needed that tool
and built it, using project funds or other project
resources. Unfortunately, tools added this way
were usually neither general nor high quality because they were oriented toward specific project
needs; as a result, we rebuilt similar tools again
and again. I do not know how many times we built
editors, compilers, and other tools that were functionally identical.

Some years ago we became aware of this constant re-
building, and the topic of environments -- that is,
the structure and the set of tools that you should
have to support software involvement -- became a
hot topic. Part of the motivation for the Ada program was to bring some focus to the development
of environments: to take advantage of a high-order
language and create a powerful set of tools oriented to a single language.

Commercial products are now beginning to emerge.
In the past, every project had to worry about its
software development environment, but now it's
even more confusing because projects have to decide whether to build or buy. In the past, you did
not have to choose; you had to build since there
was not anything to buy. Now there is an option.

With these evolving tools, or environments, we see
different directions. I'll try to characterize
some of those directions. The structure I'm going
to describe was proposed by Susan Dart, Bob El-

lison, Nico Habermann and Peter Feiler at the Software Engineering Institute[3]. One reason this topic
is important comes out in Fred Brooks's exposition
on the future of software engineering, "No Silver
Bullet"[2]. Brooks talks about areas in which we've
seen improvement in the past and where we might see
improvement in the future. His point is that there
is "no silver bullet"; but he identifies three developments that have been important: high-order
language, time sharing, and programming environments. Although the technical basis had been demonstrated, we have not yet realized the advantages
to be gained by combining these developments in
production. I think we ought to pursue those advantages first. Conceptually we have understood the
developments, but we haven't used them effectively.
If we could pull them together, we would gain a
tremendous advantage.

By way of motivation, let me talk about why we have
environments. I think there are three issues in
software engineering. (When I speak of software
engineering, I'm talking about building large systems -- I'm not talking about programming-in-the-
small.) The three issues are, in order of importance: quality, managing the software engineering
process, and productivity. Of the three, I think
the quality of the product has to be the first objective. If the quality is not in the product --
that is, if software isn't reliable and if it does
not perform the way it should -- it does not fulfill its function.

The second issue is one I think we tend to neglect:
managing the software development process. Often
we think of process in terms of its contribution to
productivity; we don't usually think of it as a
good in itself. If you consider systems such as
space stations, you realize that software is what's
going to make them run. The software is the intelligence of the system. The cost of software will
be dwarfed by the cost of all the other things that
go into the system. Yet, if the software fails --
or if it's late -- because we didn't manage the
process of software development, we have major cost
over-runs. So, even if we don't increase productivity, it is worthwhile for us to manage the process.

Finally, we want productivity -- for reasons that
we all understand. Quality and process contribute
heavily to productivity, but they are not only important in their own right; they must take precedence. Productivity is one of the things that mo-
tivates us to build tools. A good way to improve
productivity is to automate some support for the
process. However, I'm not going to talk about automation of the process because I think that's a
research topic. I'm going to talk about automated
support for the process -- that is, automating some
of the manual tasks not automating the entire process.

The original reason for building tools was to automate some pieces of the process as they apply to the coding phase. The obvious tools for doing that were compilers, linkers, loaders, and debuggers. Those were powerful tools to build on top of existing environments, and those tools got us the first real contribution to productivity: high-order language. In the early days of environments, we tended to focus on the coding phase, the part of the environment, or life cycle, that those tools address. As systems got bigger, we began to realize the need to manage the complexity of the software system, to concentrate on things like version control, configuration management, and requirement tracing. So we began to add those tools to the system. Now we're beginning to recognize the need to provide automated support for the entire software process. (Process is the latest buzz word in our industry, although it's been around since the days we started developing software.) We are now trying to understand and manage that process.

Let's discuss four environment trends in an order that is pretty much chronological but also is the order of their potential power. The four trends are: tool kits (that is, bags of tools that live on an operating system); language-centered environments; structure-oriented environments; and method- and process-based environments.

The tool kit framework -- extensions added to an existing operating system -- is the most familiar. I think all of the early environments were of this type. First, it's easy to add a new tool because there's little structure; you're simply using the underlying operating system. You have some file system that you can use. Everything is either a file or a program. Programs get applied to a file; it's very flexible. It has generality, but it offers little for management of the sequence or consistency of the user interface. You have to know how to use every tool, and your interface to each tool is enormously different. Since tool builders have their own models, there's no consistent model to interact with. You can't come to the system with an expectation of what it will do for you. Tools don't know about one another, so they don't communicate well with one another. In fact, you can apply a tool to a file for which that tool is inappropriate. For example, you can run a Fortran file against a Cobol compiler. There's no consistency checking done by the environment to preclude you from doing what we might consider silly. This was an appropriate style for the early days of environments because we didn't know how to structure an environment. We just didn't know what we needed.

These are not the only examples. Every operating system that has compilers, linkers, loaders, and all the other tools are themselves environments in some dimension. Among those where we began to think more about structure, probably the first example is Programmers' Workbench. More recently the Ada community has moved in two directions. CAIS is an attempt to define the interfaces more precisely, and PCTE is an extension to UNIX to provide standard interfaces for tools. These have evolved on the two sides of the Atlantic. CAIS is evolving in the United States, and PCTE in Europe.

The next kind of environment has a language-centered approach, with support for the semantics of a specific language. In the workbench or tool kit approach, all languages were supported. In this case, however, we're focused on a specific language. An early example of such an environment was the Interlisp environment to support Lisp. Others that have evolved are the Smalltalk environment to support Smalltalk, the Cedar environment to support MESA, and the Rational environment to support Ada. These environments support the semantics of the language and provide tools in a consistent, language-oriented model. In general, you don't need to learn a separate command language; the language that you're using is the command language. This approach promotes an interactive style. The developers usually provide a feature that enables you to make incremental changes without having to go through the complete recompilation stage. One of the trends that we see in software engineering is toward more prototyping, and this type of environment encourages that exploratory or prototyping style of development.

The next, and more difficult, type of environment is the structure-oriented approach. In this case, there's a common representation but different views; that is, you're not always working with code. You might have different views of the program for the software that you're writing, and you have an editor that manages some underlying representation. This approach was used for hardware design automation in the early to mid 1970s. The key that got us to take the step from a tool kit approach for hardware design automation was the discovery of an underlying representation, one that allowed a single representation and many different views. You could change one view; and when you changed that view, the underlying representation was changed and was reflected in all the other views. We're beginning to see that for software environments now. They provide a more semantic-driven kind of browser and other tools that are more structure-oriented. Examples of such environments are Gandalf at Carnegie Mellon, the Rational environment, and the Carnell synthesizer.

The last category of environment is method- or process-based. Method-based or process-based does not imply that the environments are also structure-based and language-centered. This is an orthogonal dimension. The advantage to a method-based environment is that it enforces a methodology or a software development decomposition -- you can pick your favorite, such as MASCOT, Jackson Structured Design, SADT, Structured Analysis/Structured Design, or state diagram. There are any number of decompostion metholologies, and the tools in this type of environment enforce or support the method. Unfortunately, we don't have a homogeneous set; not all of them will support the entire life cycle or the same part of the life cycle. Some of them will take you down to coding; some will not. Some of them will provide configuration management; some will not. Some are based on personal computers, and some are fuller environments on bigger operating systems. The process-based environments are only beginning to emerge. There aren't many examples of these, primarily because we don't have a consistent view of the process.

An interesting approach is a system called ISTAR, which is based on the notion that everything is a contract. From the beginning (defining requirements), you are establishing a contract. You are

doing so when you design (which is a contract between the designers and the coders), and so on down
the process. ISTAR enforces this model throughout
the life cycle. It's a tool that is just beginning to emerge; clearly, it has much evolution and
maturity to support it. But ISTAR, which is coming
out of Europe, and Refine, another emerging tool
that is under research at Kestrel Institute, are
the only two examples that support a view of the
process as a discipline.

One activity at the Software Engineering Institute
over the last year has been to define five maturity
levels of the process[1]. First is the ad hoc, which
is the lowest level. Ad hoc means you do what is
necessary to get the job done. At the next level
is the ability to repeat whatever you do. You repeat it successfully. The third is to have a defined process. Whatever it is, you've got a definition of it. You can go to that definition and
see where you are in the process. The next level
is to have the process managed. The final level is
to have an optimized process. You don't want to
provide automated support for the ad hoc level of
software development. If you do not have a well-
defined process, it surely won't help much to automate support for it. The introduction of technology at that level could be more harmful than helpful.

From your perspective what are the concerns? I
think the concerns about an environment occur at
three levels, that of the user, the environment
architect, and the tool builder. I am not going
to discuss the concerns of the architect and tool
builder because I do not think that many of you are
in those categories. Most of you are users or em-
ployers of users. I will focus on that level, although some of your concerns can be met only if the
architect and tool builder reflect those concerns
in design decisions.

One of your concerns is conceptual integrity in the
environment. For example, consider the Macintosh.
I'm not suggesting it's the ideal, but it represents
a consistent model. When you bring up a tool on
the Macintosh, you automatically know how to interface with it. You know there are pull-down menus
and that the menus will only allow you to do things
that are appropriate at that time. This is an example of integrity of the user interface.

You would also like conceptual integrity that supports the entire life cycle -- tools that will enforce some model of consistency, whatever that model
happens to be, and you'd like the tools to be well
integrated. That is, you do not want to apply a
tool inappropriately to a file; you'd like to apply
a tool only in the places where it is appropriate.
You want the environment to do some consistency
checking.

You also want to add new tools. You would like
third parties to add tools to whatever product you
use, and you would like those tools to enforce that
consistent model. The Apple people have done that
reasonable well. Third parties can add tools and
be consistent with their model. The method that you
choose needs to be supported. (Ideally, you choose
the method first and then choose the environment to
support it, instead of the other way around.)
Clearly, the method has to support the language or
set of languages you use, and you are also con-

cerned about the hardware base.

I am not talking about the concerns of the environment architect and tool builder, but I will summarize the current state and suggest that right
now nearly everybody is backing into an environment.
If you are doing software development, you have some
sort of environment. Your set of tools is your environment, and I suspect that you spend little time
and energy deciding what funds it should be given.
Often it is a matter of choosing an operating system and then taking whatever tools live on that
system. This is really not a viable choice any
longer. You increment your way in because you keep
adding tools, and then you find that the tool you
want to add next is no longer compatible with what
you already have. you back yourself into a box. It
is false economy to focus on hardware. You are
forced to live in a heterogeneous world today, and I
think the biggest potential difficulty is that many
of the tools you see, and what the marketing folks
call environments, are on personal computers. And
these systems do not always scale up; they do not
generalize.

Yet, if you do not design and build the environment
yourself, you are faced with having to add new hardware and new tools. You then have a heterogeneous
system and you have to worry about compatibility.
If that summary characterizes your current situation
then I suspect you should become more interested in
the developments in environment technology.

REFERENCES

[1]Humphrey, W.
<u>Characterizing</u> <u>the</u> <u>Software</u> <u>Process</u>: A <u>Maturity</u>
    <u>Framework</u>.
Technical Report CMU/SEI-87-TR-11, ESD-TR-87-112,
    Software Engineering Institute, June, 1987.
[2]Brooks, F. P. Jr.
"No Silver Bullet: Essence and Accidents of Software
    Engineering"
<u>IEEE</u> <u>Computer</u> 20(4):10-19, April, 1987.
[3]Dart, S. A., Ellison, R. J., Feiler, P. H., and
    Habermann, A. N.
<u>Trends</u> <u>in</u> <u>Software</u> <u>Development</u> <u>Environments</u>.
Technical Report CMU/SEI-87-TR-24, EDD-TR-87-187,
    Software Engineering Institute, October, 1987.
In "Software Development Improvements," Druffel
gives a preliminary, informal review of this now
formally reported work.

TECHNICAL AND MANAGEMENT
INFORMATION SYSTEM (TMIS)

Timothy R. Rau *
National Aeronautics and Space Administration
Washington, DC

Abstract

The Space Station Program (SSP) will utilize the Technical and Management Information System (TMIS) to support the design, development, and operation of the Space Station Program Elements and the management of the Program. TMIS will integrate the NASA Headquarters and all participating centers, international partners, Program contractors, and Program customers. TMIS is an integrated system of technical and management processes, automated data processing hardware and software, communications, procedures, and people supporting the design, development, operation, and management of the SSP.

Background

The next logical step in the exploration and development of space is an orbiting, manned Space Station. United States leadership in this effort was ensured on January 25, 1984, when President Reagan, in his annual State of the Union address, endorsed a national commitment to develop a permanently manned Space Station. NASA, with funding authorized and approved by Congress, has initiated the Space Station Program (SSP) (hereinafter referred to as the "Program") in accordance with the President's directive. The Space Station orbital facilities will be a laboratory, test center, manufacturing site, and observation platform that will open broad new areas of opportunity for scientific, commercial, Government, and international endeavors.

The Space Station Program is highly complex, involving the coordination of numerous organizations which are geographically dispersed, supported by heterogeneous information systems, and operated with a significant level of autonomy. In addition, the engineering and technical aspects of designing, developing, and operating the Space Station are complicated by the distribution of Space Station elements across numerous NASA centers, the involvement of numerous contractors, and the requisite extensive efforts in the areas of interfaces and integration.

---

*Senior Technical Analyst, TMIS Project Division

To facilitate Program management, a hierarchical organizational concept was established which implements overall Program policy and direction responsibilities at NASA Headquarters (HQs), Level A; Program management and technical integration at the separate Headquarters office, Level A', located in Reston, VA.; and project management responsibilities at the Work Package Centers and Supporting Centers, Level C. NASA institutional elements supporting the Program but not within the direct management control of the SSP are referred to as Level D; Level E refers to the definition, design, and development contractors.

Given this environment, it became necessary for the SSP to develop an information system which provides common capabilities to all SSP centers, facilitates the flow of technical and management information throughout the Program, and facilitates SSP decision-making processes. This system is known as the Technical and Management Information System (TMIS).

The Technical and Management Information System (TMIS)

The TMIS concept was formulated during the formative stage of the SSP when it became apparent that, in order to be successful, the Program development and management would require sophisticated information systems and support. The SSP management concept distributed initial Program responsibility for Systems Engineering and Integration (SE&I) between the work packages and thereby required significant improvements in the integration of engineering capabilities. The nature of the NASA organization, its culture, the distributed SSP responsibilities, and the challenge of the SE&I requirements clearly required operating and management systems beyond the existing environment.

In order to successfully meet these challenges, it was deemed necessary to develop an understanding of the Program's various technical and management processes and functions and to implement, where possible, integrated capabilities (i.e., organizations, systems, tools, etc.) which would serve to maximize the potential from those processes and functions. Such a system did not exist within NASA; however, the technology was available to construct one. It was clear that

NASA should exploit the opportunity and requirement to build this integrated system. TMIS emerged as the Program-wide system needed to meet the challenge. The backbone of TMIS would be its automated and integrated attributes which could achieve cost-effective productivity gains in both the management and technical aspects of the SSP.

The TMIS is to be implemented in an evolutionary and phased approach whereby the TMIS phases will support phases of the SSP itself. The present TMIS Phase 2 is a modern and efficient information management system which will support the SSP Phase C/D DDT&E activities. It will continue to be utilized as the SSP information interchange mechanism during the Space Station operational phases.

The TMIS concept calls for "common" Automated Data Processing (ADP) technologies to be implemented at Levels A, A', and C SSP organizations. These common system capabilities provide the NASA centers with the means for easily capturing, storing, transferring, manipulating, analyzing, and archiving SSP technical and management data. Program-wide databases and applications, based on information provided by all participating centers, is controlled under a TMIS database management system and made available to all authorized users. Local databases and applications will be implemented, as required, according to TMIS policies, procedures, and standards. Most TMIS applications and hardware components are to be commercially-available off-the-shelf technologies so as to reduce life cycle costs and minimize risk. TMIS provides for collecting and managing data resources from all SSP organizations so that all users can access required information to perform their DDT&E activities. Data resources submitted to TMIS are placed under configuration control so that standard applications can generate congruous information and so that Program data integrity may be insured.

TMIS Goals and Objectives

The TMIS Project has goals and objectives which were developed to support the SSP mission requirements. These goals are to:

1) Provide an integrated set of hardware, software, processes, procedures, and people to maximize the effectiveness and efficiency of Space Station Program technical and management design, development, test, and evaluation processes as they evolve through time;

2) Implement an Information Resources Management (IRM) discipline, as defined by OMB Circular A-130, to manage SSP information resources to improve productivity and cost effectiveness;

3) Create an environment for developing advances in technology, methodologies, and processes which focus on improving management systems for large-scale development projects;

4) Create the opportunity for competition in the procurement of ADP hardware and software in TMIS systems, subsystems, and components throughout the TMIS life cycle; and

5) Apply a rigorous design-to-cost methodology to implement TMIS.

Numerous associated objectives have been identified as quantifiable outcomes of each goal. These are delineated in numerous TMIS documents including the TMIS Master Plan (Ref. 1); some of these objectives (associated with each of the above goals, respectively) are to:

o Create integrated data communication links throughout the SSP which allow geographically dispersed SSP organizations to perform inter-center and intra-center data transfers;

o Increase the performance and productivity of SSP processes by ensuring accurate and timely interchange of SSP information among Program participants and easy access to TMIS ADP services;

o Create a hardware, software, and communications evaluation process for the SSP that allows the continuing evaluation and/or development of new technologies, methodologies, and processes;

o Use commercially available, off-the-shelf technologies for the development of TMIS systems, as appropriate; and

o Apply state-of-the-art modeling and evaluation tools to assess compliance with performance requirements and system life cycle cost constraints.

TMIS Scope

TMIS is an integrated system of technical and management processes, automated data processing hardware and software, communications, procedures, and people supporting the design, development, operations, and management of the SSP. It will integrate, to varying degrees, NASA HQs and all NASA centers and laboratories involved with the SSP, other Government and international participants, SSP contractors, and selected Program customers. TMIS will provide data for the support of on-orbit operations and will continue the technical and management support for the evolutionary design and development of the Space Station throughout the Program's life cycle. TMIS will archive and serve as an historical repository of SSP information after the culmination of the SSP.

The life cycle of TMIS is envisioned to be indefinite, although systems increments and enhancements will have finite life cycles. This life cycle is graphically depicted in Figure 1 and is composed of principal stages of strategic planning; design and implementation; and operations.

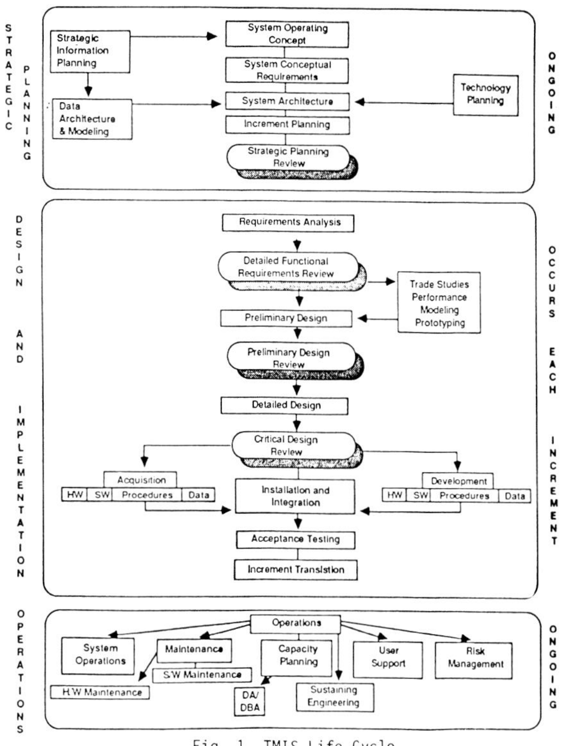

Fig. 1 TMIS Life Cycle

SSP Functionality - 28 Processes

TMIS will provide integrated system capabilities to support the major 28 SSP technical and management processes identified through the TMIS Project's Information Analysis effort. These Space Station Program processes are as follows:

o   Budgeting

Activities related to the administration of Agency funds. These are generally categorized as Program Operating Plan (POP) activities.

o   Planning

Tasks undertaken to estimate what future activities will be required to achieve Program goals.

o   Scheduling/Project Management

Activities pertaining to the development and tracking of project tasks and milestones inclusive of progress reporting and analysis;

also, early reporting of potential problems with ability to evaluate alternative solutions and resolve issues.

o   Policy Development

Activities undertaken to produce Program and project guidelines and procedures for any or all phases of the Program life cycle.

o   Performance Measurement

The monitoring, tracking, and reporting activities undertaken to confirm a contractor's compliance with schedules, budgets, and deliverables set forth in his contract.

o   Technical Contract Management

Activities associated with monitoring technical quality and direction of contract work.

o   Administrative Contract Management

Tasks associated with overseeing the managerial (e.g., budgeting) elements of contract work.

o   Program Review

Periodic SSP forums undertaken to identify, evaluate, and modify (as necessary) Program status and direction at a relatively high management level.

o   External Affairs

Activities which disseminate information outside the Agency and respond to inquiries received from external organizations (e.g., Congressional inquiries).

o   International Relations

Tasks pertaining to the official interface between the Program and the international partners.

o   Customer Relations

Activities pertaining to the official interface negotiations between NASA and potential users of the Space Station.

o   Requirements Analysis

All activities related to the definition, development, documentation, and maintenance of specifications for systems, subsystems, components, and operations.

o   Technical Analysis

Tasks undertaken to achieve quantitative or qualitative results in answer to specific technical questions.

o   Interface Control

    Activities undertaken to ensure physical interface compatibility between systems, subsystems, and components.

o   Cost/Financial Analysis

    Tasks undertaken to develop cost models, estimate system costs, or assess the potential Program gains associated with various expenditure scenarios.

o   Design

    The manifestation of requirements or a concept into an explicitly defined physical entity or representation. Design applies to hardware, software, communication, and other SSP components.

o   Design Review

    The process of evaluating and accepting the merits of a design or the relative merits of alternate designs.

o   Acquisition

    The process related to the procurement of goods and/or services.

o   Administration

    Those tasks undertaken to support the organization as opposed to the Program directly. Examples are personnel record processing, travel processing, and salary administration.

o   Implementation and Integration

    Activities related to the installation and/or activation of a system, subsystem, or component, and the physical interfacing of modules, systems, subsystems, or components. Implementation and integration includes development of hardware and software.

o   Test and Verification

    Activities supporting performance evaluation of hardware and/or software systems or components. These activities include, but are not limited to, establishing test procedures, establishing evaluation criteria, performing tests, documenting results, and presenting results/recommendations.

o   Documentation

    Tasks related to the generation and maintenance of reference materials including, but not limited to, manuals, instructions, and drawings.

o   Configuration Management

    Activities pertaining to the baselining of a design or document and the administration of changes to the baseline. These activities include the initiation, review, and approval of Change Requests (CR's).

o   Training

    Activities related to the dissemination of knowledge or skills, including the development of training materials, performing instruction, and administering training activities.

o   Operations

    Tasks undertaken to direct and control the use of a system to achieve desired performance.

o   Maintenance

    Activities undertaken to ensure continued, proper functioning of a system, subsystem, or component.

o   Prototyping

    The manifestation of a design into a physical, operational entity that can be evaluated for performance and operational capabilities.

o   Inventory Management

    Activities undertaken to maintain and control parts and/or supplies.

TMIS Capabilities

    In order to provide facilities to implement the above Space Station processes, TMIS will provide the capabilities and tools performing the following functions:

o   Computer-Aided Design (CAD)

    Hardware and software to support the creation, storage, and manipulation of engineering drawings and associated textual descriptions.

o   Computer-Aided Engineering (CAE)

    The integrated capability required to support the variety of engineering disciplines in design, development, and analysis.

o   Common User Interface (CUI)

    The CUI provides a standard man-machine interface which enables users to easily access all TMIS capabilities. The CUI facilitates the user's ability to move from one TMIS capability or application to another.

o   Data Base Management Systems (DBMS)

The capture, storage, search, and access capabilities for drawings, documentation, engineering data, financial data, and any other SSP shared information.

o   Remote System Access

The ability to access information which may be external to a user's immediate environment.

o   File Transfer

The capability to send and receive information in a defined format from a remote computer.

o   Image Processing

A tool for capturing and store text, drawings, photographs, or other hard copy material and converting it to electronic data, allowing users to later manipulate this data as necessary and incorporate it into compound documents.

o   Cost/Price Modeling

The capability to support the manipulation of financial data such as budgeted and actual costs for pricing policy analyses.

o   Word Processing

The capability to create, edit, store, and retrieve textual information.

o   Graphics

The capability to integrate graphics, including two-dimensional drawings, with text in support generation of compound documents and engineering .design documentation (e.g., Interface Control Documents (ICD's); this includes technical sketches, common business graphics, and viewgraph or slide generation.

o   Electronic Mail

The capability to send and receive text and data files at the workstation level.

o   Spreadsheets

Includes individual spreadsheets for small workstation oriented applications as well as larger applications.

o   Project Management System

Includes major project management functions as required to bring together cost, schedule, and technical performance information for analysis and review.

o   Scheduling

Includes project scheduling and associated activities as well as local or individual scheduling of organizational or programmatic meetings and conferences.

o   Computer-Aided Instruction (CAI)

CAI provides on-line interactive tutorial sessions demonstrating typical scenarios for each TMIS capability, as well as interactions between capabilities. CAI offers appropriate choices, demonstrates correct actions, provides self-paced courses, affords pertinent commentary, and provides recovery advice for incorrect choices.

The aforementioned TMIS Information Analysis resulted in the correlation of the SSP Processes with the TMIS capabilities. Figure 2 illustrates this matrix of Capabilities vs. Processes which is requisite to the support the SSP activities. This information analysis was performed to define the functional requirements for the initial increment of TMIS Phase 2 functionality. These processes and capabilities will be further expanded and defined during future TMIS increments.

WORK/CAPABILITIES MATRIX

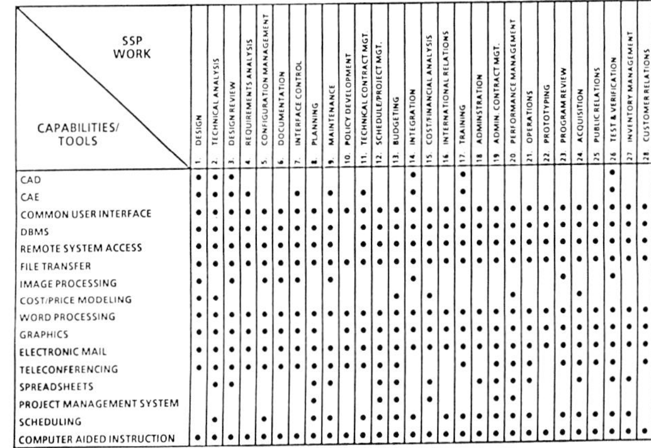

Fig. 2 SSP Work/TMIS Capabilities Matrix

<u>Implementation Approach and Acquisition **Strategy**</u>

The TMIS implementation approach and acquisition strategies were developed in light of TMIS goals and objectives. To meet the goal of providing ADP support to the SSP processes in an evolutionary manner, the implementation approach calls for phased TMIS increments. These increments consist of periodic (one to two years) implementations of system capabilities specifically required by SSP functions during the corresponding Phase C/D period. Phased implementation also supports a secondary TMIS goal of incorporating advances in technology, methodologies, and processes which improve the operation of large-scale information systems by providing both for updates of requirements associated with each increment and for a testing environment for assessing the use of new technologies within the TMIS architecture.

The procurement strategy supports the TMIS goal which calls for maximizing competition in the acquisition of ADP technology. An aspect of this strategy was to acquire the services of a Systems Integration Contractor (SIC), the selection of which was on the basis of the offering of a TMIS system architecture which is modular, capable of evolutionary growth, and which can effectively respond to changes in SSP user requirements and evolving technologies. This system architecture was required to reflect how the SIC intends to satisfy TMIS goals and objectives. The SIC is charged with the implementation of information system components via competitive acquisitions; hardware and software vendors are precluded from bidding their products if they are involved in defining TMIS requirements, system specifications, or developing solicitation materials.

The SIC issues NASA-approved Request**s for Proposals (RFP's) to acquire** hardware **and software, then evaluates the** proposals and **makes a selection which meets** NASA's functional requirements and approved design. NASA must secure a Delegation of Procurement Authority from **the** General Services Administration before the SIC **can** proceed with **the** acquisition cycle for each **TMIS** increment. This strategy supports competition among ADP vendors and reduces the Government's risk of being locked into a specific product line which may not be capable of evolving to meet SSP needs as they change through the Phase C/D period.

<u>Analysis of Implementation Alternatives</u>

The TMIS Office considered two Phase 2 implementation approaches for systems development during its Project planning activities. These approaches are referred to as (1) design driven, and (2) strategy driven. After careful analysis, the TMIS Office selected an evolutionary approach which is a combined, or hybrid, implementation approach. This approach is compatible with the evolutionary nature of the Space Station Program. It more readily accommodates system growth, changing or maturing Space Station requirements, and new technology advances. Most importantly, this approach has proven to have a higher probability of successful implementation in the cases of large, technically complex information systems.

The design-driven approach which was considered, involves completing a standard system life cycle in which specific phases are defined and executed sequentially. Comprehensive system requirements are developed, a design is prepared, and the required capabilities are acquired or developed. Implementation of the entire design follows and, upon test and evaluation, the system becomes operational. This approach is a traditional one, widely used in industry and Government. Using this approach, the full set of system requirements are defined and, when the system is implemented, all requirements are theoretically satisfied. This approach poses pitfalls for large systems because the entire life cycle is lengthy, often three to five years; by this time, requirements may have changed and the resultant operational system may not satisfy user needs or fulfill management directives.

The strategy-driven approach which was considered focuses on early development of key system functions, often via prototyping efforts, rather than fully defining all requirements "up front". Distinct system modules are identified and implemented as early as possible without the rigorous life cycle tasks required by the design-driven approach. Developers and users construct a concept for the system, then break it down architecturally into components which can be prototyped, although provide "building blocks" in developing the entire system over time. This approach strives for an "open architecture" and seeks to prevent becoming captive to a single vendor's hardware, software, or communications standards. While it may appear to be more costly for unit buys, it is ultimately more cost effective in that the total system never becomes obsolete and has to be replaced as a whole. This is especially important for long term programs such as the Space Station which has a Program life expectancy of thirty years or more. Users see results earlier and become more motivated in participating in future design, development, and operations efforts. The down side to this approach is that the "end" is never reached and a "total system buy" cannot be undertaken.

The TMIS-phased implementation, or evolutionary, approach calls for the implementation of system capabilities which are completed in multiple increments. This approach combines aspects of both the design-driven and strategy-driven approaches which meet TMIS needs and constraints. TMIS implementation of phased increments, which draws on the "strategy driven" approach, will enable TMIS to respond to changes in SSP user requirements and new available technologies over a lengthy system life cycle. In addition, the TMIS phased increments will provide the SSP users with the appropriate set of ADP capabilities required to perform Phase C/D activities. The initial system capabilities are augmented and enhanced during succeeding increments in correspondence with evolving SSP user needs. However, the TMIS evolutionary approach utilizes the design-driven methodology for completing sequential life cycle tasks. In the case of TMIS, each increment undergoes a requirement definition, design, acquisition, implementation, and operational phase. The TMIS evolutionary approach allows for flexibility in responding to evolving SSP needs, technology advances, equipment replacement, and changes in NASA's Agency-wide standards and data processing environments.

<u>TMIS Schedule Overview</u>

The TMIS Project Schedule for its phased implementation approach is based on the SSP Phase C/D schedule. As depicted in Figure 3, TMIS Phase 0 correlated to the SSP "Skunk Works" activity; TMIS Phase 1 supported to SSP Phase B effort; TMIS Phase 2 will support the SSP Phase C/D activities; and the TMIS Evolutionary Phase correspondents to the SSP evolutionary phase. The TMIS Schedule Overview also illustrates the specific SSP C/D stages to be supported by the **TMIS increments IOC' through IOC-5.**

TMIS IMPLEMENTATION PHASES

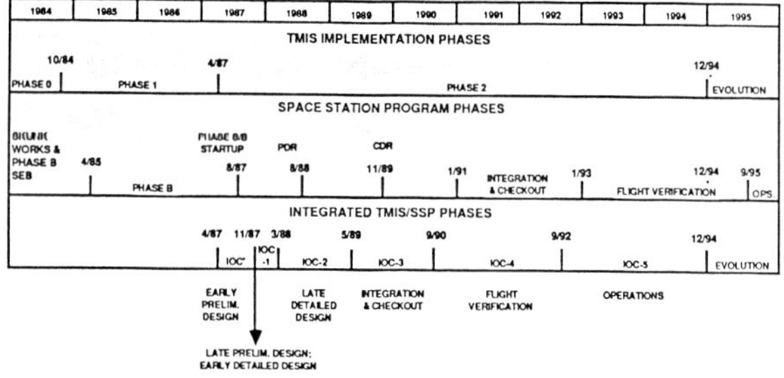

Fig. 3 TMIS Schedule Overview

Summary of TMIS Phases

TMIS Phase 0, completed in late 1984, provided limited data processing capabilities to support SSP's generation of Phase B Request for Proposals. This resulted in the Work Package procurement cycle and contractor selection for SSP Phase B Preliminary Design work. The TMIS Office produced an initial set of system requirements to be further defined in TMIS Phase 1.

During Phase 1, the TMIS Project supported the concurrent SSP Phase B activities by developing and implementing capabilities in the areas of budgeting, planning, scheduling, project management, documentation processing, electronic mail, limited CAD/CAE, and communications networks. Initial TMIS Phase 1 capabilities supported many SSP management and technical functions. Various CAD/CAE applications were available at each NASA SSP center, although these heterogeneous systems were not integrated. A TMIS Phase 1 engineering database was developed and was accessible via a dial-up capability. In addition, Phase 1 created various test beds to verify selected concepts for TMIS Phase 2 implementation. Phase 1 included all capabilities implemented prior to the Phase 2 award to the Systems Integration Contractor.

TMIS Phase 2 consists of multiple increments: an Initial Operating Capability Prime (IOC') and increments IOC-1 through IOC-5. IOC' will implement early in the Program additional capabilities which were identified during Phase 1 and which are essential to the SSP Phase C/D startup activities. This will be followed by a more thorough analysis and implementation of capabilities to support SSP Phase C/D through early 1994. Post-1994, a TMIS Evolutionary Phase will be completed on a two-year cycle to satisfy new SSP requirements and to implement emerging technologies, as appropriate.

In TMIS Phase 2 and the Evolutionary Phase, a fully developed and integrated set of technical and management functions will be implemented at SSP Levels A, A', and C/D. The system will have fully developed communications networks allowing for data transfers within NASA and between NASA and its contractors, international partners, commercial users, and scientific institutions. These phases will support the SSP Phase C/D and evolutionary phases in which the initial station configuration will be developed, assembled in space, made operational, and maintained.

<u>Phase 1-2 Transition</u>

As defined previously, Phase 1 is that TMIS capability which was in place prior to award of the SIC contract. The SIC has the option of absorbing Phase 1 capability, if appropriate, by subsequent increments. Phase 1 provided for Program-wide implementation of electronic mail and calendering, scheduling, document interchanging and archiving, budgeting, and management of engineering and customer requirements databases. Where desirable, Phase 1 relied upon institutional resources that were in place prior to the initiation of the SSP and allocated to the SSP for its use. Phase 1 also included new acquisitions to enhance and augment those capabilities to achieve required Phase B functionality. The transition from Phase 1 to Phase 2 will be effected by the SIC.

Phase 2 capabilities are to be considered available only after the SIC has implemented into Phase 2 the functional capabilities which they presently support in Phase 1. Thus, no loss in functional capability will occur when the Phase 1 systems are taken out of service or transitioned to Phase 2 in support of a functional replacement.

TMIS Project plans to transition as many Phase 1 functional capabilities as practical at the earliest possible increment. Ideally, all Program-wide Phase 1 capabilities would be transitioned into Phase 2 with implementation of the IOC' increment.

IOC' Increment

A minimum TMIS capability must be in place and operational early in the SSP Phase C/D period. This increment of capability, TMIS IOC', is based on minimum requirements defined in the TMIS Functional Requirements Summary. The TMIS Project Division, with support from the SIC, finalized the IOC' implementation plan and has begun acquire the approved TMIS IOC' hardware and software immediately following Contract Start Date (CSD).

The SIC identified those Phase 1 functional capabilities which will be incorporated into the IOC' system design and acquires, implements, and will make operational the approved IOC' system within eight months after CSD. Figure 4, IOC' Capabilities, presents the level of TMIS support which will be provided by the IOC' increment.

Remaining Phase 2 Increments

The TMIS Project Division supports SSP processes as described in the TMIS Functional Requirements Document (Ref. 2). The SIC will continue to verify these requirements, including the System Operating Concept (SOC), and will implement and refine a system architecture which fulfills these requirements and which can effectively evolve and incorporate new technology as required by the SSP.

For each increment, the TMIS Project performs an information analysis effort to define the functional requirements of the increment. These requirements are used to develop a detailed design and acquisition plan for approval by NASA. The SIC acquires/develops hardware and software, performs installation and integration, conducts acceptance testing, and transitions the increment into the Phase 2 operational capability. Figure 5, IOC-1 Capabilities, presents the level of TMIS support to be provided by the IOC-1 increment.

Figure 5. IOC-1 Capabilities

Evolutionary Phase

TMIS must evolve to support the operational phase of the SSP and accommodate changing SSP requirements. In addition, TMIS itself will be faced with new requirements. Equipment obsolescence and emerging technologies will necessitate a process of system evolution. To ensure that the TMIS Project Division addresses this challenge of evolution, the period immediately following the Phase 2 implementation has been defined as the TMIS Evolutionary Phase. The preliminary requirements for evolution shall be expanded and refined as part of the TMIS Project Division Planning function.

**TMIS IOC' INCREMENT
IMPLEMENTATION OF SYSTEM CAPABILITIES**

Minimum set of Capabilities to be procured:

- COMMON USER INTERFACE - On all TMIS Workstations
- DATA BASE MANAGEMENT SYSTEMS - REFCON, MRDB, Hazards
- REMOTE SYSTEM ACCESS } Available via TMIS Network
- FILE TRANSFER
- WORD PROCESSING - Text, Forms, etc.
- GRAPHICS - Engineering and Business Graphics
- IMAGE PROCESSING - Scanners, OCR Devices
- ELECTRONIC MAIL/CALENDARING - Correspondence, bulletins, meetings, individual appointments, etc.
- SPREADSHEETS - Individual and multi-user
- SCHEDULING - Project and Task Plans
- PROJECT MANAGEMENT SYSTEM - Project Schedules, Performance Management

Fig. 4 IOC' Capabilities

**OVERVIEW OF
TMIS IOC-1 INCREMENT**

TMIS IOC-1 Is Characterized By:

- Greater Support To SSP Technical Processes
- Development Of Additional Data Bases and Applications Specified in Strategic Information Plan
- Further Integration Of Program-Wide Data Bases
- Total Integration Of Engineering Drawing Data Base
- Expanded Integration (Bridge Building) Of Institutional Capabilities
- Integration Of Program-Wide Project Management Capability (Includes Performance Measurement)
- Integration Of Program-Wide Scheduling Capability
- Implementation Of Computer-Aided Instruction For Users

Fig. 5 IOC-1 Capabilities

<u>Evolutionary Acquisition Strategy</u>

The TMIS acquisition strategy is evolutionary-based and complements the phased implementation approach. This acquisition strategy calls for procuring a basic or "core" capability at IOC' which supports a subset of TMIS functional requirements defined in broader, all-encompassing terms. This strategy describes the framework within which evolution of the the TMIS system architecture will occur. Subsequent increments consist of system modifications and enhancements based on lessons learned from operational usage, adequacy of configuration, maturation of requirements, and new technologies.

The TMIS Phase 2 procurement effort to secure a Systems Integration Contractor was conducted at the Johnson Space Center (JSC) under NASA Source Evaluation Board (SEB) procedures. The Source Selection Official was the NASA Administrator. The TMIS Request for Proposal (Ref. 3) included detailed Functional Requirements for IOC' and phased requirements for IOC-1 through IOC-5. TMIS Functional Requirements for IOC' were generated at the Space Station Projects Offices at the various NASA centers and analyzed, integrated and documented by the Level B TMIS Office at JSC. The Functional Requirements Document was baselined by the Space Station Control Board. These requirements were incorporated into the TMIS SIC Statement of Work (SOW) in parallel with the generation of a Request for Proposal (RFP) by the SEB.

With the Program transition from Level B at JSC to the Level A' at HQs, the responsibility for TMIS was also moved from the Level B TMIS Office and is now located in the Level A' TMIS Project Division. This division is one of the organizations which comprise the Level A' Information Systems Support Group. In addition to TMIS, this group is also responsibilities for the Space Station Information System (SSIS) and the Software Support Environment (SSE).

The TMIS acquisition strategy is characterized by the development of a Request for Proposal to select a Systems Integration Contractor. As of this writing, the proposals have been received and evaluated; the announcement of the selected contractor has yet to be made. The offerors were evaluated on the basis of their proposed:

o   TMIS conceptual architecture into the Evolutionary Phase (CY 1996);

o   Detailed design and implementation plan for IOC';

o   Conceptual design for the remaining Phase 2 increments;

o   Design-to-cost methodology;

o   Transition plans for Phase 1 to IOC' and all subsequent increments of Phase 2;

o   Operations support plans for Phase 2 increments; and

o   Project management and integration plans.

<u>Contract Type and Period of Performance</u>

The contract type is planned to be level-of-effort (LOE) cost-plus-award-fee (CPAF). The commercial hardware and software acquired under the contract will be fixed-price elements of the contract for the IOC' increment.

The period of performance for the TMIS contract is planned to begin in April, 1987 and run through Space Station IOC (December 1994). However, authority to contract for an additional two years beyond Space Station IOC is essential to provide uninterrupted support during critical early operations of the Space Station. Authority to contract for this period may be requested in accordance with NFS 18-17.206-70, under a separate cover.

Upon contract award, the Systems Integrator Contractor is required to implement all IOC' hardware, software, communications, databases, processes, and procedures. The remaining Phase 2 increments will be further defined, providing the basis for design, acquisition, development, and implementation tasks.

<u>References</u>

1. TMIS Master Plan; HQs Level A' Document Number to-be-assigned; as approved by TMIS Control Board on April 3, 1987

2. TMIS Functional Requirements Document; JSC Document Number 30226; June 30, 1986

3. TMIS Request for Proposal; JSC Number 9-BF3-32-6-01P; July 7, 1986

NEW APPROACHES TO SOFTWARE ENGINEERING

Victor R. Basili
Department of Computer Science
University of Maryland
College Park, Maryland 20742

Abstract

We present in this paper approaches for software measurement and evaluation that can be applied for improving the software engineering processes and products and characterizing construction oriented Software Engineering Environments (SEEs). An automated measurement and evaluation environment (TAME: Tailoring a Measurement Environment) is currently being developed at the University of Maryland to be used as vehicle for supporting these approaches. TAME is based on the idea that the software engineering process needs to be tailored and tracked for the project productivity and specific quality goals, characteristics of project environment, and the overall organization. We present in this paper some of the basic ideas behind TAME and a brief description of the TAME system requirements and structure.

Introduction

The software engineering process needs to be tailorable and tractable. The tailorability of a process is the characteristic that allows it to be altered or adapted to suit a set of special needs or purposes[1]. The tractability of a process is the characteristic that allows it to be easily planned, taught, managed, executed, or controlled[1]. A software engineering environment (SEE) should support tailorability and tractability by automating as much of it as possbile in order to make it easy for project managers to plan, control, and modify the system for specific set of needs of the project domain. At the University of Maryland, we have been measuring and analyzing the software development processes and products for a variety of organizations for the past dozen years[8,9]. We have learned several things about the software engineering processes and software measurement and evaluation. In the following section we list experimentally derived principles about the software engineering process. Section 3 deals with measurement principles. Section 4 describes a software engineering process and in section 5 we characterize a SEE for supporting the model presented in section 4. A brief description of TAME system including its requirements and architecture are presented in section 6.

Principles

Process Principles

Some of the lessons we have learned about the software engineering process are stated in this section as principles:

-It is necessary to develop quality a priori. Quality cannot be tested or measured. We must not rely on quality assurance techniques as a substitute for development techniques. In order to develop quality a priori, we need to formalize the planning of the software engineering process.

-Development methods are often heuristic and not formal. Managers are not using their methods and tools properly. In order to improve software engineering in an organization, we need to formalize the improvement and evaluation paradigm of software engineering processes and products.

-Software engineers and managers need feedback in real time and the organization needs post mortem analysis in order to improve the process and product.

-All project environments and products are different in some way. These differences must be taken into account in the software processes and in the quality goals set for the products.

-There are many process models for software development. The process model needs to be tailored to the organization and project needs and characteristics. There is a need to formalize the tailoring of processes towards project quality goals and environments.

-We need to reuse experience, but only after tailoring it to the project.

Measurement Principles

Based on our experience in measuring the software development process and product, we present the following measurement principles:

-Measurement is an ideal mechanism for characterizing, evaluating, predicting, and motivating the various aspects of software development.

-Measurement must be taken on both the software processes and the various software products. There is a relationship between the software product and process. If there's going to be a learning process involved, we have to learn what factors in the process affect the product so we can modify these factors accordingly.

-The development and maintenance environments must be prepared for evaluation and measurement. There is a planning phase necessary for this activity and the activity must be carefully embedded in the process. As a part of the organization, measurement can't be an afterthought. It has to be a part of the entire process, in the planning and the organization. The planning phase must take into account the experimental design appropriate for the situation.

-For both definition and interpretation purposes, a characteristic set of metrics (metric vector) must be collected.
-Metrics must be associated with interpretations, but these interpretations must be given in context.
-We cannot just use other people's models and metrics as defined. They must be tailored for the environment in which they are applied and checked for validity in that environment.
-In order to evaluate and compare projects and to develop models, we need an historical data base that incorporate experience from prior projects. This data base should characterize the local environment.
-This data base could be formalized into an expert system.

Software Process

The software process principles and measurement principles described in the previous two sections suggest that software engineering needs to be concerned with planning, construction, and learning and feedback.

Planning the software engineering process is aimed at providing a basis for developing quality a priori. It includes choosing the appropriate tools and methods. It involves tailoring each of them for the project specific goals and the characteristics of the project environment and the organization. The entire planning process, the tailoring process, as well as the measurement process needs to include deriving the appropriate netrics for defined evaluation goal, experimental design, and data collection and validation. The goal/question/metric paradigm[4,5,6,7] represents a systematic approach for determining the goal of measurement (tailored to the specific needs of an organization), defining that goal in a tractable way into a set of quantitative questions that in turn define a specific set of metrics and data for collection.

Construction of the required products follows the guidelines defined as part of the planning activity in order to achieve quality a priori; the existence of construction guidelines helps in assuring that methods are being used as intended. Construction activities also includes the traditional project documents (e.g. requirements, design, code) as well as all other kinds of analytic information is supported by data collection, data validation, and the computation of metrics as prescribed during the planning phase.

Learning and feedback is based on a paradigm for evaluation and improvement.[4,7] The learning requires monitoring the engineering and management processes as well as products and feeding the lessons learned back into the ongoing project or into the planning phase of future projects.

Integrated Software Engineering Environment

The goal of a Integrated SEE (ISEE) should be to support the planning, construction, and learning and feedback activities of a tailorable and tractable software engineering process. This includes support for goal-oriented measurement as a means for achieving tailorability and tractability.

The SEE components are not just tools and methods, but people as well. People need to plan for a software engineering project by choosing a process model, a method or a set of methods for a particular project goal set, based upon information about which ones work best in this particular environment.

All the information produced and consumed in an SEE environment can be categorized according to three different schemes. According to the organizational scheme, we might be interested in the effect of one particular project or across multiple projects within the organization. According to the integrational scheme, we can support assessment, evaluation and improvement of individual methods or tools in isolation or their integrated use in an overall process model. According to the analytical scheme, we can provide support for assessment according to measurement, feedback, and planning.

Based on the different levels of support for construction oriented activities as well as the different ways for providing the needed information, we classify SEEs by the degree to which they provide constructive and analytical support. Figure 1 shows a classification of SEEs.

TAME System

The TAME (Tailoring a Measurement Environment) system supports planning activity by automating the goal/question/metric paradigm, automates as much of the measurement process as possible, provides evaluation and analysis of the development and maintenance processes, and creates and uses historical data and knowledge base that incorporate experience from prior projects. TAME is based on the software engineering process and measurement principles and designed around the process model and SEE described in the previous sections of this paper.

TAME has many requirements; we list a few of them in this paper:
-A mechanism for dealing with measurement and evaluation goals in an operational/quanitifiable way.
-A mechanism for controlling the measurement and evaluation process.
-A mechanism for interpreting analysis results in a context and providing feedback for the improvement of the process model.
-A mechanism for learning in an organization.
-An interface for construction oriented SEEs.

The basic logical structure of TAME
system is shown in figure 2. It consists
of four levels: user interface level,
evaluation level, measurement level, and
data repository level. For a detailed
architecture of the TAME system and the
first prototype, see reference 2.

Conclusions

Based upon our experience with measurement and evaluation of software engineering processes and products, we have
listed in this paper a set of empirically
derived principles about software engineering processes and measurement. We
have also presented a software engineering
process model and measurement approaches
in a SEE, and briefly described TAME system designed for supporting our suggested
ISEE. TAME is a very ambitious project
and its objectives go beyond automating
measurement and evaluation activities. It
will help us develop guidelines for the
future software engineering environment.
There is a tremendous amount of learning
and research activities involved by trying
to formalize and automate the planning,
evaluation, and improvement of software
engineering processes and products. TAME
offers a wide range of research problems
allowing software engineering to interface
with other computer disciplines such as
data base community and artificial intelligence community.

References

[1]Webster's New Collegiate Dictionary, G+C
Merriam Company, 1981.
[2]V.R. Basili, H.D. Rombach, "TAME: Integrating Measurement into Software Environments," Technical Report TR-1764 (and TAME
-TR-1-1987), Department of Computer Science, University of Maryland, College
Park, MD, 20742, June 1987.
[3]V.R. Basili, H.D. Rombach, "TAME: Tailoring an ADA Measurement Environment,"
Proc. of the Joint Ada Conference, Arlington, VA March 16-19, 1987, pp. 318-325.
[4]V.R. Basili, "Quantitative Evaluation of
Software Engineering Methodology," Proceedings of the first Pan Pacific Computer
Conference, Melbourne, Australia, September 1985 (also available as Technical Report, TR 1519, Dept. of Computer Science,
University of Maryland, College Park, June
1985).
[5]V.R. Basili, H.D. Rombach, "Tailoring the
Software Process to Project Goals and Environments," Proceedings of the Ninth International Conference on Software Engineering, Monterey, California, March 30-
April 2, 1987, pp. 345-357.
[6]V.R. Basili, R.W. Selby, Jr., "Data Collection and Analysis in Software Research
and Management," Proceedings of the American Statistical Association and Biomeasure
Society Joint Statistical Meetings, Phil-
adelphia, PA, August 13-16, 1984.
[7]V.R. Basili, D.M. Weiss, "A Methodology
for Collecting Valid Software Engineering
Data," IEEE Transcations on Software Engineering, vol. SE-10, no. 3, November 1984,
pp. 728-738.
[8]V.R. Basili, "Can We Measure Software
Technology: Lessons Learned from 8 Years
of Trying," Proceedings of the Tenth Annual Software Engineering Workshop, NASA
Goddard Space Flight Center, December
1985.
[9]F.E. McGarry, "Recent SEL Studies," Proceedings of the Tenth Annual Software Engineering Workshop, NASA Goddard Space
Flight Center, December 1985.
[10]C. Brophy, W. Agresti and V.R. Basili,
"Lessons Learned in Use of Ada Oriented
Design Methods," Proceedings of the Joint
Ada Conference, Arlington, Virginia, March
16-19, 1987.
[11]M. Zelkowitz, R. Yeh, R. Hamlet, J. Gan-
non and V.R. Basili, "Software Engineering
Practices in the U.S. and Japan," IEEE
Computer Magazine, June 1984, pp. 57-66.

AUTOMATED SOFTWARE PRODUCTION

R. J. Lauber
Institute for Control Engineering and Process Automation
University of Stuttgart, West Germany

Abstract

An automation-based software paradigm is discussed, which the EPOS system is based upon. The subjects of automated code generation (using an Ada generator as an example), of automated software verification, and of automated reuse of software are considered. Software reusability is advocated as a powerful approach to further improving reliability.

1. Introduction

Although the term chosen as the title of this paper may be, rather misleadingly, taken to mean "production of software by machines only", it is used here to imply "software production by software engineers using computers as engineering tools throughout all phases of the life-cycle and for developmental as well as managerial enginee-ring activities". It does imply, however, also that those activities which can be performed better and more cheaply by computers than by people (such as establishing documentation or coding of computer programs in a programming language) should be done by computers automatical-ly.
Furthermore, the term "software" is used in this paper rather generally, in fact as a shorthand for software/hardware systems, whether such systems are encoded as bits in a program store or, say, geometrical patterns on silicon. Thus software production is, effectively, software/hardware system production.
There are two aims to be achieved by automating the software production process (in the sense mentioned above):

- The main aim is, especially talking about space systems, very high reliability.
  It is well known that, with traditional methods of software production, many opportunities exist for injecting human error into the series of activities involved in developing software systems, and thus for jeopardizing system relia-bility, and integrity. These difficulties may be overcome through automation of engineering activities[1].

- The other aim is a substantial reduction of life-cycle cost by an order-of-magnitude.

2. Automation-based software production paradigm

In a famous article, Robert Balzer et al.[2] claimed that we need a new, automation-based software paradigm if we are to achieve orders-of-magnitude improvement. Their main idea was to concentrate the engineering activities on establishing speci-fications, and to automate the translation from specifications to implementations, thus switching the noncreative aspects of maintenance and modifi-cation from man to machine.
It seems that Balzer et al. were too pessimistic when they stated "... the technology needed to support the automation-based software paradigm does not yet exist". Experiences with the defi-ciencies of the current software paradigm had led several other software engineering teams to the same conclusion as Balzer et al. quite some time ago, and they had already developed solutions for automated software production[3,4].

One of these solutions is the EPOS system [5], which is available as a commercial product since 1980, and which is currently in practical use at more than 250 installations worldwide, being applied in industrial, government and academic projects.
In fig. 1, this environment is depicted, showing its notational means for requirements, software/hardware, and project specification, as well as its tool systems to automate requirements enginee-ring, system design, and project and configuration management.

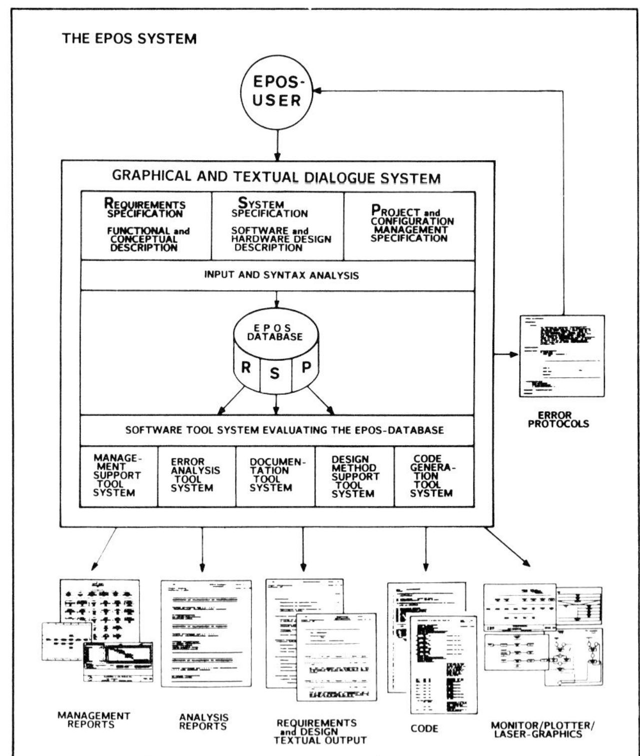

Fig. 1: The EPOS system for automated Requirements engineering, System design and Project and configuration Management.

The diagram in fig. 2 explains the automation-based software production paradigm which the EPOS system is based upon. It consists of two parts:

The upper part of fig. 2 shows the creative activities of requirements engineering and system design which are performed by software engineers supported by tool systems. The lower part contains the implementation, which is carried out almost fully automatic with the exception of some program testing, which may still be necessary.

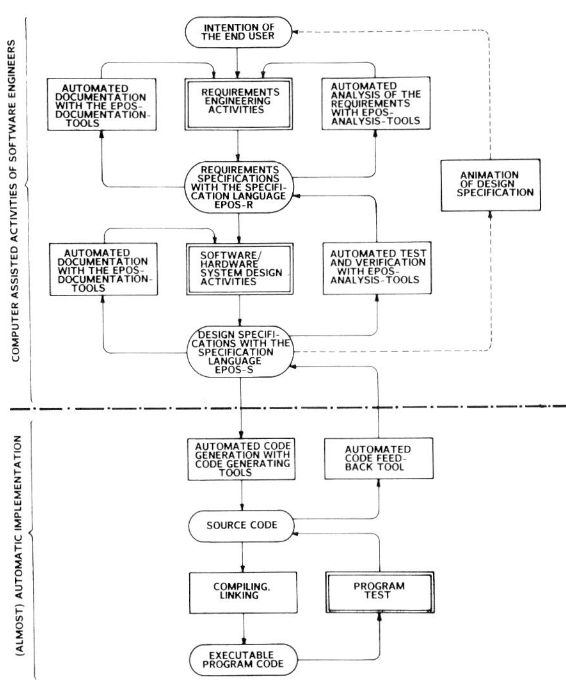

Fig. 2: The automation-based software production paradigm of the EPOS system.

Starting with the intentions of the end user, requirements engineering in the upper part of fig. 2 includes both stating user needs and finding a conceptual solution. By formulating these with the specification language EPOS-R, requirements engineering is supported by automated analysis, as well as by automated documentation using the EPOS tool systems. Therefore, there is substantial computer support already in this early phase of a project. Using the semi-formal EPOS-R-specifications, a software/hardware system design is created, which is then described with the formal specification language EPOS-S. During design activities, automated test and verification is performed on these specifications, thus immediately detecting specification errors, and tracing requirements into the design. Automated documentation provides graphical and textual representations of every design stage, giving the designer visual overview, and helping him to keep track of every design decision. In addition, there is a facility in preparation (see the dashed line on the right side of fig. 2) which provides animation of system design on each design level. Thus, the design specification itself serves as a prototype of the system. This gives the end user a means of ensuring that the specified system matches his intent. The early feedback avoids the delay and expense of implementing the software in order to find out, whether it does what it is supposed to do.

Automated implementation (see the lower part of fig. 2) starts with automated generation of code directly from EPOS-S specifications. There are EPOS code generation tools available to produce Ada, FORTRAN or Pascal, source code. Standard compilers and linkers are used then to receive executable program code.

Because of the fact that the design specification is converted into an efficient implementation by automatic means, no clerical errors can be introduced. However, in real software projects, some program testing may be necessary in the field, thus introducing changes in the source code. This could result in inconsistencies between the design specifications, from which the original source code had been derived, and the actual source code. Therefore, automatic code feedback tools are provided in order to detect such inconsistencies and to guarantee that the source code always corresponds to the design documentation.

It should be noted that fig. 2, though describing the initial development process, is valid for the maintenance process as well. This means, that maintenance is performed on the requirements and design specifications rather than on the source program. The revised specification can then be reimplemented using the code generation tools. Thus, by maintaining the specification directly, the maintenance problems are simplified drastically.

Especially for large systems, such as in the case of space projects, software maintainability may be the foremost objective (meaning not only activities to correct bugs, but also the much more important activities of enhancing the released system). Interestingly, it turns out that establishing design specifications from existing code by so-called "reverse engineering" may be the only way of achieving maintainability of those software systems which had been developed in the past without automated tool support, and which sometimes consist of FORTRAN code only without any comments. This method of "reverse engineering" by using the EPOS system was applied, for instance, to a large operational flight control software system in order to solve the enormous maintenance problems which had occurred.

3. Automated code generation from design specification

At the core of the new, automation-based software production paradigm explained above are:

- correct formal design specifications, which have been tested by automated analysis tools

- automated code generation from these specifications into a selected programming language

- automated code feedback, in order to guarantee consistency between the specification and the source code, even if the source code has been changed during program testing or maintenance actions.

As can be seen in fig. 3, the end product of the design activities using an automated development environment is a project data base, containing the design specifications. By automated code generation, a source code file is produced.

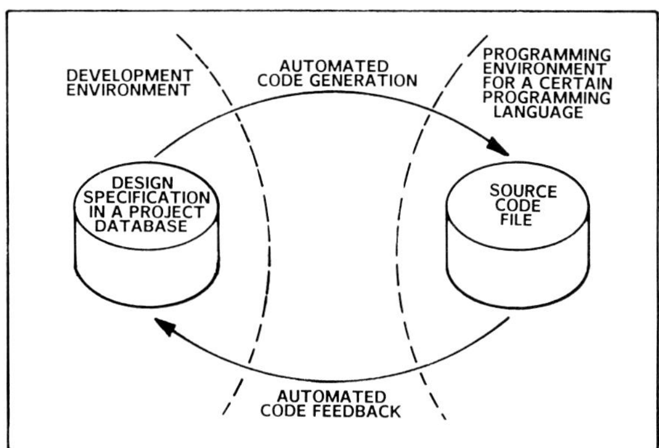

Fig. 3: Interaction between a development environment producing a design specification and a programming environment to work on a source code file.

In order to make an automated transform of a design specification into a program code feasible, there are two basic requirements:

- The specification language must be at a low enough level that purely local optimizations are sufficient.

- The level of the programming language should be close to the level of the specification language.

In the EPOS system, which was mentioned above already, two alternatives are realized:

- For certain high-order programming languages (Pascal, FORTRAN and Ada), tool systems are available which generate 80% of the total program without manual interaction.

- For any other programming language (and even assembly language!) the transition to a source file is accomplished through "code selection". Pieces of code are manually entered in the code-part of EPOS-S design objects, which are then linked into a runnable source file.

In order to explain this in more detail, the EPOS Ada Generator is used as an example[9]. It generates a complete Ada package or subprogram from an EPOS-S design specification, with correct program structure, data and type definitions, exception handling, tasking and statements for sequential and parallel control, provided the EPOS-S specification is complete and consistent. In order to check whether the specification is transformable, software tools with intensive analysis features are provided.
The only programming task left to the designer is the insertion of Ada code for certain terminal actions at the lowest level of refinement.

In order to demonstrate these code generation features, an example for the design of a real-time system is chosen. EPOS provides both very high level synchronization features and lower level features for the direct specification of rendezvous primitives. In the following example shown in fig. 4, parts of two parallel tasks named CARS-FROM-THE-LEFT and CARS-FROM-THE-RIGHT have to run exclusively as specified in the SYNCHRO-PART of the procedure NARROW-ROAD. In order to perform this requirement, a controller task with two entries is generated. This controller task synchronizes the subactions SEND-LEFT-CAR and SEND-RIGHT-CAR of the above mentioned tasks. It should be noted that even the termination of the controller task is guaranteed when it is no longer needed.

```
EPOS-S SPECIFICATION

ACTION PROCEDURE NARROW-ROAD.
DECOMPOSITION : PARALLEL      (CARS-FROM-THE-LEFT,
                               CARS-FROM-THE-RIGHT,
                               TIMER).
SYNCHRO : EXCLUSIV (SEND-LEFT-CAR
                    SEND-RIGHT-CAR).
ACTIONEND.
```

```
GENERATED ADA CODE

task CONTR1 is
   entry   request;
   entry   release;
end task;

task body  CONTR1 is
   locked:  boolean := FALSE
begin
  loop
    select
        when not locked  =>  accept request;
        locked := TRUE;
    or
        when locked => accept release;
        locked := FALSE;
    or
        terminate;
    end select;
  end loop;

end CONTR1;
```

Fig. 4: Example of automatic generation of Ada code from EPOS-S design specifications.

There are two kinds of limitations of the automatic code generation. Firstly, design features which do not have a corresponding programming language concept may not be transformed. In the case of EPOS, this applies for instance to the specification of a distributed system, where it is specified with EPOS-S on which hardware a designated piece of software will be executed. In the programming languages mentioned above, there is no concept to describe this design information.
Secondly, programming language concepts which do not have a counterpart in the specification language may not be reached. As an example, the GoTo statement, labels, renaming, overloading, and generics of the Ada programming language have no corresponding concepts in the EPOS-S specification language.

4. Automated software verification

Even if powerful analysis tools are used during automated design which guarantee almost error-free design specifications, and despite automated code generation and code feedback avoiding errors to be injected during implementation, the reliability achieved may still not be high enough considering the very stringent reliability and safety constraints of space applications. There are three different strategies which can be applied and which also may be combined to further increase reliability

- use of fault tolerance methods not only for hardware, but also for software (i.e., n-version programming or recovery block methods),

- verification and validation on code level and on the level of specification languages

- reuse of software modules which have been proved to be reliable in previous projects.

By verification we mean methods which are used to determine that the results of one development phase are complete and consistent and fulfill all requirements levied by the previous phase, and that the project techniques and procedures were followed. The term validation is used to indicate methods to determine that the completed software meets the requirements and intentions of the end user. Thus, verification is used to describe the stepwise procedures utilized to discover inadequacies during the development effort, while validation is used to describe the overall determination that the software is adequate.

A number of different approaches to software verification and validation are listed in table 1, indicating whether they are applicable to the different phases of the development process.

software verification and validation method	character- ization	applicable to the phases of		
		requirements engineering	software design	implemen- tation
formal verification (proving correctness)	M, A			Ve
symbolic execution	M, A			Ve
fault tree analysis	M, A		Ve/Va	Va
simulation	A		Va	
inspections	M	Ve/Va	Ve/Va	Ve/Va
informal testing	M, A		Ve/Va	Ve/Va
reverse design	M		Va	

Table 1: Software verification and validation methods
Abbreviations: M   Manual
                  A   Automatic
                  Ve Verification
                  Va Validation

Much research has been done in the area of providing formal proof of program correctness, and of automating this proving process. In essence it implies the use of a proof language in which mathematical axioms are formulated and against which programs are verified. Since the function of the program is defined in precise mathematical terms, there is no question of interpreting correctness. Any deviation from the proof axioms is considered wrong. The problem with this approach is that it is extremely difficult to learn and tedious to apply, and that it is restricted mostly to purely sequential programs[10,11]. Because of the complexity of the proving process, there is a certain probability that errors are made during proving. Therefore, there is still the question of the confidence one can have in the result of the verification process.

The second approach mentioned in table 1 entails testing programs with symbolic as apposed to real values. Systems for symbolically executing programs have been implemented. These systems have proven themselves to be useful in providing information for debugging programs. But, as for proving correctness, this approach is restricted to code level verification, and therefore does not fit to the automation-based approach of fig. 2. There are, however, research projects going on to apply the methods of formal verification and symbolic execution to design specifications.

Because of the weakness of automated verification tools, manual methods are still used extensively. For example, with software inspection excellent results have been obtained in the development of defect-free software. This method is claimed still to detect more defects in a software product - and at lower cost - than does machine testing. [12]
It is, however, anticipated that research efforts to build more powerful verification tools on the level of the formal design specifications will be successful eventually. We definitely need tools for an automated certification of software.

5. Automated design by reuse of reliable software

Software designers used to be - and often still are - convinced that software which they did not invent themselves is not applicable to solve their task. They are "reinventing new wheels" for every new project.

Considering the software technology available until recently, they cannot be blamed for this attitude. Available software with few informal specifications (if specification at all!) and mainly consisting of code, often without comments, really is not reusable. This is true because reuse, in most cases, means that modifications are necessary to adapt the existing software to the new requirements. [13]

As was pointed out in the sections above, automation-based project support environments (like, for example, the EPOS system), are currently available, which change the nature of software. While software used to be identical with programs, now software consists mainly of requirements specifications and of multi-level design specifications with automatically produced graphical and textual documents, whereas the final code is just a "less interesting" result of the automatic code generation process.
Based on this new nature of software, reusability is becoming feasible. [14]

One of the aims of software reusability certainly is to improve software productivity by reducing the amount of new software, that needs to be designed, tested, and implemented. But, possibly even more important, reuse of already existing software modules, which have proved to be reliable, may contribute considerably to achieving high reliability of new software systems.
There are two basic strategies to achieve reusability of software:

- the "constructive strategy" of designing reusable software components (building blocks)

- the "search strategy" of searching for "old"
  software modules, which may be existing as a
  result of previous projects, and which may be
  similar to those "new" software modules which
  are required in the new project.

Both of these strategies have a chance to be
successful, but they also have major drawbacks.

The parametrized components, which are required as
building blocks for the "constructive strategy"
need many parameters and therefore are difficult
to define and, even more difficult to use.
On the other hand, the "search strategy" also
poses difficult problems to be solved:

- The results of previous software projects must
  be accessible in project data bases
- The similarity between the "new" software modules to be designed and possibly available "old"
  modules has to be evaluated
- If similar and reusable modules have been found,
  they have to be adapted to the new requirements
  without introducing new errors.

The solution of these problems is the aim of
further research. For example, there is a research
project, which was initiated 3 years ago at the
University of Stuttgart (in cooperation with six
industrial firms and supported by the German
government), to build a tool system called "Project Advisor". The aim of this knowledge-based
tool system will be to make accessible the experience from previously completed software/hardware
projects and to provide support especially in the
area of reusability[15,16].
Though the "Project Advisor" will be independent
of a specific support environment, the EPOS system
will be used, because this system has been applied
at a number of real projects by the industrial
companies involved. Data from different phases of
these projects are available in the project data
bases.
From the requirements engineering phase are available

- formally specified requirements (i.e., decision
  tables)
- informally specified requirements and constraints
- definitions of terms.

From the system design phase result

- hierarchical structures of design objects (i.e.,
  data structures, modules etc.)
- control and data flow
- synchronization of parallel tasks
- programming code.

Concerning the project management, formal specification are available on

- time and cost planning
- work assignments
- work packages
- project progress
etc.

All of these different kinds of data may be evaluated and considered for reuse.

Fig. 5 shows the role of the "Project Advisor":
While integrated project support environments
(i.e., the EPOS system) work on the project currently under development, the knowledge-based
project advisor gives advice in finding possibly
reusable experiences and results from earlier
projects.To fulfill its task, the knowledge-based
Project Advisor is made up of the basic components
shown in fig. 6. Its central component is the
knowledge base, where knowledge about the development of software/hardware systems, criteria about
reusability etc. are stored. This knowledge is
entered by a knowledge engineer using the knowledge representation language "PATHOS".

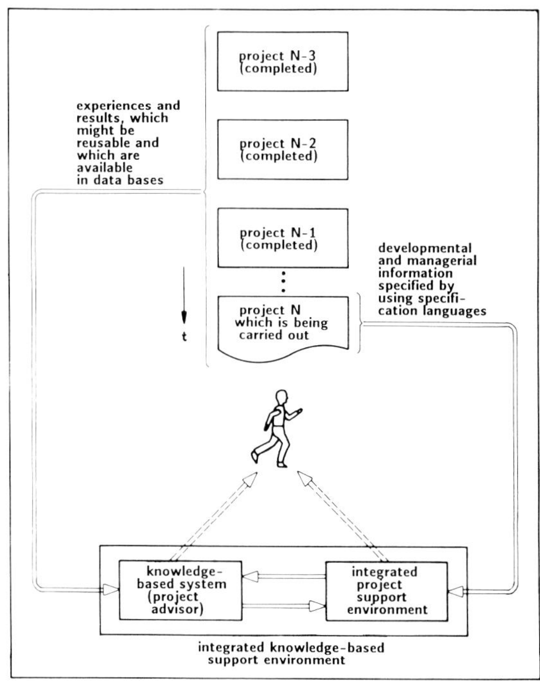

Fig. 5: Combining an integrated project support
environment with a knowledge-based system
(Project Advisor) to automate software
production by reuse of reliable software.

In order to retrieve reusable parts, the Project
Advisor has access to several project data bases.
Heuristic knowledge, i.e., experience gained
during preceeding projects, is put to use by the
inference engine. It applies the knowledge formulated in PATHOS rules on the actual project data.
To make its conclusions transparent to the developers, the Project Advisor also comprises an
explanation component.

6. Conclusion

The answer to problems of modern software systems,
such as the embedded software systems required for
space stations, is to use the computer itself to
solve them. Automated software production by using
computers as engineering tools throughout all
phases of the life-cycle is becoming a state-of-
the-art technology. Many of the difficulties that
used to plague the development of computer software - for example, high cost, late delivery, poor

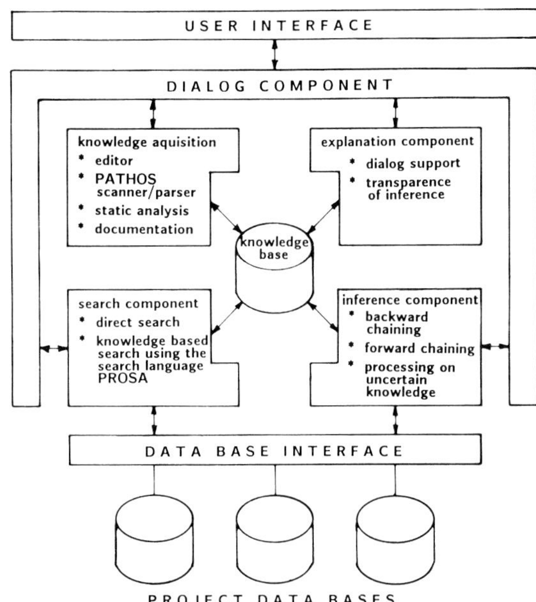

Fig.6: Basic components of the knowledge-based tool system "Project Advisor".

reliability and maintainability, nonresponsiveness to user requirements, and incompleteness and inconsistency of documentation - may be alleviated by using an automation-based software production paradigm.

At the core of this new way of producing software are requirements and design specifications, from which program code is generated automatically. Strictly speaking, "programmers" (in the meaning of people who are coding computer programs) are not necessary any more, while "software engineers" are required for the creative activities of requirements engineering and system design. As a consequence of this fundamental change, the meaning of the term "software" itself is changing:

- In the past, the term "software" was used for "programs coded in a certain programming language (i.e., FORTRAN), together with the documentation needed".
- In the future, "software" will mainly consist of "requirements and design specifications, formulated by specification languages", while program code, as well as the complete documentation will be the result of an automated generation process.

Though modern development support environments already provide powerful analysis tools to find and eliminate specification errors, verification and certification of error-free software still is an unsolved problem. Much theoretical and practical research is in progress to improve automated verification techniques.
Instead of trying to realize perfection of software systems through automated correctness proofs, another approach to further improving reliability is to reuse software modules which have been successful in operation already for some time and therefore can be considered to be error-free. Actually, this is the normal approach in all other engineering fields, where complex technical systems are being developed.

For example, if engineers develop a new automobile in a car factory, they never design everything new from scratch. Instead, they are reusing constructions and components, which have been proved to be successful, and they only modify these slightly according to new requirements. In addition, they use standard components (such as, for example, screws) for certain standard tasks.

The evolving technical area of software reusability is likely to reinvent this approach to high reliability of complex systems, which is well-known in other engineering fields. By reusing approved software constructions and software components and additionally, by reusing standard software components (which have been developed with the aim of reusability), an evolutionary software paradigm may replace the current software practice which consists of designing software new from scratch for every new project. The object oriented specification languages mentioned above are representations for design that foster this reusability-based approach.
There is evidence that in the time of the space stations, automated software production using this evolutionary paradigm will become reality, and thus the development of reliable software systems could become an ordinary engineering job.

References

1. Lauber, R.J.: Development Support Systems. IEEE Computer vol. 15 Nr. 5 (May 1982) pp. 36-46
2. Balzer, R., Cheatham, T.E., and Green, C.: Software Technology in the 1990's: Using a New Paradigm. IEEE Computer vol. 16 Nr 11 (Nov. 1983) pp. 39-45
3. Lauber, R.: Integrated Project Support Environments. In: Encyclopedia of Computer Science and Technology, Holzman, K., Kent, A., and Williams, J.G. (Eds.); New York, Basel: Marcel Dekker, Inc., 1987.
4. Sommerville, I. (Ed.): Software Engineering environments. London: Peter Peregrinus Ltd. 1986.
5. Lauber, R.J., and Lempp, P.R.: EPOS overview. Available from: SPS Software Products & Services, Inc., 14 East 38th Street, New York, N.Y.10016
6. Lauber, R. and Lempp, P.: Integrated Development and Project Management Support System. Proc. 7th Intl. Comp. Software & Appl. Conf. COMPSAC '83, Chicago, Nov. 1983. IEEE Computer Society Press, Los Angeles, 1983, pp. 412-421
7. Lempp P.: Integrated Support in the development and maintenance project "Heating control system" with the Software/Hardware Production Environment EPOS.
National Conf. on Methodologies and Tools for Real-Time Systems (R.F.Mathis, Ed.) March 2-3, 1987. The National Institute for Software Quality and Produtivity, Inc., Washington, D.C. 1987.

8. Goehner P.: Integrated computer support for
   software/hardware systems development. Pre-
   prints NAECON '86, Dayton, Ohio. 1986.
9. Lempp P. and Zeh A.: Developing Ada Software
   using the Software/Hardware Production Environment EPOS. SDI Ada Conference, Washington,
   D.C., Jan. 26/27, 1987.
10. Kley A.: Verification of processes using
    program transformation. NATO ASI Series, vol.
    F8 "Program Transformation and Programming
    Environments (P. Pepper, Ed.). Springer-Verlag
    1984.
11. Majoros, M. and Sneed, M.: Testing Programs
    against a formal specification. COMPSAC '83,
    IEEE Comp. Society Press, 1983, pp. 512-519
12. Fagan M.E.: Advances in Software Inspections.
    IEEE Trans. on Software Engineering vol. SE-12
    Nr 7, July 1986, pp. 744-751
13. Meyer B.: Reusability: The Case for Object-
    Oriented Design. IEEE Software vol. 4 Nr 2
    (March 1987), pp. 50-64
14. Shriver, B.D.: Reuse revisted. IEEE Trans. on
    Software Engineering, Vol.20 Nr 1, (January
    1987) p. 5
15. Lauber, R.J. and Permantier, G.A.: A knowledge
    representation language for process automation
    systems. To be presented at: IFAC congress
    Munich, July 27-31, 1987.
16. Wagner B.: Der Projekt-Advisor; ein wissens-
    basiertes Werkzeugsystem zur Wiederverwendung
    von Projektergebnissen.
    To be presented at: Kolloquium der Techn.
    Akademie Esslingen, Sept. 8-10, 1987.

Knowledge Servers -- Applications of Artificial Intelligence to Advanced Space Information Systems

Peter Friedland
Artificial Intelligence Research Branch
NASA Ames Research Center
Moffett Field, California 94035

Abstract

We have begun a transition from passive information systems which act only to facilitate the storage and retrieval of stereotyped data to far more active and responsive systems which can deal with widely differing forms of human knowledge. Edward Feigenbaum has coined the term "knowledge servers" to describe this next generation of active information management systems. Among the functions of a knowledge server will be: the ability to store enormous varieties of knowledge; the ability to determine, through natural discourse, the needs of its users; the ability to summarize and pursue complex relationships in its knowledge; the ability to test and critique user hypotheses and suggest previously unseen connections resulting from those hypotheses; and the ability to communicate and collaborate with other autonomous knowledge servers. Because of the complexity and variety of information relevant to future major space missions like space station, these missions will act as a driving force and testbed for the knowledge server concept.

1. Introduction

In the early days of artificial intelligence research it was thought that the key to powerful problem-solving behavior was in the reasoning process, that machines needed to utilize complex, general-purpose "inference engines" in order to emulate human performance in real world domains. This approach led to much frustration and many blind alleys until the discovery in the early 1970's (through the development of such early "expert systems" as DENDRAL, MACSYMA, and MYCIN) that it was in the knowledge, not the reasoning processes, where the power lay. This discovery was tested through the creation of thousands of such knowledge-based systems in the 1970's and 1980's. Many were capable of near-expert performance in complex domains using very simple inference mechanisms; all acted as confirmation of the knowledge is power thesis. This thesis has now become formalized as The Knowledge Principle: that a computational system can exhibit a high level of competence in understanding and action primarily because of the domain specific knowledge it contains.

An importance consequence of The Knowledge Principle is that intelligent systems must contain large amounts of knowledge (in its many and varied forms) about the domains in which they perform. Therefore, the pragmatics of building such a system dictate the need for the ability to acquire knowledge, to represent knowledge, and to manipulate knowledge; these topics have formed the core of artificial intelligence research since 1970.

Space Information Systems in the Space Station Era and beyond will be among the largest knowledge-based systems. They will need to store and effectively retrieve both structural and functional knowledge of artifacts as large as the Space Station, and will need to relate that knowledge to queries from a wide variety of other intelligent agents, both human and machine. The remainder of this paper will discuss current bottlenecks in building such comprehensive systems and will analyze the major research topics that will lead to the solution of those bottleneck problems. The paper will conclude with a vision of the "dream" space information system--the knowledge server for Space Station.

2. Current Bottlenecks

Knowledge-based systems have already been built that exhibit expert-level performance on significant problems. However, the amount of knowledge available to these systems is at least two orders of magnitude less than a comprehensive knowledge base relating to Space Station would require. Current representation mechanisms employ at least one of three basic structures: statements in the predicate calculus (logic), if-then rules, and collections of attribute-value pairs known as frames. The best current systems utilize all three structures in a hybrid approach. A typical "large" present-day knowledge-based system might contain up to a few thousand rules and/or statements in the predicate calculus, along with a few hundred frames.

There are four major bottlenecks in scaling systems up to the size we will require in the future:

- Knowledge Acquisition--Getting the knowledge into the system.

- Knowledge Consistency and Completeness-- Combining knowledge from many, potentially conflicting sources in many different forms and determining if enough knowledge has been stored for functional adequacy.

- Large Knowledge Base Manipulation--Determining if and how current methodologies for knowledge manipulation scale up to the demands of far larger systems.

- Interface Technology--Getting the knowledge out in a form that humans or other intelligent agents can understand and utilize effectively.

2.1. Knowledge Acquisition

Developing the knowledge base for a current-generation system currently requires the participation of a human "knowledge engineer" who acts as the bridge between domain expertise and computational representations. While knowledge engineers have become highly skilled at their task, they cannot avoid the introduction of significant delay and loss of accuracy into the knowledge base construction project. Major research efforts are underway to eliminate this intermediary by means of both automatic knowledge acquisition (by both inductive and deductive means) and by semi-automatic knowledge acquisition where domain experts act as their own knowledge engineers.

In a typical knowledge-based system project, knowledge acquisition can take anywhere from 50% to 90% of the time and effort; for systems two orders of magnitude larger than current systems, clearly, better mechanisms are necessary. In addition, since systems such as those which are to function as information systems for Space Station must have dynamic abilities to grow and improve their behavior, knowledge acquisition cannot be viewed as a one-time task, but a continuing one throughout the life of the system.

2.2. Knowledge Consistency and Completeness

The construction of a current knowledge systms typically involves interaction between a single (or at most a few) expert and a small team of knowledge engineers. But for the large-scale systems of the future, no single (or even small group) expert can possibly know more than a small fraction of the required knowledge. In addition, much of the expertise will come from other than human sources; it may arise directly from CAD systems during design or it may come during automated construction and testing processes. A major problem occurs in combining expertise from multiple sources. There may be disagreements both over facts (tolerances, expected lifetimes, etc.) and over heuristics (why certain design decisions were made, where components are likely to fail first, etc.). But even where disagreements are not critical, the "language" used to describe knowledge can be expected to vary significantly from source to source. Mechanisms are needed to ensure that large-scale knowledge systems are not highly dependent upon either the source or order of information entry.

A second issue arises of knowledge completeness for any significantly sized system, i.e. when is there enough knowledge to do an adequate job of problem-solving. In a sense, this problem is even more difficult than the knowledge consistency bottleneck since any notion of completeness involves "deep" knowledge of both the domain and the desired functional attributes of the knowledge system. Current systems have none of this "self-awareness," and the only present-day answers are for completeness in the formal sense for logic-based knowledge systems. The problem of knowledge completeness should be thought of as dynamic, with criteria changing during the evolutionary growth of missions like Space Station.

2.3. Large Knowledge Base Manipulation

The artificial intelligence research community has developed, over the last decade, adequate methods for reasoning with our present knowledge bases. These inference methods tend to be simple (following the Knowledge Principle) mechanisms like forward and backward chaining of rules, theorem proving in the predicate calculus, and object-oriented programming (attached procedures) for frame representations. However, we do not yet know whether our current methods will scale-up to the several orders-of-magnitude larger systems of the near future. Perhaps hardware solutions will come to the rescue and enable us to continue to rely on our present inferential methods, but it is equally likely that future problem knowledge base size and complexity will spur research into a second generation of more comprehensive and efficient reasoning methods. At any rate, much empirical experimentation will be required.

2.4. Interface Technology

Current systems are able to communicate actions and (perhaps as important) reasons for actions in a variety of formats ranging from graphical to textual. But even the best of present knowledge and information systems do not employ more than a tiny subset of true natural language nor do they employ, in any real sense, that most common of human interaction mechanisms, speech. As knowledge systems become larger, more comprehensive, and more ubiquitous, mechanisms for smooth interaction with other intelligent agents (particularly, but not exclusively humans) will become even more important.

3. Brittleness of Current Systems

A characteristic of all current computational systems (whether traditional or knowledge-based) is extreme brittleness at the edge of their performance. Even the most "user-friendly" computer programs are trivially easy to confuse with the wrong syntactic or semantic input. In the case of knowledge-based systems, even those that perform at extremely high levels of expertise within narrowly limited domains degrade rapidly to incompetence at the limits. Part of the problem comes from a total lack of self-awareness on the part of computational systems. Humans (often) know when they are near or beyond the bounds of their expertise and switch to more generic knowledge and problem-solving mechanisms or call in other, more expert consultants for advice. Part of the problem comes simply from the enormous gap between the knowledge of the least competent human and the most competent knowledge-based system. And finally, part of the problem arises from the fact that most operational knowledge-based systems are primarily heuristic or experiential; they rely solely upon compiled, mostly anecdotal experience, without a detailed model of structure and function to fall back on.

For space information systems, the brittleness problem is critical. We cannot have an information resource that simply gives up when precise, encapsulated expertise in unavailable.

But, on the other hand, a system which pretends to expertise when its information is only a very rough guess is equally dangerous. The next three sections of this paper discuss potential solutions to the brittleness problem for knowledge systems.

4. Common Sense and Analogical Reasoning

In situations where our particular expertise does not apply, we humans rely upon two, probably related, forms of reasoning. The first we label "common sense" and seems to encompass the generic knowledge which we can apply to a broad range of situations and actions. Both philosophers and computer scientists have pondered the exact nature of common sense knowledge; current thinking within the artificial intelligence community is that common sense comes from knowing a little about a lot and recognizing, from vast and continual experience with our environment, relevance and interconnectivity of knowledge. Dr. Douglas Lenat, of the Microelectronics and Computer Consortium (MCC), is currently testing this thesis in his ten-year CYC Project by attempting to build a knowledge base which contains everything in a desk encyclopedia and then determining if the resultant knowledge system responds gracefully over an extremely broad range of queries and problems.

The second mechanism is called analogical reasoning. Analogies are simply a method for relating common properties of seemingly different situations. Sometimes they are very similar situations; we may know what to expect in a Cambodian restaurant because we have already been to a Thai one. Sometimes, however, they allow us to reason about wildly different situations; understanding genetic regulation by thinking of what can slow and divert a locomotive on railroad process consists of two steps: pick a potential analogous situation, and extract out the common items. While the process sounds simple, little real progress has been made in building computational systems which reason by analogy. Many believe the lack of progress is caused by the fact that to effectively utilize analogies, as discussed for common sense reasoning above, the system needs to know a little about a lot; indeed the ability to form analogical reasoning is a major test of the CYC system. In any event, the ability to effectively analogize will add a major dimension of reasoning power to any future information management system.

5. Learning

Perhaps more than any other single characteristic, the ability to learn is what separates human problem-solvers from current computational ones. We continually improve and expand our own internal "knowledge bases" by a variety of mechanisms including explicitly from teachers, colleagues, and written material, and implicitly by experimentation and discovery from our environment. We would consider a fellow human impossibly stupid if the identical mistake was endlessly repeated, but almost all of our computational partners fall into this category. As our knowledge systems grow larger and larger, the addition of learning mechanisms will be the only hope for continuing to manage complexity.

Space Station is designed to continue and evolve over a period of at least 35 years. Even if we imagine a perfectly knowledgable information system at the initial configuration of Space Station, such a system will not be useful for very long as systems change by design or accident and missions change as we better understand the unique utility of the facility. But if each change requires specific, conscious, human-mediated changes to the information system, then humans will need to continue to be global experts in every piece of knowledge to recertify consistency of the knowledge base.

Learning mechanisms bring their own complexity to information mangement. As important as the ability to learn is the knowledge of when learning is worthwhile; clearly, every minor "discovery" need not be stored away for later use or even the large knowledge systems of the future will become overwhelmed. In addition, in the validation of computational systems, a decision must be made as to whether a learning mechanism can be validated as safe and reliable or whether each change, no matter how minor, to the knowledge system must be independently revalidated.

6. Synergism Among Intelligent Agents

Humans do not operate in an information vacuum; even experts cooperate with other humans of varying degrees of sophistication. Much of our power as problem-solvers comes from knowing when to seek out help in areas in which we are not expert. A powerful characteristic of future knowledge systems must be the ability to seek out answers from other sources, both human and machine, on the information network. Humans find that the synergism of multiple expertise causes the whole to often be much more than the sum of its parts for problem-solving. We have begun to develop the hardware mechanisms (the ethernet for distributed systems, for example) to allow such cooperation. However, the technology to know when and where to seek synergism is still well beyond the state-of-the-art. Researchers in artificial intelligence have begun to explore mechanisms, spanning a continuum from hierarchical to distributed, for achieving such intelligent behavior among intelligent computational agents, but the work is still in its infancy. In addition, much research is needed in such areas as communication of intent and abilities among intelligent agents.

7. The Dream for Space Information Systems

This paper has attempted to sketch the current state of knowledge systems along with the areas of research necessary to turn them into the kind of information utilities that will be truly useful on missions like Space Station. Let us imagine the properties of this new system, called a Knowledge Server, on our information network for Space Station.

- The Knowledge Server can determine, through natural discourse, the needs and desires (both overt and hidden) of its users. It maintains a memory of abilities and preferences of those it has served in the past, and can determine a reasonable model of the intents and abilities of new users. It is able to present answers in a form that is suitable to each particular user, whether astronaut or ground controller, specialist or generalist.

- The Knowledge Server can explain how it arrived at any particular answer and can justify its reasoning in response to user questions.

- The Knowledge Server can summarize and pursue complex relationships in its knowledge either when requested or when it believes such analysis will add to its understanding. It will actively seek out ways of improving its overall knowledge of Space Station and point out unseen and potential flaws in subsystems.

- The Knowledge Server can test and critique user hypotheses, working either autonomously or in close cooperation with a human user.

- The Knowledge Server can communicate and collaborate with other autonomous knowledge servers, both human and machine.

- The Knowledge Server can serve as an archival respository of all decisions, actions, and events that affect Space Station, and ensure that all relevant parties are informed as necessary about such knowledge.

Clearly, achieving the above abilities will be a non-trivial task; indeed, many view the research necessary as forming the core of work in artificial intelligence over the next several decades. However, building such Knowledge Servers to support the information needs of future space missions may be essential as our tasks grow increasingly complex and long-lived.

Acknowledgments

This article was inspired by Professor Edward Feigenbaum's keynote speech to the MEDINFO-86 conference, entitled "Knowledge Processing: From File Servers to Knowledge Servers." While the context of the speech was medical information processing, many of the ideas are transferable to space information processing. The author thanks Professor Feigenbaum for permission to use his speech as a starting point for his own thinking and writing on the matter.

MISSION SCHEDULING EXPERT SYSTEM
AND ITS SPACE STATION APPLICATIONS

T. Yoshioka*, T. Kato**, T. Wakaki**
M. Horio**, N. Sakamoto†, T. Jukuroki†
* National Space Development Agency of Japan, ** Fujitsu Limited
† Ishikawajima-Harima Heavy Industries Co., Ltd.
Tokyo, Japan

ABSTRACT

Japan has joined the United States space station program with its unique Japanese Experiment Module (JEM) development scheme.The space station and JEM will be launched in the mid 1990s,and many scientific experiments will be performed taking advantage of the microgravity and high vacuum.

This paper describes the mission scheduling expert system designed to prepare plans of these experiments (missions), and explains its operation .

Recently, we have developed a prototype of the mission scheduling expert system. Evaluation of the prototype has demonstrated that the system should be able to prepare an objective mission operation plan (especially a medium-short term plan or a short term plan).

This paper describes, in detail, the scheduling conditions for this prototype, as well as its functions, scheduling algorithms and heuristic knowledge, performance evaluation results, and subjects for further study.

1. INTRODUCTION

Since 1984, the National Aeronautics and Space Administration (NASA) has promoted the program of a permanently manned space station to be constructed in the mid 1990s. In response to the United States' appeal for participation, some of its allied nations including Japan, Canada, and the European Space Agency (ESA) nations joined the phase B study of this space station program in 1985 and 1986. Design, development, and utilization of space stations are in progress under international cooperation and coordination.

The National Space Development Agency of Japan (NASDA) has joined this space station program with its unique JEM development scheme to be connected to the US space station.

In the JEM, various experiments, such as material, life science, space science, earth observation, communications, and technology development will be conducted making full use of the space environment (microgravity, high vacuum, and abundant solar energy and high-energy particles).

Execution of these experiments (or operation of missions) requires consideration of the many complicated constraints (e.g. the limit to the functions of the JEM and the characteristics of each experiment or mission), which takes a lot of time and labor if done manually. To reduce this man-effort, we recently developed and evaluated the prototype expert system which has the scheduling ability and heuristic knowledge required to perform mission scheduling.

Chapter 2 discusses the applications of artificial intelligence (AI) for the JEM. Chapter 3 describes the mission operation plans for the JEM. Chapter 4 outlines the mission scheduling . expert system prototype and its evaluation, and Chapter 5 describes further problems under study.

2.APPLICATION OF ARTIFICIAL INTELLIGENCE FOR THE JEM

NASDA promotes the application of artificial intelligence technology for the JEM with the following:
Objectives
(1) Increase productivity,
(2) Lower operation cost,
(3) Improve reliability,
(4) Reduce hazards,
(5) Achieve autonomy,
(6) Perform tasks unsuited to human, and
(7) Transfer technology to other space
    development plans.
The expert system is designed to solve problems in specific domains by using the knowledge of human experts. It consists of the inference engine and the knowledge base. Table 1 lists proposed application areas for the expert system.

3. MISSION OPERATION PLAN IN JEM

This chapter describes the concepts of the mission payload operation scheduling, which is one of the proposed expert system applications for JEM (see Table 1 in chapter 2).

3.1 Types of Mission for the JEM

JEM is assumed in the following :
JEM consists of (1) the Pressurized Module, (2) the Exposed Facility, and (3) the Experiment Logistic Module.

Table 1 Expert system applications proposed for JEM

APPLICATION AREA	CONTENTS
System / payload Fault diagnosis	ONBOARD : identify the failed ORU and suggest necessary action GROUND : detailed analysis and suggest countermeasures
System / Mission Payload Operation Scheduling	ONBOARD : short-term schedule adjustments GROUND : intermediate / long term schedule planning
Crew Advisor	ONBOARD : suggest procedures for mission and system operations
System Monitoring And Maintenance Planning	GROUND : trend analysis for system conditions; planning for preventive maintenance and checkup
Logistic Management Support	GROUND : resource consumption analysis; planning for the supply and recovery

In the Pressurized Module which is supplied with one atmosphere, water and air, various experiments such as those for material or life sciences are performed under micro gravity. In the Exposed Facility, space science, earth observation, technology development and communication experiments are to be conducted in a space environment. The Experiment Logistic Module stores these experiment devices and materials. It can be accommodated in the shuttle's luggage room, and is also used as the container for return transportation to the Earth.

Execution of these experiments requires various resources such as electric power, crew, heat dissipation, and data communications.

3.2 Mission Operation Plan

This section describes the mission operation plan expected at present.

Each execution plan for a group of experiments (missions) is made in order of the long term plan, intermediate term plan, medium-short term plan, and then short term plan. Both the long term and the intermediate term plans are made on the ground, and the medium-short term plan is made both on the ground and on board. The short term plan is made on board (See Figure 1).

(1) Long term plan (strategic level)

The scheduling period is around 1 to 2 years. This plan is made in order to coordinate the resource supplies and mission candidates between nations.

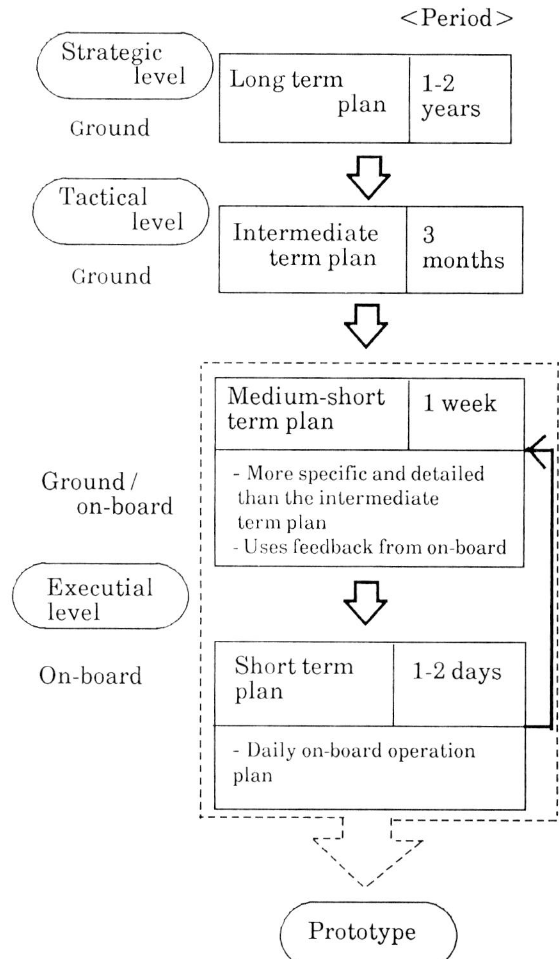

Figure 1   Planning flow for mission operation

Mission candidates and resource supplies are decided on the intermediate term plan basis.

(2) Intermediate term plan (tactical level)

The scheduling period is three months, which is the crew shift period. The main purpose of this plan is to optimize resource distribution for this period. The mission candidates and resource supplies are distributed first in units of months, then weeks.

(3) Medium short-term plan (executional level)

The scheduling period is one week. This plan is a base schedule for experiments on board. It is made on the ground taking into consideration the constraints, such as the amount of resources allocated for one week. The plan is then uplinked to the JEM and executed. If some change of the mission schedule becomes necessary on board, the plan is revised under the short term plan.

(4) Short term plan (executional level)

The short term plan is made when a change of the

base schedule is required due to unexpected events or operational situations on board. The scheduling period of this plan is one or two days.

3.3 Purpose of MISES Development

We have already manually scheduled a task, the FMPT (First Material Processing Test), which is scheduled to be performed in the space lab. at the end of 1980s. In order to reduce the man-effort required, we have performed preliminary investigations into the development of the MIssion Scheduling Expert System (MISES), which will make the above plans using the artificial intelligence technology. As a result of this examination, we recently prepared a prototype of the MISES at the executional level for making the medium-short term plan as well as the short term plan.

4. Mission Scheduling Expert System (MISES)

This chapter describes the MISES prototype developed by Fujitsu.

4.1 Specifications and Functions of the Prototype

This prototype has the following three important functions.
(1) Data editing before scheduling
(2) Scheduling, and
(3) Modifying and issuing schedules.
The scheduling function is the kernel of this prototype and is described in Section 4.1.1.

4.1.1 Scheduling function

As mentioned in Chapter 3, the scheduling term is one week for a medium-short term plan and one day for a short term plan. When making these plans using MISES, we presume that the resources and mission experiments have already been assigned.

This prototype schedules the missions taking the specified conditions into account using the specified scheduling methods.

(1) Scheduling conditions
Conditions which can be specified in the prototype are described below (see Table 2):

① Types and quantity of resources
Electric power, crew, heat dissipation, fluid and data transmission rates can be specified.

Electric power : Its supply power is specified for the JEM, modules and racks. The electric power consumed by each module (the Pressurized Module, Exposed Facility and Experiment Logistic Module) is limited, and the sum must not exceed the amount of power supplied for the whole JEM. Similarly, the total power consumed by all

racks of a module must not exceed the amount of power supplied for the module

Table 2   Scheduling conditions

Constraints		Description
Classification	Item	
Supplying capacity of JEM (resource supply)	Electric power	Power supplied for operating the missions
	Crew	Time period when a crew (one person) can work
	Heat dissipation	The amount of heat that can be dissipated
	Fluid	Fluid supply (of two types such as water, gas, etc.)
	Data communication	Transmission rate of data such as experiments /images /command data
Characteristics for to each mission	Resource consumption	Resource consumption of a mission
	Execution time	Execution time required for a scientific observation or timing of communication with the ground station
Mutual interference	Vaccum exhaustion sysyem	Mutual interference when the same line is used by multiple missions

Crew : The crew is only one person in the prototype. The working time of the crew is a resource.

Heat dissipation : There is a limit to the amount of heat which can be dissipated by the JEM.

Fluid : Two kinds of fluid (liquid or gas) can be specified. There is an upper limit to the quantities.

Data communication : There is an upper limit to the data communication rate for transmitting the experiment data , image data , and command data, etc..

② Characteristics of each mission
The following conditions are considered as characteristics of each mission.

Resource consumption : Each mission requires and consumes some of the resources in item ①.

Execution time : The start time for execution of experiments can be specified with conditions such as scientific observation or timing of communication with the ground station .

③ Mutual interference between missions

There are various interferences between missions. For example, an experiment which causes vibration or pollution, and another experiment which is damaged by vibration or pollution should not be executed at the same time. The vacuum exhaustion system described below is an example of a mutual interference in the prototype.

Vacuum exhaustion system : It is assumed that there is a common vacuum exhaust line. There would be interference between a mission which requires exclusive use of this line and another mission which can share it if they were performed together.

(2) Scheduling methods

The following scheduling methods can be specified:

① Concentrated or distributed scheduling

Concentrated mission scheduling schedules the missions as close as possible to the start of the scheduling term, and distributed mission scheduling schedules the missions as uniformly as possible.

If concentrated scheduling is used, missions tend not to be assigned in the latter part of the period, and we can then use that time for any unexpected missions.

Distributed scheduling spreads the missions in the period, and resources tend to be consumed uniformly during the period.

② Concentrated or distributed intra-vehicular crew activity

Concentrated crew scheduling schedules the missions so that the crew's work is concentrated as close as possible to the start of the scheduling period. Distributed crew scheduling schedules the missions so that the crew's work is distributed as uniformly as possible.

4.1.2 Other functions

(1) Editing resource data and mission data

This function adds, modifies and deletes resource and mission data prior to scheduling.

(2) Manual modification of the schedule and crew notification

Manual modification is available for a success case of scheduling and crew notification is useful for a failure case.

Manual modification : When it is needed to modify the executing time of some mission, the crew can change the schedule using the manual modification function. The schedule is then checked to see if the modifications satisfy the conditions.

Notification function : This function advises the crew of the reasons the schedule failed and suggests corrective action.

4.2 Knowledge and Algorithm for scheduling

4.2.1 Mission scheduling problem

The mission scheduling problem to be resolved by MISES can be modeled in the manner described below.

First, the following items are assumed to be specified.

(1) Scheduling period;

(2) Type of resources, such as electric power, crew, and gas, and their quantity and time dependency; and

(3) Missions to be scheduled and their resource consumption with time dependency.

The objective is to obtain a solution that efficiently schedules all the missions in item (3) within the scheduling period specified by item (1) under the conditions of resource supply in item (2).

It is assumed that all the resources in the space station can be represented by a time-resource supply graph, such as the following figure.

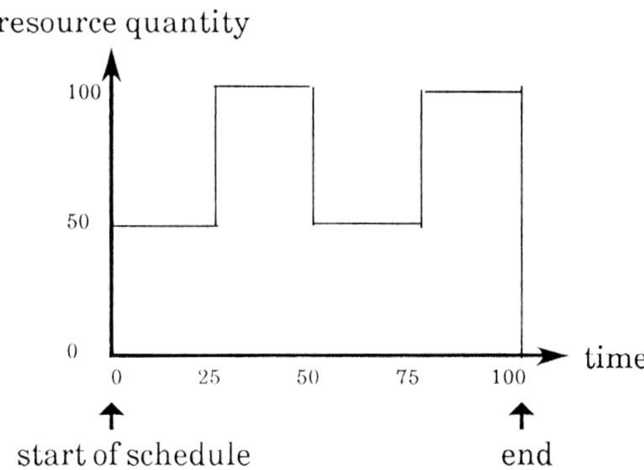

Example   Time dependency of resource supply

Also, each mission consumes resources. It is assumed that the resource consumption can be represented by a similar time-resource consumption graph.

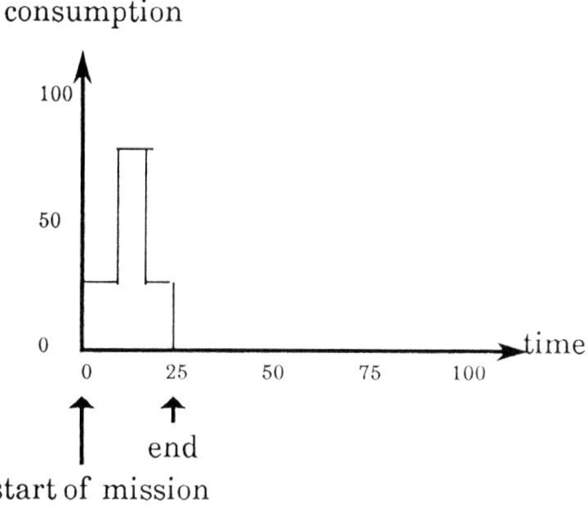

Example Time dependency of resource consumption

If the resource consumption pattern of a mission can be included somewhere in its resource supply pattern, the mission can be scheduled. Thus, the mission scheduling problem is a problem of packing the resource consumption pattern of a mission into the resource supply pattern.

<u>4.2.2 The mission scheduling algorithm as a packing problem</u>

(Scheduling of one mission)

As an example, take resource X and its supply pattern in Figure 2-a. Also, the resource X consumption pattern of a mission M is Figure 2-b. The possible schedules of one mission M can be easily found manually. Generally, a solution is not unique. As shown in the above example, a mission can be executed at any time within several time intervals. Therefore, there are infinite possibilities of selection of a mission schedule in these intervals.

Subtraction of the consumption pattern from the supply pattern makes a new supply pattern. Since the consumption pattern can be moved continuously within each time intervals, infinite new supply patterns can exist theoretically. However it is impossible to consider all of them. It is necessary to heuristically select one or several start times by some means. Figure 3 shows some new supply patterns for mission start times t_1 - t_4 in Figure 2-c.

(Scheduling multiple missions)

For scheduling multiple missions, choose one mission and schedule it in the above mentioned way. Then choose and schedule the next mission into one of the new resource supply patterns which resulted from the first scheduling. Repeat this until all missions are scheduled. If the operation fails on the way, take a backtrack (depth first search).

(Scheduling with multiple resources)

If there are multiple resources, search the

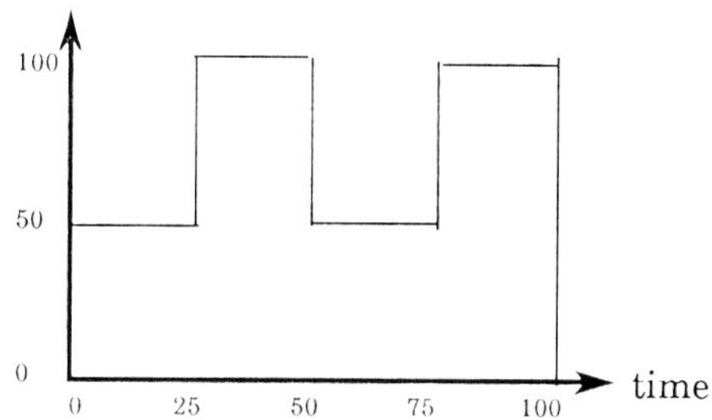

Figure 2-a   Supply pattern of resource X

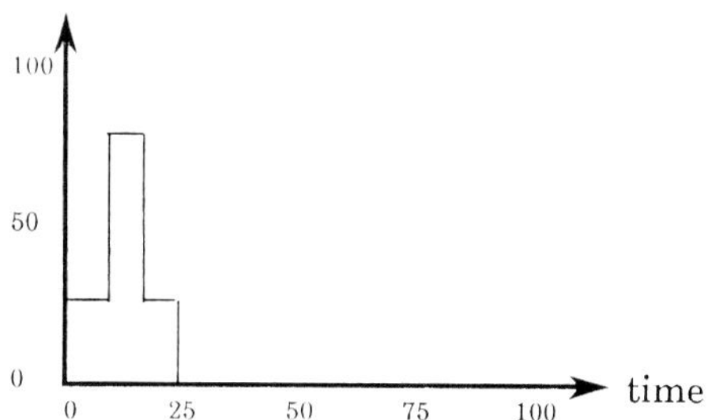

Figure 2-b   Resource X consumption pattern of mission M

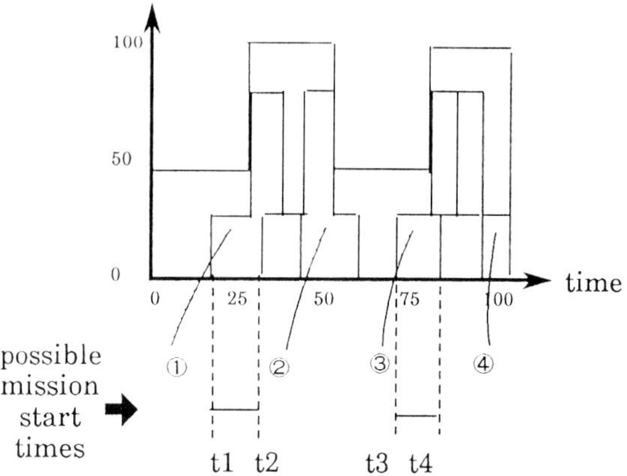

Figure 2-c
This shows the four cases in which the consumption pattern has been packed into the resource supply pattern (① to ④)

schedule time intervals for each resource and take the time interval common to them. This common time interval is regarded as the schedule time interval for the objective mission, and so this problem is now the same one as that for one resource.

Combining the above cases, the mission schedule is processed according to the following process flow in Figure 4.

<u>4.2.3 Knowledge for scheduling</u>

Acquired knowledge can be classified into the following three items:
- Order of the mission schedule

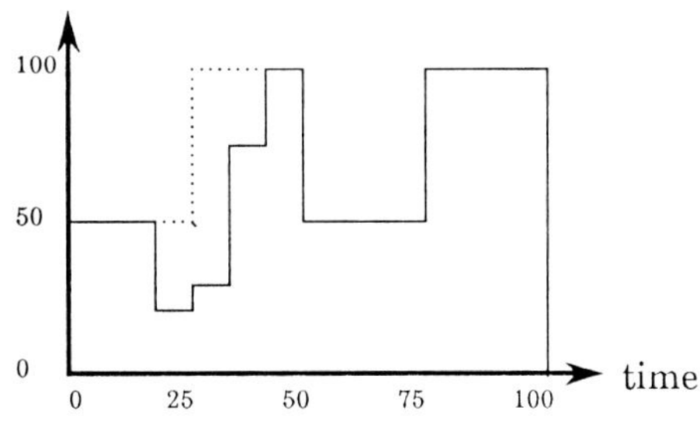

Figure 3　Generation of a new supply pattern
(case①)

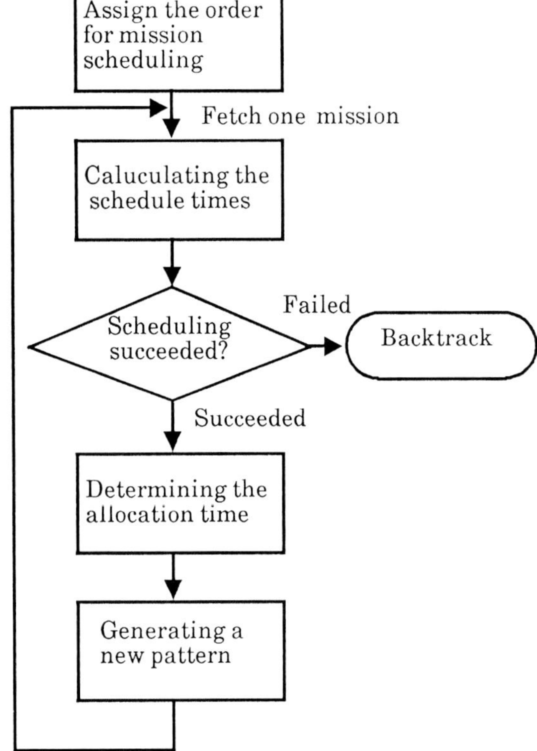

Figure 4　Mission scheduling sequence

- Method for finding the schedule time intervals for one mission
- Method for selecting a scheduling time from the schedule time intervals

(1) Order of the mission schedules
① Generally, to pack boxes of various sizes into a space, it is well-known that it is more successful to pack the largest first. Thus, an attempt is made to schedule the missions with the larger resource consumption (greater area of the resources consumption pattern) first. If the missions require multiple resources, mission scheduling is tried in order of greater area of the driver resource consumption pattern of each mission, where the driver resource is defined as the resource whose consumption rate is highest of all. The resource consumption rate of some resource A is defined as follows :

consumption rate of resource A

$$= \frac{\sum^{all\ the\ missions} (area\ of\ consumption\ pattern\ of\ resource\ A\ in\ each\ mission)}{Area\ of\ the\ supply\ pattern\ of\ resource\ A}$$

② For concentrated or distributed mission scheduling, the following points ,as well as the area of pattern mentioned in ①, are considered.
(For concentrated missions)
Since missions should be allocated as close as possible to the start of the scheduling period ,the residual resources become more scare at the start as more missions are scheduled. Accordingly, it would probably be difficult to schedule missions having a long duration of consumption if they are scheduled later. Therefore, in ordering the missions, the following two items are taken into account in the prototype.
-- Area of consumption pattern
-- Duration of consumption

(For distributed missions)
Since missions should be distributed evenly within the scheduling period, resources become consumed evenly. Accordingly, it would probably be difficult to schedule missions having a consumption pattern with a high peak if they were scheduled later. Therefore, the following two items are taken into account in ordering the missions.
-- Area of consumption pattern
-- Height of consumption pattern

(2)Computation of the schedule time
In the prototype, both the resource supply pattern and the mission consumption pattern are decomposed into their respective sets of rectangles. After that, the schedule time intervals can be computed using the conditions that the rectangle belonging to the consumption pattern should be included in the rectangle belonging to the supply pattern.

(3) Selection of the scheduling time points from the schedule time intervals
As described in section 4.2.2 , it is impossible to

perform infinite trials for the timing of the objective mission within its schedule time intervals. Therefore, we must select some finite points of time (candidate solutions) by some means. We call this selection allocation.

On the other hand it is indispensable to obtain better solution within the time required from the actual operation by specifying an scheduling method. We tried to find some quasi-optimum (approximate) solutions for concentrated and distributed scheduling useful for the actual operation as soon as possible and with as little memory as possible on the basis of heuristic knowledges and scheduling know-how, rather than to obtain the optimum (exact) solution, which would consume enormous amounts of time and memory.

Our approach has been partly heuristic. For concentrated schedules,the candidates whose schedule time is closer to the start of the scheduling period is evaluated sooner. For distributed schedules, the candidates with greater residual of driver resource at its schedule time is evaluated first.

Solutions for concentrated or distributed crew schedules have also been obtained by a similar approach.

4.3 System Configuration

This prototype was developed using ESHELL, which is a tool to construct expert systems, and currently operates with about 4MB memory under the UTILISP environment on a FACOM M382 Computer System.

ESHELL supports the blackboard model. The knowledge representation offered by ESHELL consists of knowledge sources (KS, sets of production rules), frames, a blackboard(BB) and user defined functions(UDF), which are described as a function of UTILISP. The man-machine interface of the prototype has been made mainly using FORTRAN77 and software packages IPF and GSP for screen generation. Figure 5 shows the system configuration and the development environment of the prototype.

4.4 Execution Results and Performance Evaluation

4.4.1 Execution results

Figures 6 and 7 show the results of the short-term plan (for one day) for a concentrated and a distributed scheduling method respectively.

4.4.2 Performance evaluation

Current planning for the JEM, allows for an average of about ten missions per day for a short term plan and an average of about thirty missions per week for a medium-short term plans. The medium-short term plan has a smaller number of missions per day than the short term plan, because the medium-

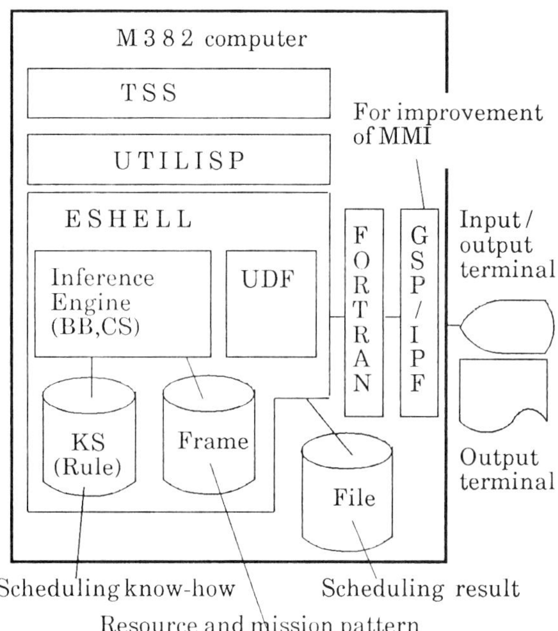

Figure 5   System configuration

short plan includes some experiments covering several days or several weeks.

We tried to evaluate whether or not the number of missions expected for a short term plan or a medium-short term plan can be scheduled, considering a typical resource supply and the material testing equipment expected to be mounted on the space station .

The following conclusions have been obtained as the result of scheduling experimental cases:
(1) Almost all the expected missions can be scheduled if the missions do not include crew work.
(2) If the missions include crew work, it becomes fairly difficult to schedule the expected missions. The number of missions that can be scheduled will decrease about 30 percent.

The causes of problem (2) are as follows:
-The crew work as a resource is supplied only for one third of a day (8 hours). Furthermore, the crew is limited to one person in the prototype and he cannot perform more than one mission operation at a time. Therefore, the resource pattern of the crew work does not allow much packing.
-These missions request the crew work too often.

This is the problem peculiar to the present system operations or the mission operations to be scheduled. This situation may cause a failure to meet mission requests. It is necessary to find a way to escape this problem. A solution may be to give resource and mission data some flexibility; that is, it may be possible to find a solution by making good use of the knowledge interactively acquired. For instance the system (i.e. MISES) could inquire to its operators interactively about the possibility of the crew's overtime work or the possibility of a change of the mission's request for crew's work.

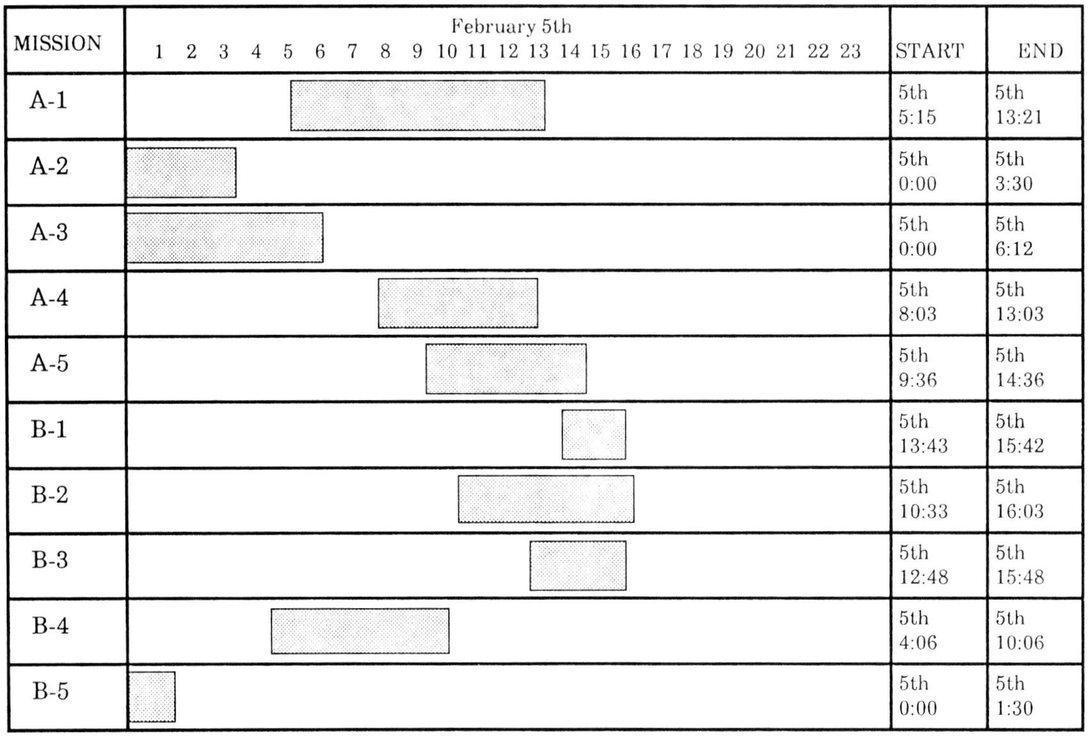
MISSION
February 5th
1 2 3 4 5 6 7 8 9 10 11 12 13 14 15 16 17 18 19 20 21 22 23
START
END
A-1 5th 5:15 5th 13:21
A-2 5th 0:00 5th 3:30
A-3 5th 0:00 5th 6:12
A-4 5th 8:03 5th 13:03
A-5 5th 9:36 5th 14:36
B-1 5th 13:43 5th 15:42
B-2 5th 10:33 5th 16:03
B-3 5th 12:48 5th 15:48
B-4 5th 4:06 5th 10:06
B-5 5th 0:00 5th 1:30

Figure 6 Result by concentrated scheduling

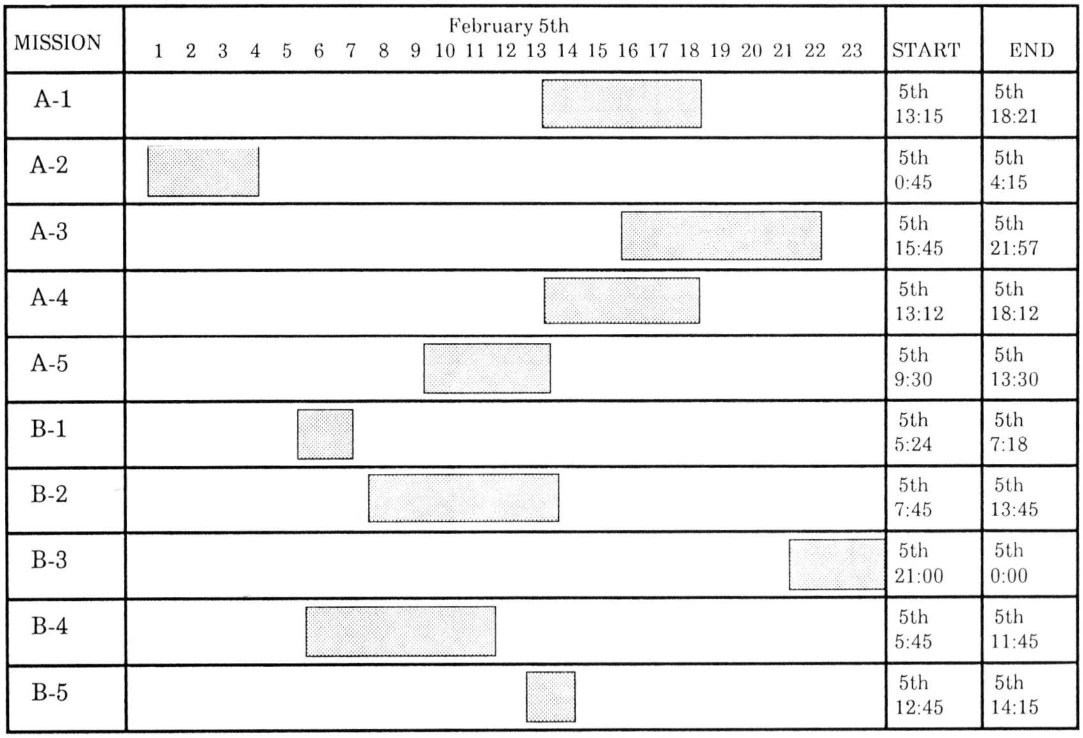
MISSION
February 5th
1 2 3 4 5 6 7 8 9 10 11 12 13 14 15 16 17 18 19 20 21 22 23
START
END
A-1 5th 13:15 5th 18:21
A-2 5th 0:45 5th 4:15
A-3 5th 15:45 5th 21:57
A-4 5th 13:12 5th 18:12
A-5 5th 9:30 5th 13:30
B-1 5th 5:24 5th 7:18
B-2 5th 7:45 5th 13:45
B-3 5th 21:00 5th 0:00
B-4 5th 5:45 5th 11:45
B-5 5th 12:45 5th 14:15

Figure 7 Result by distributed scheduling

Our prototype of MISES can generate the short term plan, which is now expected to schedule an average of ten missions per day, with several seconds of CPU time. We can compare this scheduling task to the similar case of FMPT, in which it took about a week to manually make a schedule for a 1 day period with about ten missions. Accordingly, we are convinced of the feasibility of MISES in actual operation after launch, since the CPU time is fast enough for a crew to make a daily schedule (i.e. short term plan) every morning in JEM.

The significant subject of the mission scheduling expert system (MISES) is to solve one of the search problem which has been an old theme of the artificial intelligence rather than to construct an expert system or use the sophisticated techniques of knowledge engineering.

As described in section 4.2, the scheduling of one mission includes the infinite possibility in the schedule time intervals. Then, trying to search the possibilities thoroughly for the solution candidates will cause such a combinatorial explosion of the number of them to be searched as even very fast super-computer could not process them. We overcame this problem of the combinatorial explosion by heuristically selecting a finite number of scheduling times as the candidates.

Therefore, even if the message "can't schedule" is displayed as a result of searching all of the candidates in this prototype, it cannot be positively asserted that there is really no solution. That is, a problem still exists that a solution may exist.

In the near future, we intend to develop a search method to find a practical effective solution quickly as well as within the allowable limit of memory by growing up these heuristics and knowledges more intelligently.

Furthermore, we are going to develop the system on a workstation, to make a more precise evaluation of its performance in the system environment similar to the onboard computer in JEM, and to improve its man-machine interface by using a multiwindow technique.

ACKNOWLEDGMENTS

We would like to thank the Manager, Mr. Sohma, Mr. Mikami and Dr. Okada of Space Development Promotion Group in Fujitsu for their valuable suggestions and helpful support in the prototype development.

REFERENCES

(1) Donald A. Waterman: A Guide to Expert Systems, Addison-Wesley Publishing Company, 1985

(2) Elaine A Rich : Artificial Intelligence, McGraw-Hill Book Company, 1983

(3) Advanced Technology Advisory Committee (ATAC) Aaron Cohen: Advancing Automation and Robotics Technology for the Space Station and for the U.S. Economy, NASA Technical Memorandum 87566, 1985

(4) Takashi Chikayama: Utilisp Manual, METR81-6, Dept. of Mathematical Engineering, University of Tokyo, 1981

Management 2: Plans and Approaches

Co-Chairmen: Robert E. Smylie and Claude Honvault

LESSONS LEARNED FROM PAST PROGRAMS: POLICY AND MANAGEMENT ISSUES

Stuart J. Evans
NASA Headquarters
Washington, DC

The whole theme of this session is "lessons learned." An important question to ask is: have they been learned? The answer I don't think is yes. It is no. I was reviewing some of the famous quotations on this subject such as, "those who ignore history are doomed to relive it." One of Churchill's quotes from 1940 also came to mind: "if the present sits in judgment of the past it will lose the future." A number of your sons, when they get to be about thirty years old, will probably say to you that it's remarkable how much you've learned in the last fourteen years. And then finally, visibility and communication in transmission of what one learns in a program is quite important. Very often it's completely missed. It is like an Alaskan husky dog sled. Only the lead husky has a good view of things. There are, however, some lessons that people have learned. In the last year there has been a lot in the papers on spare parts such as six hundred dollar ashtrays, eight hundred dollar toilet seats and a whole range of such things. That cycle was repeated around twenty years ago when Porter Hardy was chairman of the House Government Operations Committee. There was a whole range of hearings on exorbitant spare part pricing and the reasons for it, as well as books written on the subject. Now twenty years later, it comes up again, like some of the concepts of the 60's: fixed price development, fixed price research and development, and total package procurement. Bob Charles, when he was the Assistant Secretary of the Air Force, initiated the total package concept of competition for design, development and production of a program on a fixed price basis. Many volumes have been written on the case known as the C5A where it was applied. That was in the mid-60's. In 1970 David Packer came to Washington as a Deputy Secretary of Defense. He put out quite an edict, a portion of which was "thou shalt not fix price development programs." Take the 80's with the rise of a new fad; I'll call it competition advocates and fixed pricing. Look at what the Navy is doing today in fixed pricing programs. Look at what the Air Force did in the early 80's under General Slay, the AF Systems Commander. All the lesson are there. Are the learned? No. No, they're not.

I came to NASA with the conviction that most of what I learned in the Navy I would never let happen at NASA. This was because I inherited more difficulties trying to fix programs that had been put together in the wrong way, and with foreseeable consequences of just disastrous courses, whether it was in shipbuilding, or in aircraft. I think part of the problem is, in a way, one of memory; and it

struck me when I was in the Navy and seeing more of the kind old gentleman Admiral Richover than I cared to do. His office reception area had what looked like a bullet proof window with closed circuit TV so that people could be monitored as they entered. You couldn't get in the office without an escort. Over this bullet proof window was a little sign and it said, "You can't trust anybody over 30 or under 82." There's a lesson in that.

There are also some profound lessons concerning the Tracking Data Relay Satellite Program. We ignored them. The same lessons that we learned there, were learned in the 60's in the services. But what was the concept? We took things farther than DOD had done. The concept was this: we would compete for the tracking and data satellite system to replace the ground stations. We would do it on a shared basis, government and commercial. We would do it on a fixed-price, ten year service contract basis. Consequently, we did not directly enter a development program. We signed a fixed-price service contract for ten years, under which, through a subcontractor, a very advanced type of spacecraft had to be developed. It was no more than perhaps ninety days after signing the contract that trouble began. It has continued in one form or another for perhaps ten years. We have learned a great deal.

Again, it is a fallacy to think that one can enter into a major developmental effort, forecast its course, schedule and cost, and price it on a fixed-price basis. It is an utter, total fallacy. We went, as they say, a step farther. We didn't fix the price of the development contract. We fixed the price of a ten year service contract following that development. This meant that for a period of time things began to get difficult: the development did not go as anticipated, the design began to mature and cost began to become clear. We were in a fixed price mode, prime and subcontract. Because of that we had no privity of contract with the developer, TRW, because it was fixed price on a service performance type specification. We had no contractual means of sitting down with the contactor, discussing technical matters, making technical changes, or even seriously monitoring performance. Every time there was a difference of opinion, the position would be taken that we have a service contract and what we're developing, we'll provide that service. All the difference of opinion in the world on the subject doesn't do any good.

The other aspect of this was a commercial and a government service, a shared system. Part of it was a tracking and data relay satellite, part of it was an advanced C-band commercial spacecraft. We shared systems. The pricing of it was constantly a tug-of-war. The override of one on the other was constantly a tug-of-war. As it became closer to the time of completion and launch (initially it was to be completed and launched in September 1979, but it was not launched until 1982 or 1983), what happened? As it came closer to launch, the focus on relative control and priorities under this contract between a commercial service, which was one main interest of the company, and the service required by NASA, came into continual conflict.

Interpretations of contract provisions led to differences of opinion, differences in position, and concern in terms of "you interrupt commercial service and you do us great damage from the standpoint of revenue." From our standpoint, we needed immediate response if something went wrong. We needed an immediate override. So what has occured? We have done two things. Very late in the contract and at great cost (cost in terms of relationships, people, wear and tear, and money), we have converted this contract to one of reimbursement. We have essentially bought out the C-band commercial transponders. Thus, after eight years we have finally come around to where we should have started in the first place: on a dedicated spacecraft, on a cost reimbursement basis, recognizing that the development of the spacecraft was a major effort. We didn't do it. There are a lot of lessons there.

Many of us are still around to recall these things. Many of us. Ten years from now we won't be, just like what has happened in DOD. The people in DOD have not learned the lessons of the 70's. I really hope that this does not happen to us. But people change. Ideas come in cycles. If you look at it, you centralize, you decentralize. You organize, you reorganize, you unorganize. It's simply in cycles. Concepts have a way of coming up. There's not much new under the universe, but cycles do come up. And they're brought up by people. New people who have not experienced the cycle maybe ten years or fifteen years earlier. So they're doomed to try again in a lot of ways. Again, I hope that what we have learned from this type of contract stays with us. It's not the type of thing you can put in a book. I have read a very thick case of what was known as the lockheed C5A total package, a package not unlike this tracking and data relay system. You can write all you want but people won't read or rarely read. They simply won't. You have to experience it, like the sixteen year old who becomes 30 and tells you how much you've learned in the last fourteen years. It is not

just true in this business. It's true across the board, in essentially everything we do.

Perhaps the only lesson learned that is still with us, I suppose, is the one learned in the 30's on appeasement, and we overreacted in some ways to that for maybe fifteen or twenty years. But 50% of today's populace were born after World War II, and those things dim. You can read about it in books, but you have to experience it yourself. And it seems to me that is human nature. The one thing I would leave in this area is whenever one embarks on a program, particularly a major program, to do so with one's eyes fully open, wide open, to look back, to look at comparisons, to see if approaches have been tried before. If they have, then one must find the people that experienced them.

This is true in the Space Station. It was true in the organization of the Space Station. I thought one of the greatest lessons we could learn in terms of Space Station management and development was in the Polaris/Poseidon/Trident Program. This program was continually evolving, growing, and going through concept formulation of evaluation; one building on the other. One had a very, very complex integration job: putting nuclear power in a submarine hold with ballistic missiles in it. And the one person around that was there the entire time was Vice Admiral Levering Smith, a walking encyclopedia of knowledge on lessons, on how to do and how not to do. I stongly suggested we get Levering in, talk to him and get his experience from the standpoint of organization, management, delegation, direction, and management system across the board. We need to take advantage of people like that, get their advice and heed that advice. If we do not, we are doomed to relive some of the failures of the past.

You will see it in the Defense Department in the next four years. The fallacies of fixed pricing are indeed what took place in the late 70's and will come home to roost in claims, and worse than that, in significant delay and adverse relationships. The worst thing the government or its industry partner can do is to enter a form of communication, a contract, which has the seeds of formulating an adverse relationship. Where pricing is fixed and risk is high, the motivation is to cut corners, to protect your business, to protect costs, to protect the profit, minimize the loss and work on the very minimum required. Communications dry up, postures are taken and the program gets hurt. If it is not on a basis of partnership, a common interest, the common proceeding, and if the means of communication, the contract, doesn't express that then I think we're doomed to again learn a lot of

the lessons of the past. This is the message of overriding importance to all of
us: to look back, to learn from the past.
The lessons are there. The record would
indicate that human beings do not learn
from the past; but it is critical that we
start.

LESSONS LEARNED FROM PAST PROGRAMS-AIR TRAFFIC CONTROL

B. N. Etheridge* and R. W. Peak, Jr.**
Martin Marietta Corporation
Information and Communications Systems
Air Traffic Control Division
Washington, D. C.

Abstract

This paper discusses the classical approach to system engineering and a modern adaptation of that approach necessitated by a major modernization to a system that began as a loosely coupled set of subsystems, procured independently over a long time, and which consequently did not have the usual system documentation, nor unified system engineering management disciplines. This adaptation is seen as a potential pathfinder for future large system modernizations which have similarly evolved.

Introduction

The development of any large system is a complex engineering effort which usually costs millions or billions of dollars, takes years of time, and requires the dedicated efforts of hundreds or thousands of personnel.

History has shown that the prudent management of such heroic undertakings requires both management and engineering disciplines that provide an orderly documentation of requirements, a comprehensive integration of design, a rigorous standardization and control of key development processes, and detailed reporting and control of cost and schedule.

Among the expected disciplines for a classical system engineering development environment are integrated cost, schedule, and technical program/project management systems; centralized hierarchical configuration management; formal design review process; interface management process; integrated logistics support; formal verification of all levels of engineering requirements; centralized software and hardware development; and comprehensive engineering standards for development, documentation, and various aspects of management which are universally applied and enforced. (see Fig. 1)

These disciplines form the basis of system engineering management, which is the subject of this paper. The particular aspect of system engineering discussed here is the adaptation of classical methods to allow application of these disciplines to the modernization of large deployed systems which were not the product of classically applied system engineering. In addition to the inherent risk in any large scale development, a major modernization adds additional risk as each new element is integrated into the existing system. System engineering helps to reduce or manage such risks. (see Table 1)

The Classical Approach

The classical system is a group of hardware, software, and personnel or procedures which has a defined mission or set of functions to accomplish to a defined performance level, in a defined context or environment, including interfaces with other systems. It is designed and developed from the top down, from the beginning, by a unified system engineering organization.

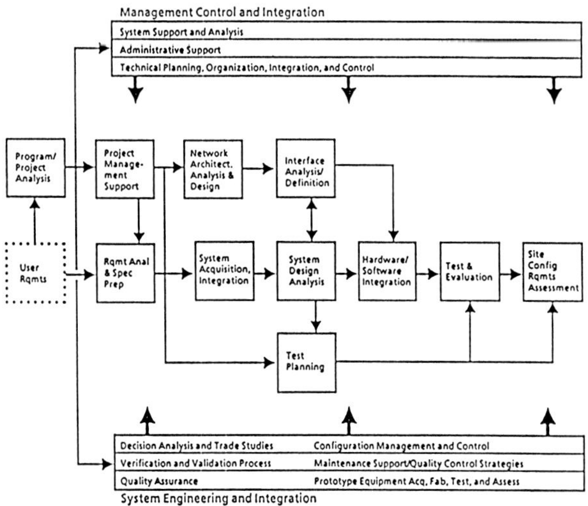

Fig. 1 Systems Engineering Process.

Table 1 Modernization Risks for a Large System

Development	- Modern Computer Technology - Advanced Technology Workstations - Software - Communications - Interfaces - Facilities - Support Systems
Transition / Implementation	- Configuration Management - Integration Testing - Installation

---

*Vice President and Program Manager, Member AIAA

**Director, Advanced Automation Program, Member AIAA

The classical approach to system engineering, documented in MIL-STD-499A, is to define the mission to be accomplished, the operational requirements to be satisfied in the accomplishment of the mission, and the system requirements that satisfy the operational requirements. Concurrently, with this requirements definition, the system design concept and operational concept are being evolved.

After the system requirements and design have been established, the allocation of system functions and performance is made to the system segments or subsystems. The same process repeats at this next lower level to establish the segment or subsystem functions, performance, operations concept, and interfaces.

This approach is designated as top down, since the hierarchy of requirements and design is developed from the mission statement, downward to the system, the segments or subsystems, and finally the components which when properly integrated into the system, satisfy the mission.

Engineering specialties such as human factors, reliability/maintainability/availability,security, safety, and electromagnetic compatibility are integrated into this process as the requirements and designs mature. Also, during this process, Integrated Logistic Support (ILS), which provides for maintenance, supply support, training, technical documentation, and special tools and test equipment, is integrated to influence the system design to ensure cost-effective supportability over the entire life cycle.

The classical approach is to accomplish this top down development before any detailed component or subsystem design is begun. This approach ensures that the requirements drive the design and that the resulting system can accomplish its intended mission.

The classical approach is for the prime system developer or a single system engineering and integration contractor to be responsible for the entire system design, development, and deployment process. This facilitates management of the engineering activities and assures consistency of approach, analytical techniques, design style, and documentation at all levels.

<u>A Modern Adaptation</u>

The National Airspace System (NAS) consists of existing hardware and software some of which has been fielded for 20 years or more, and some of which is in every stage of development--from conceptualization to installation and activation. Some of the new equipment is required to replace aging, unsupportable devices with new solid state electronics that accomplish the same function. Other equipment and software is being developed to advance the capabilities of the NAS in terms of response, capacity, accuracy, and additional functionality.

The System Engineering and Integration Contractor for the NAS was selected three years after the publication of the NAS Plan, and after the beginning of procurement of subsystems for the modernization.

The history of NAS development began with the development of communication systems and radar systems to allow controllers to better communicate with each other and with pilots, and to allow controllers to more accurately detect aircraft location during any weather conditions. Initially, there was little connectivity among these systems, and they were treated as systems in their own right, rather than as subsystems of a larger system.

As the number of flights increased and aircraft acquired all-weather capability with the advent of radar and more sophisticated navigation avionics, the pilots' dependence on the air traffic control system grew. With this dependence came the need to expand or upgrade the air traffic control system, and to set standards of performance at the system level that could be expected by pilots wherever they flew in the American airspace.

The formal notion of a National Airspace System emerged in the late 1970's, as the connectivity and interoperability of the air traffic control subsystems increased. At this time, the FAA recognized that it must also expand the system engineering applied to the development of the NAS, and extend it to the NAS level to tie all the subsystems together. In 1981, the first National Airspace Plan for Facilities and Equipment (NAS Plan) was developed. The FAA conceived a four-level scheme for the design of the NAS, which laid out top level functional requirements for a 1995 end point design. These requirements were first published in the Level I Design Document three years later.

The Level I Design Document tied all the previously defined "systems," or "facilities" together into one National Airspace System. It documented the top level functional requirements for the NAS, and identified the interface relationships among the "systems"/facilities. It began the formal extension of the system engineering process to the NAS level.

The Level II Design was to add performance requirements, and the Level III design was to add interfaces and define implementation packages of subsystem groups that could be concurrently transitioned into the field and integrated with the existing subsystems. Finally, the Level IV Design was to tailor the standards approach established in Level III to the individual sites and facilities into which the new subsystems were to be installed. (see Fig. 2)

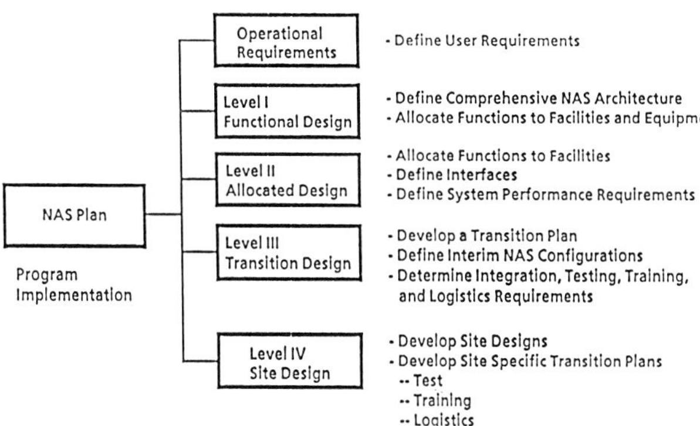

Fig. 2 Systems Engineering Approach.

Two years after the publication of the Level I Design Document, the Martin Marietta team was chosen to be the System Engineering and Integration Contractor for the NAS, and joined the FAA in this extension and expansion of system engineering for the NAS.

In order to minimize the time required to deploy new subsystems, the FAA had proceeded with the NAS modernization effort at the component and subsystem level. Much of this engineering had to be incorporated into the system level design which started later, in order to preserve the ongoing subsystem designs, and to minimize subsystem cost and schedule growth.

The difficulty of this adaptation is that the individually developed subsystems were not governed by a consistent set of system level requirements. Various design standards were applied, but not rigorously or consistently, and these were not always mutually compatible.

Therefore, this situation necessitated a tradeoff balancing the preferred "top-down" system approach and the pragmatic "bottom-up" approach required to accommodate on-going programs with the accompanying full range of political and economic factors that bear on such a system development. In some instances, the subsystem requirements will be revised; in others, the system requirements will be relaxed, to achieve a practical solution.

Mission/Operations

Overview

Classically, the mission statement for a system is analyzed and specific operational requirements are developed. These operational requirements indicate what products and/or services the system will provide the user, and to what level of performance it will provide them.

For example, to land an astronaut on Mars, a system that can escape Earth's gravity and support the astronaut during the journey must be hypothesized. It must provide a pressurized environment, a survivable temperature range, radiation protection, food and water for the duration of the trip, and a relatively soft landing at the end of each trip.

A design reference mission is developed from these requirements, and that mission is used to gage the efficacy of the system designs proposed to meet the operational requirements. The design reference mission is the set of operational requirements and scenarios against which the system analysis will proceed. It defines the system products and services required by the system user, in terms of the top level functionality and performance.

Failure to develop such a design reference mission can produce a spaceship design with too much or too little propulsion, too little space for samples, or too many "bells and whistles" to ever get off the ground.

The interrelationship of the hardware, software, and personnel in carrying out the system operational requirements is defined in a concept of operations which establishes operational scenarios that guide the designers in subsystem development and verification. This concept also includes keeping the system operational; this aspect is called the maintenance or support concept.

Application to the NAS

The NAS mission is evolving as the air traffic control environment evolves. Traffic projections indicate significant growth in commercial and general aviation during the next 15 years. As the number of flights and the

performance levels of aircraft increase,
the NAS must respond with greater
capabilities. The goal of the NAS Plan
is to reduce technical staff required to
maintain and operate the modernized and
expanded system by one-third by the year
2000. At the same time, the goal is to
maintain the overall air traffic and
airway facilities maintenance costs of
field operations of the NAS at the 1980
level, when adjusted for inflation,
excluding the capital cost of
modernization.

The basic mission of the NAS is to
safely control aircraft flying under
instrument flight rules in designated
airspace. In the same sense that NASA
might have a mission to safely land an
astronaut on Mars and return with
samples of the Martian soil, this
mission must be defined to be achieved.

The NAS mission concept was well
understood by many engineers in the FAA
when the new computers and radar were
envisioned, but the details of the
mission were not sufficiently well
documented to ensure that the hardware
and software that were being procured
would meet the mission requirements.

For the 1995 NAS, the operations
concept is still under development. As
the concept matures, further tradeoffs
must be accomplished to determine
whether to modify subsystems under
development, or to modify the concept
of operations.

<u>System Requirements and Design</u>

<u>Overview</u>

When the system design concept is
known, system requirements should be
defined and documented in a system
specification that will govern the
development of all subordinate
subsystems.

This process begins with a system
requirements analysis to convert the
operational requirements into functions
and concomitant performance levels.
These are then allocated from the
system down to the constituent
subsystems. Requirements include
interfaces to other systems and methods
of verifying that the desired functions
and performance levels have been
achieved in the selected design.

The requirements analysis also
identifies interface, support, and
verification requirements that will be
documented in the system specification.

ILS involvement in the
requirements analysis process is
important, since the chief value of
early logistics analysis and planning
is influence on the evolving design to
ensure that it is supportable over its
life cycle at the lowest cost. (see
Fig. 3)

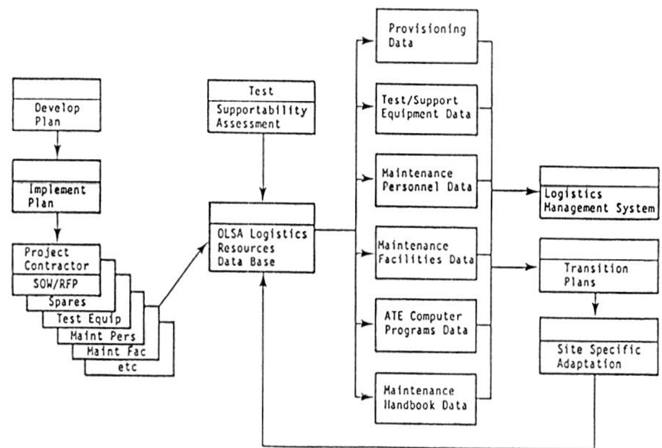

Fig. 3 Logistics Process.

The system specification consists
of an orderly and rigorous listing of
the system level functional,
performance, interface, support and
verification requirements necessary and
sufficient for the system development
in order to satisfy the operational
requirements of the design reference
mission.

In most systems, the system
specification is developed from the top
down to the provisions of MIL-STD-490.
It forms the top document in the design
control hierarchy, and all other
specifications for the documentation of
segment, subsystem, equipment, or
software requirements, are subordinate
to and must comply with the
requirements contained in the system
specification.

When the system specification has
been approved, the subsystem
requirements specifications can then be
developed to satisfy the allocated
functions from the system level, and the
subsystems can be developed.

<u>Application to the NAS</u>

The system requirements analysis
activity was compromised in the
development of the NAS system design to
incorporate subsystem requirements and
analyses that existed when the system
level analysis began.

The possible outcome of this
approach is that when the system is
integrated, some of the subsystems will
not perform at the levels required to
achieve the system level performance
requirements. Those subsystems will
either have to be upgraded, or the
system requirements will have to be
relaxed. In some instances, the system
requirements may be impractical to
achieve with current technology or with
available funding, and they should be
reduced. In other cases, safety
mandates that the subsystem
requirements be strengthened to achieve
the overall performance objective.

System Integration

Overview

In any system, it is important that subsystems interface properly to exchange data necessary for the system to meet its objectives. It may also be important that the system itself interface with other systems.

Interfaces are boundaries among subsystems, or between subsystems and facilities or other systems, at which these entities interact in some physical or functional way. When design and development of interfacing entities is concurrent, and is accomplished by different organizations, it is important to obtain early agreement among these organizations on the requirements at the interface and the responsibilities of the organizations in meeting a common interface.

These agreements are published in interface requirements documents (IRD) negotiated between the interfacing entity design organizations. Later, when the designs are implemented, these requirements are used to assure that the designs indeed provide the common interface. The implementation of these requirements are specified in interface control documents (ICD), also developed by the entity design organizations.

Application to the NAS

System integration and, in particular, interface management, is critical to the development of a system as complex as the NAS. For example, its radar systems must feed flight and weather data to the central computers that assist the air traffic controllers. Parallel development of upgrades to these individual systems required early identification and control of requirements at the interfaces.

The NAS must also interface and interoperate with elements of Department of Defense and National Weather Service systems. Thus, it is important to achieve early definition of the interfaces and the phasing of the interfacing subsystems so that the agencies can negotiate interface requirements and design implementations in time to avoid costly design changes.

The FAA/SEI team developed a comprehensive and system approach to this complex interface management problem. System-to-system interfaces within the NAS and between the NAS and external systems were identified through detailed analysis of the "end state" system design and the planned transition strategy. Individual system procurement packages were assessed for adequacy of the interface requirements. Comparison of these analyses provided a set of needed IRDs. The FAA/SEI team developed these IRDs through a vigorous process that ensured proper coordination between individual design organizations and proper compliance to the "end state" system requirements.

System Verification

Overview

The assurance that the many dollars invested in new hardware and software for a large system have yielded a system that meets the users needs cannot be left to chance. Thorough planning and analysis is required to develop requirements and methods of verifying that the functions and performance levels required in a system or subsystem specification have been achieved.

This verification can be accomplished many ways, and is controlled by the quality assurance provisions of the system specification, which contains a matrix that indicates how each functional requirement of the specification is to be verified. Acceptable methods of verification include test, analysis, inspection, and demonstration. The matrix also indicates whether a requirement is to be verified at the system level, or at a lower level. System level verification requires either full or partial system models or integration of all or some subsystems to prove distributed performance and cooperating functional capability.

It is difficult and costly to perform extensive system level testing of distributed systems, but the risk of fielding a system that does not meet its performance requirements can be significantly reduced by an intelligently structured system verification program that includes lower level verification of those requirements that can be allocated entirely to a subsystem or component, but requires system level verification of distributed performance. Such a program usually includes analysis, factory testing, and integrated subsystem testing at a system testbed or operational site.
(see Fig. 4)

Application to the NAS

As for any very large, deployed system, the entire NAS is not amenable to full scale tests, but since it replicates functionality at each major facility of kind, and the existing subsystems remain functioning until the replacement has been satisfactorily proven, this level of test is not required to gain a high confidence level that the system functions properly.

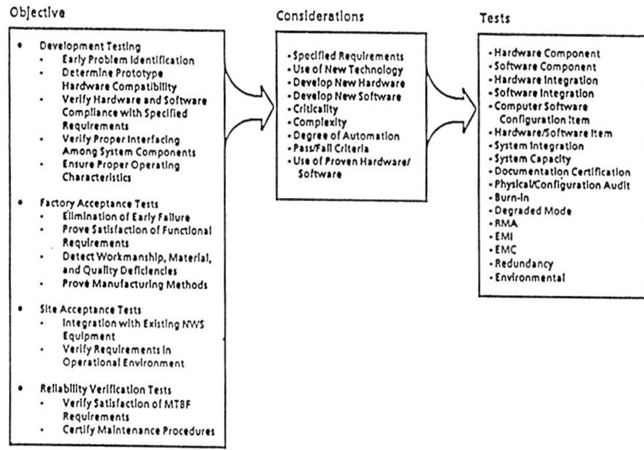

Fig. 4   Key Elements of Test and Verification

In the NAS verification program, unit and integrated subsystem tests are being performed at the development contractor facilities, and most subsystems undergo further testing integrated with other subsystems either at the FAA Technical Center or early operational sites to reduce the risk of not achieving system level performance requirements.

<u>System Deployment and Activation</u>

<u>Overview</u>

The real proof of the system design occurs when system elements are successfully deployed into the operating environment and the entire system plays together as intended.

When some of the system elements are deployed prior to the system design, field modification is usually necessary to incorporate additional requirements or interfaces not foreseen by the subsystem developer.

The phasing of subsystems may also need adjusting, or may cause need for additional hardware and/or software to correct for the fact that some subsystem schedules slip during development and planned interfaces may not be present when some elements are deployed. Otherwise, some subsystems will not function properly or may not be able to exchange data until the requisite interface arrives.

Facilities that have been engineered in advance of the system design may not easily accommodate all subsystem requirements or may not tolerate the grouping of subsystems required by the system design due to space, environmental, or electromagnetic conflicts. Facilities may not be standard, and the site variant differences may further complicate installation engineering of identical subsystems at different locations.

The national level deployment planning for a system establishes the sequence, schedules, and methods for the implementation of the entire system in its intended operational environment. This planning establishes the template for more detailed planning and implementation of the constituent elements and subsystems in the specific locations where they will operate. (see Fig. 5)

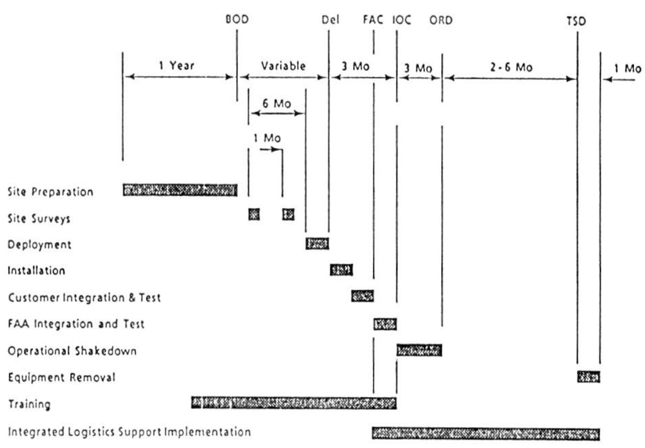

Fig. 5   Critical Transition Planning Milestones - Example of Analysis of Computer Replacement Program.

Where the same system elements and subsystems are deployed in geographically diverse areas, it is often prudent to provide a detailed level of planning for the implementation of these elements and subsystems on a regional basis, so many elements can be deployed simultaneously. This top down process ensures that the new equipment is being installed in a standard manner in the same locations in each facility, so that the overall NAS will have a high degree of standardization that minimizes training and support during the operational life of the system, where the cost is usually the greatest.

Transition Engineering proceeds from the system level planning. It defines the system configurations that exist at each stage of the implementation process and the requirements for each configuration in terms of added functionality, interfaces, or performance to accommodate the implementation. (see Fig. 6)

Installation and test engineering defines the specific tasks that must be performed to activate the system elements/subsystems in their facilities/environments, to integrate these into the system, and to verify by formal procedure and measurement that the system functionality and performance have been achieved at a specific site.

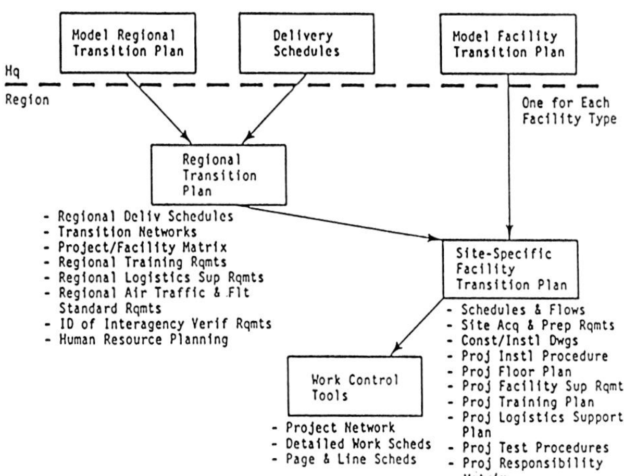

Fig. 6 Transition Design Process.

This activity includes the necessary facility modifications to accommodate the new subsystems, including electrical power, HVAC, grounding/shielding, and structural changes.

Application to the NAS

For the NAS modernization, new equipment is being deployed into thousands of facility sites across the nation. Nine FAA regions manage the operation and maintenance of the equipments in these facilities. In addition to a national level transition plan, a regional level transition or implementation plan has been developed for each of these regions that takes into account the exact numbers and facility locations of the installations and the regional resources required to support the installations.

Further, for each standard facility type, including Air Route Traffic Control Centers (ARTCC), Automated Flight Service Stations (AFSS), and Air Traffic Control Towers (ATCT), an implementation plan has been developed that can be tailored to the specifics of each individual facility.

The NAS transition was originally conceived to be accomplished in packages of subsystems that could be simultaneously installed in the field. This concept proved to be impractical because of subsystem phasing and different production rates, and a different approach was adopted.

The current approach lays out activity networks for each subsystem to be deployed and a standard order of installation for all regions and facilities. These networks are the basis of an integrated schedule developed in Martin Marietta's ARTEMIS-based program scheduling

system. This schedule phases each deployment activity to achieve early installation with the least operational impact.

To assure that current levels of service are maintained, the FAA/SEI team will model the performance of the existing and interim configurations by tracking field performance reports and specification requirements for existing and new subsystems to see if interim configurations provide more or less performance.

With this information, instances that threaten the existing service levels can be identified before they are implemented, and management actions can be initiated to preclude service degradation.

In addition, standard facility layouts are being developed, reserving common locations for each remaining and new subsystem to be installed in standard facility types throughout the country. These will be tailored at the regional and facility levels to account for variations from the standard designs.

This approach facilitates the eventual standardization of all NAS facilities for certain configuration aspects. Local variants caused by environmental factors including topography and climate will always exist.

Extensive modifications have already been made to the 20 major en route control centers throughout the nation to accommodate all the planned subsystem equipment and personnel required to operate and maintain it. Other facility designs have been modified, or new designs generated for the remaining locations.

New electrical power systems have been defined; raised floors and water cooling systems have been installed, and for some new installations, new sites have been developed.

Detailed drawings have been prepared indicating cable routing, power and grounding access, and office and administrative space.

System Engineering Management

Overview

The classical system engineering approach has been developed around several technical management disciplines that are assumed to be present at the outset of any system development. If these are not present, additional modification of the system engineering process is required.

This entire approach depends upon certain system engineering management concepts and disciplines relating to cost, technical, and schedule management for success. Among these are standards for design, development, documentation, central configuration management, testing, quality assurance, electromagnetic compatibility, reliability/maintainability/availability, human factors, integrated logistics support, interface management, and cost management. (see Table 2)

Table 2 Standardization in System Engineering Management

Necessary Standards	- Software - Electronics-Commercial Equipment - Drawing Practices - Microfilming - Design and Constuction - Power and Energy - Electronics - Design Reviews - Procurement Requests - Statement of Work - Lightning Protection - Transient Protection - Communications Protocol - Specifications - Spare Parts - Technical Documents - Interface Control - Test Plans and Procedures

Configuration management rigorously identifies the functional and physical configuration of the system hardware and software at any stage of development and implementation, and provides a formal process for managing changes to the configuration. (see Fig. 7)

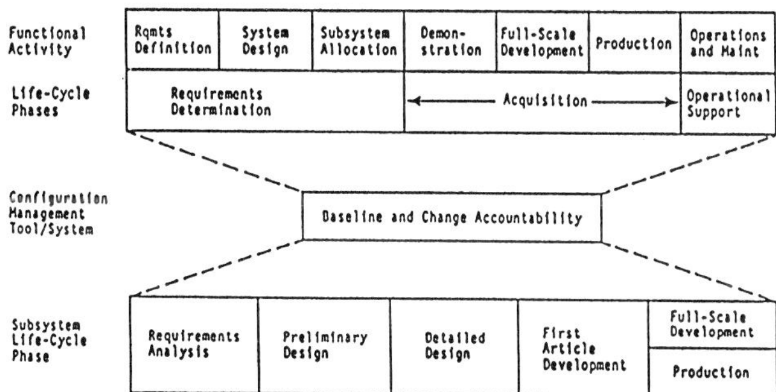

Fig. 7 Configuration Management.

Configuration management is classically applied at both the system and subsystem levels, for both hardware and software configuration items.

Cost management establishes formal identification of the cost aspects of development, deployment, operation and support of a system, and provides a formal process for managing changes to the cost of each of these phases.

The usual structure for collecting costs against the development of a system is the Work Breakdown Structure (WBS), which assigns numerical identity

to work packages associated with the system products that are to be developed.

For large system developments, a performance measurement system (PMS) allows tracking performance against cost and schedule targets.

Schedule Management provides rigorous identification of timing and phasing relationships of each activity necessary to the development and deployment of a system. It provides a formal process for controlling changes to the scheduled performance of each of these activities that requires program management approval.

Classically, a program management process requires several levels of schedules. The top program schedule depicts major program level milestones activities and some visibility into the program structure.

System level engineering and management activities are contained in program level schedules, while subsystem schedules are generally detailed in project level schedules and summarily reported in the program level schedules.

Application to the NAS

In the case of the NAS, some of these disciplines existed, but needed to be extended from the subsystem level to the NAS level of application. Others had to be jointly developed by the FAA/SEIC team at the same time the NAS system development was being accomplished.

The FAA had practiced configuration management since the 1960's, for the subsystems that had been developed and fielded. When only one subsystem at a time was being fielded or modified, this was a comparatively simple process. With the advent of the NAS Plan modernization, and the increased interconnectivity and interoperability of the subsystems, it became necessary to extend the process across the entire NAS.

A NAS configuration control board (CCB) hierarchy has now been established, and baselines have been approved for major design controlling documents such as the NAS operational requirements, Level I Design, and System Specification.

Subsystem specifications are being placed under configuration control as well and are being formally changed to meet the system level requirements via a Project Compliance Plan (PCP) process pioneered on this program. The air traffic control center floor space for operational areas has also been baselined.

Software patches to the national release of operational software are limited at each center, and the trend is toward establishing comprehensive national level control of all operational software.

All this has been accomplished in three years with the advent of a NAS configuration management policy and the necessary vehicles to carry it out.

The NAS is being developed and implemented using FAA funding for Facilities and Equipment (F&E). The NAS Plan for Facilities and Equipment is the top planning document used to formulate requests for this funding from the Congress.

A joint FAA/SEIC effort has established a baseline for the NAS cost that is being used to manage the entire program. Any change proposed to that cost baseline must be justified to and approved by the Deputy Associate Administrator for the NAS Program. This provides sufficient cost visibility across the entire NAS to make informed management decisions on tradeoffs among projects.

A schedule baseline for each of the NAS projects has been approved and is under configuration control. Any requested change to a schedule milestone requires approval by the appropriate level of project or program management. Key project milestones such as contract award, and first and last activations are controlled by the Associate Deputy Administrator for the NAS Program.

The NAS cost and schedule status/ control and configuration management are supported by automated processes provided under the SEI contract. Periodic computer-generated status reports provide the basis for effective NAS management.

<u>Summary</u>

The application of System Engineering to the modernization of an ongoing system that was not totally "system engineered" from the beginning requires a departure from the classifical approach. Exclusive use of top down analysis and design principles that would be practical in "new-start" systems must be tempered with "bottom up" constraints of the existing portions of the system. For each design issue, iterations must be made and tradeoff studies performed between the existing and the desired to convert solutions that may be technically optimal for a "clean slate" to the most cost-effective answer considering all of the technical, fiscal, programmatic, organizational, and political aspects of each issue.

The implementation of system engineering technical and management disciplines allows this to be done in a more orderly, reasoned, NAS system basis than would be realized by attempting to individually optimize individual subsystems.

The congressman who compared the NAS modernization to attempting to fix an automobile while it is speeding down the highway may have been correct when he forecast that the approach the FAA/SEI team is following in the NAS may well set precedents that will lead the way for similar upgrades to other systems in the future.

INTEGRATED MISSION OPERATIONS
LESSONS LEARNED

P. Piotrowski and H. Wanke

DFVLR - GSOC

Oberpfaffenhofen, Fed. Rep. Germany

ABSTRACT

The national space program of the Federal Republic of Germany is carried out
by the German Space Operations Centre
(GSOC) of DFVLR. In the execution of
these activities many valuable lessons
have been learned with regard to the
effective performance of all phases of
satellite operations.

This paper describes briefly the various activities involved addressing par-
ticulary experiences gained and showing
how these experiences will be incorporated into the data processing systems
to be implemented for future missions,
especially those in the Space Station
Era.

1. INTRODUCTION

The Federal Republic of Germany's space
program is planned and initiated by the
Federal Ministry of Research and Technology (BMFT). The program is executed by
the industry under the supervision of
ESA or DFVLR.

For national projects DFVLR provides
various services to BMFT, mainly project
management and planning, along with, in
many cases, the realisation of mission
operations.

DFVLR is an aerospace research institution and operates several technical facilities for both internal and external
users.

The German Space Operations Centre (GSOC)
consists of a ground station complex at
Weilheim and a Control Centre at
Oberpfaffenhofen, both near Munich.
The national program has up to date been
operated using these facilities.
While the responsibility for the entire
data processing systems and mission
operations lies within GSOC, additionally
research work in the fields of satellite
dynamics and orbital mechanics is done.

The national space program experienced
an intensive phase in the late 60's
and 70's with the successful executive of a series of scientific and experimental communications satellite
missions. These programmes were often
bilateral, together with the USA or
France.

Following the establishment of ESA in
the mid 70's the national program
came to a virtual standstill.

Since the beginning of the 80's the
national space program has received a
new impetus. Today the list of planned
projects is even more imposing than
during the 70's and provides an inter-
resting mix of project types, ranging
from cooperation activities with the
Indian government in the area of experimental scientific satellites
through the positioning and operation
of geostationary communications satellites up to manned space flight operations.

1.1 Past and planned projects

The following tables provide an overview of the missions which GSOC has
performed or are currently in preparation.

As the list of achieved and planned
activities shows, GSOC is dealing with
projects of various types, as

. Manned and unmanned

. Near earth and deep-space

. Scientific and commercial

and is cooperating with nearly all
major space agencies in a cooperative and/or commercial manner.

The necessity to cooperate with other
agencies is forced due to two major
factors:

SCIENTIFIC SATELLITES
- Mission responsibility within GSOC

| PROJECT | TIME FRAME | COOPERATION WITH | | | | |
| | | NASA | | CNES | ESA | ISRO |
		DSN	GSTDN			
AZUR	1969-1970		X			
AEROS-A	1972-1973		X			
AEROS-B	1974-1975		X			
AMPTE	1984-1986	X				
ROSAT	1989-1992	X				
HELIOS-A	1974-1986	X				
HELIOS-B	1976-1983	X				
GALILEO	1989-1995	X				

- Network support provided by GSOC

| PROJECT | TIME FRAME | COOPERATION WITH | | | | |
| | | NASA | | CNES | ESA | ISRO |
		DSN	GSTDN			
GIOTTO	1985-1986				X	
SROSS-1	1987					X
SROSS-2	1988-1989					X
IRS-1A	1987-1990					X

GEOSTATIONARY SATELLITES

| PROJECT | TIME FRAME | COOPERATION WITH | | |
| | | NASA | | CNES |
		DSN	GSTDN	
SYMPHONIE-A	1974-1984		X	X
SYMPHONIE-B	1975-1985		X	X
TV-SAT-1	1987-1994		X	X
TV-SAT-2	1989-1996		X	
TDF-1	1988-1995			X
KOPERNIKUS-1	1988-1995	X		
KOPERNIKUS-2	1989-1996	X		
EUTELSAT-1	1989	X		
EUTELSAT-2	1990	X		
EUTELSAT-3	1991	X		

MANNED MISSIONS

| PROJECT | TIME FRAME | COOPERATION WITH | |
		NASA/JSC	ESA
SPACELAB-1	1983	X	
D-1	1985	X	
D-2	1991	X	
COLUMBUS	1997	X	X

. Many programmes are carried out in
  bi-lateral or even multi-lateral
  manner.

. A trend towards international cooperation exists to minimize costs
  associated with ground support
  systems.

Therefore, intensive contacts are maintained with NASA, CNES and ESA relationships and of a cooperative nature
have been established with the space
agencies of several other countries.

1.2 MEANING OF INTEGRATED MISSION OPERATIONS

In this paper the word "integrated" is
used for integration at each of the
following levels.

. Integration of management activities
  in cooperative projects.

. Integration of the support of various
  space agencies in the performance of
  the operations.

. Integration of the various requirements on the data processing systems
  and provision of services for
  different types of users.

  .. scheduling and mission
     planning support

  .. housekeeping support

  .. experimenters services

  .. provision of access possibi-
     lities to data.

Each of these integration topics is
illustrated in the following section
using as examples 3 missions which have
been performed by GSOC.

2. TYPICAL SCENARIOS

Due to the fact that GSOC owns only one
Ground Station (with however several an-
tenna`s) and does not have any Launch
Facilities it has always been necessary
to cooperate with:

. A Launch Vehicle Agency

. An agency with Tracking Facilities
  to provide visibility when satellites cannot be observed/accessed
  from Weilheim.

. Other institutes for experiment
  data processing.

In the following, the experiences
gained in the execution of mission
operations, three different missions,
out of the many projects supported,
are presented with regard to the
`integration` aspect and the lessons
learned will be shown.

. The first example is the interplanetary mission HELIOS

. The second example is from
  missions having a high degree of
  integration. This is the French -
  German cooperation in the field of
  geostationary satellites for communications and direct broadcasting
  satellites.

. The third example is from the
  spacelab Missions, with the establishment of the first Payload
  Operations Control Center in Europe.

2.1 HELIOS OR THE INTEGRATION OF THE DEEP SPACE NETWORK

The interplanetary mission HELIOS was at
that time a gigantic challenge to the
German Space Operations Center. Two
missions, HELIOS-A and HELIOS-B were
executed successfully. The missions were
conducted in a coope- rative manner
between Germany and the United States of
America, with Ger- many providing the
space probes and the U.S.A. providing the
launch vehi- cles. Experimenters from
both countries contributed with scientific instruments. This had the following
consequences for operations:

. Launch operation was done by NASA

. Mission operations (with the exception of orbit determination)
  were performed by DFVLR.

The network for the mission consisted of the NASA Deep-space-network
and one deep-space dish implemented
in Germany. While for the first
launch (HELIOS-A) early mission operations were conducted by German personnel using U.S. control center facilities the second launch (HELIOS-B) was
operated from GSOC.

In order to reach the project
objectives, DFVLR implemented

. Its own deep-space facility

. Its own control center, consisting of all necessary hardware
  and software.

All these German facilities had to
be combined with the existing U.S.
facilities to an integrated network.
The fact, that German personnel operated the spacecrafts (HELIOS-A and
HELIOS-B) at two different control
centers with different stations
required a high degree of:

. Interoperability at the management and organisational levels.

. Interoperability at the technical level.

It is clear that at that time implementation in Germany benefitted
highly from the existence of standards and methods used within the
NASA JPL & DSN complex. At that time
integration for DFVLR meant the adoption of the following standards:

. Communication methods and protocols such as NASCOM

. Ground station standards as used
  by DSN

. Data formats and structures used
  by NASA-JPL.

At the start time of HELIOS operations
in 1974 and having essentially only
one interface to one other space
agency, the technique of adopting
existing standards and methods directly fron NASA was certainly adequate and suitable, bearing especially in mind that the mission duration was planned so as to not exceed
18 months and that later activities
could not be accurately forseen.

As is also goes with other successful
missions, HELIOS was not terminated
after the planned mission duration of
18 months but continued to perform very
very successfully after this date.

At that time it was fully realised
that the NASA-JPL system was a living
system, which meant in practice changes and upgrades to the system, largely imposed by new missions (more
important for NASA than the continuation of the HELIOS support during
its extended mission period) and the
need to replace old systems introduced changes in the communications
interface which DFVLR/GSOC.

This was GSOC's first lesson in integrated operations.

2.2 INTEGRATED OPERATIONS FOR GEO-STATIONARY SATELLITES

The positioning and control of geostationary satellites has been conducted within Germany for almost two
decades in cooperation with France:

. SYMPHONIE with two satellites
  operated successfully by Germany
  and France in the time frame
  1974 to 1985,

. TV-SAT / TDF-1 each a dedicated
  satellite for Germany or France
  to be launched and operational by
  the end of 1987 and beginning of
  1988 respectively.

These cooperations included

. Common use of ground stations

. Mutual operations back-up

. Commonalities in the data processing systems regarding

  .. data structures
  .. data exchange protocols
  .. software modules,

allowing the necessary cross-support.
The fact that for TV-SAT the positioning phase will be performed utilising the ground stations of four different space agencies, namely:

. NASA GSTDN network (26-meter)

. Ground stations of CNES in Kourou
  and Toulouse

. ESA-Stations in Malindi (Kenya)
  and
. DFVLR Station at Weilheim,

implies much work in the integration of
the networks in the areas of

. Telemetry data

. Radiometric data

. Command data exchange and adaption
  to the functionality of foreign networks.

This requirement for increased interoperability and the need to replace
the existing systems at GSOC, led in
1979 to the decision to implement a new
data processing system for the forthcoming projects in the scientific area
for both manned and unmanned missions
as well as in the commercial field for
geostationary satellites. This system
should replace.aced

The main idea behind the new concept
was to build up a system which would
allow GSOC to meet the challenges of
the 80`s was based on the experiences
acquired with the cooperation in the
previous decade.

The system design goals were:

. modular and highly flexible
  architecture

. all computers to be acquired from
  a single manufacturer

. high level programming language
  to be used throughout

. distributed processing concept

. use of standards wherever
  possible.

In order to allow a high degree of
flexibility and cooperation with
other agencies, so-called front-end
processors were introduced. The
main task of these front-end processors are:

. to convert protocols and formats
  of foreign agencies into the
  DFVLR-GSOC inhouse format and
  rice versa

In this manner the Control Centre software was made independent of the:

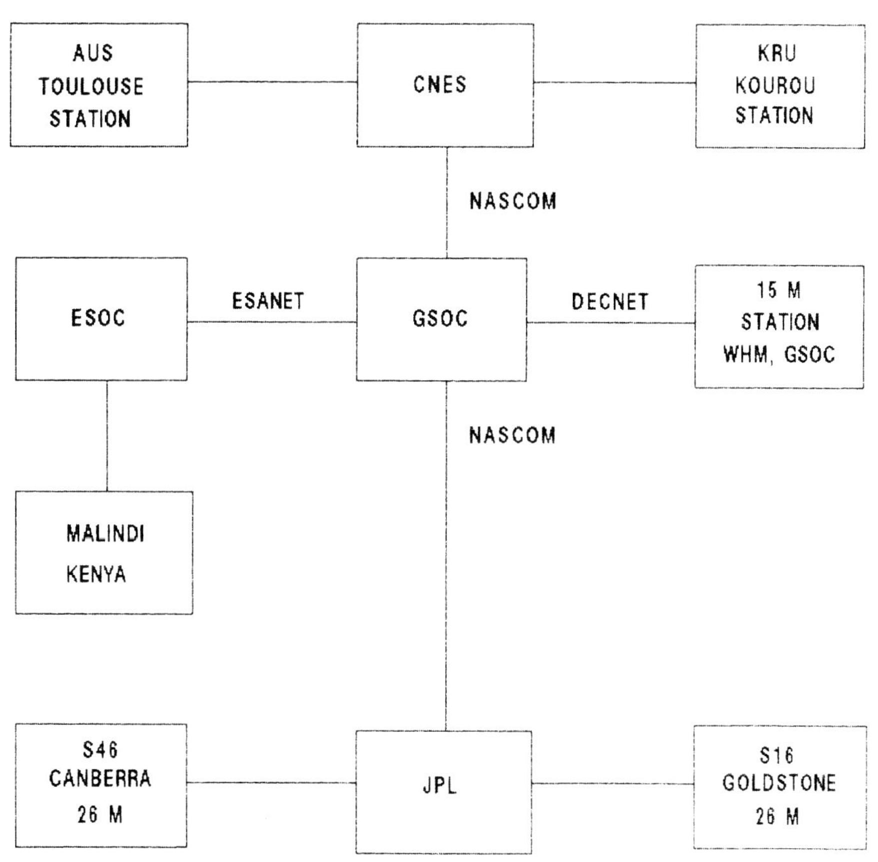

NETWORK CONFIGURATION TV-SAT / TDF-1 FOR SPACECRAFT POSITIONING

. Protocols for data exchange

. Data structures

. Data formats

of the other agencies. Using this technique, if a change was made to the data processing system of the foreign agency relevant for GSOC, only the front-end of the overall data processing system must be modified.

In order to cooperate with our partners, the following system conversions have been implemented in both directions.

. NASA-NASCOM - DFVLR-DATANET

. ESA X.25 - DFVLR-DATANET

. CNES-NASCOM - DFVLR-DATANET

. ISRO - DFVLR-DATANET

2.3 MANNED MISSION OPERATIONS

DFVLR as the first space agency in Europe to play an active role in the field of manned missions. The decision by the Germen government to built SPACELAB was logically lowed by the decision of DFVLR to implement Control Center facilities at Oberpfaffenhofen to conduct payload operations. At a first step, the mission with the first Spacelab payload served as a qualification for GSOC to conduct manned missions. However, this fact implied many changes to the typical sccnarios of mission operations with which GSOC was dealing. The characteristics of manned missions are.

. Intensive use of the Control Center facilities throughout the relatively short mission duration

. High data rates

. Accomodation of many scientists from different fields

These characteristics and the introduction of a payload Operations Control Centrer remote from the mission Control Center led to the following tasks:

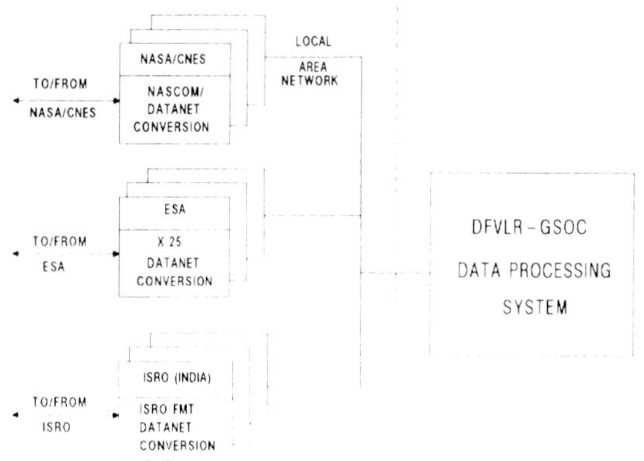

. Direct the overall payload operations system including the coordination and control of all payload and experiment related activities in many different locations.

. Development and operation of a communications system (JSC-KSC-GSFC-Experimenters) including 4 56KBit Datalines, and one video line.

. Development and operation of a data handling and processing system to serve 5 control rooms.

. Operation of voice communication between astronauts and scientists for guided payload operations and trouble-shooting.

. Analyze non-nominal experiment operations, adjust crew activity plans and change timelines by dedicated mission planning and scheduling computer programs.

. Coordination and provision of optimal payload data flow from SPACELAB to ground elements with extremely short coverage periods.

. Generation data products (tapes, prints, etc.) in near realtime.

. Provision of infrastructure for 150 operations and science representatives.

. Monitoring of the data acquisition and selection process at Houston.

. Provision of sufficient redundancy in the data processing system along with the neccessary back-up switching criteria and procedures.

For FSLP as well as during the D-1
Mission, the data interface for
telemetry data was complex.
GSOC had access to all telemetry
data at Houston to select there the
data relevant for the payload Operations Control Center and for real-
time operations. This was done with
its own computer system located at
at JSC to new telemetry data blocks
according to the NASCOM standards so
that the NASCOM services between
Houston and Goddard (effectively the
Gateway to Europe) could be utilised.

While this procedure was defined to
save communication costs and was indeed effective, the staffing problems
associated with maintaining communication equipment at another agencys
site have caused GSOC to decide not
to use this strategy for the forthcoming D-2 mission. GSOC is currently
aiming for a "cleaner" and operationally simpler interface; that means
that NASA should provide and multiplex selected data as are neccessary
for payload operations in Germany,
This scould be done at Houston JSC
or possibly other sites using NASA
facilities.

D-1 was effectively the middle stage
in a 3-stage integration process
where FSLP was the first stage in
which the remote POCC concept was
checked-out:

. With regard to the communication
  system.

. As to how much support could be
  provided for experimenters.

. Data processing system capabilities.

This first stage reqired integration
primarily from the communications
aspect although certain procedural
aspects had to be accommodated.

In the second phase of the integration in the execution of the D-1
mission the entire remote-POCC concept
was exercised. The only exception made
was to exclude some life-science experiments which had a very high data
rate or volume (cost ineffective) but
the major part of these will be incorporated into the D-2 mission.

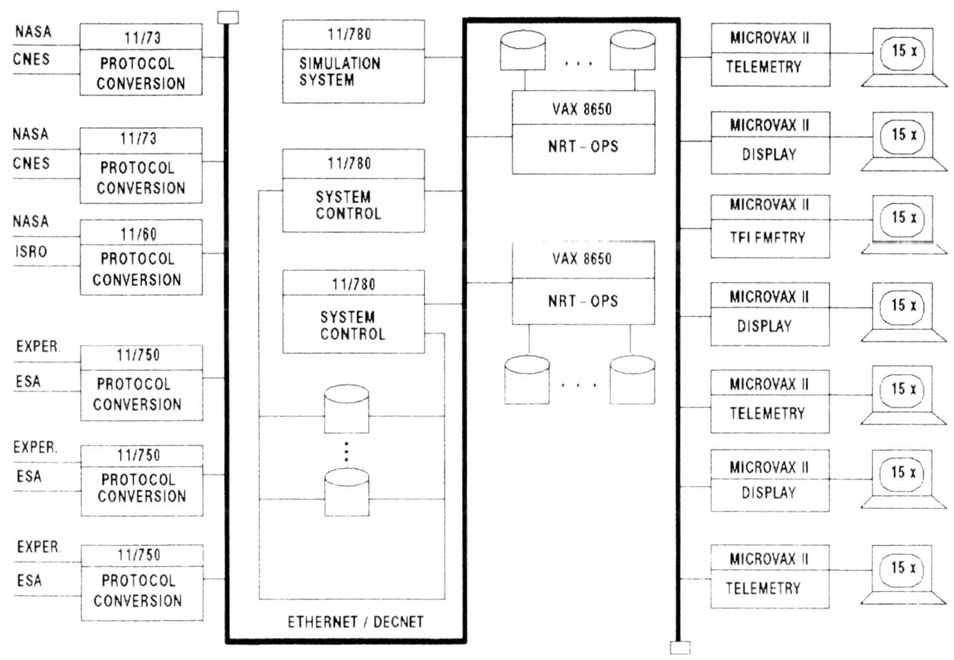

3. LESSONS LEARNED

The most important lessons learned from
the mission experiences sketched above
were:

. The necessity to provide a distributed processing capability

. The essentiality of a network to
   support this distributed processing capability in an orderly fashion.

. The need for standardisation in many
   aspects of the space business particularly with regard to cross support.

. A high degree of flexibility to accomodate the multiplicity of requirements and experimenter wishes associated with modern space operations.

3.1 NECESSITY TO PROVIDE A DISTRIBUTED
     PROCESSING CAPABILITY

Due to the often conflicting nature of
the priorities established by the various groups involved eg. Payload operations director, experimenters, etc.,
it is essential to provide a degree
of autonomy to any processing subsystem, allowing each function to be
independently without influencesari-
sing from other subsystems.

3.2 ESSENTIALITY OF A NETWORK TO
     SUPPORT THE DISTRIBUTED PROC-
     ESSING

Distribution of processing implies that
a high flexibility is required with regard to where specific tasks are to be
accomplished. Such flexibility may only
be provided if all processing is largely independent of its immediate environment and to provide this capability on today`s computers the technolo-
gies required are networks or clusters
of computers where peripheral equipment
and resources may be efficiently shared.

3.3 NEED FOR STANDARDISATION

In view of the high impact of maintaining effective gateways to the networks operated by other space agencies
we feel there is a need to standardise
this area in order to reduce the high
costs associated with cross support
in the area of

. Communication methods

. Communications protocols

. Provision of standardised services.

3.4 HIGH DEGREE OF FLEXIBILITY

To accomodate the multiplicity of requirements established by the mission
operations teams and to satisfy the
wishes of experimenters involved in the
evaluation of experiment behaviour in
both realtime and near-realtime, a system
allowing evolution and reconfiguration
using standard tools, is of paramount importance.

4. FUTURE DEVELOPMENTS

4.1 COLUMBUS

In COLUMBUS preliminary planning the
notions of distributed processing have
been captured and the following paragraphs give some idea of the impact
that the lessons learned have had on
the design of the data processing system.

4.1.1 DATA PROCESSING ARCHITECTURE

A data processing system architecture
is shown below. The architecture is
based on DEC`s VAX cluster. This confi-
guration provides an efficient network
implementation, and good system software
support for real time applications.

The architecture comprises a minimum of
three processors within the cluster
sharing the central filing. Two are
loosly coupled sharing mission workload
and providing `hot backup` for each other. The third processor would be uncoupled from the other two and would provide
software development facilities, system
testing and a cold backup for the other
two processors.

External communications are handled
using dedicated communications processor
controlled from the main cluster. These
processors control the input/output and
routing of data for the EPOCC and will
provide long and short term storage of
raw data. These are further described
under Communications Front End, below.

Workstations, terminals, low rate communications interface and any other internal low speed interfaces required for
the EPOCC are served via a Local Area
Network from the main processing cluster.
This is further described under the Main
Processing System, below.

4.1.2 COMMUNICATIONS FRONT END

The functions of the communications
front end are as follows:

. Receive data destined for the
  EUROPEAN Payload Operations
  Control Centre (EPOCC)
  archive it, copy it to short term
  log and route it internally within
  the EPOCC.

. Receive data destined for users,
  archive it, copy it to short term log
  and route it directly to the appropriate user Note that it will not enter
  the EPOCC processing system.

. Upon request from the main processing
  system retrieve data from the archive
  or shorter term log and route it as
  required.

. Output data originating from within
  the EPOCC to users and
  Mission Control Center (EMCC)
  as required by the main processing
  system.

The communications processors will be
simple devices. Their main task is to
provide protocol drivers for the communications lines, after that they will
carry out the archiving and routing functions under direction of the main processors.

The communications processors chosen for
the architecture can be configured to
control up to twenty one external lines
each one supported by a dedicated,
6 MIPS processors.

Three processors are used in order to
provide redundancy.

4.1.3 THE MAIN PROCESSING SYSTEM

The architecture has been based on existing equipment. It is too early to settle
on specific equipment now, however a
brief note of some of the main points
within the architecture will be useful
for later design.

The chhluster provides efficient networking between processing nodes and to the
filing system. Internode communications
in a distributed processing system can be
an expensive use of processing power, but
adopting a vendor's own tailored archi-
ticture, for the latter, rather than
`foreign` networks, is a more efficient
solution.

Three coupled processors two are proposed
for the cluster. In the loosely coupled
configuration the processors will share
the real time workload rather than running in parallel on the same data.
Machine goes down then the others will
take over. This has the advantage
in normal operation of maximising the
availability of processing power, one
machine is not `wasted` by duplicating
the work of the others. However, each
machine must be sized to take on the
entire real time workload if necessary.

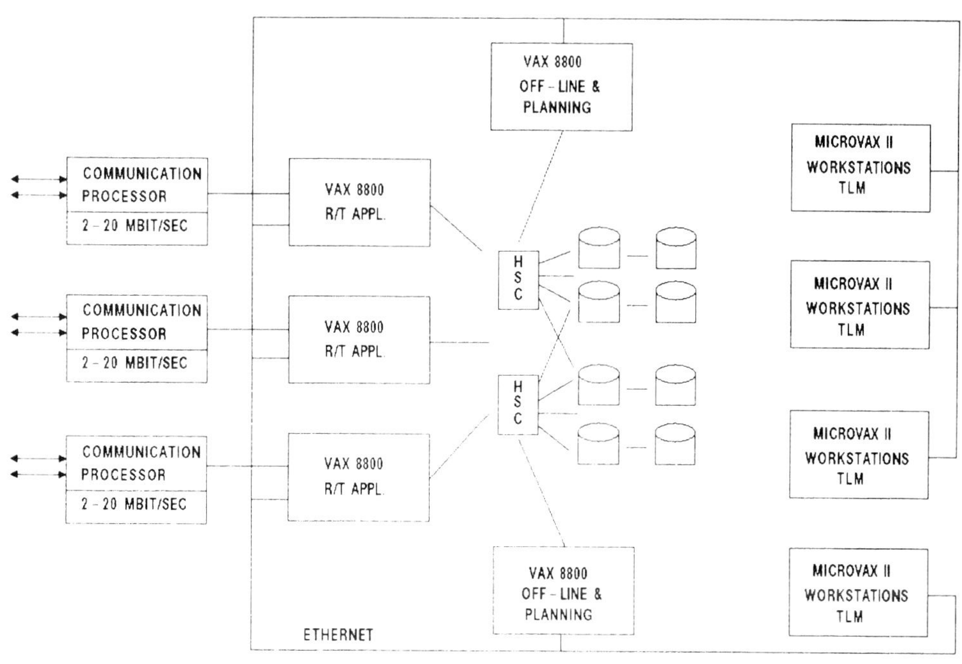

This is achieved as follows:

. The real time processing software
  will be duplicated in each of the
  loosely coupled processors.

. In normal operation the incoming
  data to be processed will be shared
  between the machines according to
  the load on each machine.

. If one machine goes down the whole
  of the incoming data stream is then
  directed to the other machines.

The uncoupled processors in the
cluster are primarily required for software maintenance and testing where they
must be isolated from the realtime
system.

Workstations and other low speed interfaces will communicate with the main
processing system via a Local Area Network. The only path to the central
filing system will be via the LAN and
then via software in the main processors.
Currently it is not possible to interface Micro VAXs directly to the main
processing cluster, however it is expected that this may be possible shortly. If this does become possible communication with the main processors
and also the filing system will be
very much more efficient. The `power`
of workstations using Micro VAXs would
then be significantly increased.

4.1.4 RELIABILITY AND REDUNDANCY

The architecture has been designed with
both reliability and redundancy in mind.

. The use of a three processors in
  the communications front end is
  forseen.

. Recovery from the main processor
  failure within the cluster has been
  taken into account.

. Communications within the cluster
  are provided by redundantly configured interfaces.

. The generalised software services
  will be designed to provide access
  control to prevent unauthorised
  memory access by a failed or fai-
  ling applications software task.

given to recovery from other failures
such as storage media, LAN, and support facilities, and to the results
of `external` events such as fire.

4.2 CCSDS

DFVLR has been a member of CCSDS since
its inception and is very keen to encourage the definition of standards
since this is in fact a very effective
from of integration. We have been contributing to the work in all panels and
feel that the work in these panels will
provide in the long-term a very strong
basis for standardised products and
services throughout the space community.

Each of the four panels of CCSDS provide
space operations oriented standards
which will encourage the provision of
standardised services and data products
by each space agency, allowing improved
inter-operability and achieving the integration so neccessary for mission
operations in the Space Station Era.

In conjunction with CCSDS we are working in Panel 3 in an effort to provide some degree of standardisation
in the are of cross support. This work
will obviously be motivated strongly
by our need to cooperate closely with
other agencies and by our realisation
that the costs of mission operations
intergration on a one-off project
basis are too high to allow this activity to continue in this inefficient manner.

EVOLUTION OF DATA MANAGEMENT
SYSTEMS FROM SPACELAB TO COLUMBUS

Guenther Brandt and Hans-J. Pospieszczyk
Project Manager Spacelab Mission D1/D-2 and
Columbus Systems Engineering Manager at
MBB/ERNO in Bremen, West-Germany

Abstract:

This paper describes the evolution of data
management systems, starting with the
generic Spacelab design, followed by its
utilization during the missions FSLP, D1,
D-2, it describes the EURECA System and
finally outlines the Columbus plans. It
discusses the experience gained in particular from Spacelab development and
mission preparation. An attempt is made to
formulate Columbus guidelines respectively.

1. S p a c e l a b

1.1 General

In June 1974 an industry consortium led by
ERNO won the contract for the design and
manufacture of the Spacelab, the European
contribution of the STS program and one of
the most important, reusable Orbiter payloads.

The Modular Spacelab concept did foresee
two principle flight configurations:

- a manned one with a habitable
  cylindrical pressure shell for
  accommodation of subsystems, scientific instruments, and up to three
  astronauts,

- an unmanned one with a pressurized
  container (so called igloo) for
  subsystem equipment and up to five
  pallets which expose instruments to
  the external space environment.

Spacelab required on orbit time was in
accordance with the nominal Orbiter
scientific missions duration of seven days
with growth provision up to 30 days.

1.2 Data Processing Concept

Different to satellite programs realized
by that time in Europe Spacelab had to
provide a multipurpose reusable space
laboratory satisfying a set of generic
requirements derived from a multitude of
experiment requirement analyses. Especially in respect to data processing no
specific experiment requirements did
exist.

Therefore a concept had to be found
providing a maximum of flexibility for
instruments interfacing the data processing resources.

Different to todays state-of-the-art a
computer with acceptable data processing
capabilities on one hand and being able to
satisfy the unique requirements of a
space-laboratory was not easy to find at
that time. The task was even further
complicated by the political requirement
to base the concept on European technology, by which the Orbiter computer (IBM
AP 101) was ruled out.

Out of several alternatives finally the
French Mitra 125 MS has been selected
which was based on a military aircraft
computer.

The Control and Data Management Subsystem
(CDMS) was divided into two assemblies,

a)   for subsystem data processing,

b)   for payload processing

with a maximum of commonality between the
two.

This split-up provided the feature to
configure and verify the subsystem portion
for a particular mission independent from
the payload integration process. The CDMS
blockdiagram is shown in Figure 1.

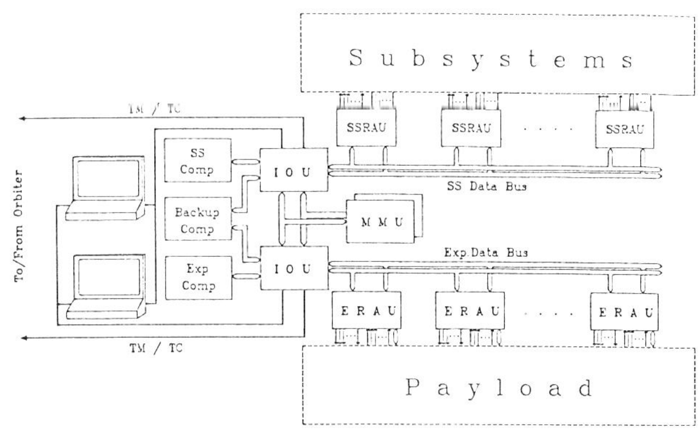

Figure 1:   Spacelab Control and Data
Management Subsystem (CDMS)

The central point of each CDMS assembly is a dedicated Input/Output Unit (IOU) managing the different data streams to/from the

- computer,
- Mass Memory Unit (MMU),
- Display/Keyboards (one located in the Orbiter Aft Flight Deck, the other one in the pressure shell for the manned configuration),
- Data Bus with Remote Acquisition Units (SSRAU = Subsystem Remote Acquisition Unit),
- Telemetry/Telecommand links via the RF-Equipment of the Orbiter.

The difference between the subsystem and payload section is to be found within the higher capabilities of the Experiment Data Bus and the Experiment Remote Acquisition Units (ERAU).

The major characteristics of the CDMS and its units is visualized in Figure 2.

Computer	Mass Memory
Word Length: 16 bits	Type : Tape
Memory Type: core	Capacity: 800 kwords
Capacity : 64 kwords	

Data Bus	Display/Keyboard
Clock Rate : 1 Mbps	Screen : CRT, 14 inch
Max. Length: 50 m	Type : 3 colours
No.of RAU's: up to 30	Keyboard: Full ASCII

Operat.System SW
Synchronous with overlays

Subsystem RAU	Experiment RAU
Inputs : 128	Inputs : 128
Outputs : 64	Outputs : 64
	Serial : 4 inputs
	4 outputs
Note:	Time : 4 Clock
Inputs configurable	4 Update
by SW (Analog or discrete)	

Figure 2:   Mayor CDMS Characteristics

In course of the various project Requirements and Design Reviews the selected concept was criticized for not providing adequate payload resources and many changes were implemented as the Spacelab design matured.

The most important one was the addition of a High Rate Multiplexer (HRM with a capability of up to 48 Mbps output rate and an associated tape recorder with a capability of up to 30 Mbps input/output rate and 20 min. recording time at highest speed.

In addition the operational requirements were changed drastically upon availability of first operational analyses and scenarios leading to extensive modifications of the software concept.

1.3 Design Responsibilities

ERNO divided the Spacelab System design into several subsystems allocated to so-called co-contractors which were fully responsible for the design and qualification of their subsystem under ERNO technical supervision and management.

The CDMS co-contractor MATRA in France placed several contracts with subcontractors responsible for CDMS unit design and development. The industrial organization is depicted in Figure 3.

Subsystem qualification took place at Matra with a worst case CDMS configuration and all interfaces to other subsystems simulated as far as necessary.

In accordance with the original model philosophy ERNO should have received the individual CDMS hardware and software items as well as other subsystem units and should have integrated them into the complete engineering model of Spacelab for the first time, and the second time into the flight of model Spacelab.

However, derived from first test results on subsystem level it turned out, that too many incompatibilities would show up late, in the project schedule if this original plan would have been maintained. Therefore an Electrical System Integration Model (ESI) was introduced into the program.It covered all units from CDMS and other subsystems which were considered critical in respect to interface compatibility and functional performance.

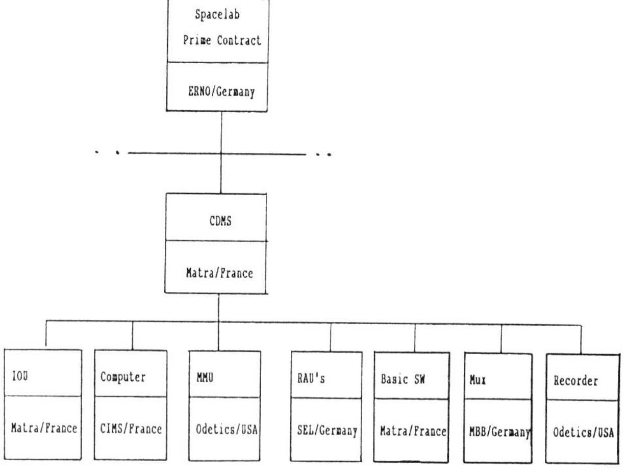

Figure 3:   Spacelab CDMS Industrial
            Organization

This decision was not made easily as the activities for manufacturing of additional hardware and testing had remarkable effects on cost and schedule without knowing if this effort would be compensated by the results received now earlier from the ESI, instead of waiting for the engineering model integration of Spacelab.

Looking back, however, it can be stated clearly, that the ESI test results achieved were extremely important as they did allow the necessary modifications on the units to be integrated into the engineering model without causing further delays. In addition the ESI turned out later on to be a valuable facility for trouble-shooting during engineering and flight model integration and even for assessment of some abnormal Spacelab flight events in Europe as all engineering hardware had to be delivered to NASA. For payload integration and testing the ESI is in usage as of today.

2. S p a c e l a b   U t i l i z a t i o n

2.1 First Spacelab Mission

Spacelab utilization started in Europe already in 1977 with preparation of the First Space Lab Payload (FSLP) which was the European complement of the NASA MSFC Spacelab mission 1 (SL 1). This demonstration mission was devoted to basic science out of various scientific disciplines such as material science, life science, earth observation, plasma physics, solar physics, etc.

One of the most important experiment interface is the one to the CDMS because it controls: the data transmission to ground and/or storage, the commanding capabilities by the on-board crew and/or ground personnel, and the on-line monitoring capability by the on-board crew and/or the experimenter on ground. Designs of most FLSP instruments required an extensive usage of the CDMS services. In other words experiment application tasks had to be executed by the Spacelab experiment computer, extensive display requirements had to be implemented on the central display unit and telemetry data of up to 32 Mbits/sec had to send to ground. Some instruments however had Dedicated Element Processors (DEP) with an communication link to the Spacelab computer.

FSLP was data wise a very complex payload which led to a difficult integration and verification sequence. The application software had to be developed and tested by a centralized team utilizing as simulation facility with a ground version of the Spacelab MITRA computer. The requirements or this application software had to be discussed and defined between the scientists and the programmers well in advance, the actual hardware/software testing

(instrument and application SW) upon completion of individual software package verification of individual software package verification. This approach extended the payload integration process and was cost intensive.

2.2 German Spacelab Mission D1

The German Spacelab Mission D1 was the first complete Spacelab mission under German mission management responsibility. The mission was dedicated to basic and applied research mainly utilizing effects resulting from absence of Earth gravity in the disciplines material research, biology and medicine.

Experience gained from FSLP led to a changed approach for the data system concept of the D1 payload configuration. Early in the process of mission definition the following groundrules were imposed on instrument design:

-   minimize instrument dependance upon Spacelab subsystems to become as autonomous as practical

and

-   no application software shall run in the Spacelab experiment computer.

Implementation of these groundrules, however, evolved to a great variety of instrument data management designs as shown in Fig. 4. In some cases no data interface at all was foreseen to the Spacelab computer, which constrained very much the on-line monitoring during the various test steps and during the mission itself. Another disadvantage of stringent interpretation of the a.m. groundrules was the resulting limited capability of programs and parameter sets change flexibility during the system test phase and during the mission. Only those instruments which had their programs and parameters on the central Spacelab memory were in the position to introduce changes easily during the mission.

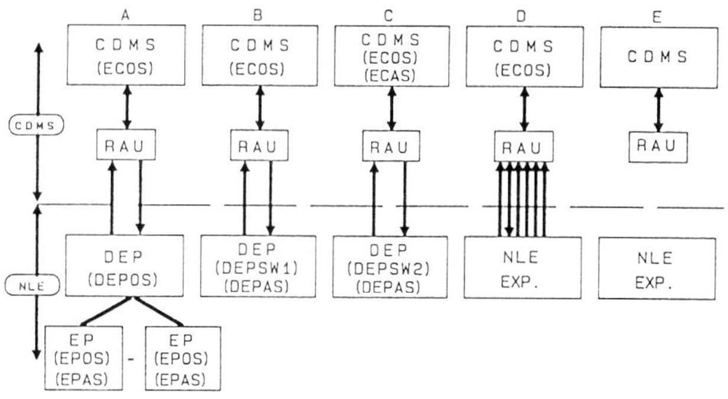

Figure 4:   D1 Payloadelement Data Management Interfaces to Spacelab System

The D1 test concept was structured such that it followed the mechanical step by step build up of the payload. The goal was:

- Performance and interface verification as much as possible on the lowest integration level possible

with

- a test environment as close to the flight as possible.

The latter means, usage of flight type software and flight representative hardware for test purposes to the greatest extent practical and feasible. By this means the risk of a malfunction during the final payload verification process prior to launch at KSC was minimized, and by the way proven to be valid.

Figure 5 depicts the D1 integration and test sequence. In addition to the steps shown in Figure 5 software tests with an instrument breadboard model and a payload system software model were performed with the test payload facility at MBB/ERNO to verify early in the project schedule software interface design. This approach minimized the software problems during integrated hardware/software testing. Figure 6 depicts a typical configuration of an instrument breadboard model test.

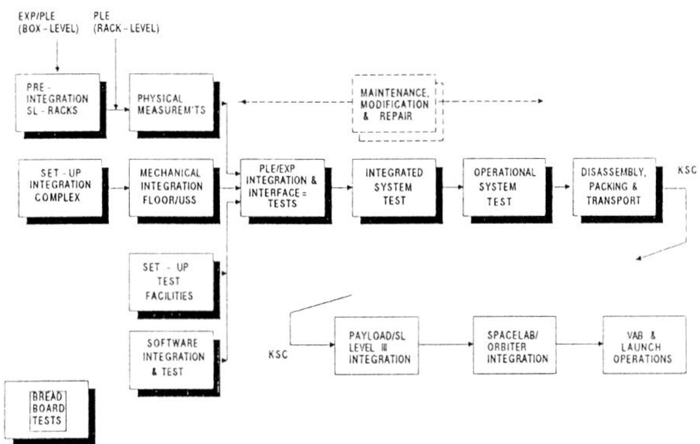

Figure 5:   D1 Integration & Test Sequence

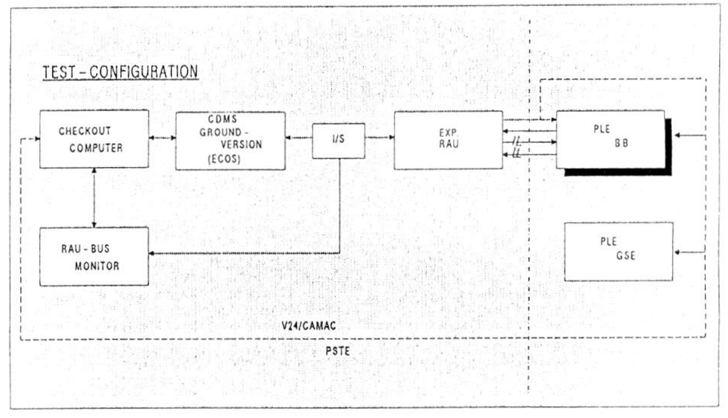

Figure 6:   Typical Instrument Breadboard
             Test Configuration

2.3 German Spacelab Mission D-2

The German mission D-2 will be the logical next step after D1 and will be again dedicated to basic and applied research. The payload composition will be different from D1 however some of the D1 instruments will fly again. The D-2 mission is in the conceptual design phase right now, the payload configuration is not finally selected. However as far as the data management design is concerned an approach will be implemented which will take both lessons learned, FSLP and D1, into account. The now valid guidelines can be defined in the following way:

- experiment application tasks shall be executed in decentralized processors on the lowest level possible.

- instrument upstream to the Spacelab computer system the standard provided interface services shall be utilized to guarantee

  1. full transparency of experiment status to the on-board crew and ground personnel, and

  2. easy change of programs and parameter sets during test and mission.

- downstream from a dedicated element processor to individual task controllers (microprocessors) a standard serial interface shall be used (RS232) to ease software development and test and checkout.

- decentralized display capability shall be foreseen were required to ease operations. The Spacelab display shall be the backup.

A new aspect within the D-2 payload will be the image processing or video data management. Video data was available in previous Spacelab mission also, however with the CCD-technologie as of today the capabilities of installing CCD sensors at multiple locations can be extended by an order of magnitude. Within the D-2 payload several CCD cameras will be integrated within instruments to monitor experimental processes and will provide on-line the images to the scientists on ground, even in 3-D technique. The first step towards telescience will be made. MBB/ERNO is in the process of developing a CCD-camera family for STS and Spacelab application.

The D-2 test concept will follow very much the D1 approach taken, however stream-lining of individual steps and upgraded user services in terms of display and quick-look processing will be implemented.

2.4 Payload Design Responsibilities

Payload design integration and systems engineering is a complex task, since not all instrument developments are the responsibility of a sole contractor. They are funded by various agencies (ESA or NASA) and/or national authorities and designed by many independent companies, in some case by the institute of a particular scientist. The only way to impose system level requirements are specifications, handbooks and guidelines defined by the system integrator and be made binding by the mission manager. Very often facts like available hardware flown on other missions, commercially available equipments etc. govern the real world and waivers and exceptions have to be granted throughout the mission preparation process.

3. E U R E C A

3.1 General

Beginning 1985 MBB/ERNO got the prime contract for the design and manufacture of the <u>Eu</u>ropean <u>Re</u>trievable <u>Ca</u>rrier (EURECA).

Eureca is a platform to be put into its initial orbit by the Orbiter from where it will boost autonomously into its operational orbit to perform for half a year microgravity experiments.

At the end of the mission it will descend to the rendezvous orbit where it will be picked-up by the Orbiter again to be brought back to earth.

3.2 Data Processing Concept

Different to Spacelab MBB/ERNO was given this time also the responsibility to integrate the first payload as well as the task to ensure that adequate generic resources are available for later payloads.

Due to the overall responsibility which included mission preparation and payload integration and due to the experiences gained during Spacelab utilization, a more decentralized payload data processing has been selected. The overall platform data management, however, is more centralized, that is to say subsystem tasks and payload tasks are executed by one computer. The configuration is shown in Figure 7.

It is based on the data bus system and the Remote Acquisition Units developed for Spacelab. The new computer is equipped with the integrated CPU 80C86 and relies on the standard INTEL computer architecture (equipped with a 128 kbytes memory) and RMX based operating software.

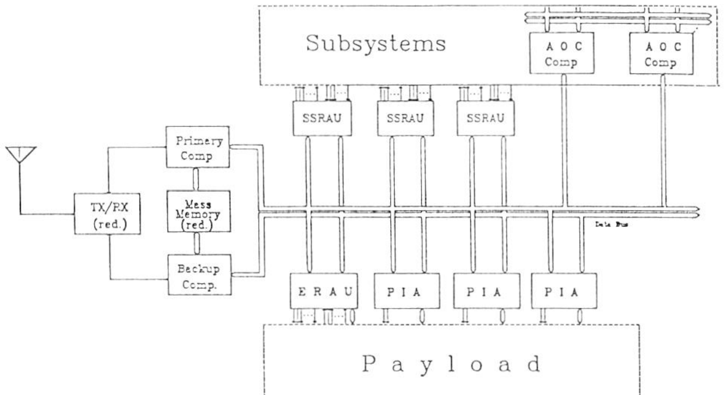

Figure 7:   EURECA Data Handling Subsystem Configuration

Already during Spacelab System development it became obvious, that the future trend will go clearly towards decentralized data processing and therefore the Spacelab contract had been extended to develop a further peripheral called "Payload Interface Adapter" (PIA) allowing a more elegant data exchange between the central system management computer and the payload computers via an IEEE 488 bus interface compared to the serial RAU channels.

For program and intermittent data storage a Bubble Mass Memory with internal redundancies and a capacity of 128 Mbits will be implemented.

The Data Handling Subsystem (DHS) will perform overall platform management (incl. failure detection, isolation and recovery) and payload initialization. The specific function of attitude and orbit control will be performed by a dedicated, cold redundant computer communicating with its peripherals via its own real time bus.

3.3 Design Responsibilities

The Spacelab C/D contract covered the development of the Spacelab hardware and software only and excluded payload design integration and detailed operational analyses covering the end-to-end utilization.

As mentioned before many changes had to be implemented during the C/D phase and work-around solutions had to be found during the Spacelab utilization phase.

Set up of the EURECA contractual structure is different to Spacelab. Contract responsibility covers also payload design integration, ground segment involvement and spacecraft operations. Also the continuous discussions with the investigators developing the instruments for the first Eureca flight allow to gain a high confidence on the design and therefore a "Protoflight" model philosophy. (Protoflight means: - functional qualification with the Flight Model) will be implemented. For early Software integration and verification an Eureca simulator is used.

267

4. Columbus

4.1 General

Since 1983 studies were performed (initially on German/Italian initiative only) to reuse Spacelab design concepts for future programs, especially in cooperation with NASA's manned space station scenario. The results of these studies led to the decision to make Columbus one of the most important European space programs for the next decade under the leadership of the European Space Agency (ESA).

Presently the phase B2 is running to specify the concept in such a detail that an industrial proposal covering the implementation phase C/D can be submitted in beginning 1988.

The current concept foresees three flight configurations:

- A pressurized, 4-segment laboratory (double in length of the long Spacelab module) permanently attached to the ISS, shall provide basic research capabilities under continuous supervision of astronauts

- A Man - Tended Free - Flyer (MTFF) shall perform microgravity experiments during its half year unmanned operation and then being serviced/reconfigured by astronauts at the ISS or by docking with a servicing vehicle (Orbiter or Hermes),

- A Polar Plat Form (PPF) shall perform earth research in an 900 km orbit and shall descend to a lower orbit to be serviced by a servicing vehicle every 4 years.

4.2 Data Processing Concept

Concerning payload data processing the situation is quite similar to Spacelab, that is to say no specific data processing requirements can be derived from the user side.

In addition there are other requirements and constraints which will influence the data processing concept quite strongly, such as:

- On-orbit repair and reconfiguration for new payloads,

- Infinite (30 years) life time, with adequate on-orbit grow-up and implementation of improved resources as new technology becomes available and is qualified for space application (e.g. artificial intelligence).

Analyzing the requirements related to the various flight configurations it turns out that the concept driving requirements are identical and can be satisfied by a common design ,which has to be extended by man - machine related functions for the manned configurations.

The principle concept of the Columbus Information Management System (CIMS) for the MTFF, which is the most complex one, is shown in Figure 8.

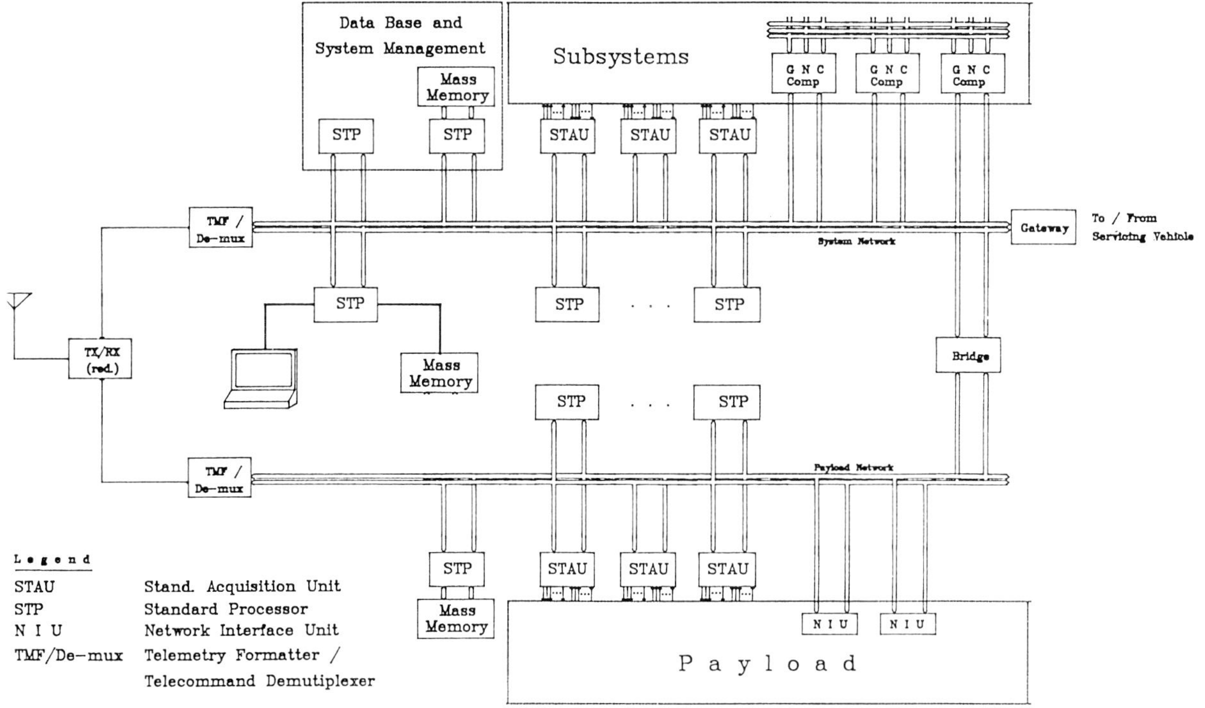

Figure 8:   Columbus Information Management System (CIMS)

A fully decentralized data processing concept is proposed with an open architecture of data processing nodes connected by a Local Area Network (LAN).

Data will be gathered and distributed by Standard Acquisition Units (STAU's) comparable with the previous Spacelab RAU's but with own local intelligence for preprocessing (limit checking etc.). The main data processing will occur in Standard Processors (STP's) which can be allocated to a single, so-called "intelligent" subsystem advantageous, that is to keep complex functions isolated from others and in addition to allow independent qualification.

For the Guidance Navigation and Control (GNC) computer a standard processor will be equipped with an interface for driving a real-time bus, which is necessary to fulfill the time homogeneity requirements for precise attitude control. For the two - failure tolerance requirement 3 GNC computers and 3 busses will be provided.

The overall system management (including data base management) will be performed by the system node.

For the manned configurations a Crew Work Station will be added with display/keyboard and voice recognition and synthesis for efficient communication between crew and the system.

4.3 Design Responsibilities

For the three flight configurations covered by the Columbus program commonality and standardization is to be implemented wherever advantageous for minimization of the overall mission-life-cycle costs.

Especially for the data processing hardware and software this principle is considered being applicable. Therefore a working group has been established composed of the flight responsibility system engineers of the particular configuration under the leadership of the anticipated Prime Contractor MBB/ERNO coming up with a.m. described CIMS concept.

Due to the key role of the data processing hardware and software for the timely development of the various flight configurations and the anticipated commonality to minimize operational costs, a special responsibility scheme has been defined by MBB/ERNO, which is the implementation of a Common Subsystem Contractor (COSCO) under one contract to the Prime Contractor MBB/ERNO. This role will be given to MATRA, who will control the design and qualification of the data processing hardware and software and will deliver it as "Prime Contractor Furnished Equipment" to the companies being in charge for a particular flight configuration.

The overall system model philosophy copes for one engineering model and one flight model per flight configuration.

As mentioned above for early design assessments, software integration and qualification a Columbus Simulation Facility (CSF) is foreseen, which will be available to all Columbus consortium members, and will contain in the final implementation phase real CIMS hardware, so that all software functions can be exercised by influencing the simulation models of the subsystems which could run on a host computer as software environment.

5. S u m m a r y

The past development of Spacelab and its utilization, the ongoing development of Eureca and the present design and specification phase of Columbus with their different design concepts and responsibility distributions allow to setup some ground rules, which should be considered when planning/defining space programs.

1. Data processing concepts:
    the development of special architectures, interfaces and protocols should be avoided and performed only, if existing commercially available solutions cannot be used. The architecture design shall be such that technology progress can be accommodated efficiently.

2. Model Philosophy
    The more and more complex data processing concepts need powerful simulations not only for early testing but for in-depth and efficient assessment and monitoring of the data processing system performance. A flight model philosophy on system level will not pay off over a long utilization period of the system, since engineering models and/or qualification models can be utilized later on as perfect test beds for payloads.

3. Responsibility Distribution:
    A design responsibility structure could be set up which corresponds mirrorwise to the data processing architecture. However care must be taken to keep directly related hardware and software developments in one hand.
    User- and operational requirements change system design very often as they mature. Distribution of design responsibilities shall take this into account.

MANAGING LARGE COMPLEX SYSTEMS IN THE FUTURE: SUPPORT SERVICES

LeRoy Allen
Boeing Computer Support Services, Inc.
Vienna, Virginia

Introduction

First of all, let me declare my bias. My work over the past 20 years has been mostly in the support services business. I have been involved in managing, operating and maintaining systems that other people built, so if my remarks betray a certain bias towards "developers," I want to assure you it is intended.

Now that I've shown my colors, I'd like to begin with a personal story. A few years ago my wife bought a Honda Civic. She bought the lowest priced model and has been very pleased with the car. But, last year, backing up under a tree, she bent the antenna. I took her car by the shop for them to repair it. They told me they didn't normally do that sort of thing and suggested I purchase a kit for $7.50 and repair it myself.

Understanding my limitations on such a project, I finally coaxed an estimate of $75.00 for doing the work if I failed. So I bought the kit. Frankly, I procrastinated for a few days, knowing that I would never be able to figure out how to get the antenna cable through the windshield post to the roof. But, finally I started.

Opening the plastic package I found the antenna, cable, appropriate fasteners, instructions and a 6-foot length of string. The instructions read as follows:
1. Unscrew antenna cable from rear of radio.
2. Tie string to cable.
3. Unscrew antenna fastener from roof.
4. Pull cable until string appears.
5. Tie string to end of new cable.
6. Pull string until cable is in position inside car and attatch.

Now, I suspect that the point of that story is as obvious to you as it is to me; indeed, I suspect many of you have had a similar experience. The story illustrates the business end of a very important principle: all companies or institutions make their money where the product or service meets the customer. I brag about that Honda because someone was making my life easier, and incidentally, a lot cheaper. When customers see that they are getting a problem solved or a need met at a reasonable cost, the business prospers. If not, then sooner or later, the business will fail. Simple as that.

Another way to look at it--one that brings us a little closer to my basic message to you today. The point of my little story is nothing more than a down-to-earth way of talking about the ultimate outcome of three critical phases that describe the life-cycle of any product, or service, or indeed, of any business.

There is, first of all, the development phase, which is short on time but with high leverage on life-cycle cost; then comes the implementation or production phase, which is crucial to determining the success of the endeavor. And, finally, there is the operational and maintenance phase, which has, relatively speaking, lower levels of units or absolute cost, but spreads them out over the longest possible time.

Now, you don't have to be a management genius to figure out the most important things about these phases: First, that it is vitally important to provide, and, secondly, that the customer is satisfied, or, even better, enthusiastic about the performance. Success comes down to the customer's view of "COST" and "QUALITY." And that brings me to a basic question: if a large, complex system is going to be operated and maintained in a way that satisfies both quality and cost objectives, then what do we have to do to accomplish that?

At Boeing we produce and manage extremely large, complex systems. We are well known for our airplanes, and I'm sure if you think about it, you are as grateful as I am for the quality in those systems. We build other aerospace systems for the government that have served the country well: the B-52, Minuteman, KC-135, the ACLM or cruise missile, and the AWACS system, to name a few.

Less well known than our airplanes and aerospace products is the fact that we operate and manage some complex support systems for Boeing and the government. I'm part of Boeing Computer Services which provides computing and communications services to the Boeing companies under a Cost Plus Incentive Fee contract which is over $750 million a year and employs over 6000 professional and operational personnel. We also provide similar services to the government in excess of $150 million per year. These projects involve the management of large complex computing and communication facilities for the Department of Energy, NASA and other agencies that provide a complete spectrum of services to multidisciplined users.

In managing these contracts, we at Boeing have learned to trust in a triad of elements that work together to create a powerful "field of force," which we have learned to count on to help us manage any large, complex system. These three elements, we have learned, are inseparable. You cannot have any one of them without the other two and expect to manage any system successfully. The three legs of this management stool are:

First, the prudent application of technology.

Second, disciplined management methods, with an emphasis on performance measures.

And third, the link that binds them together and makes them work--a commitment to total quality.

Technology

Let's begin with technology. I think technology is one of the most seductive elements in systems management today, because in the end, it promises more than it can deliver when it comes to operations and maintenance.

Let me explain what I mean by that. When you consider what happens in the three processes described by the three phases of the product or service cycle-- that is development, implementation and O&M--you see very quickly that technology is a more prominent feature of the development cycle than of the other two. New products and services, above all else, require new technologies and result from their application. In short, technology is the driver of the development part of the cycle.

Within the Boeing Company, for example, our biggest customer is the Boeing Airplane Company, and we are heavily involved in applying the new technology of artificial intelligence capabilities to both military and commercial aircraft. In modern aerial combat, for example, the ability of the pilot to make split-second decisions is the margin that separates victory from defeat. As the capability of weapons systems increases, the decision-making time for pilots decreases. And even though we can acquire real-time information about enemy aircraft and missiles, we are pressing the limit of the ability of even the best pilot to make effective use of that information.

Artificial intelligence techniques, particularly the use of expert systems, can expand the pilot's margin. Pilots equipped with the AI technology fo symbolic problem-solving techniques, that is, with expert systems, can prioritize threats instantly, determine the course of action accordingly. Of course, given enough time, the pilot would probably make the same decision--but the key fact is, that the time is NOT given.

The application of new technologies to these kinds of problems is essential to achieving some very laudable purposes, but they are not necessarily cost-effective-- and we should not allow ourselves to be lulled into thinking they are. My experience with my wife's Honda being the case in point. There is nothing inherent in AI technology at the moment, for example, that says by the time we get to the operations and maintenance phase, that that steep investment curve will have been worth climbing. In other words, technology alone is not enough; it is not self-justifying when it comes to systems management.

The fact that technology alone is not enough can be demonstrated by comparing Ford with other U.S. car manufacturers over the last few years as they have fought to recover their competitive position against the Japanese. Some of these manufacturers poured enormous amounts of money--hundreds of millions--into new technologies. Some of those technologies were essential to more up-to-date manufacturing. But some of them, unfortunately, were more of the gingerbread variety. A six-way power seat costs money, and many consumers considered it a poor trade, opting instead for better gas mileage and cars that would last longer--features they knew were readily achievable.

Ford, on the other hand, decided they could build better cars using less technology if they concentrated on something else. Naturally, they adopted and adapted some new manufacturing technologies. But rather than let their development cycle be driven by technology alone, they decided to put the concept of customer-driven Quality in the driver's seat. The result, as we all know, has been extremely gratifying to the people at the Ford Motor Company.

My point about technology is a simple one. It is essential in complex systems of the future, but if it is going to pay off, it has to be controlled by a set of ideas and principles that have equal application throughout the life cycle of processes that make up any system.

Management Discipline and the Use of Standards and Measures of Performance

A second way to improve management of complex systems in the O&M phase is through the use of quantitative standards and measures of performance. Typically, this happens in the implementation phase of a system's life, the time when the ideas of the development cycle are being infused into the product or service being offered --that is, during production, or installation, or both.

Here, the conventional wisdom goes, is where you begin to recover the costs of the development phase. You develop lists of evaluation criteria to test the quality of products, to drive down the costs of production processes, and to speed up scheduling, you introduce performance assessments from top to bottom, throughout the organization, and M.B.&O. become every employee's middle initials.

But come to find out, these very management methods, which have been taught for years in our best business schools, don't work. In fact, as Dr. Edwards Deming has been pointing out for some time now, these methods actually inhibit the very things we want to achieve.

Quality control measures are a prime example. As with motherhood and apple pie, no one is against quality control, but we need to be clear that managing outcomes by detecting defects is far too expensive and wasteful. Modern methods of production are too complex and variable for such a system to enable a company to remain competitive. Process feedback in manufacturing and customer feedback in product market testing are much more reliable.

Performance appraisal is another example--it can destroy teamwork, foster mediocrity, increases variability, and sacrifices long-term success on the altar of short-term results.

System managers are often judged on the basis of the results they can achieve now, or at the latest, tomorrow. But achieving short-term objectives is not the same as making long-range improvements. Too often, the improvements one manager makes during one quarter come back to haunt him or another manager who depends on his output in the next quarter, when, having received his terrific evaluation, he is told that obviously since he did so well with X resources, he can continue to improve, though perhaps not so much, with half the resources to work with. Do you think that manager will ever again be as open about how well he as done? You cannot improve a system at the expense of the people in it.

The problem here--both with the defect-management method of quality control and with the performance appraisal method of people control--is simply the wrong application of a good idea. Standards and measurements are too often applied at only one level--to manage operations--instead of bringing them to the level where they will do the most good--to manage for improvement in the system's processes. To do that, you have to focus on and create a total quality program which is the third and balancing leg of the stool.

A Total Quality Program

By now, the idea that we are in a new economic age being fashioned primarily by the Japanese has approximately the same news value as a headline reading "Seattle Slew Wins 1977 Kentucky Derby." There are all kinds of people who know what the Japanese are doing and understand it very well. But there are not so many who are running their lives and their operations as if what they know were true.

The old philosophy continues to prevail. That is the philosophy that says the way you manage is to set up the system, direct work through subordinates, make crisp, unambiguous assignments, set goals, and rate performance in accord with those targets. And, finally, what you learn from this exercise is that the more quality you want--whether in a product, a service, a process, or an organization--the more it is going to cost you. Whereas if you want to drive costs down, you produce more units. Production and quality are incompatible, says the old business manual. End of story.

What the Japanese have done, however, is simply to write a new story. What they have learned, and what they prove every day, is that higher quality costs less, not more. And when the clipboard brigade started investigating this and asked the line workers why it was that productivity went up as more attention was paid to quality, the answer came back in two words: "less waste."

When the emphasis is on improving processes instead of rejecting completed work, uniformity of product output is multiplied by the cost saved by fewer mistakes, wasted labor, less machine time, and fewer wasted inputs. Not only that, workers are happier, have a bigger stake in their jobs, and they improve the competitive position of the company.

Come to find out, the way it actually works is exactly the other way around. High quality drives the costs down, not up. In other words, you build quality into the processes and the system from the onset by creating opportunities to eliminate mistakes up front.

At Boeing, we have developed a total quality program, initially called "Operation Eagle" and now called "TQC," or "Total Quality Commitment." TQC is a structured process that continually improves quality, increases productivity, and achieves cost effectiveness throughout the life of a program, while at the same time providing an innovative environment.

But TQC is more than a way of increasing productivity, reducing costs, eliminating defects, or reducing inefficiency and rework. It focuses on the way all employees do their work every day, from senior management to the person in the factory. TQC works because it focuses on processes. It enables us to do what is necessary to meet our customer's needs by doing our job right the first time. And, by integrating total quality planning into everything we do, Boeing has achieved significant process improvement and has realized considerable cost savings.

As we applied these concepts across the company, we at Boeing are proud of the results of the first two full years of this commitment. The results show a tripling of the number of suggestions per employee and more than a doubling of the savings as a percent of cost from 3.5% to 8.1%. Although our employees' achievement of almost one suggestion for each employee is impressive, we still fall short of the average Japanese industrial worker who turns in 25 suggestions per year. However, our employees will take an increasing personal interest in the welfare of our companies when we give them an opportunity, an encouragement and a reason to do so.

My business is managing support services facilities for the government and Boeing. Then, why so much about this total quality concept when we allknow that it only applies to producing products? Well, I'm not convinced; these techniques apply equally to support services as they do to products!

We are aggressively pursuing that concept and have some results--but before I

describe examples, I'd like to relate the
three elements of the previously described
management stool to our support services
business:

Technology

The biggest leverage technology has in
large-scale systems management is in the
development phase. There, it produces a
system with the lowest O&M costs. But
lowering those costs requires that a formal feedback system be designed in. That
worker at Honda understood my concern
about an easy and inexpensive way for even
me to replace that antenna; he even provided the technology--a string. The continued infusion of technology into an ongoing operation is important. It can be
done if we listen to the people who are
doing the job and provide them the tools
and training to improve the system and the
product. The only caution here is that we
can become preoccupied with technology and
ignore the improvement of the process using the current technology.

Performance Measures and Standards

Quantified performance standards that
measure not just costs and schedules, but
the quality of the service a customer receives are key to making that customer
happy...and to making a profit. Such measures and standards are also required if a
feedback system is going to do what it's
supposed to, namely, to tell the developer
or designer what to change to make it better. Here's how it's done. We relate the
customer's requirements and his objectives to quantified measures and standards
of performance. These measures are assigned to the lowest possible level in the
organization and tracked. At that point
in the table of organization people can
really see if the organization is meeting
the customer's needs or not. Indeed, we
have learned that, on their own, people
participate in setting new standards, often suggesting that the current ones be
tightened to improve quality and lower
costs. And it is no accident that these
measures and standards are the same ones
that the customer uses to determine how
well we have met his needs.

A Commitment to Total Quality

As I said earlier, this commitment
provides balance for these elements. In
our support services projects our goal is
continuous improvements in the processes,
because we have learned that improved processes yield improved systems, and improved systems deliver better service. Inter-
disciplinary teams and training infrastructure make it possible for technology
to really improve processes and systems,
not merely embellish them. And that's
true when the technology is a string or a
robot. Eventually, as technology, performance standards, and the commitment to

quality join together to create a feedback
loop, they become self-reinforcing, and
the system becomes not only stronger but
self-renewing.

Managing Large Complex Systems In The Future: Support Services

If you have doubts about what I've
been saying, they probably take the form
of a question like, "Okay, but where can
you show me these principles have worked
and produced results in a service organization?" to answer that question, I want
to tell you one last story. Since people
are one of the primary drivers of quality
and cost in support services projects, the
measure of effectiveness is how much work
you are able to accomplish versus the number of people it requires to do the work.
In our contract with NASA at the Marshall
Space Flight Center, using these principles, we were able to improve the quality
of service by every measure while reducing
manpower by 14% over the five-year life of
the contract. TQC principles, including
the identification and resolution of problems, and statistical quality control,
contributed to this overall increase in
quality and productivity.

The heart of the system is a Management Information Center (MIC) with an automated information system. This system
contains project schedules, problem identification and resolution data, and other
performance standards, which keep both
Boeing people and the customer informed of
contract goals and progress. All personnel have access to the room. If I may be
permitted to boast just a little, during
the contract period, we registered some
very gratifying indications of quality and
productivity improvement:
--Reruns were reduced from 16% to 1%.
--Down Equipment decreased from 23 units
  per day to 1 unit per day.
--Response time improved form the 3.0 second range to 0.5 second range in 1986.
--We increase the effectiveness of Help
  Desk personnel in resolving initial user
  problem inquiries from 15% to 40%.
--Our data entry group doubled output and
  consistently meets 97% of its schedule.
--And finally, costs have been cumulative-
  ly underrun during the contract.
These results earned those people the recognition of being selected as one of the
seven finalists in NASA's Excellence Award
for Quality and Productivity.

That's what I mean by total quality!
By creating the system, that itself creates the possibility to make changes in
both product and services, we were able to
bring standards and technology together to
create a new atmosphere in our company.
This chart summarizes what I mean (Fig. 1).
An envelope of improvement comparisons for
six programs including the one I just described over a five-year time-frame begins
to border on fairly compelling evidence
(Fig. 2).

As we move toward a future of increasing complexity, that evidence will become, I believe, increasingly significant due in no small measure to the growth of both the number and size of large scale systems. But I am neither deterred nor disturbed by that challenge because we have already mapped out the royal road to managing those systems, and the first signpost along the road is quality. That I am convinced, will take us where we want to go.

Boeing Experience With TQC/Operation Eagle

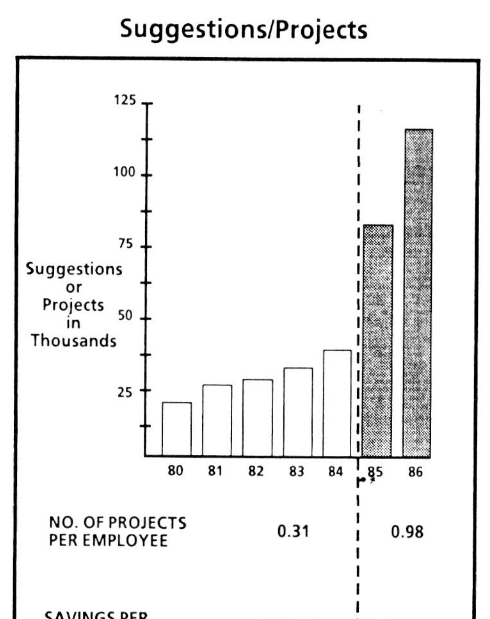

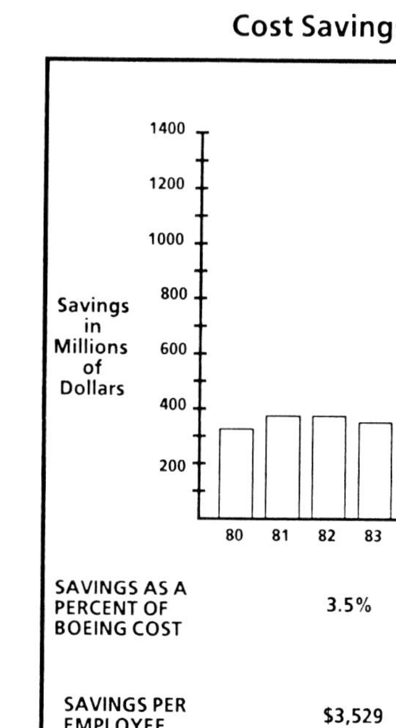

Figure 1

Improvement Comparisons
on Six Boeing Support Services Projects

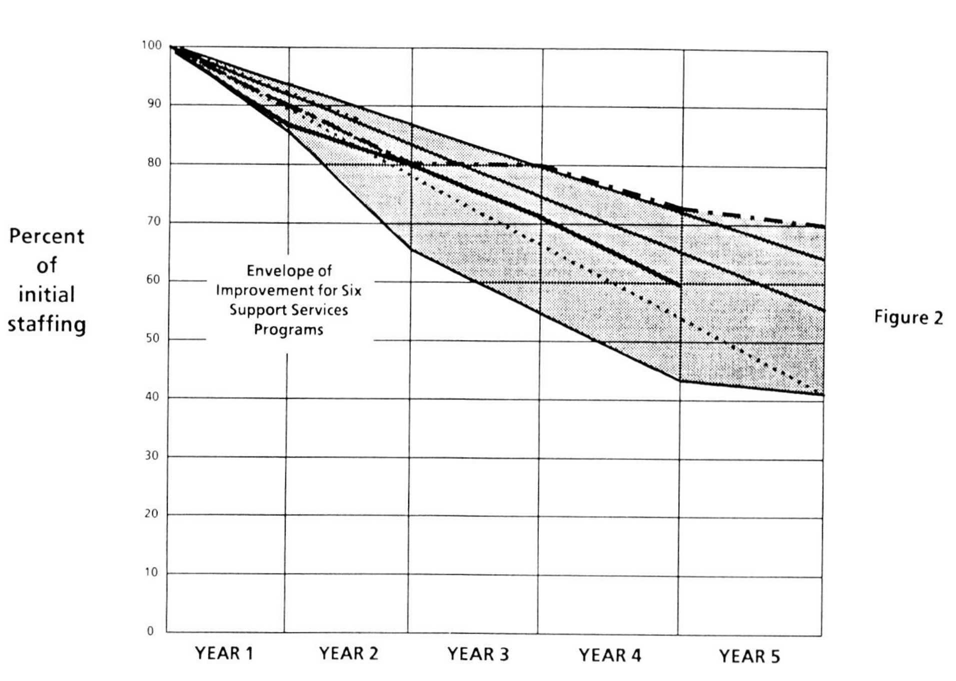

Figure 2

Space Information Systems Workshops

Workshop 1—*Chairman*: Michael J. Wiskerchen
Workshop 2—*Chairman*: C. E. Velez
Workshop 3—*Chairman*: Kurt Debatin

AIAA/NASA INTERNATIONAL SYMPOSIUM ON
SPACE INFORMATION SYSTEMS IN THE
SPACE STATION ERA

SPACE INFORMATION SYSTEMS WORKSHOP I

How can space information system
design best meet customer needs?

Michael J. Wiskerchen
Stanford University
Stanford, California

Abstract

The purpose of workshop 1 was primarily to
bring together key communications and information system (CIS) design engineers and
managers, system users, and technologists to
discuss the design and development process,
the management, and the operations of Space
Station era systems in terms of efficient
functional utilization and user productivity.
The necessity for the discussions was driven
by the decisive evolution of NASA responsibility from one dealing primarily with R & D
to one that concentrates most of its resources on operations and user services.
Workshop 1 has formulated a set of recommendations which deal with systems engineering and operational issues pertinent to the
development of Space Station Era information
systems which are functionally efficient and
productive. Although no specific reference is
made to international impacts in the recommendations, the panel assumed that most, if
not all, Space Station era information systems would be designed and developed as international projects and that CIS services
would be provided to an international user
community. The following key recommendations should be viewed as applicable to
both U.S. and international CIS projects.

**1. A NASA-wide System Engineer-
ing and Integration (SE & I) function must be established to coordinate all CIS activities across
organizational matrix elements
(Codes S, E, T, R). Each major program office (i.e. Space Station)
should also have an adequately
staffed and budgeted SE & I activity
for the CIS area.**

**2. NASA should take the appropri-
ate steps to insure that its systems engineering methodology is
structured to be responsive to the
operations and users services re-**

**quirements of Space Station era
information systems.**

**3. NASA must provide formal mech-
anisms for active and continuous
involvement of system users and
technologists in all phases of Space
Station CIS projects. These mechanisms should promote cooperative
working arrangements involving
personnel from the government, industry, and university sectors.**

**4. NASA must take steps to coordi-
nate and integrate the Space Station CIS testbed activities currently underway within Code S,
Code E, Code T, and Code R. Formal
mechanisms must be established to
incorporate, in a timely way, the
testbed results into the phase C/D
contractor efforts.**

Workshop Panel

Chairman: Michael J. Wiskerchen (Stanford
              University)
Recording Secretary: Allan Jaworski (Ford
              Aerospace & Communications Corp.)

Panel Members: Jack Garman (NASA), Richard
Costa (NASA), James Weiss (NASA), Erwin
Schmerling (NASA), Elaine Hansen (Univ. of
Colorado), Lee Holcomb (NASA), Barry Leiner
(USRA - RIACS), Larry Young (MIT), John
Dalton (NASA)

I.   Introduction

The operational mode of scientific endeavor
on Space Station must emulate the adaptive
science methodology used in terrestrial research laboratories. The iterative, trial and
error process, which is inherent to the scientific method, must be recognized and supported by the Space Station communications

and information system. The statements above reflect a functional requirement presented to Space Station by the Task Force for Scientific Uses of Space Station (TFSUSS). To address this issue, the TFSUSS has coined the term telescience, which they describe as the interactive acquisition of new scientific knowledge through remote observations and experiments. The Office of Space Science and Applications (OSSA) as part of the Science and Applications Information System (SAIS) activities has further defined telescience as follows:

Telescience is the direct, iterative and distributed interaction of users with their instruments, data bases, specimens and data handling facilities, especially where remote operations are essential.

The distributed interaction is meant to include all members of a user team, in space and on the ground, and may involve either manned or unmanned operations. It is the general desire of the user community to conduct their operations from their home institution by on-line computer networking. Telescience can be divided into three portions, centered on pre-flight, flight, and post-flight:

1. **Teledesign**

   The ability to send drawings, documents and specifications, to plan, manage, and coordinate science investigations among geographi-cally distributed investigators, to perform interactive design with remote facilities, and to conduct interface and other tests of instruments by remote computer access.

2. **Teleoperations**

   The ability to conduct remote operations by making rapid adjustments to instrumental parameters and experiment procedures in order to obtain optimum performance.

3. **Teleanalysis**

   The ability to access and merge data from distributed sources and to perform analyses and studies on computers that may be located at other institutions.

Although the telescience concept was formulated for the space science and applications community, it is applicable to all potential Space Station users (science, technology and commercial). Telescience, as defined above, clearly defines the functional needs of the Space Station user community but it does not identify the engineering methodology needed to implement it. You might also ask whether the functional needs expressed by the SAIS telescience definition are new or similar to ones expressed for previous space missions (i.e. free-flyers and Shuttle). The answer to this question is that telescience is a new name for functional requirements that have existed for every past space science mission flown, including Shuttle missions. Space Station era CIS projects must give these functional needs a high priority and and provide the means to translate them into system specifications such that they are included into the operational design. The workshop will probe and analyze this problem and attempt to formulate a constructive methodology which, for Space Station, insures that the Space Station CIS is functionally productive and operationally efficient.

The Workshop participants were asked to address a number of questions which would assist in assessing the problems and help to formulate recommendations for dealing with CIS design and development.

QUESTION 1:

Does NASA's linear system engineering approach (phased approach - phase A, B, C/D) lend itself to system designs which support efficient telescience operations?

Question 2:

How do we engineer telescience-oriented systems in the context of resource and communications limitations?

Question 3.

What impact will telescience have on existing systems and working methods?

Question 4.

What services and interfaces are needed and what impact does telescience have on their design?

Question 5.

How is telescience to be supported through all life-cycle phases of an experiment?

The above questions formed a framework for the discussions although often the discussion covered several questions simultaneously. The Workshop discussion, both from the panel and the audience, grouped itself into three distinct areas: Organizational issues, the needs/requirements/specification process, and the user/designer/contractor interaction. It is important to briefly summarize the discussion points in each of these areas and state the consensus recommendation that followed.

II. ORGANIZATIONAL ISSUES

It was generally recognized that NASA was structured in a matrix fashion with the various office codes taking responsibility for different segments of the CIS design and development. The responsibility matrix was seen as follows: Code T (OSTDS) has responsibility for all communications and tracking along with data flow management, Code E (OSSA) has responsibility for all science and applications payload unique CIS requirements, Code S (OSS) has responsibility for Space Station unique systems (core systems), and Code R (OAST) has responsibility for advanced CIS technology including advanced automation and robotics R & D. Many statements were made by the workshop participants that indicated that the NASA matrix structure often operated such that the matrix elements did not have the same coordinated mission objectives and goals, had independent budgetary priorities, and engineered CIS systems based on either poorly stated or misunderstood system requirements. This was seen, by the panel, as not being due to the matrix organizational structure but to the fact that an Agency-wide system engineering and integration (SE & I) function did not exist. OSS (Code S) has addressed this issue for the Space Station by establishing an SE & I function within the Space Station Program Office. Whether this office will give a high priority (adequate budget and personnel) to CIS activities still remains an open question. The panel concluded that that the matrix organizational structure was the optimum structure for dealing with Space Station era CIS tasks as long as an overall SE & I organizational function was established. Also, for the matrix elements to interact and coordinate adequately, each element must commit adequate resources with a high enough priority so that the resources will not be used to fund other urgent engineering problems. Specific recommendations concerning organizational issues are as follows:

RECOMMENDATIONS

1. NASA should establish an Agency-wide System Engineering and Integration organizational function to coordinate all CIS activities across the functional matrix elements. The newly established SE & I office within the Space Station Program Office should be adequately staffed and funded to carry out the CIS coordination function for Space Station.

2. All matrix element responsibilities should be divided along functional lines. Historical boundaries for responsibilities should be reassessed in terms of Space Station era system functions and service responsibilities.

3. Each major functional element should establish a single point of contact for coordination and ease of information flow across the system.

III. NEEDS/REQUIREMENTS/SPECIFICATION PROCESS

It was generally agreed by all of the workshop participants that the CIS needs - requirements - specification process was the most difficult to manage and be successful with. It was observed by the panel that most engineering systems have their designs driven by specific product delivery while most CIS designs are driven by required user services. This makes CIS activities considerably more challenging in their design, development and operation. It requires a close interaction between system designers/developers, the end users, and technologists during all program phases. Service requirements will evolve continuously as users become familiar with new technologies and capabilities. It must be recognized that compromises and tradeoffs on various aspects of the system will need to be made throughout the design and development phase and that the entire engineering - user - technologist team will be required to efficiently make these changes.

RECOMMENDATIONS

1. A mechanism must be established to actively involve users and technologists in all phases of CIS design, development and operations.

2. CIS engineering methodology must be structured such that it is responsive to user service requirements.

IV. USER/DESIGNER - TECHNOLOGIST - CONTRACTOR INTERACTIONS

Due to the dynamic and evolving nature of user services requirements and their ultimate impact on system design, development, and operations, new means need to established which allows an iterative assessment of new concepts and technologies throughout the life-cycle of a CIS project. To respond to this need, Codes S, E, T, and R have established Space Station CIS testbeds to test and evaluate new technologies and concepts. The Science and Applications Information System (SAIS) activity within Code E has the most detailed methodology for dealing with the evolving and dynamic nature of CIS requirements. SAIS has established a University Consortium for Space Station Telescience Testbeds which combines users, system engineers, and technologists into a team to rapid prototype new CIS concepts and technologies. Although university and government teaming is mandatory, industry participation is strongly encouraged and sought. The major shortcoming of the SAIS program is that it does not have a recognized, formal way to input its testbed results into the Space Station phase C/D contractor activities. All of the Space station CIS testbed activities suffer from a lack of a central focus and coordination. From these discussions came the following recommen-dations:

RECOMMENDATIONS

1. NASA must take the appropriate steps to coordinate the Space Station CIS testbed activities.

2. Formal mechanisms should be established such that CIS testbed

results can be input into the Space Station phase C/D activities in a timely manner.

3. NASA should establish mechanisms for universities, government, and industry to work cooperatively on rapid prototyping testbeds.

WORKSHOP 2: HOW CAN GROWTH AND NEW TECHNOLOGY BE ACCOMMODATED?

Chairman
Tom Velez
President
Computer Technology Associates
Englewood, Colorado

Recording Secretary
Thomas E. Ryan
Data Systems Study Manager
NASA/Goddard Space Flight Center
Greenbelt, Maryland

Introduction

The overall theme of Workshop 2 was established by the question: How can growth and new technology be accommodated? Specifically, this Workshop addressed the following five questions:

1. How is new technology introduced into operational systems?

2. How do we engineer systems to grow gracefully?

3. Can we predict advances in commercial/military technology?

4. Can we engineer hardware-transparent software?

5. How do we reconcile modularity/layering with efficiency?

The eight panelists for Workshop 2 were:

Chairman	Tom Velez - President, Computer Technology Associates, Inc.
Recording Secretary	Tom Ryan - Data Systems Study Manager, NASA/GSFC
Man-Machine Systems	Bill Rouse - Professor, Georgia Institute of Technology, and President, Search Technology, Inc.
Data Storage	Don Waddell - Senior Scientist, BDM Corp.
Software	Frank McGarry - Head, Systems Development Branch and Director of SEL, NASA/GSFC
Communications	Pat McClure - Technical Representative, Information Systems Division, JPL
NASA Standards	Ray Hartenstein - Head, Flight Data Systems Branch, NASA/GSFC
FAA Systems	Larry Culhane - Associate Technical Director, Air Transportation Systems, MITRE

Overview

The challenge to NASA and the aerospace community of introducing new technology into the Information Systems of the Space Station Era is clearly enormous. It was timely that a symposium and workshops were held at this time to address this and related questions.

Space Station itself represents a new market with a planned lifetime of 20 to 30 years. By 1995, it has been estimated that NASA will generate 4.5 x 10.13 bits per day to be captured, processed, archived, displayed, and analyzed. This data volume is equal to the capacity of all but the very largest Mass Store Systems in existence today. Since the data volume is expected to continue to increase at an accelerated rate, care must be taken to plan for this growth.

At the same time, technology is continuing to grow at an increasing, but not always linear, rate. Planning is required so that we can take advantage of the technology as it becomes available. One trap to be avoided is the belief that technology alone will solve our problems. Experience shows that technology indiscriminately applied results in managers, operators, and maintainers of information systems being overwhelmed by data volume, the number of operational modes, and the complexity of the systems. In an effort to solve the problem, more technology is sometimes infused into the system. The result of this technology spiral is often less productivity.

To avoid this trap, we need to understand the functional requirements and ask ourselves why we are building the Station and other advanced systems, and what we want them to do. NASA has been working these problems with the Space Station Operations Task Force and other study teams and needs to continue to ask itself these questions in today's ever-changing environment.

A major factor to be recognized is that requirements for the Station are evolutionary in nature. They will continue to change and, as past experience indicates, to grow until--and after--the Station is operational.

Evolving functional requirements will result from asking the five questions addressed in this Workshop. It is these functional requirements that are needed. Engineering requirements are derived from the functional requirements and will flow from them.

It is recognized that determining the requirements is not easy. But in an effort to do this, and to optimize the design, NASA must continue to have user involvement in all phases of the Space Station Information System (SSIS) program.

Several basic themes emerged a number of times during the Workshop. The essence of them was:

- SSIS is a very large, complex system with many interfaces. It is easy to miss the forest for the trees. NASA needs to continue to step back and ask, "What are we trying to build and what do we want it to do?"

- Don't try to do the SSIS job all at once. Break it up into doable packages that can be implemented as requirements are better known and technology is available.

- Have the design driven by consideration of how the system will support the Space and Ground personnel.

- Have the objectives drive the technology used, not vice versa.

- Plan for the SSIS that we ultimately hope to have.

- Look for and identify problem areas. Apply extra resources to those problems.

Recommendations

The following are the Recommendations from the Workshop:

<u>Provide a Common Execution Environment Throughout the Space Station Information System Architecture.</u>

Because of the large number of subsystems in the SSIS and the wide diversity of software developers involved, a common execution environment for all elements of the SSIS is mandatory.

A common execution environment will ensure that all functions available to one element of the SSIS will be accessible and available to all other elements of the SSIS. A strong driver for this is the need for commonality between the Space and Ground segments. Many of the functions, and much of the software, will start on the ground and migrate onboard later. A common execution environment will facilitate this migration.

The Space Station Software Support Environment (SSE) facility will provide the means to effectively develop software for SSIS. The SSE is a repository of tools available to all SSIS software developers. It is a technology that in itself will provide the mechanism for infusing/transferring new technology into the software development world for NASA. As new tools evolve from research, NASA will incorporate them into the SSE, thus allowing them to become widely used both in Government and in Industry.

Workshop participants agreed that NASA should continue to vigorously support the SSE. The SSE RFP does not specify that ground versions of the Station onboard computers are to be provided. Once these computers have been defined, it is recommended that they be provided in the SSE.

It is also recommended that NASA provide overall system engineering for the SSIS to include not only the Code S elements, but also Codes E and T subsystems. A focused end-to-end development management system is necessary to achieve operational status of the Station. It is important that overall system engineering be applied to the SSIS. If this is not done, it is possible that the individual elements could each be developed in such a way that the total system would be less than optimal.

A test bed facility should be established with the SSE for rapid prototyping of new requirements and software. It would provide prototype demonstrations and solutions to evolving operational requirements, and serve as the basis for the generation of formal requirements and hardware/software specifications.

<u>SSIS Should Be Designed to Meet Future Requirements.</u>

Because the requirements for the Station are not well-known, and because history shows that requirements tend to grow with time, it would be ideal if the SSIS could be designed with a projected 20-year life and a comfortable computational margin. SSIS architecture should be periodically analyzed in an effort to identify potential problems. When bottlenecks are foreseen, resources should be focused on the areas that will solve the problems.

For example, technology should be infused into the design and specification areas of software where the need is the greatest and the opportunity for productivity increases are large. In the past, technology was often applied to the coding of software simply because that was easiest to do.

It is important that up-front analysis be used to identify problems and that available resources be used to solve them.

282

In the best of all worlds, NASA would be planning to develop SSIS on three levels using a common architecture:

(A) What we are sure can be delivered.

(B) What we would like to have delivered.

(C) What we hope to have eventually.

This would provide a minimum risk system that can be periodically updated to handle changing requirements and take advantage of new technology. This approach is feasible, and requires thorough development and implementation planning. However:

1. The initial cost of such an approach may be higher than a "closed-end" system.

2. RFPs tend to be written based on poorly understood requirements. Contractors tend to treat them as unchanging fact and produce systems that meet the specification but do not perform the intended operational mission, nor provide room for growth. The contractor has no incentive to do otherwise. Is it possible that creative procurement policies could be developed that would change this?

3. It may not be practical to try to keep the same static architecture over the 20- to 30-year span involved. Can a dynamic architecture be developed that can periodically be revisited and updated as required?

Establish a "Software First" SSIS Architecture with the Family of Standards That Will Facilitate This.

Because software is the prime cost driver, it is recommended that it drive the SSIS architecture rather than having the hardware drive it. It is felt that a commitment to an Open (Public Bus) architecture system is required. A dynamic architecture that could be revisited periodically would be ideal to deal with new requirements and technology.

Commit to the Use of Ada for the SSIS.

Because Ada appears to be the most viable transportable high-level language, NASA should use it for the SSIS. In addition to the Code S elements, Codes E and T subsystems should be examined closely to see if Ada should be specified for all or some of them. It is further recommended that NASA institute a vigorous coordinated intercenter training effort for Ada.

Develop a Risk Mitigation Plan for Ada.

The state of Ada maturity is such that it may not be ready to support SSIS until the early to mid 1990s. A risk mitigation plan is needed to help ensure that Ada is ready to support SSIS in time. However, it is important that NASA have a commitment to Ada, and follows through appropriately. Experience indicates that if the "contingency plan" is too easy to use, people will not use Ada, but will go to older existing technologies.

Use Existing Opportunities to Test/Validate Technology.

Flights that occur prior to the Station, e.g., Space Lab and HST, should be used when possible to test and validate new technology such as Ada.

Since the SSIS job is so large and requirements not well known, it is important that existing opportunities be exploited as soon as possible to test and validate new technology.

Revisit the SSIS Architecture Definition Document.

The SSIS functional requirements should be revisited and the reference architecture revalidated or modified. A lessons-learned study of what went wrong on previous systems should be made. NASA should determine if a poll needs to be made of people who have previously flown payloads to solicit their recommendations.

On a continuing basis over time, NASA should ask the questions: Why are we building Space Station? and What do we want it to do?

Proposed Designs Should Be Analyzed Against Evolutionary Requirements.

Requirements on systems tend to change and evolve throughout the life of the systems. Proposed designs should be evaluated on this basis. A dynamic architecture that can be revisited periodically and updated based on changing requirements is needed. Test beds should be exploited for this purpose. Bottlenecks should be identified and resources applied to solve the problems. A priority structure should be put in place and the requirements projected as to how they might evolve over a 20-year period.

Exploit Modularity.

Because of the long life of SSIS and its great diversity, modularity is required to support efficient growth to meet changing requirements and new technology opportunities. A commitment should be made to ISO/OSI and IEEE Reference Models and to ANSI Standards as applicable so that modularity can be fully exploited. An interagency group could facilitate this process. A review should be made of existing IEEE standards to determine if they should be adopted for SSIS.

Maintain Commercial Compatibility and Use Existing Technology When Possible.

Commercial compatibility and existing technology should be used to reduce cost

whenever possible. For example, it was
mentioned that digital tape recorders
capable of record rates of 240 mbps are
presently available.

Develop a Meaningful Life Cycle Cost Model.

A Life Cycle Cost Model that takes into
account all elements of the SSIS should be
developed. This would include Codes E and
T elements in addition to Code S elements.
A goal would be to set better growth
boundaries and to be up front with projected SSIS evolution goals. Such items
as logistics, sustaining engineering,
operations, and maintenance need to be
considered in the Life Cycle Costs.

Investigate If an Appropriate Budget Cycle Is Possible.

Due to the lack of firm functional requirements for the SSIS, an iterative
approach is required. This would involve
Preliminary Requirements Reviews (PRR),
System Requirements Reviews (SRR), Preliminary Design Reviews (PDR), and Critical Design Reviews (CDR). At each of
these junctures the requirements/system
costs could change. Today's budget cycles
have difficulty dealing with this. NASA
may have to live with this constraint.
However, if it is possible to have a more
flexible approach to the budget process,
the Space Station program would benefit
from it. Innovative RFPs that require a
contractor to meet certain requirements
but reward him for features that provide
additional benefits to SSIS users could
reduce Life Cycle Cost on this long-term
program.

Provide for User Involvement During All Phases of the Acquisition.

In order to ensure that the SSIS evolve
creatively and iteratively, the users must
be involved in the requirements and design
phases and continue to be involved for the
life of the system.

Provide Commonality Between Space and Ground Systems.

Because the functions and software on SSIS
will, in many cases, migrate from the
Ground to the onboard Data Management
System, it is essential to have commonality between Space and Ground Systems.

Discussion on Theme Questions

The following is a summary of the discussion associated with each of the five
questions:

1. How is New Technology Introduced into
Operational Systems?

Perhaps we can learn from other agencies.
The Federal Aviation Administration (FAA)
has been developing, operating, and updating the Air Traffic Control system (ATC)

for over 50 years. Portions of the approach and methodology may be beneficial
to NASA in the Space Station Era. Some of
these are listed below.

(a) Detailed living plans are prepared
     for upgrades to operational sys-
     tems. Projected increases in the
     number of flights, radar, etc.,
     are factored into the planning.

(b) Development is staged into techni-
     cally and operationally doable
     packages.

(c) Provision is made for the major
     involvement of users.

(d) Incremental development/produc-
     tion/test milestones are estab-
     lished.

(e) Adequate engineering and opera-
     tional tests are conducted.

(f) Interfaces are carefully engi-
     neered.

The FAA's experience in upgrading existing
critical operational systems is potentially applicable to Space Station. In the
early 1980s, FAA developed a master plan
for updating the ATC system. For the En
Route Centers, FAA was able to reuse most
of the existing ATC applications software.
The old IBM 9020 computers were replaced
with IBM 3083 computers. By hosting the
old software on an instruction compatible
set of computers, 97% of the old software
could be used unchanged. The second phase
of the upgrade is to replace the air traffic controllers' displays with more modern
displays. This will be done only after
the new host computer systems are operational. The displays were chosen to present the same type of data to the controllers that the old system presented.
As will occur on Space Station, it is
important that the impact on operations
personnel be minimized. The final upgrade
scheduled for the mid to late 1990s will
add new computers and software and implement advanced automation functions. The
same display systems will be used with the
new computers.

The principles of long-range, multiyear
planning; the use of the system evolution
concept previously discussed; an iterative, staged approach; modularity; and
maximizing the reuse of software while
minimizing the impact on critical operational systems contain many lessons for
the Station.

A related example cited by one of the
panelists was a situation when six expert
systems were required to work together.
The approach taken was to have them all
"think alike." From an individual element, this was not optimum but it was
better from an overall system standpoint.
In a sense, this is a microcosm of the
Station. System engineering is required

to avoid optimization of the subsystems in
such a way that the overall system is
suboptimized.

A major software technology that NASA
should continue to commit to is the Space
Station Software Environment. It was felt
that the SSE should have the ground equivalent of the flight computer when these
computers are defined. It was felt that
the SSE should be the conduit for getting
software development technology to all
practitioners.

A recurring theme was that requirements
are continually iterated on any system and
a means must be provided on the Station to
manage this. Several recommendations were
that the users continue to stay involved
throughout the life of the system. In
addition, the requirements should be revisited and reviews conducted periodically.

It was agreed that it is difficult to keep
the same architecture for a long-life
system like the Station, particularly
since the requirements are not well-known
at this time. It was suggested that when
the Station requirements are periodically
revisited that the reference architecture
should be reviewed and revalidated.

2. How Do We Engineer Systems to Grow
Gracefully?

Several approaches were suggested:

  (a) By using distributed systems.

  (b) By the use of standards.

  (c) By the judicious choice of com-
       puters.

2(a). Distributed Systems

An effective way of introducing new technology into operational systems is by
using a distributed architecture. This
implies that overall systems engineering
practices have been applied. It is important that the interfaces be well thought
out and defined up front.

A distributed system offers a number of
advantages:

  (1) Each element of the system can
       grow at its own rate. As require-
       ments change, a specific part of
       the system can be updated without
       impacting the complete system.
       For example, if the quantity of
       data to be archived increases, the
       mass store unit can be changed out
       relatively easily if it is func-
       tionally separate from the rest of
       the system. Conversely, if there
       is a major breakthrough in optical
       disk technology, a different type
       mass store could be added with in-
       creased capability to handle fu-
       ture requirements.

  (2) New functions that were not origi-
       nally envisioned can be accommo-
       dated relatively easily by adding,
       for example, another processor.

  (3) Reliability can be increased. If
       there are several systems tied
       into the same bus they can back up
       one another.

Space Station, by its very nature, is a
"distributed system." It is distributed
between Centers, between many contractors,
and between the Space and Ground segments.
Many of the Data Management System functions that are initially done on the
ground will be migrated onboard as the
functions and their implementations are
better understood.

It was pointed out that there are limitations on distributed architecture. On
very large systems the problem of integrating all the elements of the system
together may outweigh the advantages of
distributed architecture.

2(b). Standards

Standards can be used effectively in defining interfaces and facilitating the use
of distributed systems. It is best to use
common Industry Standards for hardware,
software, and interfaces. Standards need
wide agreement and acceptance to be effective. The Consultative Committee for
Space Data Systems (CCSDS) which involves
NASA, the European Space Agency (ESA), and
Japan is an excellent forum for producing
effective, well-accepted standards. It
was emphasized that manufacturer-developed
proprietary standards and the in-house
"better" standards are to be avoided.

NASA has an excellent example of standard
systems and well-defined interfaces. The
Deep Space Network (DSN), the Tracking and
Data Relay Satellite System (TDRSS), and
the NASA Communication Network (NASCOM)
all have well-defined and publicized interfaces that users of those systems must
design to in order to use them.

2(c). Choice of Computers

Several ideas were put forth for the selection of ground system computers that
also have merit for flight hardware.

Use state-of-the-art computers. If we can
choose a computer that is new but somewhat
proven, it's likely to remain a viable
system for a longer period. Flight computers, as evidenced by NASA's NSSC-1 and
RCA's onboard computer for meterological
spacecraft, have remained in production
for a longer time than originally planned.
Two reasons for the longevity:

  (1) The NSSC-1 was designed with extra
       margin.

  (2) The RCA computer has evolved ex-
       tensively over the years, but

still retains its original architecture and instruction set.

A widely used family of computers can provide better future support and the possibility of upgrading to a bigger machine.

3. Can We Predict Advances in Commercial/Military Technology?

It was generally agreed that it is difficult to predict advances in technology beyond about 5 years. An exception to this may be software. Studies indicate that software systems have reasonably predictable rates of maturation. It was pointed out requirements will continue the push for more technology, and when the technology comes, it will be "needed."

4. Can We Engineer Hardware-Transparent Software?

For flight systems and real-time ground systems, it is difficult to engineer hardware-transparent software. Those computer systems tend to be designed around the strengths and weaknesses of the machine.

An ongoing effort was reported of trying to develop hardware-independent software for the A6 flight computer.

Transparency is enhanced by the use of transportable languages such as Ada and transportable operating systems such as UNIX.

In distributed systems, the problem can be fenced by changing out hardware and software together--hopefully in a small part of the system.

It was pointed out that it may be counterproductive to try too hard to engineer this transparency, particularly for ground systems. Changing requirements tend to outdate software as fast as the hardware becomes outdated.

It was also mentioned that it is productive to design when possible in such a way that hardware dependency is isolated.

5. How Do We Reconcile Modularity/Layering with Efficiency?

It was felt that modularity and layering do not significantly lower efficiency--the ratio is of the order of 2:1. For ground systems this isn't much of a penalty. Hardware is cheap compared to software and usually can more than offset this penalty. If the system is properly layered, less software will have to be thrown away. An example of this is the initial step of the FAA's en route computer upgrade where, on a system of 1 1/2 million lines of source code, 97% were used unchanged on the new host computer. It was pointed out, however, that particularly in the past, flight computers have not had the luxury of a 2:1 margin.

The point was made that modularity can extend to the hardware, software, operating system domain. If properly layered, each of these elements can grow independently at its own rate as the requirements dictate and the technology becomes available.

Chairman
Kurt Debatin
Head, Computer Department
European Space Agency
Darmstadt, West Germany

Recording Secretary
Macgregor S. Reid
Manager, Program Design Planning
Jet Propulsion Laboratory
Pasadena, California

Objectives

The overall theme of Workshop 3 was established by the question: How can information system interoperability be achieved? Workshop attendance was open to all interested parties, but system designers and implementers in particular were encouraged to participate. Specific objectives of the Workshop were to analyze the role of standards in simplifying operations or achieving interoperability; to identify relevant existing international standards that should be used; to define areas where new standards need to be developed; and to identify other elements (besides standards) needed to achieve interoperability. An additional, related objective was to analyze approaches for developing systems that are compatible with those of our international partners.

Approach

Participants from the European Space Agency and the NASA/Johnson Space Center each submitted position papers on interoperability (attachments 1 and 2). These papers were presented; the results were then summarized and points of agreement/disagreement noted. This process led to the identification of key issues, and the ensuing discussion resulted in the formulation of the Workshop Recommendations. Before these Recommendations could be developed, however, it was necessary for the participants to agree upon a number of definitions and technical approaches to interoperability.

Interoperability was defined as the capability to build an end-to-end system from parts, and to replace parts with functionally equivalent parts. This would allow a variety of tasks to be performed using interoperability, such as: (1) controlling a payload carried by another Agency's satellite, (2) acquiring data from another Agency's payload, (3) acquiring data from another Agency's archive, or (4) controlling a payload from another Agency's control center. The Workshop agreed that the major purposes of interoperability are to reduce cost and risk, to increase flexibility, and to promote cooperation.

Areas of interoperability were identified within Agencies as well as between them. For example, multisatellite support facilities and telescience opera-

tions (both data exchange and command services) are areas where interoperability may be valuable within a single Agency. Between agencies, a number of suggested areas of valuable interoperability were added to the general functions noted above, including shared software development, functional services (e.g., tracking), and satellite data relay. Several barriers to interoperability were noted and discussed, including problems of technology transfer, "up-front" costs, desire to utilize "own" designs, lack of enforcement, and a perceived low priority of interoperability vs. autonomy.

Next, the technical basis for interoperability was established. This was accomplished by defining sets of services that generally fit into either the upper three or lower four layers of the Open Systems Interconnection model (defined for Workshop purposes as Upper and Lower Layer Services, respectively), then discussing specific issues based upon developing a set of these services to achieve interoperability. The topics of discussion included whether a simple set of services (lower layer) would be sufficient to allow useful interoperability, or if a more comprehensive set including upper layer services would be necessary; whether upper layer services should be optional or mandatory; how to standardize on user interfaces and languages; how to provide standard computer hardware while meeting requirements for competition; and the extent to which Space Station Information System interoperability needs should drive the standards process.

During the discussion period, general agreement on a number of important points was reached, and these agreements formed the basis for the Recommendations.

Recommendations

The Workshop recognizes the fundamental importance of interoperability both in major projects of a single agency as well as in multiagency projects.

The Workshop believes that the following activities are necessary, and that appropriate interagency bodies need to be charged with each task:

1. A unique operations concept has to be defined and agreed for the Space Station and other individual programs.

This must take into account an end-to-end concept of each task requiring interoperability.

2. Agencies should jointly agree on interoperability requirements (derived from the operations concepts) in at least four technical areas, namely: user interface, applications, equipment, and communications.

3. Agencies should establish joint conformance testing for development and operations.

4. For communications/data standards, Agencies should

   (i)   Evaluate the CCSDS/ISO/CCITT/CCIR as a <u>basis</u> of interoperability

   (ii) Agree on and select subsets of lower layer standards to achieve a worldwide network; develop any missing standards.

   (iii) Define, adopt, and/or recommend standards for higher layers.

5. Agencies should use joint bodies to define interoperability standards for applications and equipment similar to the way the CCSDS is defining interoperability standards for user interfaces.

6. Agencies should prototype and validate in interagency test beds all newly developed interoperability standards before adopting these for program use.

The Workshop acknowledges that the Space station project is at present one of the major drivers for interoperability and recommends that the International Space Station partners initiate a joint study of interoperability in an urgent manner without waiting for the Space Station MOU signatures. The Workshop notes that the CCSDS Panel 3 has already begun a similar study, and suggests that the partners consider using this resource to examine Space Station interoperability.

INTEROPERABILITY: AN ANALYSIS
WITH EXAMPLES

M. Jones
M. Clendining
European Space Agency
Darmstadt, West Germany

Introduction

This paper makes an exploratory examination of the concept of 'interoperability.' The word 'interoperability' is a new
one, its provenance lying in the growth of
large space projects which because of
their size and cost necessitate the pool-
ing of resources of several Agencies. This
applies not only to the provision of prime
facilities, but also to backup facilities
which can be used at short notice in an
emergency. In this latter case it should
clearly be more cost effective to use
another Agency's facilities than for one
Agency to develop duplicate, backup, systems.

The paper proceeds to study the definition of interoperability in an iterative fashion. A preliminary definition is
given. Some practical examples of interoperability drawn from ESA's past experience are examined, proceeding from the
restricted domain of ESA to a wider domain
involving two Agencies. Specific areas
where interoperability will become important in the future are also examined.

Interoperability: Meaning and Scope

Interoperability is characterized by:

1. The capability to build end-to-end
systems from elements or subsystems provided from different sources (e.g. Agencies)

2. The capability to replace one
system/element in such a system by another
from a different source, the substituted
element providing the same services and
products or a known subset thereof as that
of the element which it replaced

Interoperability can reduce the total
effort and cost involved in developing and
operating an end-to-end system in a multi-
agency environment. However achievement of
interoperability requires 'up-front' investment of effort for:

- Agreement and common understanding
  of products and services provided
  by elements of a given type

- Agreement and definition of interfaces between elements

- Test/checkout of those interfaces;
  this involves checking that the
  elements work together correctly
  and in effect that each element
  provides its agreed services and
  products in the correct formats

It is obvious therefore that interoperability is desirable from a point of
view of cost effectiveness and in certain
cases may be essential since with 'interoperability' certain projects may remain
impossible. In addition interoperability:

- Increases data return and raises
  safety by removing 'zone of exclusion' periods when flight elements
  are out of contact with the ground
  network of a single agency

- Improves scheduling flexibility by
  enabling off-load of peak demands,
  e.g. for ground stations, relay
  satellites, operations centres
  (especially during launch and early
  orbit phases), etc.

- Allows real time 'multi-drop' of
  data to many users increasing scientific return

The meaning and usefulness of interoperability is thus clear. The principal
problem is how to achieve it.

In fact a surprising number of examples can be found of cooperation between
Agencies, most of which involve a certain
degree of interoperability. The following
two chapters will concentrate on some real
examples of interoperability covering
radically different domains. For each
example, following a description of the
scope of interoperability in each case,
the following are identified:

- Any conditions (such as common
  observance of standards, use of
  particular equipment type/manufac-
  ture, etc.) which eased interoperability

- Problems encountered in achieving
  interoperability

Interoperability Within an Agency

Satellite Checkout and In-Orbit Operations

Satellite Control. Satellite checkout
and in-orbit satellite control involve the
following broad areas of commonality:

- Acquisition of telemetry from a
  satellite

- Commanding the satellite and verifying correct receipt and execution
  of those commands

- Support of the operator interface
  associated with the foregoing functions

Satellite checkout and in-orbit satellite control would therefore appear to
be good candidates for interoperability.
Specifically, can one not reuse satellite
checkout equipment in the control centre
for in-orbit operations, or alternatively,
develop the operational system to a schedule which allows its use for satellite
checkout?

The matter has been studied extensively in ESA, one driver being the large
investments ESA makes in both checkout
equipment and control centre hardware and
software which make the possibility of
reuse of the checkout equipment in the
control centre attractive. To this end a
study was carried out in 1980-81 by ESA's
Board for Software Standardisation and
Control (BSSC). This confirmed that there
are indeed a number of common areas of
similar or even identical functionality.
However the study identified some significant areas of difference, which it divided
into two categories:

(i)   Functionalities existing in one area
       but not the other

(ii)   Implementation differences

The differences of type(i) result
from the different operational environments of pre-launch testing and in-orbit
control; they include for example: (1) the
need to carry out manoeuvres during operations, which may impose additional re-
sponse-time constraints; (2) larger volume
of data acquired and processed during
operations; and (3) greater reliability
needed during operations. On the other
hand Check-out needs more flexibility
since the level of readiness of the satellite varies during checkout and unforeseen
test sequences may have to be run at short
notice, whereas in operations, procedures
are normally fairly stable, there being a
strong emphasis on pre-planning, principally for reasons of safety.

On the other hand, implementation
differences are organisational in origin,
resulting from the fact that check-out
systems and spacecraft missions operations
systems are developed by different, independent, units within ESA. Consequently
different decisions were made on matters
such as computer types, peripherals (in
particular displays), basic software, computer languages.

The report recommended increasing the
level of commonality between check-out and
operations, in particular by avoiding
unnecessary disimilarities such as those
in category (ii). At the same time it
rejected the idea of modifying the then
current systems to achieve commonality,
since the investment would have been too
large in relation to the lifetimes of the
systems concerned.

Since the time of the study a number
of areas of commonality have been or are
being exploited thus:

- The work stations used to support
  the man-machine interface; sophisticated devices with keyboard input, multiple (three per workstation) screens with colour and
  graphics capabilities have been
  used in both applications, although
  in the past different types of
  devices were used in each. ESA now
  intends to use the same type of
  workstation for both checkout and
  satellite operations.

- Checkout involves, like spacecraft
  control, putting together a so-
  called satellite database (SDB)
  comprising in effect a description
  of the telemetry parameters under
  test. This description includes for
  each parameter its name, position
  in telemetry format, details of
  what its 'normal' range should be,
  and, where applicable, calibration
  tables to convert to engineering
  values. The checkout process will
  also result in corrections to this
  data base in the light of actual
  experience with the satellite.
  Checkout will also set up tables of
  command data. These tables form an
  extremely valuable input to the set
  up of the corresponding database
  for satellite operations; their
  availability in computer-compatible
  form avoids the need for tedious
  and error-prone manual entry of
  such data from satellite documentation and automatically carries over
  experience gained by checkout staff
  about actual satellite behaviour.
  This approach has been taken recently during development of the
  Hipparcos spacecraft control system
  at ESOC, a computerised SDB having
  been provided by the satellite
  Prime Contractor (MATRA). In practice difficulties are still encountered because of differences
  between checkout and operations
  infrastructures.

<u>Reuse of Payload Checkout Equipment for
In-orbit Operations</u>. The GIOTTO project
presents an interesting example of reuse
of Experiment Checkout Equipment (ECOE)
for 'quick look' science data processing
post-launch. This processing was directed
at checking of instrument health and performance: it had the greatest importance
during the payload commissioning phase
(in-orbit checkout). During the encounter
itself all operations were pre-defined.
The set-up which was used is shown in Figure 1, from which it can be seen that the
checkout equipment was in fact integrated
into the ESOC control centre facilities.
The checkout equipment itself is structured as an Overall Checkout Computer
(OCOE) to which are linked the 10 ECOEs
(one for each experiment). The OCOE acts

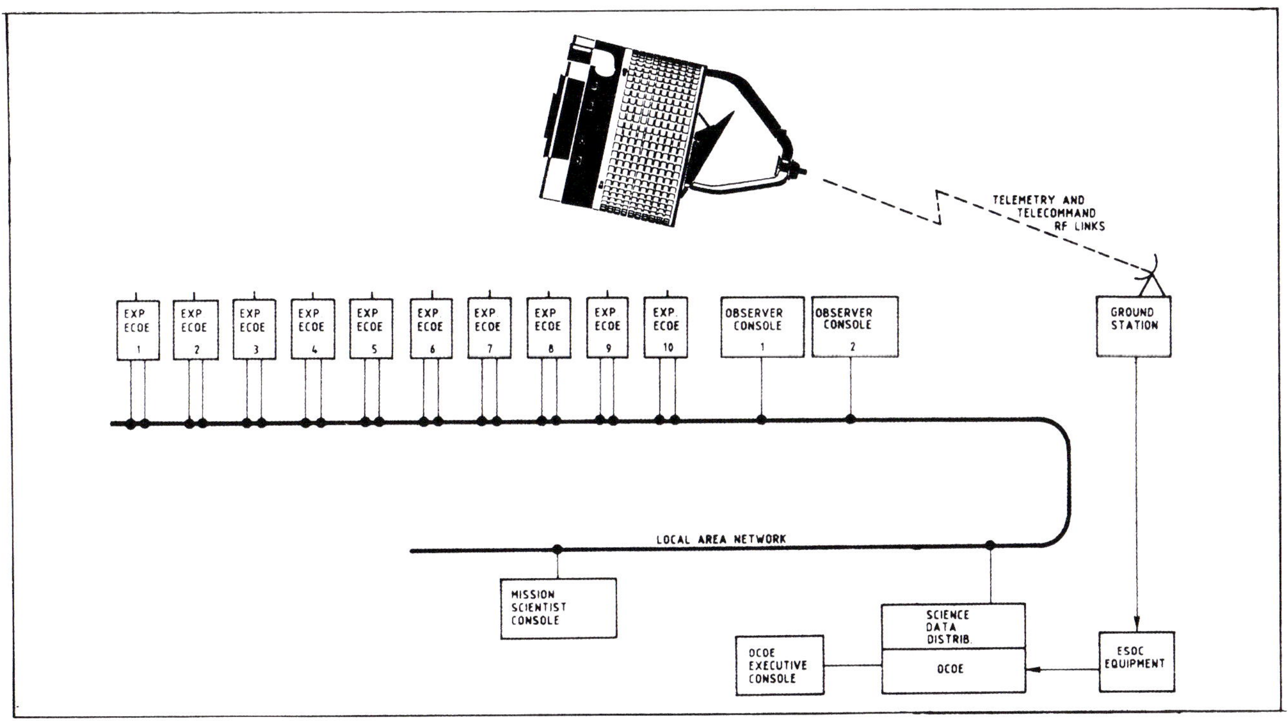

Fig. 1. Quick look science data processing configuration in the GIOTTO ground system

in effect as a data bus, acquiring telemetry data from the ESOC real-time computers [the Multi-Satellite Support System (MSSS)] and distributing it to the ECOEs. During GIOTTO's four hour encounter with comet Halley, on-line assessment of the data from the individual experiments ensured real-time availability of scientific results. It should be pointed out that this quick-look configuration had no direct capability for payload commanding and verification, which was handled centrally on the MSSS, together with determination of payload instrument status. In practice the science teams on the OCOE/ECOE in effect took an important responsibility for directing instrument commanding activity; thus during any OCOE/ECOE outage payload commanding and operation was generally forbidden.

The main lessons to be learnt from this experience are:

- The equipment (OCOE/ECOEs) was not available for installation at the Control Centre until after launch, since it was used for experiment check-out until quite shortly before launch. The installation, testing and integration of this equipment into the control centre was done during the four-month cruise phase. This was a potentially risky operation in that interface incompatibilities might have occurred during the integration which could have endangered overall system readiness. For a mission requiring commissioning and operation of the payload shortly after launch such a solution would only be possible if the checkout-supplied equipment were available prior to launch, e.g. by duplicating the equipment or advancing the payload development and testing schedule during science operations

- Incompatibilities between the MSSS and the OCOE/ECOE which were not removed because of the late delivery of the latter impacted overall system reliability and in fact caused the majority of outages

- Inevitably, because of the separate development of the MSSS and the quick-look system, some duplication of functions occurred which however were not implemented in the same way. Thus for example real-time telemetry which was distributed to experimenters (at their ECOE's) turned out (post encounter) to have had incorrect time-labelling although correct timing data was available on the MSSS. This shows the need for a careful separation of functions avoiding duplication where possible and ensuring that key auxiliary data such as timing is computed in only one place, thus avoiding inconsistently labelled products

- In view of the important influence which the experiment teams had over payload commanding, it would have been very helpful for them to have had the capability to produce experiment command sequences off-line on the OCOE/ECOE

Multi-satellite Support System (MSSS)

The MSSS developed at ESOC was specifically designed to accomplish the control of several spacecraft at the same time and to accommodate additional spacecraft as they were developed and launched with a minimum of new software and with no down-time of the operational system. This was achieved by the following:

- Standard (single) interface to the ground stations using a packet-switched network.

- Table driven telemetry definitions (such that the format of any spacecraft's telemetry can be defined by operational staff using database editing facilities, without any software changes).

- Table driven monitoring rules (as above, allowing definition of limits, expected status, alarm messages, mode-dependent checks, etc).

- Table driven command definitions (as above).

- Table driven display format (page) definitions allowing operators to define new page contents at will, without software change.

- Standard History filing facilities (CMD, TLM, etc.) allowing any operator access to all spacecraft history.

- Standard workstations which can be allocated to _any_ supported spacecraft. In particular, a greater number of workstations are allocated during launch compared to routine ops.

Maintaining the above principles the MSSS has been in operation since 1977, during which time it has supported many different spacecraft, including both scientific spacecraft (e.g. GEOS, EXOSAT and Giotto) and communications spacecraft (OTS and the MARECS series), LEOPS for METEOSAT spacecraft and one non-ESA spacecraft (GOES 2). During its lifetime the MSSS has undergone several major modernisations, starting its life as a distributed computer system based on CII10070 mainframes and Siemens 330 minicomputers to handle all time-critical tasks and evolving to its present form consisting of two SEL/Gould 32/67 computers running in warm standby.

It has also been possible to install reduced versions of MSSS in separate sites for support of particular missions (ESA stations at REDU and FUCINO). This has been made possible by use of compatible (i.e. SEL/GOULD) hardware, in conjunction with the ESA DPS25 network for station communications.

The main lessons learned from the MSSS experience are that:

- The idea of a table driven system capable of handling different satellites is viable

- The extent to which table-driven architecture is successful also depends on the extent to which the interfaces to the spacecraft are standardised; in practice some dedicated software had to be implemented for each spacecraft (or spacecraft family) to cope with individual spacecraft idiosyncrasies

- Multi-satellite support on a _single_ hardware configuration can lead to conflicts between different missions, particularly as concerns allocations of resources which are in general fixed (disk space, CPU power, I/O capacity), particularly if one of the missions is in a critical stage (e.g. LEOP)

Telescience

Telescience is understood to cover the real-time or near real-time distribution of science data from satellites to users at sites remote from the control centre or payload operations centre, associated with a capability of those users to submit requests or commands for payload operation, normally using the same data distribution infrastructure.

Telescience is best developed in the area of data exchange. The example given here, that based on transmission of data from the ICE spacecraft encounter with the comet Giacobini-Zinner is described in more detail in ref. 2. This was a successful experiment involving ESANET (the communications network linking ESA establishments) and SPAN (the Space Physics Analysis Network, which links many US Universities, Research Institutes and NASA Centers). A link between the networks was established for the first time to support Investigators involved in a European experiment onboard the ICE spacecraft. On the day of the encounter, the data from the encounter, which lasted from 0800 to 1200 UT was made available to experimenters at 1400 UT. Encounter data comprising some 100 kbits of data was transferred from computer of the NSSDC (National Space Science Data) Computer at GSFC to ESOC via SPAN and from thence via ESANET to ESTEC, where the data was processed. The first processed data appeared in the form of plots some 30 min. after the beginning of the transmission of the data from GSFC.

These plots were transmitted back to the NSSDC via facsimile well in advance of the first Press Conference held at 1600 UT, five hours after the encounter.

The general conclusion which can be drawn from this experiment is that it was greatly facilitated by computer hardware commonality: SPAN almost exclusively uses VAX computers, so a VAX at ESOC was chosen to be the European node. (See ref. 2 for further details.) This hardware commonality not only assists data transfer, but because of the availability of common software tools and office automation facilities on the VAX computers, related activities, such as joint preparation of papers by authors at different sites, are made easier.

<u>Interoperability in a Multiagency Environment</u>

<u>Cross Support</u>

In this Chapter some examples of interagency cross support are examined. Three examples are taken, both involving support of operations of one Agency by Ground Facilities of another namely;

- ULYSSES (operations of an ESA satellite using NASA facilities)

- GIOTTO (Provision of data acquisition and back-up commanding facilities by the US for an ESA satellite)

- IUE (joint operation of a spacecraft by ESA and NASA)

<u>EXAMPLE 1: ULYSSES.</u>

<u>Background.</u> ULYSSES is a joint ESA-NASA mission. It was planned for launch in May 1986 using the shuttle and a Centaur upper stage. All system integration and checkout was performed before the Launch was delayed by the Challenger failure. Launch is now planned for October 1990 using the shuttle and a Boeing IUS upper stage.

For the ULYSSES mission, ESA provides the spacecraft while NASA provides all launch facilities plus use of necessary ground stations, communications networks and the mission control centre infrastructure at JPL Pasadena. The ULYSSES-dedicated control and monitoring system is provided by ESA and integrated into the existing NASA facilities at JPL. It is this integration of an ESA-provided control and monitor system (H/W, S/W and staff) into a NASA infrastructure which is the subject of this section.

<u>Split of Operations Functions between Agencies (Fig. 2).</u>

NASA (JPL) were responsible for:
- Telemetry Acquisition and frame synchronising
- Long term TLM Filing and Processing
- Command transmission
- Ranging, Doppler processing and Navigation

ESA (ESOC) were responsible for:
- Command Schedule Generation
- Command event checking
- Real time TLM processing (limit checks, calibration, display)
- Near Real time TLM processing (performance analysis etc.)
- Science Printouts
- Flight Dynamics

<u>Interfaces.</u> Human interfaces are not considered here. The prime data interfaces are:

- Telemetry Frames (Real-time or recorded and replayed from spacecraft or from ground)
- Command Schedule
- Navigation/Flight Dynamics Data

<u>Constraints.</u>

- To minimise cost, the existing ESA MSSS concept (and software) was used as far as possible.

- The data acquisition system provided by JPL was a version of their 'standard' Data Acquisition and Control System (DACS) (Based on VAX hardware).

- The existing JPL commanding system was to be used, with minor extensions to accommodate ULYSSES specific features such as block commanding. This system was based on a standard format magnetic tape input to the command system (cf. the MSSS real time command, transmission and verification concept).

- To minimise cost (particularly for hardware maintenance during the 5 year mission life) the ESA-supplied spacecraft control system called the RTTP (Real Time Telemetry Processor) was to be hosted on a MODCOMP Classic computer, a computer type which was already in common use at JPL, and for which there was on-site support; in addition an existing machine could be used as backup.

<u>Approach.</u> The <u>telemetry</u> interface (DACS (VAX) to RTTP (MODCOMP)) was implemented at RS422 at the electrical level, X25 level 2 at the communications protocol level, and an application-specific protocol based on SFDUS (Standard Format Data Units) at the transfer level. This interface was fully defined down to the bit level in a mutually agreed and approved Interface Control Document (ICD).

The <u>telecommanding</u> interface utilised the existing magnetic-tape based interface to the JPL multi-mission command system;

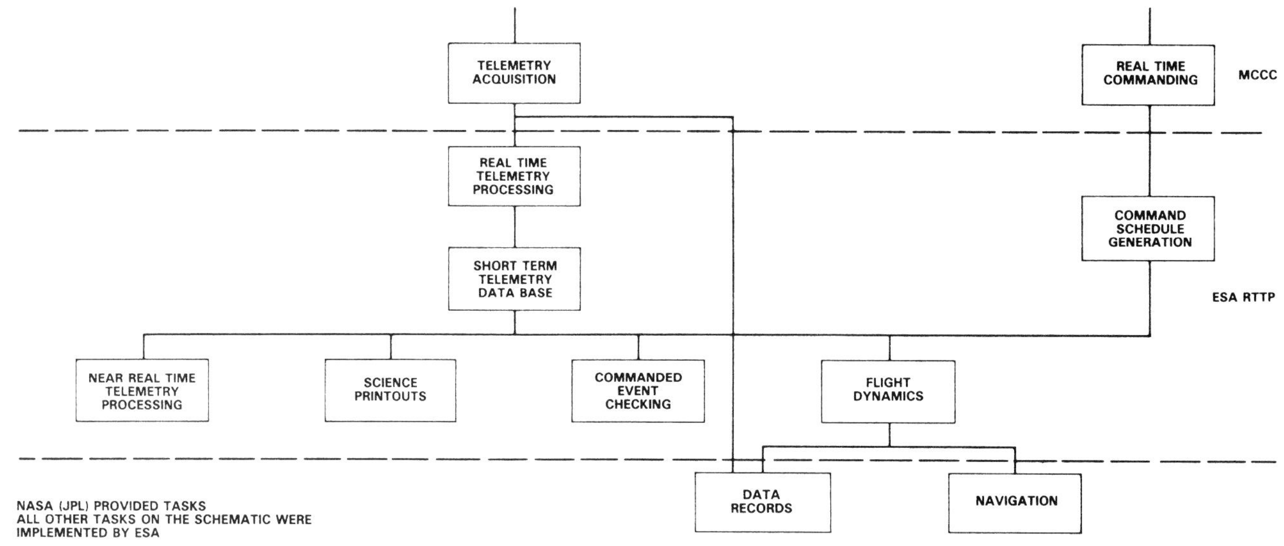

Fig. 2. Split of functional responsibility between ESA and NASA

the necessary extensions to accommodate ULYSSES-specific command structures did not affect the existing tape/record structure, only the record contents. A mutually approved ICD was generated for this interface.

An interface for navigation/Flight Dynamics data involving a ULYSSES-specific tape format was agreed and controlled by ICD.

During <u>development</u> the ESA-supplied spacecraft control and monitoring system (the RTTP) was initially implemented at ESOC (Germany) and shipped to JPL for integration into the NASA infrastructure some 6 months before the scheduled launch date. Some 4 months prior to this (i.e. 10 months before launch) the RTTP software system was taken to JPL on tape and installed on the (JPL provided) backup RTTP machine to enable exhaustive testing of the TLM interface between DACS and RTTP. The procedures for these tests being fully defined, documented and approved between JPL and ESOC during working sessions over the preceding months.

The command and Flight Dynamics tape interfaces were verified, using agreed test procedures, by exchange of magnetic tapes between ESOC and JPL.

<u>Lessons</u>. Although the Modcomp hardware was initially selected as compatible with NASA machines, during the time span of the project NASA 'standardised' on VAX hardware leaving the ULYSSES system as potentially 'non/standard' hardware with resultant maintenance problems.

The VAX-MODCOMP interface (X 25 level 2) used hardware in both machines which was not part of the suppliers' nor-

mal range (because no 'standard' interface between the machines was available). This was the source of some integration problems.

The pre-testing of interfaces between ESA and NASA systems proved invaluable and led to a smooth subsequent integration of the full system.

The only significant transfer level interface problem arose because the agreed and signed-off interface control document was transcribed within JPL into a different document for their implementation department. This transcription contained a non-trivial error.

<u>Example 2: Giotto</u>. Ref. 2 gives a good introduction to the Giotto Mission Operations systems, describing in some detail the complex ground system set up to support the mission. The discussion here will be restricted to highlighting the various areas of interoperability within this mission, principally:

(1) TT&C and X-band acquisition support, both for backup and during the encounter phase to maximise data recovery as well as provide a sort of 'hot-standby' security.

(2) Pathfinder, involving interoperability between Giotto and the USSR spacecraft Vega1 and Vega2

<u>TT&C and Science Telemetry Acquisition</u>. During the Giotto Mission, the NASA Deep-Space Network (DSN) was used to provide:

- Backup TT&C support during the first days of the mission

- Additional tracking

- High-speed telemetry backup at
  encounter: parallel 24-hour support
  was provided so that maximum data
  coverage was ensured

- Telecommand capability for spacecraft emergency

DSN stations at Canberra, Madrid and
Goldstone were used. It should also be
noted that the backup so provided was
based on 64m. antennas and had in some
respects a greater capability (performance) than the prime TT&C facility at
Carnarvon which had a 15m. antenna. This
meant that the backup could handle contingencies in which the telemetry signal
level from the spacecraft was too weak to
be acquired by the 15m. antenna at Carnarvon.

During the Cruise Phase use of the
DFVLR station at Weilheim to backup the
prime TT&C station at Carnarvon, W. Australia. Weilheim was modified by ESA to
provide compatible ranging support and
Reed-Solomon decoders for telemetry reception becoming in effect part of the   ESA
S-band Network for the duration of the
Giotto Mission

The operation with NASA DSN deserves
further study here, since it contains some
practical examples of the interoperability
defined in (2). Turning first to telemetry, it is noted that

- it was received as a bit stream
  over NASCOM

- ESA provided equipment to transform
  this stream into formats compatible
  with ESA facilities i.e. spacecraft
  formats with Reed-Solomon decoding
  performed

- similarly data tapes provided by
  DSN stations for off-line processing were bit-stream structured,
  which had to be taken into account
  in the software which provided
  final experimenter tapes

Although considerable effort had to
be put into the checkout of this interface, this aspect of operation with NASA
worked rather well, possibly because the
interface on the NASA side was a rather
simple one, so the protocol conversions
were relatively straightforward. The only
problematic area was the organizational
one of scheduling of the usage of DSN,
particularly during the test phases where
the absence of direct ESA participation on
the DSN scheduling meetings (for geographical and logistical reasons) made the
process somewhat cumbersome.

Regarding telecommanding, setting up
the facilities for backup commanding involved finding a solution to the problem
that the ESA and NASA commanding systems
are based on radically different philoso-
phies, the former being far more orien-

tated to automation, in particular as
concerns the spacecraft controller's man-
machine interface. Broadly two choices
were open to ESA in using NASA facilities
for backup commanding:

(i)   install a remote terminal to the MCCC

(ii)   integrate the interface into the ESA
commanding system so that commanding via
(NASA 'looks like' (as far as possible)
manual commanding on the ESA system. The
spacecraft controller would then be offered the usual standard ESA facilities of
automatic pre-transmission command validation, command execution verification, and
command logging via the standard man/ma-
chine interface.

Solution (ii), which would have involved developing complex NASCOM interfacing software 'behind' the manual stack
handler on the MSSS was rejected on
grounds of cost. Solution (i) was therefore adopted, but turned out not to be
totally without problems, the first of
which was that the MCCC terminals of the
type used for commanding were no longer in
production. The solution to this was to
implement an emulation of the MCCC terminal on an IBM PC, which was then connected to GSFC via a 9.6 kbaud link to
GSFC via INTELSAT. The resulting man machine interface was workable but uncom-
fortable since the command files were set
up on the terminal and then had to be
manually subjected to a series of operations (passage to ground station, configuration of ground station for uplink
etc.). Verification if needed would have
involved feeding the commands into MSSS as
dummy   commands   (i.e.   with   uplink
suppressed) so that verification against
Giotto telemetry received on MSSS could be
done. In practice this was not necessary
since it was mainly used for commands not
requiring verification (ranging commands,
on-off sequences, commands without verification).

The conclusion on commanding is that:

- A backup was successfully implemented

- In practice it only had to be used
  a limited number of times

- From an operational viewpoint it
  was unsatisfactory because (a) it
  looked radically different from the
  usual ESA system (alternative types
  of operator interfaces for achieving a giving functionality are
  generally undesirable); (b) it was
  heavily manually biased; (c) it had
  an exceedingly low throughput (c.
  one command every 2 min.); and (d)
  it would create increased mission
  risk when used under contingency
  conditions, since untried sequences
  and files had to be created. It
  would therefore have been unacceptable for heavy operational use

- The successful implementation of this backup commanding was in part made possible by the familiarity with the NASA commanding system gained as a consequence of the ULYSSES Project; this shows the obvious but important fact that one attempt at achieving interoperability may make further attempts easier to accomplish

Pathfinder Activities. Pathfinder involved the reevaluation of comet Halley's trajectory based upon cometary ephemeris data from the encounters of the USSR spacecraft Vega1 and Vega2 with the Comet 3 and 5 days respectively prior to the encounter. This involved three Agencies-- NASA, ESA and Intercosmos:

- NASA performed very precise tracking of the Vega spacecraft in order to achieve the accuracy of the orbit determinations for the Vega spacecraft necessary for improvement in cometary orbit determination

- ESA and IKI used Vega data for the refinement of the comet's orbit. This involved establishing a communications 'hot-line' between ESOC in Darmstadt and the Institute for Space Research (IKI) in Moscow. This supported not only exchange of data but also remote access to ESA computers from Moscow, interactive dialogue (via terminals) and voice communications between the teams concerned

It should be noted that these activities were the culmination of an international campaign of analysis of astrometric observations of the comet involving twelve observatories and associated observers (see for example ref. 3).

Example 3: International Ultraviolet Explorer (IUE). IUE, a joint development by ESA, NASA and SERC (GB) was launched 78.01.26. Since then it has been operated jointly from NASA's Goddard Space Flight Centre (GSFC) Greenbelt, Maryland, USA and from ESA's operations facility at Villafranca, Spain (VILSPA).

NASA has three principal areas of responsibility in IUE operations: firstly, overall responsibility for monitoring and maintaining the health of the spacecraft; secondly, providing a backup system for the purpose of spacecraft safety during the shift controlled by ESA's ground station; and thirdly, operating IUE for two eight/hour shifts each day, designated US1 and US2. The spacecraft-related tasks are carried out on a 24-hour per day basis by GSFC.

ESA provides spacecraft control and science operations for one shift, designated VILSPA, and in a second shift the standard data processing is carried out.

Because the ESA station has limited backup facilities, e.g. only one computer capable of controlling IUE, GSFC maintains readiness to take control and keep the spacecraft safe in the event of a failure in the VILSPA station.

In the case of IUE the ability to operate the spacecraft from two sites was achieved by contracting out the development of the operations system (H/W and S/W) to a single firm which supplied systems to both GSFC and VILSPA (in fact VILSPA is a subset of the GSFC system) thus interfacing and compatibility problems were minimised, but still existed between antenna and control system.

Interoperability in the Immediate Future

European Space Software Development Environment (ESSDE)

ESA is proposing to use a single ESSDE environment for future projects. The need for this development is driven by:

- The amount of software to be developed for future projects such as Columbus, Hermes.

- The consequent large number of software engineers involved (both in ESA and European Industry).

- The higher S/W reliability requirements for manned missions.

- The need to minimise S/W development costs.

- The need to achieve S/W portability between projects, between contractors, between mission phases (e.g. space segment development and operation) and between agencies.

- The need to harmonise S/W development activities between projects.

ESA proposes a standard hardware and software set to be used for the entire S/W development life-cycle from user requirement to production of project history document. The ESSDE proposed is based on the European Portable Common Task Environment (PCTE) and is aimed at supporting ADA plus one efficient language for time-critical applications.

Discussions have taken place between ESA and NASA with respect to commonality between ESSDE and the NASA SSE. So far it is thought impractical to try to ensure full commonality, the current aim is to support sufficient of the same tasks/languages to allow software developed on one environment to be maintained on the other.

Combined operations: Safety and Redundancy

Future ESA missions (especially manned missions) will have more stringent

safety requirements than previous missions. To achieve the necessary levels of
safety will require careful attention to:

- Procedures for analysing safety
  risks

- Assessment of proposed methods of
  prevention or containment of problem impacts

- Safety problem detection and diagnosis

- Recovery from unsafe (or potentially) unsafe situations ('safety
  problem recovery')

This analysis of operations systems
safety will undoubtedly indicate the need
for redundant/backup configurations. Such
backup systems can be most cost effectively implemented by use of existing
systems which are normally dedicated to
other routine operations but can be made
available at short notice in contingency
situations. In order for such backup systems to be certified they must first be
developed in such a way as to be capable
of supporting the spacecraft for which
they act as backup (i.e. possess interoperability) and must be subject to the
same QSA procedures as the prime system to
ensure required levels of safety are
reached.

This implies common development approaches and common (or at least compatible) QSA Standards.

<u>Discussion and Conclusions</u>

The paper has discussed a number of
areas in which the degree of difficulty in
achieving interoperability varies greatly.
From this it follows that it is perhaps
best to systematically approach the subject by classifying it into application
areas, for example telemetry acquisition,
telecommanding, telescience and data communications, checkout and spacecraft control systems,...In this way the problems
in each area can be localised and perhaps
made more manageable.

From the examples studied, three
principle conclusions may be drawn as
follows:

1. Interoperability depends on the
   interfacing of substituted elements
   into a system; this interfacing can
   be eased by the observance of commonly agreed protocols and interface procedures

2. Achieving common agreement of protocols and interfacing is made
   easier by the existence of standards; however standards on their
   own will not achieve compatibility
   (a necessary condition for interoperability) on their own because
   they have to leave a certain flexi-

bility if they are to have a useful
domain of applicability. This
leaves room for implementation differences ('dialects'), which can
only be avoided by organisational
measures, e.g. common interagency-
developments, agreements to use
particular proprietary products
etc.

3. Successful examples of interoperability appear to have been based on
   use of relatively simple or low-
   level protocols, e.g. Giotto DSN
   interface (NASCOM), ULYSSES DACS/
   RTTP interface (X25 level 2); thus
   it would appear advisable to base
   interoperability on the use of
   protocols corresponding to the
   lower layers of the OSI (Open Systems Interconnection) model of ISO.
   The use of the higher layers is not
   recommendable at present since (a)
   for the highest layers there is no
   common agreement (b) for the intermediate layers (e.g. layers 4 and 5
   they will not be commonly available
   on the variety of computer types
   likely to appear in a multi-agency
   environment; in particular they are
   likely to appear in different 'dialects' on different computers, and
   (c) compatibility between peer
   processes at each level is necessary making overall compatibility
   between independently-developed
   system increasingly more difficult
   to achieve as more layers are incorporated.

It might seem from (3) that the authors are recommending a backward-looking
approach. In fact this is not the case,
since we believe that in the future use of
new standards such as packet telemetry and
packet telecommand will considerably facilitate interoperability, providing a
basis for much cleaner interfaces than
those based on NASCOM. However it is fair
to point out that these standards actually
correspond to the lower to intermediate
layers of the OSI model (ref. 4). Thus for
e.g. packet telemetry a suitable interface
level between say the Control Centre of
one Agency and a ground station provided
by another could be telemetry transfer
frames rather than telemetry source packets, thus leaving the problem of "opening"
the transfer frames and routing the source
packets (to end users) to one Agency only
and keeping the interagency interface as
simple as possible. The reader is referred
to ref. 4 for more details.

A number of subsidiary conclusions
can also be drawn thus:

- Interagency agreements are obviously necessary when interoperation
  takes place. Single versions of
  such agreements must be dissemi-
  nated and maintained; transcription
  must be avoided, since errors and
  subsequent misunderstandings result.

- Operational concepts underlying the
implementation of a facility will
affect interoperability between
facilities performing in principle
similar functions. The case of
back-up of Giotto commanding operation station did not provide an
equivalent level of service to the
ESA system (validation, verification, 'transparency' of ground
station, response times, etc.).

- Hardware, and basic software compatibility between different facilities can ease interoperability
see e.g. (1) the telescience example of connection to the SPAN
network (2) exploiting commonalities between checkout and in-orbit
operations by using a common infrastructure (computers, special peripherals, basic software). If standard proprietary products are used
for interfacing it is possible to
ensure compatibility and, in simple
cases such as data exchange, also
ensure interoperability. However a
warning must be issued namely that
such interoperability can be endangered by uncoordinated modernisation of hardware/software without due regard to forwards compatibility. (ref. the case of ULYSSES).
A suitable modernisation policy
would require rigorous management
agreements, e.g. to adopt one particular supplier, who himself guarantees forwards compatibility. In
the view of the authors, in practice, this will be difficult or
impossible to achieve between two
or more agencies since considerations other than interoperability
have a strong influence on infrastructure procurement policies.
Indeed such coordination can even
be difficult within a single Agency
(witness the case of checkout and
operations). It could perhaps be
feasible in a limited applications
area, as the SPAN example shows.

- Interoperability can be designed
into a system from the start, although this will normally only be
the case for specific 'one-off'
projects. One way of doing this
(ref. IUE) is to contract out the
development of the operations sys-

tem to a single supplier, who du-
plicates subsystems as necessary
and ensures overall integrity.

- Successful interoperability requires exhaustive verification of
interfaces well in advance of that
of the rest of the system (ref.
Giotto, ULYSSES).

- Easier exchange of software may be
facilitated by use of standard
software development environments
within Agencies (e.g. ESSDE within
ESA and European National Agencies,
SSE for the NASA Space Station
Program). Obviously efforts should
be made to ensure as much commonality as possible between ESSDE and
SSE, with the view of making software exchange possible.

- Similarly, standard computer access
to data bases would greatly aid
cooperative projects; such data
bases could be used for coordinated
compilation of requirements, spacecraft data, ground segment infrastructure characteristics etc. Such
standardisation may need to be
extended to other tools for e.g.
documentation, so that joint preparation of interface agreements
etc. becomes easier and less dependent on bulky intercontinental
paper exchanges.

References

1. Wilkins, D.E.B., 'The Giotto Mission
   Operations System', ESA Journal,
   $\underline{10}$(3), 1986.

2. Sanderson, T.R. et al., 'Near Real-
   Time Data Transmission During the ICE-
   Comet Giacobini-Zinner Encounter', ESA
   Bulletin, $\underline{45}$, pp. 21-23, 1986.

3. Morley, T.A., 'Astrometric Observations of Comet Halley', ESA Bulletin,
   $\underline{43}$, pp. 19-23, 1985.

4. Jones, M. and Head, N.C., 'The Stan-
   dardisation of On-Board Data Manage-
   ment Systems and its Impact on Ground
   Systems', presented at the EUROSPACE
   Symposium on Standardisation of On-
   Board Management Systems, Madrid, 1-5
   June 1987.

INTEROPERABILITY IN THE SPACE STATION INFORMATION SYSTEM

Virginia Whitelaw*
Walter Marker**
John Muratore+
Jim Bigham++

National Aeronautics and Space Administration
Lyndon B. Johnson Space Center
Houston, Texas

Abstract

Goals for interoperability in the Space Station Program are described. Various levels of interoperablity are identified, including communication services, user interfaces and computer application programs and equipment. Potential barriers to interoperability are also described, as well as the types of actions required to overcome them. This paper reviews some successful steps that have been made toward achieving communications interoperability and outlines future work needed to make international interoperablity possible in this and other areas.

Goals for Space Station Interoperability

The Space Station Information System (SSIS) will comprise a wide variety of networks and networked resources, built by many different vendors in many parts of the world. Future users of SSIS have repeatedly emphasized their needs for interoperability within this widespread environment. For example, a recent issue paper on interoperability from the Space Station Users Working Group (1) states, "The Data Panel believes that the goal of interoperability should apply to all areas in which there is an interface between the users and Space Station Project personnel." Moreover, the operational concepts emerging for the Space Station virtually take as a given that international interoperability will be present, for example, to coordinate software actions and human procedures in real time operations.

Interoperability - whether it's allowing two ends of a communication channel to exchange meaningful information, whether it's allowing flight or ground controllers at different workstations to see the same user interface, or whether it's providing a standard service interface to payload or system components across all SSIS elements, is regarded as one of the intrinsic "goods" of the Space Station Program. Everybody agrees it is good to have interoperability. Everybody agrees that mechanisms to achieve interoperability are needed within NASA, since the development of NASA's portion of the SSIS and other related information systems will be a multi-center, multi-vendor affair. It is agreed that even greater mechanisms, from a political perspective, are needed internationally, such that the portions of SSIS provided by NASA and the international partners function as an integrated whole. So strong is the sentiment that this interoperability is necessary that, from the perspective of system and payload operations, no alternative is being seriously entertained.

Achieving interoperability is far more difficult than desiring it. It will require a thorough understanding of the levels of interoperability and the nature of the agreements that need to be put in place to achieve each one. The following section describes the levels of interoperability that are implied in current operational concepts. Achieving interoperability will further require a thorough understanding of its sizable obstacles - both technical and political - and conscious organizational actions by all users and providers of the SSIS to overcome each one. Section 3 exposes some of these barriers and the kinds of actions that will be required to overcome them. Through the work of the SSIS Data Flow Working Group (DFWG), building on the efforts of the Consultative Committee for Space Data Systems (CCSDS), some positive steps have been in understanding what is required to achieve one type of interoperability within the SSIS, namely successful end-to-end communications services. The latter part of this paper summarizes that effort, and outlines some follow-on work needed to make international communications interoperability, as well as other levels of interoperability, possible.

Levels of Interoperability

There are many types of interoperability, many of which build on one other. These levels of interoperability are supported by distinct equipment or service interfaces at which agreements are reached to develop to common specifications. This relationship between interoperability levels and interfaces is depicted in Figure 1 for the four categories of interoperability described in the following subsections.

Communications Interoperability

The capability to provide SSIS end-to-end communications services is captured in this type of interoperability. This would allow, for

* Research Engineer, Avionics Systems Division, Member AIAA
** Research Engineer, Tracking and Communications Division, Member AIAA
+ Flight Controller, Mission Operations Directorate
++ SSIS Project Integration Manager, Space Station Project Office, Member AIAA

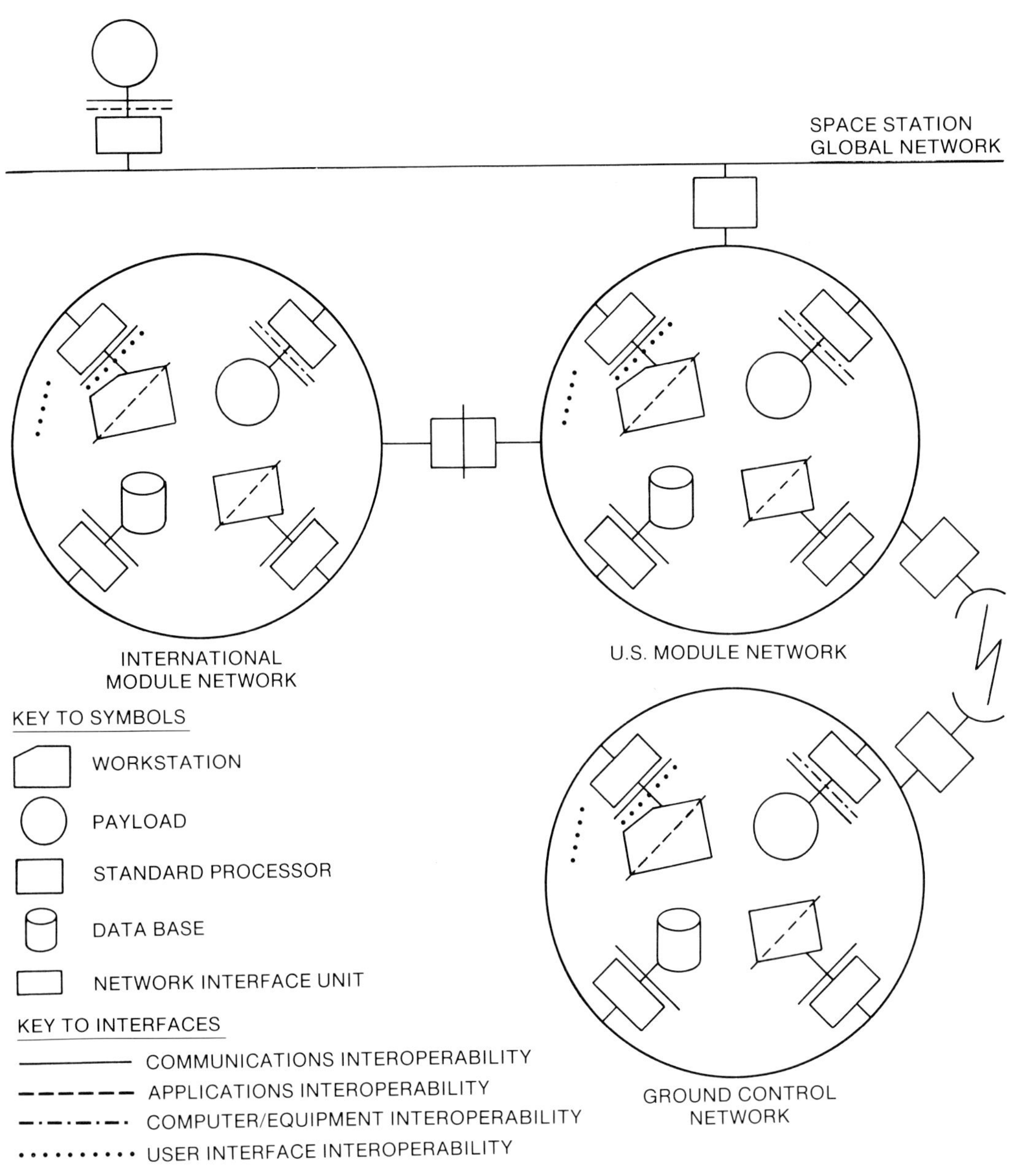

Fig. 1 Interfaces involved in various levels of interoperability.

example, two application programs in different parts of SSIS to exchange a meaningful dialogue or to move a file. It would allow people at workstations to communicate with electronic mail and to access remote applications and networked resources elsewhere in SSIS. In its most general sense, this type of interoperability allows whatever a user puts in at one SSIS endpoint to "pop out" at the appropriate endpoint for the intended recipient.

This type of interoperability, like the communications architectures that make it possible, can be achieved in layers. For example, with one set of agreements, we may achieve routing interoperability, by which we can physically reach any endpoint in SSIS. This is the level of interoperability agreements currently being worked within CCSDS. Further, higher level agreements are needed in order to, say, exchange a file between those endpoints. Full communications interoperability, as

discussed further in Section 4, will basically require developing a very detailed specification of the SSIS communications protocols and management functions and obtaining widespread agreements to implement it. A SSIS-wide conformance/ interoperability testing authority would have to be established to certify each portion of SSIS as it's developed. Full communications interoperability, while an extensive task in itself, is necessary but not sufficient for the levels that follow.

Applications Interoperability

There will be many parts of SSIS that provide computers in which system and payload application software runs. The U.S. onboard Data Management System (DMS), the Japanese and European modules' Data Management Systems, the Space Station Training Facility, and the Space Station Control Center are examples of SSIS end systems that will support application processing. Applications interoperability would allow a program (e.g., source code) written for any one of these systems to run in any of the other systems or to migrate between them. Beyond having access to common communication services, this would require that the applications have a common interface to the operating system/run time environment of the computers in each system. Furthermore, if the application uses data base services, application interoperability requires common data base interfaces; if the application interacts with users at terminals, interoperability requires common user interfaces; if the application draws on sophisticated artificial intelligence services, those interfaces, too, must be common. As with communications interoperability, each level of interoperability would require development of and adherence to a detailed interface specification, as well as thorough conformance testing.

Computer/Equipment Interoperability

The onboard data management systems - U.S., Japanese, and European - as well as several other end systems in SSIS, will provide standard network interface units (NIU) by which computers and payload equipment are attached to their respective networks. This level of interoperability would allow computers or payload equipment designed to interface to any one of these systems' NIUs to interface to the other systems as well. Some of these systems also define interfaces to instrumentation-type buses, called Bus Interface Adapters (BIAs). Computer/equipment interoperability requires standardizing these NIU and BIA interfaces, and developing detailed specifications and conformance test arrangements. This type of interoperability (and/or applications interoperability) is required if payload design is to be independent of the Space Station module in which it's designed to fly.

User Interface Interoperability

This level of interoperability describes the capability for a person at a workstation to control and monitor space station operations with a consistent user interface, regardless of the location of the workstation (U.S. or international module, onboard or on the ground) or the location of the system or payload being controlled. This level of interoperability requires agreements not only on the presentation of information to the user at various workstations (e.g., human factors), but also on system and payload responsibilities for interacting with the various workstations. Agreements on the human factors for workstations can be approached through the specification process described for earlier levels of interoperability. Agreements pertaining to system and payload responsibilities will be more subtle and difficult; to a large extent they will depend on how successfully other levels of interoperability have been achieved.

As an example of user interface interoperability, consider a payload that is sometimes controlled from a workstation in the U.S. Lab Module and other times from a different workstation in the Japanese Module. For starters, much about the basic input/output capability of the two workstations would have to be common: display resolution, use of color or black and white, common high level menus, icons, and input devices, user interface language, etc. Beneath the common display capability, the display programs for this particular payload would have to run in both workstations (application interoperability) and interact with the payload in the same way from either location (communications interoperability). If the control sequence for this payload required, for example, data from a U.S. data base, both workstations would require the communications services that provide access to that data base (communications interoperability). Clearly if other forms of interoperability cannot be achieved, user interface interoperability - which goes well beyond the veneer of what the user sees and does - will be difficult to impossible.

The Barriers to Interoperability

There is no doubt that interoperability is desirable, but the forces acting against it are significant, both technically and politically. Some of the major forces are identified below:

Lack of Mature Standards

In all of the areas in which interoperability is needed, use of standards provides the start of a solution, but far from the whole solution. Standards for communications protocols, operating system interface definitions, user interface languages, etc. are emerging, but we cannot yet look to countless examples of where their use has achieved multivendor interoperability. In the communications realm, perhaps the greatest progress has been seen in the Manufacturing Automation Protocols (MAP) and Technical Office Protocols (TOP) programs. But current specifications for these programs are built on still-evolving standards, such that products built to different versions of the specifications do not interoperate. Moreover, conformance to a particular version of the MAP or TOP specification does not yet guarantee product interoperability, though interoperability conformance tests and a testing authority are currently being established. SSIS specification and conformance testing will have to track a similar path to these major programs.

Learning Curve on Standards That Do Exist

If interoperability is difficult to achieve with standards, it is impossible without them. Lack of widespread knowledge of, and implementation experience with, the standards that do exist hampers progress toward understanding the nature of the agreements that have to be achieved for interoperability. This barrier can be overcome through pronounced organizational and individual efforts to climb the standards learning curve.

Autonomous Development and Verification

Interoperability requires that we agree up front, at a high degree of detail, exactly what we're going to build before we build it. Writing detailed specifications and achieving agreement on them takes a good deal of time. This time impinges on every individual vendor's

schedule; the push to "get on with development"
intensifies as schedules slip. Within the Space
Station Program, there is a popular concept of
autonomous system development. This concept is
incompatible with having to hold up or
continually re-direct development based on an
interoperability specification. There is also a
concept of system or element verification
responsibility. But if interoperability is to
be achieved between systems and elements, the
verification process must extend beyond the
domain of any single system or element.
Interoperability will only be achieved where it
is consistently placed at a higher priority than
autonomy in the development or verification of
systems that have to interoperate.

Cost

For reasons such as those outlined above,
it should be obvious that interoperability is
not inexpensive. The alternative is not
inexpensive, either. But the monetary commitment that buys interoperability tends to be
needed up front, whereas, the money that pays
for incompatibilities can come later. In
addition, the up front costs of interoperability
tend to be borne by a very few funding sources
(e.g., the funding sources for the providers of
the system), whereas incompatibilities tend to
be paid for from a much broader base, e.g., all
of the users who have to develop multiple
specialized interfaces for their equipment.
There is uniform agreement that interoperability
is less expensive in the long run, but programs
tend not to be funded "for the long run".

There are several approaches to overcoming
the cost barrier to interoperability. One is
for a concerted monetary commitment from the
funding sources for both the providers of SSIS,
as well as the users of SSIS (e.g., scientific
and commercial users) towards achieving
interoperability. This is suggesting that some
of the cost for achieving interoperability can
be defrayed by the funding sources for the users
who will otherwise bear the brunt later.
Moreover, this would give the user community
more control over the levels of interoperability
achieved. This also implies that the providers
of the system could specifically fund for
interoperability up front. It could be treated
as a separate budgetary line item, along with
the recognition that it is the responsibility of
a given area to bring it about.

A second approach is to reduce duplication
of development among the international partners.
Right now, for example, everyone wants to build
their own workstation and their own networks.
These tasks could be divided up among the
international partners, rather than duplicated
by each one of them.

If these suggestions for attacking the cost
barrier seem radical or highly controversial,
that simply underscores the magnitude of the
cost barrier.

Building Local Capability/Industry

Combating the suggestion above, that costs
could be reduced with less duplication of
equipment among the international contributors

to SSIS, is the reality that every country wants
to boost its own industrial capabilities in all
of the areas relevant to SSIS interoperability.
Challenging as interoperability is, it would
probably be easier to achieve international
agreements on user interface interoperability,
than on which one country produces the
workstations.

Technology Transfer

Concerns about technology transfer from one
country to another clearly inhibit open discussion on how to "do business" the same way.
Significant technical exchanges on interoperability among the international partners cannot
proceed until technology transfer concerns are
behind us.

Installed Equipment Base

The SSIS will not be built from scratch.
Even if the newly developed parts of SSIS are in
compliance with interoperability specifications,
those parts shared with, or interfacing to,
other programs will not be. Moreover, most
users of the SSIS will have some installed base
of equipment that, ideally, they'd like to make
maximum use of in interfacing to SSIS. Compromises will have to be made between the
dimensions of interoperability and the use of
non-compliant equipment.

Enforcement Authority

If the barriers to interoperability are to
be overcome, an internationally recognized Space
Station Program authority must be established
that forces the compromises and enforces the
agreements that interoperability entails.

SSIS Data Flow for Communications

Interoperability

Building on the work of the CCSDS, the
activities of the SSIS Data Flow Working Group
(DFWG) have made positive inroads toward
achieving communications interoperability within
SSIS. This group was formed to address the
problem of identifying what communication
services were needed end-to-end in SSIS and how
they could be provided. The work of the DFWG is
discussed in two companion papers in these
proceedings. The paper of Muratore et al. (2)
discusses the Data Flow Model. This model
analyzes SSIS data flow according to the seven
layer Reference Model for Open Systems Interconnection (OSI) and defines key SSIS concepts and
services. The paper of Marker etal. (3)
discusses the requirements that followed from
the Data Flow Model and their impact on specific
deliverables within SSIS, particularly on the
DMS and Communications and Tracking (C&T)
System.

The following definitions are taken from
those papers, as they are germane to the subject
of international interoperability.

* SSIS Primary Domain - consists of SSIS
  subnets that agree to exchange
  information according to SSIS standard
  protocols and management functions.

* SSIS Secondary Domain - consists of all
  other subnets in SSIS.

* SSIS endpoint - where the user is
  located; it can be either in the
  Primary or Secondary domain.

* SSIS services are only guaranteed by
  SSIS between endpoints in the Primary
  SSIS domain.

* SSIS services are accessible from the
  Secondary domain, but the quality of
  service cannot be guaranteed by SSIS,
  since services are provided, in part,
  by networks over which SSIS has no
  direct control.

The span of the SSIS Primary Domain is not
known at this time. In particular, there are
not agreements yet with the International
Partners for them to develop at least their real
time operational networks within the SSIS
Primary Domain (Subsequent discussion focuses on
the onboard real time networks, however, it is
largely applicable to other international
subnets of SSIS as well). If those agreements
are achieved, it will be a major step toward
communications interoperability; anything else
would be a major departure from it.

If the international partners' onboard LANs
are in the Primary SSIS Domain, the gateways
they share with the U.S. DMS can be low layer
relay points (i.e., at Layer 2 or 3 of the OSI
model). This is highly advantageous from the
perspective of the performance and power
requirements of the gateway. The lower the
interface can be pushed (with respect to the OSI
Reference Model layers), the higher the
performance and/or the lower the power needed by
the gateway. In network communications, nothing
is so inefficient as a full 7-layer gateway.

Another key point on being in the Primary
SSIS domain has to do with the guarantee of
services. SSIS only guarantees services between
endpoints in the Primary SSIS domain. If those
services are involved in safety critical
functions, as some of them will be, they have to
be verified and guaranteed by some authority.
If they cannot be guaranteed by SSIS, serious
impacts are likely, either on the operational
concepts for the Space Station or on its program
management concept. Conceivably, the Secondary
domain could guarantee the service between the
user and the gateway to the Primary domain,
while SSIS guarantees it within the Primary
domain; however, some authority must still
guarantee the gateway.

A third advantage of being in the Primary
SSIS domain has to do with service transparency.
If the international partners are in the Primary
SSIS domain, the interoperable upper layer
protocols between international endpoints and
U.S. endpoints would allow, for example, a file
transfer between the ESA module and the U.S.

lab module to work the same as between the two
U.S. modules.

If the International Partners' onboard LANs
are in the Secondary SSIS domain, the gateways
between the DMS and the international LANs would
potentially be full 7-layer gateways. At these
gateways, each side's non-interoperable
protocols would be terminated and translated to
the other's. At very least, these gateways
would entail a sizable power, weight, and
performance penalty, relative to a low layer
relay. The magnitude of these penalties would
depend on the degree of difference between the
U.S. and International Partners' protocol
implementations. In the the worst case, the
gateways would be functionally inoperable.

In addition to incurring run time
penalties, if the International Partners' LANs
are in the Secondary SSIS domain, they will
impose additional requirements on the integration, test, and verification (IT & V) process.
Since the successful completion of communication
services cannot be guaranteed by a single
program administration (i.e., SSIS), a more
extensive, combined IT & V process will have to
be arranged. Additional tests for verifying the
service integrity of each network will have to
be established, as well as verification tests on
the gateways themselves. In total, these
agreements and arrangements would probably be
more difficult than up front interoperability
agreements. The rub, however, is that interoperability agreements tend to be made by choice
early in the program, while laborious IT & V
arrangements are made when there really is no
other choice late in the program.

In summary, the Primary SSIS Domain is the
span over which communications interoperability
can become a reality. Currently, high level
requirements have been specified for what membership in the Primary SSIS domain would entail.
From here, a detailed functional specification
is required that describes the protocols,
parameters, and options, as well as the
associated management procedures by which the
SSIS Primary domain will operate. Finally,
detailed implementation conformance tests must
be specified. Test procedures and a testing
authority should be established or designated
for all vendors' implementations. To the extent
that the SSIS functional specification has
overlap with existing specifications (e.g., from
the MAP program or from Government OSI Procurement specification), SSIS may be able to make
use of some of the conformance testing
"machinery" in place by that time.

Achieving Further Interoperability:

Issues and Recommendations

The previous section outlined a course of
action for achieving communications interoperability, following from the work of the DFWG.
Clearly such working groups, if they are to be
international in impact, must become international in membership. The CCSDS, while not
tackling the full range of communications

interoperability issues, is in an excellent
position to achieve initial international
agreements on a subset of protocols and
management functions needed for Space Station.
International working groups could be spawned
from this effort that extend the breadth and
depth of the forthcoming CCSDS recommendations.
Similarly, international working groups need to
be established for other areas in which
interoperability is needed. Progress in this
direction is being made in the areas of the User
Interface (i.e., User Support Environment),
Network Management, and Data Base Services. A
sequence of activities for these working groups
that the DFWG found very fruitful involved (1)
defining user requirements in terms of
scenarios, (2) developing a model that supports
the scenarios, (3) identifying SSIS-wide
requirements from the model, and (4) deriving
system-specific requirements for program
deliverables.

Not only do requirements on all vendors'
designs need to be established, but also an
international enforcement authority needs to be
established that has responsibility for assuring
that design, development, test, and evaluation
is in conformance with these requirements and
their follow-on specifications. If this work is
to proceed on an appropriate scale, it must be
solidly funded up front and recognized within
the Space Station Program.

Full interoperability is an excellent goal,
but as we proceed in our international negotiations on the Space Station Program, we need to
be realistic. We need to continually assess

where interoperability fits in our priorities.
We may have to compromise on building a local
capability in a particular technology or on
supporting an installed base of equipment if
interoperability is, indeed a higher priority.
To the extent that our conscious or unconscious
choices put interoperability at a lower
priority, we MUST re-think safety and
reliability implications in the non-interoperable Space Station environment that will result.

Acknowledgements

We wish to thank Bill Mallary and Dave
Pruett for their careful review and helpful
suggestions on this paper.

References

1. N. N. Spencer and A. J. Fleig,
"Interoperability With International
Partners' Modules - Issue Paper",
March 4, 1987

2. J. Muratore, et al., "Space
Station Information System Integrated
Communications Concept," AIAA/NASA
International Symposium on Space
Information Systems, June, 1987.

3. W. Marker, et al., "Space Station
Information System Requirements for
Integrated Communications," AIAA/NASA
International Symposium on Space
Information Systems, June, 1987.

Maryam ABOLVAFA – Programmer Analyst, MCI
Barbara A. ACKERMAN – Program Development & Systems Application Manager, TRW
Rick ADAMS – Product Manager, Central Research Labs
A. W. ADKINS – Flight Control Element Manager, IBM Corporation
Lottie AEPLI – Manager, Military and Data Systems – GE
Gary T. ALCOTT – Electronics Engineer, NASA Goddard Space Flight Center
LeRoy ALLEN – President and General Manager, Boeing Computer Support Service
Michael A. ALLEN – Group Leader, The MITRE Corporation
Robert O. ALLER – Associate Administrator/SFO, NASA Headquarters
Robert E. AMBROSE – Communications Management Specialist, NASA Goddard Space Flight Center
Barbara ANDERSON – Member Technical Staff, Jet Propulsion Laboratory
Seth ARENSTEIN – Reporter, Defense Daily
Gary P. ASPESI – Member Technical Staff, GTE Strategic Systems Division
Janene L. AUMILLER – Senior Consultant, Booz, Allen & Hamilton, Inc.
H. AWAZAWA – Assistant Senior Engineer, NASDA Jet Propulsion

David L. BAHRS – Senior Member Executive Staff, Computer Sciences Corporation
Sidney C. BAILIN – Chief Scientist, Computer Scientist
Robert M. BAKER Jr. – President, West Coast University
Melvin D. BANKS Jr. – Head, Mission Support Section, NASA Goddard Space Flight Center
R. C. BARBIERE – Senior Engineer (SSIS), TRW
Quinton BARKER – Mathematician, NASA Goddard Space Flight Center
J. BARSKY – Manager, MOSS, Allied Bendix Field Engineering Corporation
Lisa BASILE – Computer Scientist, NASA Goddard Space Flight Center
Victor BASILI – Department of Computer Science, University of Maryland
Peter C. BELFORD – Executive Vice President, NYMA, Inc.
Miroslav BENDA – The Boeing Company
Anderson BENNETT – Manager, C&T Space Station, Jet Propulsion Laboratory
Gregory R. BENNETT – Group Engineer, Space Station Operations, McDonnell Douglas Astronautics Company
Pawan K. BHARTIA – Project Manager, ST Systems Corporation
Graham T. BIDDIS – Columbus Utilization Group OPS, DFVLR
James P. BIGHAM Jr. – SSIS Project Interaction Manager, NASA Johnson Space Center
Fred C. BILLINGSLEY – MTS, Jet Propulsion Laboratory
George BIRUTIS – Manager Business Development, Harris Corporation – GCSD
Joseph L. BISHOP – Data Processing Program Manager, NASA Headquarters
William BISIGNANI – Principal, Booz, Allen & Hamilton, Inc.
David C. BLACK – Chief Scientist, Space Station, NASA Headquarters
Lyman BLANCHARD – Data Systems Engineer, NASA Kennedy Space Center
Fredrick C. BLUMBERG III – Director S/W Technologies, Planning Research Corporation
Angelo BODINI – Head Technical Operations Division, European Space Agency – ESRIN
Jerome R. BOHSE – Engineering Group Manager, Computer Technical Association, Inc.
J. David R. BOULDING – Space Station Project Office, National Research Council of Canada
James P. BOYLE – Senior Systems Engineer, RMS Technologies, Inc.
William E. BRADLEY – GAO Evaluator, U. S. General Accounting Office
Charles BRADY – Information Systems Engineer, Space Station Program Office, NASA Headquarter SSI
Daniel L. BRANDEL – Manager TDRSS Continuation Program, Office Space Operations, NASA Headquarters
Gunther BRANDT – Project Manager Spacelab Missions, MBB-ERNO
Edward J. BRAZILL – Manager Advance Development, NASA Headquarters
Robert BREKKE – Senior Systems Engineer, Ford Aerospace
Francis BRETHERTON – National Center for Atmospheric Research
Fred BROSI – Vice President, Computing Technology Association
Ronald C. BRUNO – Manager Advance Development, NASA Headquarters
Robert J. BRYER – Major Account Manager, Hewlett Packard Company
Joseph M. BUCH – Senior Staff Engineer, Contel Technical Services
R. K. BUREK – Senior Engineer/Applied Physics Laboratory, The Johns Hopkins University
William C. BURKHART – Computer Specialist, NOAA/NESDIS
Robert L. BURNS – Flight Mission Simulation Director, NASA Goddard Space Flight Center
Kristine BUTERA – Deputy Director Manager, SAIC
Madeline J. BUTLER – Head, Digital Systems Section
Thomas E. BUTLER – Supervisor, Astro Telecommunications, NASA Goddard Space Flight Center

Ralph CACACE – Electrical Engineering Supervisor, United Technology, Inc.
Ralph H. CAGER – Manager Communications & Electronics Division, Systematics General Corporation
David E. CALLENDER – Technical Staff, Jet Propulsion Laboratory
William M. CALLICOTT – Deputy OSDPD (NOAA/NESDIS), Environmental Satellite Service
John W. CALVERT – Manager Program Development, Hughs Aircraft Company Defense Information Systems
Divison Jack CAMPBELL – Project Manager, UNISYS
William J. CAMPBELL – Geographer, NASA Goddard Space Flight Center
Roy T. CANTRELL – Space Station Services Manager, Boeing Computer Support Services
Richard D. CARPER – Engineer, NASA Goddard Space Flight Center
T. S. CASSELL – Senior Engineer (SSIS), TRW
Gian C. CASSISA – Avionics Systems Manager, Aeritalia Space Systems Group
Mike CERNECK – Senior Washington Representative, TRW
Lloyd W. CHAMBERLAIN III – Astronautic Data Systems, NASA Kennedy Space Center
Scott CHAMBERLIN – GAO Evaluator, U. S. General Accounting Office
Robert P. CHASE – Research Specialist, Advance Engineering Laboratory
Cynthia CHEUNG – Senior Scientist, The BDM Corporation
Lynn CHILSON – Consultant, Booz, Allen & Hamilton, Inc.
Barbara A. CHRISTOPH – Department Manager, Space Sciences, SRA Corporation
Paul A. CLARK – Consulting Scientist, Diesel Kernel Systems
Laura CLARKE – Director, Program Development, Planning Research Corporation
Micheal CLENDINING – Analyst, ESOC
Paul COLEMAN Jr. – Space Sciences Department, University of California
Cynthia COLLINS – Electronic Engineer, NASA Goddard Space Flight Center
Edward B. CONNELL – Head, Data Systems Application Branch, NASA Goddard Space Flight Center
James M. CONOVER – Staff Member, TMIS Panel, McDonnell Douglas Astronautics
Pierre L. CONTRERAS – Chief of the Division of Program & Support, CNES
Earl COOK – Space Research and Applications Laboratory, 3M Company
John COPECK – Systems Engineering Operations, SED Systems, Inc.
Robert CORNETT – Photographer, National Space Science Data Center
S. Richard COSTA – Director, Communications and Data Systems Division, NASA Headquarters
Jack T. COULSON – Vice President, Space Science IDEAS, Inc.
R. CREASEY – Space Information Systems WS, ESA/ESTEC
Philip J. CRESSY – Science Operations Study Manager, NASA Goddard Space Flight Center
David R. CRISWELL – Director Consortium for Automation and Robotics/CALSPACE, University of California
Charles J. CROTEAU – Program Manager, NYMA, Inc.
Larry CULHANE – Associate Technical Director, MITRE Corporation
Bernard CURBELIE – CNES

Pamela A. D'ANGELO – Reporter, Federal Computer Week
Edgar DALKE – Project Manager, NASA Johnson Space Center
John DALLAT – Software Development Manager, DFVLR-GSOC
John DALTON – Chief, Data Systems Technology Division, NASA Goddard Space Flight Center
Ronald R. DAPICE – Director, A&RMB, NASA Headquarters
Ranjit DASGUPTA – Senior Engineer Specialist, Ford Aerospace & Communications
Leonard DAVID – Gateway Terrace
Dale R. DAVIDSON – Manager, Program Development, Computer Science Corporation
Donna J. DAWSON – International Group, NASA Office of Space Station
Kurt DEBATIN – Head, Computer Department, European Space Agency
Micheal DEESE – Business Development Manager, Computer Technology Associates, Inc.
John A. DEFERRARI – GAO Evaluator, U. S. General Accounting Office
Claude Del FOSSE – Program Development Manager, Software Productivity Consortium
Richard desJARDIN – Associate, Computer Technology Associates
Roberto F. Di COCCO – Electrical Systems Engineer, Societa' Italiana Avionica
Anthony DICKINSON – DRPP Project Manager, European Space Agency – ESTEC
William B. DICKINSON – Assistant Director/Space Station, NASA Goddard Space Flight Center
Donald L. DICKSON – Director, Business Development, UNISYS Corporation
James E. DIEUDONNE – Associate Department Head, The MITRE Corporation
Simon E. DINWIDDY – DRS Communications Systems Manager, European Space Agency
Dennis D. DOE – Vice President, Soft Ware Product Consortium
Edward V. DONG – SSIS Project Manager, Rockwell International
Mary Kay DOUGLAS – Computer Scientist, NASA Goddard Space Flight Center
Toshio DOURA – Senior Assistant, Systems Engineering Department, National Space Development Agency/Japan
William H. DOYLE – Program Development Manager, TRW
Robert S. DROSDZAL – Marketing Manager, NASA, Boeing Computer Services
Larry E. DRUFFEL – Director, Software Engineering Institute, Carnegie Mellon University
Carroll G. DUDLEY – Chief, Spacetelescope GSO, NASA Goddard Space Flight Center
Roland L. DUPREE – Staff Engineer, Martin Marietta
Ronald W. DURACHKA – Assistant Division Chief/Space Station, NASA Goddard Space Flight Center
Robert E. DWYER – Systems Engineer Supervisor, Ford Aerospace & Communications

D. R. EARNEST – Ford Aerospace & Communications
Pete EASON – Applications Engineer, Honeywell/Sperry Space Division
Charles R. EASTON – Staff Engineer, McDonnell Douglas Astronautics Company
Paul EBERT – Director, Communications Satellite Corporation
Jim EDELMANN – Department Manager, Science Applications Research
Ivan EGRY – DFVLR
Iwao EGUCHI – Manager, NEC Corporation
Curtis EMERSON – Data Systems Engineer, NASA Goddard Space Flight Center
Norman ENGELBERG – Director, User Interface Systems, Century Computing, Inc.
John L. ENGVALL – Principal Engineering – AI Laboratory, Ford Aerospace
Don ERAT – Marketing Representative, IBM Federal Systems Division
Brian ETHERIDGE – Vice President, Air Traffic Control, Martin Marietta Corporation
Stuart EVANS – Assistant Administrator for Procurement, NASA Headquarters

Dale L. FAHNESTOCK – Deputy Director, MODS, NASA Goddard Space Flight Center
Lonnie W. FARMER – Senior Software Engineer, Ford Aerospace & Communication
V.J. FARMER – Principal Associate, CACI, Inc.
A. Frederick FATH – Manager, Architecture & Development, Boeing Computer Services
Micheal R. FEAZEL – Managing Editor, Satellite Week
Ronald R. FELICE – Chief Engineer, NASA Goddard Space Flight Center
Eugene FERRICK – Director, Tracking Data Relay S.S., NASA Headquarters/Office of S.O.
B. FEUERBACHER – Director, Institute Space Simulation, DFVLR
Marco FLAGG – Development Engineer, PC/M Corporation
Morton FOXE – Staff Engineer, RMS Technologies
Peter FRIEDLAND – Chief, AI R&D Center, NASA Ames Research Center
Stuart W. FRYE – Member, Technical Staff, The MITRE Corporation
Yasuo FUJIWARA – NASDA

Lois M. GADDIS – Senior Marketing Representative, IBM
Richard A. GAJEWSKI – Systems Analyst, Grumman Data Systems
John A. GARMAN – Space Station Operations Office, NASA Headquarters
James P. GARY – Manager, Computer Networking Division, NASA Goddard Space Flight Center
James M. GATLIN – Head Engineer, McDonnell Douglas Corporation
Martha GAY – NASA National Representative, IBM
George T. GERGORA – Program Director, Harris Government Communication System
Kam GHAFFARIAN – Systems Engineer, Ford Aerospace & Communication
Richard A. GIBBY – Staff Consultant, UNISYS
Moury GILLES – Centre Nationale Etudes Spatiales – CNES
Gerd GOELZ – Deputy Director, Space Station Program Office, DFVLR
Nancy GOODMAN – Mathematician, NASA Goddard Space Flight Center
Charles GOOREVICH – Senior Consulting Engineer, Computer Science Corporation
Marc GORDON – Senior Manager, MCI
Tony GRANDI – Associate Chief Flight Mission Support Officer, NASA Goddard Space Flight Center
Edward GREENBERG – Section Technologies, Jet Propulsion Laboratory
Richard C. HAHN – Senior Research Engineer, Rensselaer Polytechnic Institute
Dana HALL – Accounts Manager, Information Systems Management Office, NASA Headquarters
Bud HAMILTON – Staff, General Accounting Office
Barbara N. HANCOCK – Member Technical Staff, The MITRE Corporation
D. H. HANCOCK – Senior Staff Engineer (SSIS), TRW
Elaine HANSEN – Research Associate, Laboratory for Atmospheric & Space Physics
Randy HARBAUGH – Ground Software Systems, NASA Goddard Space Flight Center
David W. HARRIS – Member Technical Staff, The MITRE Corporation
Ray HARTENSTEIN – Head, Flight Data System Branch, NASA Goddard Space Flight Center
Jonathan B. HARTLEY – Mission Operations Division, NASA Goddard Space Flight Center
Micheal G. HERLIHY – Foreign Affairs Officer, U. S. Arms Control/Disarmament Agency
Scott E. HERMAN – SDI Program Coordinator, Comsat Technical Services
George E. HEYLIGER – Manager of Technologies, Computer Technology Association, Inc.
Kiyoshi HIGUCHI – Director, Space Utitlity Promotion Office, NASDA
Paul D. HILL – Electronics EngineerLight Center, NASA Goddard Space Flight Center
Kent HILLS – Group Manager, Science Applications Research
Noel W. HINNERS – Director, NASA Goddard Space Flight Center
Ronald HIRSCHFIELD – Technical Director, GE
Jack HISEY – Logistics Engineer, Raytheon Service Company
Robert W. HOBBS – Manager NASA Programs, Computer Technology Associates
Hans E. HOFFMAN – Managing Director, INTOSPACE GmbH
Claude HONVAULT – Head, Systems & Project Department, European Space Agency
Adrian J. HOOKE – Member of Technical Staff, NASA JPL
Ivy HOOKS – President, Bruce G. Jackson & Associates

Grace M. HOPPER – Rear Admiral, USNR (Ret.), Digital Equipment Corporation
Masanori HORIO – Systems Section, Space Systems Department, Fujitsu, Ltd.
Eugene HORN – Project Manager, Informatics, Inc.
Kent W. HUGHES – Director, NOAA/National Oceanographia DC
Richard C. HUGHES – Simulation and Software Manager, National Research Council/Canada
Wayne HULLETT – Senior Software Engineer, NASA/Jet Propulsion Laboratory/CALTECH
Kelley E. HUNTER – Engineer, IBM

M. IKEUCHI – Director, NASDA
Inger M. ILDSVAD – MTS, TRW
Isao ILZUKA – NASDA
Chu ISHIDA – Systems Manager, NASDA
Yukio ISHIDA – Newspaper, Asahi Shimbun
Frank ISLAM – Chief Systems Engineer, Science Applications Research
Roland IVARNEZ – SW Development Manager, CNES – Satellite Control
Weber E. IVY – Member Technical Staff, The MITRE Corporation/NASA Headquarters

John T. JACKSON – Head, Operations Management Branch, NASA Goddard Space Flight Center
Allan JAWORSKI – Manager, Advanced Technical Engineering, Ford Aerospace & Communications Corporation
George JEAMBRUN – Deputy Director, Operation Systems Division, CNES
Demetrius JELATIS – Senior Consultant Scientist, Central Research Laboratory
Alan T. JOHNS – Computer Scientist, NASA Goddard Space Flight Center
Bradley A. JOHNSON – Advanced Programs, Logicon
James W. JOHNSON – Manager Advanced Development, NASA Langley Research Center
Kurt P. JOHNSON – Director, Strategic Analysis, McDonnell Douglas Astronautical Company
Michael JONES – Head Scientific Branch Data Processing Division, ESA, ESOC
Robert L. JONES – AST, Electronics Engineer, NASA Goddard Space Flight Center
Takuo JUKUROKI – System Engineer, Space Station, Ishikawajima – Harima Heavy Industries

Ralph KAHN – Visiting Senior Scientist, Office of Earth Science & Applications, NASA Headquarters
Makoto KAJII – Assistant Senior Engineer, National Space Development Agency/Japan (NASDA)
S. J. KALLAOS – Manager (DA35), Rockwell International
Yodoh KAMEO – Manager, Mitsubishi Space Software Company
Michio KATO – Assistant Senior Engineer, T&DA Department, National Space Development Agency of Japan
Takaji KATO – Manager Systems Security, Space Systems Department, Fujitsu Limited
Julian A. KATZ – Manager Business Development, General Electric Company
Mark S. KAUFMAN – Senior Engineering Specialist, Ford Aerospace & Communication
Lori KEESEY – Managing Editor, Space Commerce Bulletin
Jochen KEHR – Columbus Ground System Operations Manager, DFVLR
John KELLER – Editor, Advanced Military Computing
Angelita C. KELLY – Head, High Rate Data Hand UPG, NASA Goddard Space Flight Center
William KELLY Jr. – Section Head, Information Processing, NASA Goddard Space Flight Center
Linda S. KEMPSTER – Optical Disk Technolgist, Computer Technology Association
Richard KESSINGER – Manager, NASA Business Development, SOFTECH, Inc.
Jeffrey KEYS – MTS, Jet Propulsion Laboratory
James F. KIBLER – Ast, Applications Data Management, NASA Langely Research Center
Ludie M. KIDD – Aerospace Technologist, SS, NASA Goddard Space Flight Center
Frank S. KING – Data System Manager, NASA Ames Research Center
Joseph H. KING – Head, Central Data Service Facility, NASA Goddard Space Flight Center
Philip S. KIRBY – Supervisor, Electronics Engineers, NASA Goddard Space Flight Center
Russell KOFFLER – Deputy Assistant Administrator for SDIS, NOAA
Gerald KOWALSKI – Senior System Engineer, Harris Government Aerospace Systems
Rolf B. KRAEHMER – Task Leader, The MITRE Corporation
George N. KREBS – Senior Engineer Specialist, Ford Aerospace & Communication
Charles J. KRESCH – Supervisor, Hardware Systems Section, Ford Aerospace & Communication
Peter R. KURZHALS – Director Utilization & Operations, Space Station Division, MDAC
Michirou KUSANAGI – Special Advisor to the President, NASDA T&DA Department

Jerry K. LAKE – Manager Data Systems, RCA Government Services
Nand LAL – Computer Scientist, NASA Goddard Space Flight Center
E. W. LAND Jr. – Deputy Director, NASA/Space Station Integration
C. LANGTON – President, Shartel Systems, Inc.
Bruce W. LARSEN – Asr, Technical Engineering Operations Management
Rudolf J. LAUBER – Director, Institute for Control Engineering & Process Automation
D. A. LAURENZIO – Electrical Systems Engineer, Spar Aerospace Limited
Michael LAURIENTE – Technical Program Manager, NASA Goddard Space Flight Center
Andrew LAWLER – Associate Editor, Pasha Publications, Inc.

Bill LAWSON – Task Leader, Science Applications Research
Doug C. LEE – Program Manager, PM3, Harris Government Communication System
James S. LEGG Jr. – Systems Analyst, Ford Aerospace & Communication
Barry LEINER – Senior Scientist, RIACS
Peter LEMPP – Teamleader Software Engineer, SPS Software Products Service
Klaus LENHART – Data Relay Ground Segment Study Manager, ESA/ESOC
Claude LENSEIGNE – Programme Hermes, Aerospatiale
K. C. LEUNG – Senior Project Manager, Computer Sciences Corporation
Arnold LEVINE – Senior Reporter, Federal Computer Week
Gene LEW – Engineering Department Manager, Honeywell Sperry Space System
J. LICHTENFELS – Technical Staff, EER, Inc.
Howard C. LINDSEY – Senior Marketing Representative, IBM
Gregg LINEBAUGH – Countdown Manager, Washington Correspondent
Richard H. LIU – MITRE Technical Staff, The MITRE Corporation
Gene LONG – National Manager, Market Development, Odetics, Inc.
B. J. LONGFELLOW – Group Director, Computer Sciences Corporation
Mario LOPRIORE – Head, RF Systems Division, ESA
Edward C. LUCZAK – Consulting Engineer, Computer Sciences Corporation
John LYON – Data System Manager Space Station, NASA Goddard Space Flight Center
William H. LYTLE – Operations Manager, UNISYS MD Telecomm OPs

Janet MACKEY – Senior Consultant, Booz, Allen & Hamilton, Inc.
M. L. MACMEDAN – EEIS Standards Engineer, NASA/Jet Propulsion Laboratory
William O. MACOUGHTRY – Assistant Head, SCPB, NASA Goddard Space Flight Center
Don MADILL – Systems Engineering Integration Test, SED Systems, Inc.
Michael MAHONEY – Associate Chief Information Processing Division, NASA Goddard Space Flight Center
Richard C. MAINS – Principal, Mains Associates
Foachim F. MAJUS – DFVLR
Gary MANTUEFFEL – Program Development Manager, Eastman Kodak Company
Walter S. MARKER – Research Engineer, Tracking and Communications Division EE8,
   NASA Johnson Space Center
Michael D. MARTIN – PDS Project Engineer, Jet Propulsion Laboratory
Warren L. MARTIN – Manager, TDA Advanced Projects, Jet Propulsion Laboratory
Stephen MASLEN – Associate Director, Martin Marietta Laboratories
Joseph S. MATNEY – Department Head, The MITRE Corporation
Kazuo MATSUMOTO – Director, Tracking & Data Acquisition Department, NASDA
Cario MAZZA – Head, Data Processing Division, ESA-ESOC
Frederick W. McCALEB – Electronics Engineer, NASA Goddard Space Flight Center
Debora McCALLUM – Photographer, NASA Headquarters
J. P. McCLURE – Program Representative, Jet Propulsion Laboratory
Gail McCONAUGHY – Task Leader, Science Applications Research
Kenneth McDONALD – Mathematician, NASA Goddard Space Flight Center
David R. McELROY – Group Leader, MIT Lincoln Laboratory
Henry W. McELROY – NASA Technical Accounting Manager, Digital Equipment Corporation
Frank McGARRY – Head, Systems Development, NASA Goddard Space Flight Center
Merle McKENZIE – Management Science Group Supervisor, Jet Propulsion Laboratory
John L. McLUCAS – President and Chairman of the Board, Questech, Inc.
Mark McNICKIE – Senior Systems Analyst, Planning Research Corporation
Stephen McREYNOLDS – Technical Director, Applied Technology Associates, Inc.
Juergen MENZEL – Messershcmitt-Boelkow-Blohm-GmbH, Space Systems Group
Barry D. MEREDITH – Aerospace Engineer, NASA Lewis Research Center
George O. MEYERSON – Vice President, Support Operations, Computer Sciences Corporation
Arshad MIAN – Senior Systems Engineer, RCA Government Services
Giovani MICA – Head, Payload Technology Department, ESA
James L. MICHAEL – Engineer, NASA Goddard Space Flight Center
Gerhard R. MICZAIKA – Director, Europe, International Operations,TRW, Inc.
S. MILANI – Manager, AI Department, FIAR
Betty Jane MILLAR – Electronics Engineer, NASA Goddard Space Flight Center
J. MILLER – Manager-Laser, Allied Bendix Field Engineering Group
Richard B. MILLER – Manager, Information System Program Office TRW, Inc.
Francois MILLON – Telecommunication Engineer, MATRA
Marilyn MILLS – Chief Engineer, NYMA, Inc.
M. MITTAL – Project Manager, EER, Inc.
Karen L. MOE – Head, Systems Technology Section, NASA Goddard Space Flight Center
Alan MONTGOMERY – Director, Aerospace Systems, NYMA, Inc.
Carol J. MOORE – Director of Products, Sterling Software
Otis C. MOORE – Manager Business Development, Martin Marietta Information & Communication

Glenn MORRIS – Senior Staff Advisor, STX
W. C. MOSLEY – Data & Information Systems, Manager, General Electric Company
Robert R. MOYLAN Jr. – NASA Marketing Program Manager, Control DATA Corporation
Glenn H. MUCHKLOW – Systems Architect, NASA Headquarters
John F. MURATORE – Flight Controller, Mission Operations Directorate, DF2, NASA Johnson Space Center
Kenneth L. MURPHY – Consultant, Control Data Corporation

Michael C. NEELY – Senior Engineer, Computer Technology Associates, Inc.
William J. NEFF – Program Manager, Project Engineering, Inc.
Phillip H. NELSON – Manager, NASA Information Systems, Boeing Computer Services Company
R. L. NELSON – President, PC/M Corporation
John NEWBAUER – Administrator, Scientific Publications AIAA
David A. NICHOLS – Assistant Manager, Information Systems Program, Jet Propulsion Laboratory
Brian E. NIMMO – NASA Program Development Manager, Digital Equipment Corporation
Bob NOBLITT – Senior Washington Representative, TRW, Inc.
James H. NOLES – Group Leader, The MITRE Corporation
Douglas NORTON – Chief, International Program Support, NASA Headquarters
J. L. NORTON – Senior Staff Engineer, TRW, Inc.
Dale NUSSMAN – Member of Staff, Dynamics Research Corporation
Patricia A. NUSSMAN – Deputy Assistant for Information Systems, The MITRE Corporation

David OKERSON – Senior Scientist, Science Applications
John F. OTRANTO – Staff Engineer, Computer Technology Associates, Inc.
Howard K. OTTENSTEIN – Chief, Public Affairs, NASA Goddard Space Flight Center
Gunther H. OTTO – DFVLR
Robert F. OVERMYER – Director Mission Success/Space Station, Martin Marietta Corporation

Daniel PALMER – Electronics Engineer, NASA Goddard Space Flight Center
Mark E. PARKIN – Member Technical Staff, The MITRE Corporation
Robert PEAK – Director, Advanced Automation, Martin Marietta Corporation
Dorothy C. PERKINS – Head, Software Engineering Section, NASA Goddard Space Flight Center
A. PESSEMIER – Space Information Systems WS, ESA-ESTEC
William A. PFEIFFER – Director, Business Development, The ORI Group
Melvin A. PIERCE – Principal Engineer, Fairchild Space Company
Paul B. PIERSON – Manager, Intelligent Systems, RCA Advanced Technology Lab
John V. PIETRAS – Lead Engineer, The MITRE Corporation
Peter PIOTROWSKI – Manager, Control Center & Data PD, DFVLR
David A. PORTNOY – Washington Representative, GE/RCA
Hans POSPIESZCZYK – Columbus Systems Engineering, Manager
John Scott POUCHER – Distinguished Member Technical Staff, AT&T Bell Laboratories
James A. PRITCHARD – Electronics Engineer, NASA Goddard Space Flight Center
David PRUETT – Acting Chief/Data System Services, NASA Space Station Program

John J. QUANN – Deputy Director, NASA Goddard Space Flight Center

Marie A. RAYBOR – Supervisory Electronics Engineer, NASA Goddard Space Flight Center
James L. RANEY – Chief, Space Station Office, NASA Johnson Space Center
Robert W. RANSOM – Electronics Engineer, NASA Goddard Space Flight Center
Daryl RASMUSSEN – NASA Ames Research Center
Timothy R. RAU – Senior Technical Analyst, NASA Headquarters
Bill RAWSKY – General Manager, StorageTek Integrated Systems
William REDISCH – Engineer, NASA Goddard Space Flight Center
Macgregor S. REID – Manager, Program Design Planning, Jet Propulsion Laboratory
Siegfried REINHOLD – Diplom-Ing., Messerschmitt-Boelkow-Blohm
Charles R. ROBERTS – SSIS Program Manager, TRW
John C. RODGERS – Program Manager, NASA Headquarters
Paula RODKEY – Reporter, Aerospace Daily
John H. ROEDER – Deputy Director, Communication & Data Systems Division, NASA Headquarters
Ruth ROGOFSKY – Computer Specialist, NASA Goddard Space Flight Center
Andrew J. ROLINSKI – Engineering Specialist, Ford Aerospace & Communication
C. J. RORIE – Assistant, Director for Planning, Johns Hopkins University
Alex ROSENBERGE – Advanced Project Manager, Lockheed/EMSCO
H. ROSENTHAL – Senior Engineer (SSIS), TRW
Joseph H. ROTHENBERG – Chief, Mission Operations Division, NASA Goddard Space Flight Center
Alan M. RUFFNER – Manager, MCI
Fidel R. RUL Jr. – Chief, Information Systems Configuration Office, Space Station Program Office,
  NASA Headquarters

Richard RUPP – Software Development Engineer, Hewlett-Packard
Thomas E. RYAN – Data Systems Study Manager, Flight Mission Support Office,
  NASA Goddard Space Flight Center
Lisa M. RYKOWSKI – Computer Scientist, NASA Goddard Space Flight Center

Phillippe SAINT-AUBERT – MATRA – Space Branch
Marc SAINZ – MATRA Espace
Nobuhiko SAKAMOTO – System Engineer, Space Station, Ishikawajima – Harima Heavy Industries
John SAKSS – Chief, International Program Support Office, International Relations Division, NASA Headquarters
Motoatsu SAKURAI – Deputy General Manager, Mitsubishi International Corporation
Joan SALUTE – Research Scientist, NASA Ames Research Center
Michael J. SANDER – Manager, Flight Program Support Office, Jet Propulsion Laboratory
Gregory SANGER – Director of Technology, UTC/Space Station Program Office
Lornezo SARLO – Software Systems Engineer, Aeritalia Space Systems Group
Donald M. SAWYER – Standards Manager, National SSDC, NASA Goddard Space Flight Center
Ashok SAXENA – President, Informatics, Inc.
Richard G. SCHELL – NASA/NOAA Business Area Manager, Computer Technology Associates
Patrick SCHENNING – Senior Associate, Booz, Allen & Hamilton, Inc.
Erwin R. SCHMERLING – Chief Data System Scientist, NASA Headquarters
Lawrence C. SCHOLZ – Division Fellow, RCA Astro-Space Division
Stanley, J. SCHRETTER – Project Manager, Stanford Telecommunications
Curtis A. SCHROEDER – St. Payload Operations System Implementation Manager, NASA Goddard
  Space Flight Center
Billy W. SCOTT – Auditor, U. S. General Accounting Office
Patricia A. SCOTT – Marketing Manager, IBM
Ed SEIDEWITZ – Aerospace Engineer, NASA Goddard Space Flight Center
Marc J. SELIG – Manager, Network Support Group, Allied Bendix Field Engineering Corporation
Moury SERGE – CNES
Anthony E. SERSEN – Systems Engineer, RMS Technologies
J. Clayton SHADECK – Seas Program Manager, Ford Aerospace
Lisa R. SHAFFER – NOAAPORT Project Manager, NOAA
Richard SHARUM – SS-SEI S/W Manager, NASA Kennedy Space Center
Charles E. SHAW – Director, Advanced Programs, Century Computing, Inc.
Nick SHORT – Computer Scientist, NASA Goddard Space Flight Center
Jesse R. SILVERS – Principal Engineer, RMS Technologies, Inc.
Charles T. SIMONDS – Director, HQ Information Systems & Technology Office, NASA Headquarters
M. SIMOPONS – Engineer, CNES
R. P. SINHA – Chief Scientist, STX
Susan D. SITKO – Electrical Engineer, NASA Kennedy Space Center
Gene A. SMITH – Senior Systems Engineer, NASA Goddard Space Flight Center
Paul A. SMITH – Data Management Systems Facility, NASA Goddard Space Flight Center
Robert E. SMYLIE – Vice President, Government Services, RCA American Communications, Inc.
Stanley SOBIESKI – Assistant Chief, Network Planning, NASA Goddard Space Flight Center
Claudio SOPRANO – Engineer, ESTEC
John Y. SOS – Chief, Systems Management Office, NASA Goddard Space Flight Center
Tom SPARN – Research Assistant, Laboratory for Atmospheric & Space Physics
Robert E. SPEARING – Director, Mission Operations & Data System, NASA Goddard Space Flight Center
James R. SPIEGEL – Systems Engineer, Ford Aerospace & Communication
John K. STEEDMAN – Electronics Engineer, NASA Goddard Space Flight Center
M. J. STEINBACHER – MTS, JPL/NASA
Robert STEPHENS – Manager Program Development, Systematics General Corporation
James M. STEVENS – Space Station Operations Manager, NASA Office Space Operations
Henry STEWART – Data System Engineer, NASA Kennedy Space Center
Rona B. STILLMAN – Chief Scientist, U. S. General Accounting Office
Gerald STILLWELL – Member Technical Staff, California Institute of Technology/JPL
Eric STOCKER – Principal Member Technical Staff, Science Applications Research
Andrew STOFAN – Associate Administrator/SS, NASA Headquarters
John H. STRAITON – Information Systems Manager, NASA Kennedy Space Center
Robert H. STRATMEN – Lead Engineer, The MITRE Corporation
David P. STRUBA – NASA Spectrum Manager, NASA Headquarters
Don E. STUCKLE – Manager, TRW
B. L. SULLIVAN – FRE Development Manager, IBM Corporation
Daniel S. SUROWIEC – Manager, Program Development – Space Operations, Bendix Field
  Engineering Corporation
Dunham T. SWIFT – Ford Aerospace & Communications
Martha R. SZCZUR – Computer Scientist, NASA Goddard Space Flight Center

Richard C. TAGLER – Associate Chief, System Management Office, NASA Goddard Space Flight Center
Edward TAGLIAFERRI – Director Special Projects, Business Development, UNISYS
George TAKAHASHI – Space Systems Team, Aerospace, Mitsubishi Corporation
T. TANAKA – Senior Engineer, NASDA
Michael A. TESSMAN – Software Engineer, MCI
Roger V. TETRICK – Acting Assistant Director/Space Station, NASA Goddard Space Flight Center
John A. THOMPSON – Senior Software Engineer, Ford Aerospace & Communication
Thomas H. THORNTON Jr. – Manager, Information Systems Division, Jet Propulsion Laboratory
Paul G. TIERNEY – Manager, Advanced Development Programs, GE – FESD – Ground Systems Department
Claudia T. TOM – Electronics Engineer, NASA Goddard Space Flight Center
Steven P. TOMPKINS – Electronics Engineer, NASA Goddard Space Flight Center
James L. TRACY – System Engineering Manager, EDS
Mark TRUEBLOOD – Program Manager, Ford Aerospace & Communication
Walt F. TRUSZKOWSKI – NASA Goddard Space Flight Center
Y. TSUJINO – Assistant Senior Engineer, NASDA
Knox W. TULL Jr. – Principal, Jackson and Tull
H. Blair TYSON – Manager, EDS

David E. ULMER – Mission Manager, Lockheed/EMSCO

Paul VALE – Senior Marketing Representative, Storage Technology Corporation
Robert R. VALLENI – Line of Business Manager, TRW
Dennis C. VANDER TUIG – Head, Logistics Management Section, NASA Goddard Space Flight Center
Douglas W. VANDERMADE – Group Leader, The MITRE Corporation
Joseph C. VASSALLO – Member Technical Staff Supervisor, AT&T Bell Laboratories, Inc.
Edwin T. VAUGHAN – Mathematician, NASA Goddard Space Flight Center
P. A. VAUGHAN – Data Handling Engineer, BNSC, Rutherford Appleton Laboratory
C. E. VELEZ – President, Computer Technology Associates, Inc.
Arthur P. VERBIN – Vice President, NYMA, Inc.
Francois VIGNES – Director Space Stations, MATRA
A. VILLASENOR – Program Manager, NASA Headquarters
C. W. VOWELL – Manager/Systems Architect/Analysis & Technology, NASA Johnson Space Flight Center

Don WADDELL – Senior Scientist, BDM Corporation
Michael WAHL – Coordinator for Special Applications, W.L. Gore & Company GmbH
Tokshiko WAKAKI – Systems Section, Space Systems Department, Fujitsu, Ltd.
Barbara A. WALTON, – Space Station Platform Inst Manager, NASA Goddard Space Flight Center
Hubertus WANKE – Head Mission Operations Department, DFVLR – GSOC
Michael T. WARD – Senior Systems Engineer, Ford Aerospace & Communication
William M. WATT – Engineer, NASA Goddard Space Flight Center
Frederick W. WEBER – Vice President, NASA Programs, Contel Technical Services
Louie G. WEED – Supervisor, TMIS, Boeing
Aaron WEINBERG – Director, Communication Science, Stanford Telecommunications
Roland G. WEISS – Senior Engineering Specialist, Ford Aerospace & Communication
Robert A. WEISSMAN – Senior Member Technical Staff, GTE Strategic Systems Division
Division Bill WELLER – Senior Systems Engineer, Ford Aerospace & Communication
Phillip C. WHEELER – Special Assistant for Technology, TRW S&TG Federal Systems Division
Virginia A. WHITELAW – Research Engineer, Avionics System Division, NASA Johnson Space Center
Chris WILKINSON – Electronics Engineer, NASA Goddard Space Flight Center
David A. WILKINSON – Manager, Civil Programs, GE/RCA
Michael WILLIAMS – Project Manager, NYMA, Inc.
Pete WILLIAMS – Director, SSIS Development Division, Space Station Program Office, NASA Headquarters
Robert F. WILLIAMS – Director, Space Station Program Office, Digital Equipment Corporation
Kelli WILLSHIRE – A&R Coordinator, NASA Langley Research Center
Ed WILSON – Group Manager, Science Applications Research
Michael J. WISKERCHEN – Senior Research Associate, Stanford University
Janet K. WOLFE – Chief, Public Affairs, NASA Goddard Space Flight Center
Drew WOLOSHYN – Manager Data Systems, RCA Corporation
Alice WONG – Electronics Engineer, Federal Aviation Administration AST 100
William H. WOOD – Consultant, Bendix Field Engineering Corporation
Charles E. WOODYARD – Associate Chief, Flight Dynamics Division, NASA Goddard Space Flight Center
Robert W. WOOLFOLK – Director, Business Development, SRI International

Takahiro YAMADA – Assistant Professor, The Institute of Space & Astronautical Science
Akahiko YAMAMOTO – Manager, Nippon Oil & Fats Company, Ltd.
Masaru YAMBE – Fujitsu Ltd.
D. YANNUZZI – Technical Support Manager – NLTN, Allied Bendix Field Engineering Corporation

Ben YEHUDAIFF – Electrical Systems Group Leader, Spar Aerospace Ltd.
Nancy YELVERTON – Systems Development Manager, Control Data Corporation
Takeyasu YOSHIOKA – Assistant Senior Engineer, SEG, NASDA
Susumu YOSHITOMI – Manager, Technical Information Section Regional Planning & Development, Space
  Communication Research Corporation
Howard L. YUDKIN – President & Chief Executive Officer, Software Productivity Consortium
Russell W. ZEARS – Group Leader, The MITRE Corporation
Andrew D. ZEITLIN – Associate Department Head, The MITRE Corporation
Donald J. ZUREK – Staff Engineering, Honeywell/Sperry Space System

Author Index